Lecture Notes in Computer Science 4964

Commenced Publication in 1973
Founding and Former Series Editors:
Gerhard Goos, Juris Hartmanis, and Jan van Leeuwen

Tal Malkin (Ed.)

Topics in Cryptology – CT-RSA 2008

The Cryptographers' Track at the RSA Conference 2008
San Francisco, CA, USA, April 8-11, 2008
Proceedings

 Springer

Volume Editor

Tal Malkin
Columbia University
Department of Computer Science
514 CSC, 1214 Amsterdam Avenue MC 0401
New York, NY 10027-7003, USA
E-mail: tal@cs.columbia.edu

Library of Congress Control Number: 2008924621

CR Subject Classification (1998): E.3, G.2.1, D.4.6, K.6.5, K.4.4, F.2.1-2, C.2, J.1

LNCS Sublibrary: SL 4 – Security and Cryptology

ISSN   0302-9743
ISBN-10   3-540-79262-7 Springer Berlin Heidelberg New York
ISBN-13   978-3-540-79262-8 Springer Berlin Heidelberg New York

Springer is a part of Springer Science+Business Media

springer.com

© Springer-Verlag Berlin Heidelberg 2008
Printed in Germany

Typesetting: Camera-ready by author, data conversion by Scientific Publishing Services, Chennai, India
Printed on acid-free paper   SPIN: 12257843   06/3180   5 4 3 2 1 0

Preface

The RSA Conference is the largest regularly-staged computer security event, with over 350 vendors and many thousands of attendees. The Cryptographers' Track (CT-RSA) is a research conference within the RSA Conference. CT-RSA began in 2001, and has become one of the major established venues for presenting cryptographic research papers to a wide variety of audiences. CT-RSA 2008 was held in San Francisco, California from April 8 to April 11.

The proceedings of CT-RSA 2008 contain 26 papers selected from 95 submissions pertaining to all aspects of cryptography. Each submission was reviewed by at least three reviewers, which was made possible by the hard work of 27 Program Committee members and many external reviewers listed on the following pages. The papers were selected following a detailed online discussion among the Program Committee members. The program included an invited talk by Shafi Goldwasser. The current proceedings include a short abstract of her talk.

I would like to express my deep gratitude to the Program Committee members, who volunteered their expertise and hard work over several months, as well as to the external reviewers. Special thanks to Shai Halevi for providing and maintaining the Web review system used for paper submission, reviewing, and final-version preparation. Finally, I would like to thank Burt Kaliski and Ari Juels of RSA Laboratories, as well as the RSA conference team, especially Bree LaBollita, for their assistance throughout the process.

February 2008   Tal Malkin

CT-RSA 2008

The Cryptographers' Track of the RSA Conference

Moscone Center, San Francisco, CA, USA
April 8–11, 2008

Program Chair

Tal Malkin, Columbia University, USA

Program Committee

Masayuki Abe	NTT Corporation, Japan
Feng Bao	Institute for Infocomm Research, Singapore
Dario Catalano	Università di Catania, Italy
Orr Dunkelman	Katholieke Universiteit Leuven, Belgium
Nelly Fazio	IBM Almaden Research Center, USA
Marc Fischlin	Darmstadt University of Technology, Germany
Michael Freedman	Princeton University, USA
Stuart Haber	HP Labs, USA
Danny Harnik	IBM Haifa Research Lab, Israel
Susan Hohenberger	Johns Hopkins University, USA
Aggelos Kiayias	University of Connecticut, USA
Eike Kiltz	CWI, The Netherlands
Ilya Mironov	Microsoft Research, USA
Kobbi Nissim	Ben Gurion University, Israel
Satoshi Obana	NEC, Japan
Elisabeth Oswald	University of Bristol, UK
Kunsoo Park	Seoul National University, Korea
Rafael Pass	Cornell University, USA
Josef Pieprzyk	Macquarie University, Australia
Tal Rabin	IBM T.J. Watson Research Center, USA
Matt Robshaw	France Télécom, France
Rei Safavi-Naini	University of Calgary, Canada
Alice Silverberg	UC Irvine, USA
Adam Smith	Pennsylvania State University, USA
François-Xavier Standaert	UCL, Belgium
Eran Tromer	MIT, USA
Yiqun Lisa Yin	Independent Consultant, USA

External Reviewers

Joel Alwen
Nuttapong Attrapadung
Lejla Batina
Amos Beimel
Come Berbain
D.J. Bernstein
Guido Bertoni
Olivier Billet
Carl Bosley
Chris Charnes
Jiun-Ming Chen
Lily Chen
Joo Yeon Cho
Baudoin Collard
Scott Contini
Glenn Durfee
Jean-Charles Faugère
Martin Feldhofer
Dario Fiore
Craig Gentry
Benedikt Gierlichs
Rob Granger
Matthew Green
Johann Großschädl
Iryna Gurevych
Shai Halevi
Jason Hinek
Dennis Hofheinz
Shaoquan Jiang
Antoine Joux
Marc Joye
Marcelo Kaihara

Seny Kamara
François Koeune
Hugo Krawczyk
Mario Lamberger
Tanja Lange
Kerstin Lemke-Rust
Tieyan Li
Benoit Libert
Huijia (Rachel) Lin
Jennifer Lindsay
Stefan Mangard
Krystian Matusiewicz
Alexander May
Pradeep Kumar Mishra
David Molnar
Tal Moran
Shiho Moriai
Phong Q. Nguyen
Antonio Nicolosi
Svetla Nikova
Miyako Ohkubo
Yossi Oren
Dag Arne Osvik
Serdar Pehlivanoglu
Chris Peikert
Pedro Peris-Lopez
Ray Perlner
Duong Hieu Phan
Krzysztof Pietrzak
Benny Pinkas
Gilles Piret
Mario Di Raimondo

Raj Rajagopalan
Nalini Ratha
Vincent Rijmen
Thomas Ristenpart
Tom Roeder
Guy Rothblum
Tomas Sander
Bagus Santoso
Dominique Schröder
Jean-Pierre Seifert
Nicolas Sendrier
Hovav Shacham
Siamak Shahandashti
Zhijie Shi
Igor Shparlinski
Michal Sramka
Ron Steinfeld
Tamir Tassa
Stefan Tillich
Dustin Tseng
Vinod Vaikuntanathan
Muthu
  Venkitasubramaniam
Shabsi Walfish
Zhenghong Wang
William Whyte
Jin Yuan
Hong Sheng Zhou
Hong-Sheng Zhou

Table of Contents

Hash Function Cryptanalysis

Cryptographic Building Blocks

Fairness in Secure Computation

Message Authentication Codes

Improved AES Implementations

Public Key Encryption with Special Properties

Side Channel Cryptanalysis

Cryptography for Limited Devices

Invited Talk

Key Exchange

Cryptanalysis

Cryptographic Protocols

Security of MD5 Challenge and Response: Extension of APOP Password Recovery Attack

Yu Sasaki[1], Lei Wang[2], Kazuo Ohta[2], and Noboru Kunihiro[2]

[1] NTT Information Sharing Platform Laboratories, NTT Corporation[*],
3-9-11 Midoricho, Musashino-shi, Tokyo, 180-8585, Japan
sasaki.yu@lab.ntt.co.jp
[2] The University of Electro-Communications,
Chofugaoka 1-5-1, Chofu-shi, Tokyo, 182-8585, Japan
{wanglei,ota,kunihiro}@ice.uec.ac.jp

Abstract. In this paper, we propose an extension of the APOP attack that recovers the first 31 characters of APOP password in practical time, and theoretically recovers 61 characters. We have implemented our attack, and have confirmed that 31 characters can be successfully recovered. Therefore, the security of APOP is completely broken. The core of our new technique is finding collisions for MD5 which are more suitable for the recovery of APOP passwords. These collisions are constructed by employing the collision attack of den Boer and Bosselares and by developing a new technique named "IV Bridge" which is an important step to satisfy the basic requirements of the collision finding phase. We show that the construction of this "IV Bridge" can be done efficiently as well.

Keywords: APOP, Challenge and Response, Password Recovery, Hash Function, MD5, Collision Attack, Message Difference.

1 Introduction

MD5 [13] is a hash function that is designed to obtain high efficiency in terms of computation time. MD5 is widely used all over the world, so its security is very important.

The first attack against MD5 was proposed by den Boer and Bosselaers in 1996 [2]. They found that collisions could be generated when a pair of initial value (IV_1, IV_2) exhibits some specific difference. After that, in 1996, Dobbertin found a collision with the common initial value IV', where IV' is different from the real IV of MD5 [5,6]. Since these attacks didn't work for the real IV of MD5, the real MD5 collision could not be generated. In 2005, Wang et al. proposed the first collision attack that could handle the real IV of MD5 [21]. This attack can generate a collision for any initial value. Since the proposal of [21], several papers have improved its result [1,9,12,18]. At the present time, the fastest attack is the one proposed by Klima [9]. It generates a collision of MD5 in 1 minute on a standard PC.

---

[*] A part of this work was done while Yu Sasaki was a master student of UEC.

T. Malkin (Ed.): CT-RSA 2008, LNCS 4964, pp. 1–18, 2008.
© Springer-Verlag Berlin Heidelberg 2008

From the above results, we can easily get many collisions for MD5. Therefore, in recent days, many studies have attempted to utilize the MD5 collisions in developing an attack for other protocols that are based on the collision resistance of hash functions [3,4,8,10,20].

Password recovery attack against APOP

APOP is a challenge and response authentication protocol; it is used by a mail server to authenticate a user who tries to access e-mails for that person.

In 1996, Preneel and van Oorschot proposed the key recovery attack against Envelop MAC [15] whose protocol is partially the same as APOP. In order to recover the key, they use collisions of MD5 which were generated according to the birthday paradox. In FSE 2007, Leurent proposed a password recovery attack on APOP that recovers the first three password characters [11]. In this attack, the attacker impersonates the server. This attack is based on Preneel and van Oorschot's attack, but generates collision by utilizing Wang et al.'s MD5 collision attack. In this paper, we write "APOP attack" to denote the attack that recovers an APOP password. So far, the messages generated by collision attacks are considered to be totally random. However, Leurent shows how to generate a collision where some part of the messages is fixed, and apply this technique to the APOP attack. In March 2007, Sasaki et al. independently proposed a similar attack [19].

In the APOP attacks of Leurent and Sasaki et al., the number of recoverable password characters depends on the location of message differences for MD5 collision attack. Both methods use MD5 collision attack proposed by Wang et al [21]. Since the message differences proposed by Wang et al. are not optimized for the APOP attack, but for finding a collision, the number of maximum recoverable characters is limited to three.

Our contribution

In this research, we succeed in practically recovering more APOP password characters by constructing a new MD5 collision attack whose location of message differences is optimized for the APOP attack.

Our collision attack uses a two-block collision, a pair of 1024-bit messages. First, we found that the collision attack proposed by den Boer and Bosselaers employs good differences for the APOP attack. However, their attack needs specific differences on IV, and thus, it cannot be applied to the real MD5. We overcome this shortcoming by using the techniques introduced in Wang et al.'s collision attack; that is, we construct a differential path that yields initial value differences suitable for den Boer and Bosselaers's attack from the real IV of MD5. We call this differential path "IV Bridge."

In the APOP attack, whenever we make a guess on a target password character, we need to generate a collision. Therefore, to recover a character (1-octet), we need to generate 255 collisions in the worst case[1]. When the password is less than or equal to 31 characters, the proposed attack generates a collision in

---

[1] 256 collisions are not needed. If we fail to recover the target character with 255 trials, we know that the last candidate is the real password.

practical time, so such password can be recovered in practical time. Theoretically, we can recover up to 61 characters of the password faster than using the birthday paradox. Moreover, by combining the proposed attack and exhaustive search, the number of characters that can be recovered in practical time is extended by 5 or 6 more characters.

For confirmation, we implemented the proposed attack and recovered the first 31 characters of APOP passwords. As a result, we confirmed that for up to 11 characters, a collision can be generated within 1 second, and for up to 31 characters, a collision can be generated approximately in 5 seconds.

This paper is organized as follows. Section 2 explains the specification of APOP and MD5. In section 3, we summarize the previous MD5 collision attacks and the APOP attacks on which we based our extension. In section 4, we explain how to construct the IV Bridge, which is a new MD5 collision attack that extends the number of recoverable password characters. In section 5, we propose efficient algorithms for finding the IV Bridge. In section 6, we show the result of an experiment on recovering APOP passwords. In section 7, we conclude this paper. Finally, we show detailed data of the IV Bridge in the appendices[2].

2 Preliminaries

2.1 APOP Algorithm

APOP (Authenticated Post Office Protocol) [14] is a challenge and response authentication protocol between the mail server and the user. Let the pre-shared Password be *Pass*. APOP is as follows.

1. The user connects to the server. (The protocol is triggered by the user.)
2. The server generates a challenge nonce C, which is a random string with some limitations shown below, and sends it to the user.
3. The user concatenates his password to C, and computes $R=\mathrm{MD5}(C\|Pass)$. He sends the result R to the server.
4. The server itself computes $\mathrm{MD5}(C\|Pass)$ by using the password stored in the server. It authenticates the user by comparing its result and R.

Restrictions placed on APOP challenge string

1. Challenge must start from '<' and include at least one '@'.
2. Challenge must end with '>'.
3. Challenge must not include 'NULL', '\n'and '<','>' in the middle.

2.2 Description of MD5

MD5 [13] employs a Merkle-Damgård structure, which takes an arbitrary length message M as input, and outputs 128-bit hash value $H(M)$. First, M is padded

---

[2] After the review process, we found that SIP [16] and Digest Authentication [7] were vulnerable to the APOP attack. We mention this in the appendices.

and divided into 512-bit block messages $(M_0, M_1, \cdots, M_{n-1})$. These messages go through compression function (CF) with a 128-bit chaining variable. The initial chaining variable (H_0) is set as follows: $a_0 = $ 0x67452301, $b_0 = $ 0xefcdab89, $c_0 = $ 0x98badcfe, $d_0 = $ 0x10325476. The procedure of MD5 algorithm is as follows:

$$H_1 = CF(M_0, H_0), H_2 = CF(M_1, H_1), \cdots, H_n = CF(M_{n-1}, H_{n-1}).$$

H_n will be the hash value of M.

The MD5 Compression Function

The compression function of MD5 takes M_i and H_i as input, and outputs H_{i+1}. First, the message block M_i is divided into sixteen 32-bit length messages $(m_0, m_1, \cdots, m_{15})$. The hash value H_i is divided into four 32-bit length chaining variables (a_0, b_0, c_0, d_0). The compression function consists of 64 steps. Steps 1-16, steps 17-32, steps 33-48 and step 49-64 are called the first, second, third and fourth rounds, respectively. In step j, the chaining variables a_j, b_j, c_j, d_j are updated as follows.

$$a_j = d_{j-1}, \qquad b_j = b_{j-1} + (a_{j-1} + f(b_{j-1}, c_{j-1}, d_{j-1}) + m_k + t) \lll s_j,$$
$$c_j = b_{j-1}, \qquad d_j = c_{j-1}.$$

f is a Boolean function which depends on the round number. m_k is one of $(m_0, \cdots, m_{15})$, and the index k depends on the step. t is a constant defined in each step. $\lll s_j$ denotes left rotation by s_j bits, where s_j changes each step.

3   Related Works

3.1   Weakness of MD5 Shown by den Boer and Bosselaers

In 1993, den Boer and Bosselaers showed that when the IV of MD5 had the following differences, one can compute M such that MD5(IV_1, M)=MD5(IV_2, M) with practically computable complexity [2].

$$\Delta IV = (IV_1 \oplus IV_2) = (\text{0x80000000}, \text{0x80000000}, \text{0x80000000}, \text{0x80000000}).$$

Hereafter, we call this difference Δ^{msb}. In their attack the MSB of b_0, c_0 and d_0 of each IV must be the same value.

3.2   Wang et al.'s Collision Attack on MD5

Wang et al.'s attack [21] generates a collision with complexity of 2^{38} MD5 computations[3]. Let m and m' be a pair of messages that yields a collision. The difference Δ is defined to be the value yielded by subtracting the value for m from value for m'. The attack procedure is as follows.

1. Find a "Message Difference (ΔM)" that yields a collision.
2. Determine how the impact of ΔM propagates. The propagation of the differences is called a "Differential Path (DP)".

---

[3] In the original paper [21], the complexity of the attack was estimated to be 2^{37} MD5 computations. However, in [12], it was shown that there was a mistake in the complexity analysis, and computed the correct complexity of 2^{38} MD5 computations.

3. To realize the DP, generate "Sufficient Conditions (SC)" on the value of the chaining variables.
4. Locate a message that satisfies all SCs by randomly generating messages and applying "Message Modification (MM)". Let the located message be M_*.
5. Compute $M'_* = M_* + \Delta M$. Finally, M_* and M'_* become a collision pair.

The attack finds collisions using 2-block messages. First, the attacker finds a message block M_0, such that $\text{MD5}(M_0 + \Delta M_0) - \text{MD5}(M_0) = \Delta H_1$[4]. Then, the attacker searches for a message block M_1, such that $\text{MD5}(M_1 + \Delta M_1) = \text{MD5}(M_1)$ assuming the input difference ΔH_1. Specifically, for the APOP attacks of [11,19], the used differences were of the form:

$$\Delta M_0 = (\Delta m_0, \ldots, \Delta m_{15}) = (0,0,0,0,2^{31},0,0,0,\quad 0,0,0,2^{15},0,0,2^{31},0),$$
$$\Delta M_1 = (\Delta m_0, \ldots, \Delta m_{15}) = (0,0,0,0,2^{31},0,0,0,\quad 0,0,0,-2^{15},0,0,2^{31},0).$$

It is important to note that m_{15} of ΔM_1 is without a difference, and that m_{14} contains a difference.

3.3 Previous APOP Attacks

In the APOP attack, the attacker impersonates the server. The attacker uses "Chosen Challenge Attack," that changes a challenge C into a convenient C' and makes the user compute the corresponding response R'.

In 2007, Leurent [11] and Sasaki et al. [19] independently proposed the attacks that recover the first three characters of APOP password by using MD5 collisions. These attacks are based on the key recovery attacks against Envelop MAC proposed by Preneel and van Oorschot [15]. The generation of the collisions in [11,19] and [15] are different and their attack complexities are also different. However, algorithm for recovering key is almost the same. To begin with, we show the key recovery algorithm proposed by [11,19] in Table 1.

Methods for Generating MD5 Collisions
Preneel and van Oorschot's attack [15] generates MD5 collisions by using the birthday paradox. As a result, this attack requires $2^{67.5}$ offline computations. Since $2^{67.5}$ is too heavy to compute at the moment, this attack is impractical.

On the other hand, Leurent and Sasaki et al.'s attacks [11,19] generate collisions based on Wang et al.'s attack. As a result, each collision is generated in approximately 5 seconds, and thus, APOP passwords are practically recovered.

3.4 Summary and Problems of Previous Works

Preneel and van Oorschot attack generates the collisions by using the birthday paradox. Since its complexity is too high, the attack cannot be practical.

The attacks proposed by Leurent and Sasaki et al. practically recover the first three APOP password characters by generating collisions based on Wang et al.'s attack. Generated collision forms $\text{MD5}(C||known||p) = \text{MD5}(C'||known||p)$.

---

[4] $\Delta H_1 = (2^{31}, 2^{31} + 2^{25}, 2^{31} + 2^{25}, 2^{31} + 2^{25})$, which is not used in this paper.

Table 1. Algorithm for recovering a password in the APOP attack

1. Let *known* be already recovered characters. Initialize *known* to be **NULL**.
2. Let p be an octet that is used to guess one character of the password.
3. **for** $p=0$ to 254 {
4.   Generate (C, C') such that MD5($C\|known\|p$) = MD5($C'\|known\|p$). Here, The lengths of $C\|known\|p$ and $C'\|known\|p$ are identical and multiples of 512.
5.   Send C to the user, and get the corresponding R.
6.   Send C' to the user, and get the corresponding R'.
7.   **if** $(R = R')$ {
8.   Password is p.
9. $known \leftarrow (known \ll 8) + p$
10.   **goto** stage 15.
11.   }
12. }
13. Password is $p + 1 (= 255)$.
14. $known \leftarrow (known \ll 8) + (p + 1)$
15. **if** (Three characters are recovered ?) {
16.   Output *known*, and halt the algorithm.
17. } **else** {
18.   **goto** stage 3.
19. }

Here, $known\|p$ is a fixed value and identical for both C and C'. Since $C\|known\|p$ is a multiple of 512-bit, the fixed part is located in m_{15} first. When we recover many characters, *known* will be long, and the fixed part is located in not only m_{15} but also in $m_{14}, m_{13}, \cdots$.

Both of Leurent and Sasaki et al.'s attacks use Wang et al.'s collision attack, which needs a difference in the MSB of m_{14}. Therefore, only m_{15} can be the same between C and C'. This is why these methods cannot recover more than three characters[5]. No other attack that can practically find collisions of MD5 is known. Therefore, it is impossible to practically recover more than three characters.

4   Construction of MD5 Collision Attack That Is Efficient for the APOP Attack

As explained in section 3, previous works cannot be used in practice to recover more than three characters since they generate collision by using birthday paradox or Wang et al.'s message differences. In this section, we extend the number of practically recoverable APOP password characters by proposing a new MD5 collision attack that uses a different approach from that of Wang et al.

4.1   Conditions for Extending the APOP Attack

To extend the number of recoverable APOP password characters, we need a new MD5 collision attack. First, in order to recover many characters, long identical

---

[5] Since C must end with '>', m_{15} can contain at most three other characters.

values need to be set in the last part of messages. Second, in order to recover each additional character, we need to generate 255 collisions (in the worst case). Considering this background, the following conditions are necessary for a new collision attack.

Condition 1: There exists no message difference in the last part of messages.
Condition 2: Many collisions can be generated in practical time.

4.2   Applying den Boer and Bosselaers's Attack to APOP

den Boer and Bosselaers's attack generates collisions by using a common message and (IV_1, IV_2) such that $\Delta IV = \Delta^{msb}$. This property is optimal for Condition 1 in section 4.1 since no message difference is needed. Furthermore, if the message modification technique proposed by [9,18,21] is applied, a collision can be generated within a few seconds. Therefore, den Boer and Bosselaers's attack also seems to be efficient for satisfying Condition 2 in section 4.1.

However, den Boer and Bosselaers's attack requires a ΔIV difference in the input hash value, and so cannot be applied to APOP directly. Therefore, in our approach, we construct an IV Bridge; that is, a differential path that produces ΔIV from the real IV of MD5.

Remarks
den Boer and Bosselaers's attack has already been utilized by Contini and Yin to attack HMAC [3]. However, they did not tackle the problem that den Boer and Bosselaers's attack could not work for the real IV of MD5. Therefore, their attack is related-key attack.

4.3   Constructing Differential Path to Produce $\Delta IV = \Delta^{msb}$

By using various message modification techniques, it is known that the cost for satisfying sufficient conditions for the first and second rounds are smaller than that for the third and fourth rounds. Consequently, we need a differential path that holds with high probability in both the third and fourth rounds.

Constructing the differential path for the fourth round
To archive $\Delta IV = \Delta^{msb}$, we need to have Δ^{msb} at the end of the fourth round. In the fourth round, if the input chaining variables in step i have Δ^{msb} and Δm_k is 0 or 2^{31}, the output chaining variables have Δ^{msb} with probability $1/2$ because of the f function. Therefore, we make initial chaining variables for the fourth round have Δ^{msb} with message differences $\Delta m_k = 0$ or $2^{31}, 0 \leq k \leq 15$.

Constructing the differential path for the third round
In the third round, once Δ^{msb} appears, all the remaining steps have Δ^{msb} as long as $\Delta m_k = 0$. Since the f function in the third round is XOR, this happens with probability of 1. To increase the success probability of the whole differential path for the third round by utilizing this property, we make several differences in the initial step of the third round, and make Δ^{msb} within few steps by only using $\Delta m_i = 0$ or 2^{31}. As a result of the analysis, we determined the following differences. Here, the notation $*$ means that the sign does not have to be considered.

$$(\Delta a_{32}, \Delta b_{32}, \Delta c_{32}, \Delta d_{32}) = (*2^{31}, *2^{31}, 0, 0),$$
$$\Delta m_{11} = *2^{31}, \Delta m_k = 0 \text{ (for other } k).$$

Constructing the differential path for the first and second rounds
We constructed the differential path for the first and second rounds by hand. We omit our differential path search algorithm since automated path search algorithm has been proposed by [20]. We outline the generated DP in Table 3 and the sufficient conditions in Table 4.

5   Finding an IV Bridge

Section 4 explained how to derive the differential path and the sufficient conditions for the IV Bridge. In this section, we explain how to find the IV Bridge.

5.1   Overall Strategy

In order to get the IV Bridge, we need to locate a message that satisfies all the sufficient conditions shown in Table 4. The search strategy of Wang et al. cannot be applied to the IV Bridge. This problem is mainly caused from the huge number of conditions in steps 17-24 of the IV Bridge. In Wang et al.'s attack, all the conditions in the first round and a part of conditions in steps 17-24 are satisfied by message modifications. However, since the IV Bridge has 101 conditions in steps 17-24 as shown in Table 4, partially satisfying the conditions in steps 17-24 is not enough to find the IV Bridge in practical time. Therefore, we need other strategy for finding the IV Bridge.

In Wang et al.'s attack, message modification is applied to satisfy all the conditions in steps 1-16 (16 steps in total). On the other hand, In the IV Bridge, we do not need message modification in steps 1-8 since there is no condition. This enables us to satisfy all the conditions in steps 9-24 (a total of 16 steps). There are 42 conditions left after step 24. Therefore, if 2^{42} messages that satisfy all the conditions up to step 24 are generated, one of them is likely to be an IV Bridge. Our search process consists of the following three phases.

Phase A: The goal of Phase A is setting chaining variables in steps 9-23 to satisfy all the conditions up to step 23. Simultaneously, the chaining variables in steps 1,2 and 8 are fixed. Phase A is executed only once. It takes approximately 1 second to finish.

Phase B: The goal of Phase B is determining chaining variables in steps 3-7 so that the values for steps 2 and 8 determined by Phase A are connected. Simultaneously the chaining variables in step 24 is fixed. Here, we also guarantee that all the conditions in step 24 are satisfied. As a result of Phase B, all fixed messages that satisfy all conditions up to step 24 will be output.

Phase C: From one result of Phase B, generate messages that also satisfy all the conditions up to step 24 using the techniques described in [9,18]. Compute after step 24, and check if all the 42 sufficient conditions are satisfied.

5.2   Phase A

Phase A takes as an input the sufficient conditions, which are shown in Table 4. Phase A outputs chaining variables in steps 1,2,8-23 that satisfy all the sufficient conditions up to step 23, all the extra conditions that are used in Phase C, and $m_0, m_1, m_5, m_6, m_{10}$-$m_{15}$.

The step update function can be written using only the chaining variable b.

$$b_j = b_{j-1} + (b_{j-4} + f(b_{j-1}, b_{j-2}, b_{j-3}) + m_k + t) \lll s_j$$

In this case, the IV is represented as $(b_{-3}, b_0, b_{-1}, b_{-2})$. From the above expression, if $b_{j-1}, b_{j-2}, b_{j-3}$ and two of b_j, b_{j-4}, m_k are determined, the last variable is uniquely determined. We detail the three expressions as follows.

Standard-b for b_j: $b_j = b_{j-1} + (b_{j-4} + f(b_{j-1}, b_{j-2}, b_{j-3}) + m_k + t) \lll s_j$,
Inverse-m for m_k: $m_k = ((b_j - b_{j-1}) \ggg s_j) - b_{j-4} - f(b_{j-1}, b_{j-2}, b_{j-3}) - t$,
Inverse-b for b_{j-4}: $b_{j-4} = ((b_j - b_{j-1}) \ggg s_j) - f(b_{j-1}, b_{j-2}, b_{j-3}) - m_k - t$.

The search algorithm is shown below.

1. Choose randomly b_{13} to b_{22} under the condition that they satisfy all the sufficient conditions. Compute *Inverse-m* for $m_1, m_6, m_{11}, m_0, m_5$ and m_{10}, which are used in steps 17-22.
2. m_0, m_1 are also used in step 1 and 2. Compute *Standard-b* for b_1 and b_2.
3. Choose randomly b_{12} under the condition that it satisfies all the sufficient conditions. Compute *Inverse-m* for m_{15}. Since m_{15} is also used in step 23, compute *Standard-b* for b_{23} and check whether all the sufficient conditions for b_{23} are satisfied. Repeat this stage until all conditions for b_{23} are satisfied.
4. Choose randomly b_{11}, b_{10} and b_9 under the condition that they satisfy all the sufficient conditions. Compute *Inverse-m* for m_{14}, m_{13} and m_{12}.
5. Compute *Inverse-b* for b_8 and b_7, output the result and halt this algorithm.

5.3   Phase B

The input of Phase B is the result of Phase A. The chaining variables input to step 3, the chaining variable output in step 7 and messages in step 6 and 7 are especially important. Phase B searches for m_2, m_3, m_4 that connect the differential path from steps 3 to 7, and satisfy all conditions in step 24. Then, m_7, m_8, m_9 are computed. The output of Phase B is all the fixed messages that satisfy all the sufficient conditions up to step 24.

As shown in Table 4, steps 3-7 do not have any sufficient condition. Therefore, any values are acceptable for chaining variables in these steps. From steps 3-7, there are 96 free bits in messages, and only 32 bits of a chaining variable are fixed. Therefore, we can expect that 2^{64} results of Phase B exist for a result of Phase A. However, 2^{64} results are too many for finding IV Bridge. Therefore, we reduce the search space and raise the efficiency of the algorithm. The idea is removing the dependency of the input of the f function in step 6 by fixing the value of the chaining variable. This ensures that when the output of f in step 6 is determined, we can immediately get input chaining variable of f that will successfully connect step 3 to 7. The algorithm of Phase B is shown below. A graphical explanation is given in Figure 1 of the appendices.

1. In order to remove the dependency of f function in step 6, we fix the value of b_5, which is one of the chaining variables input to f, to `0xffffffff` [6].
2. `for` $(c_7 = 0$ `to 0xffffffff)` {
3.   Compute the output of f in step 6. Due to the removal of dependency of f in step 6, the value of c_5 is uniquely determined.
4.   Compute the output of f in step 7, then, the value of a_6 is uniquely determined.
5.   All chaining variables from step 3 to 7 are determined now. Compute *Inverse-m* for m_2, m_3 and m_4.
6. m_4 is also used in step 24. Compute *Standard-b* for b_{24}.
7.   `if` (all conditions for b_{24} are satisfied) {
8.   Compute *Inverse-m* for m_7 to m_9. Then goto Phase C.
9.   }
10. }
11. Halt Phase B, then goto stage 1 of Phase A.

The number of conditions for b_{24} is 11, therefore, all conditions for b_{24} are satisfied with probability of 2^{-11}. Since the maximal number of iterations for stages 2 to 10 is 2^{32}, Phase B can generate 2^{21} results. As shown in the above procedure, every time we get a result of Phase B, we move to Phase C. However, if Phase C cannot find IV Bridge, we go back to Phase B, and compute another result.

5.4   Phase C

The input of Phase C is the output of Phase B, namely, all fixed messages that satisfy all the conditions up to step 24. Phase C generates 2^{21} new messages that also satisfy all the conditions up to step 24 from each result of Phase B by using the message modification technique proposed by [18] or "Q9 Tunnel" proposed by [9]. In this paper, we omit the details of these techniques.

Since Phase B can output 2^{21} results and Phase C generates 2^{21} messages for each result, we can try 2^{42} messages in total. Since there are 42 conditions after step 24, we have enough messages to find an IV Bridge.

6   Evaluation and Trial of the Extended Attack

We implemented the extended APOP attack using the IV Bridge. The attack's process is based on previous works [11,19]. Instead of stage 15 in Table 1, we check whether 31 characters have been already recovered. Our APOP attack generates three blocks challenges that yield a collision with the real password[7].

Block 0: Common messages satisfying the restrictions explained in section 2.1.
Block 1: The IV Bridge.

---

[6] The f function in step 6 is $f(b_5, c_5, d_5) = (b_5 \wedge c_5) \vee (\neg b_5 \wedge d_5)$. If all bits of b_5 are fixed to 1, then $f(b_5, c_5, d_5) = c_5$.

[7] The challenge is three-block, however, users compute a four-block message since the unknown part of the password will be pushed into the fourth block.

Block 2: den Boer and Bosselaers's attack, where last part of messages is fixed.

Blocks 0 and 1 are the pre-computation part, where one result is used repeatedly. Block 2 is computed many times with the last fixed part of messages.

6.1 Collision Search for Blocks 0 and 1

The message for Block 0 is determined to satisfy APOP restrictions. This is finished in negligible time. Block 1 is the IV Bridge. Finding the IV Bridge costs 2^{42} MD5 computations (and the IV Bridge can be used again). We ran 18 PCs whose CPUs are Pentium 4 2.0GHz, and found the IV Bridge in 3 days.

6.2 Collision Search for Block 2

According to [2], finding a collision in block 2 requires 46 sufficient conditions. Therefore, a collision is generated with complexity of 2^{46} MD5 computations using a naive search. This complexity is reduced if the message modification techniques described in [9,18,21] are applied. If all the bits are not fixed, and thus, we can freely modify all bits, the complexity is sufficiently low to be practical. However, as we recover more password characters, the longer the fixed part of the messages becomes, and some message modifications cannot be used. The most effective modification for Block 2 is the technique named "Q4 Tunnel" by [9], which modifies m_7. Therefore, if m_7 is fixed, we cannot use "Q4 Tunnel." Given that each message consists of 4 bytes and the last character of the challenge must be '>', guessing up to 31 characters does not fix m_7. If "Q4 Tunnel" is available, the complexity of finding Block 2 is 2^{23} MD5 computations, which can be quickly performed.

In our experiment, we implemented the APOP attack for 31 characters, and confirmed that 31 characters were successfully recovered. The computation time to generate a collision for recovering up to 11 characters was less than one second, and for recovering 31st characters was 5.86 seconds on average. In table 2, we show examples of collisions, where we use to guess the 31st character.

Remarks
The complexity of generating collisions for recovering more than 31 characters is estimated as follows.

     35 characters: 2^{38} 39 characters: 2^{39} 43 characters: 2^{41}

     47 characters: 2^{42} 51 characters: 2^{43}

To recover more characters, the message space in Block 2 is too small, and we need to search Block 1 again. In theory, our attack is more efficient than finding collisions using the birthday paradox (up to the first 61 characters).

6.3 The Speed of Password Recovery Attack

In most cases, speed of password recovery attack is limited by the frequency of interruption by impersonation. According to the assumption of [11], a password is leaked character by character every 1 hour. This estimation is also applicable to our attack. Therefore, 31 password characters can be recovered in 31 hours.

7   Conclusion

In this paper, we extended the APOP password recovery attack. Previous attacks can recover at most three characters, whereas our attack recovers 31 characters in practical time, and theoretically can recover up to 61 characters.

The core technique of our improvement is a new MD5 collision attack that offers more advantages for an APOP attack. Our approach uses den Boer and Bosselaers's attack, and we introduced the IV Bridge, which is a differential path producing den Boer and Bosselaers's ΔIV from the real IV of MD5.

We experimentally confirmed the practicality of the extended APOP attack. As a result, we successfully recovered up to 31 characters of APOP passwords.

References

1. Black, J., Cochran, M., Highland, T.: A Study of the MD5 Attacks: Insights and Improvements. In: Robshaw, M.J.B. (ed.) FSE 2006. LNCS, vol. 4047, pp. 262–277. Springer, Heidelberg (2006)
2. den Boer, B., Bosselaers, A.: Collisions for the Compression Function of MD5. In: Helleseth, T. (ed.) EUROCRYPT 1993. LNCS, vol. 765, pp. 293–304. Springer, Heidelberg (1994)
3. Contini, S., Yin, Y.L.: Forgery and partial key-recovery attacks on HMAC and NMAC using hash collisions. In: Lai, X., Chen, K. (eds.) ASIACRYPT 2006. LNCS, vol. 4284, pp. 37–53. Springer, Heidelberg (2006)
4. Daum, M., Lucks, S.: Hash Collisions (The Poisoned Message Attack) The Story of Alice and her Boss. In: Eurocrypt 2005 (2005),
   http://th.informatik.uni-mannheim.de/people/lucks/HashCollisions/
5. Dobbertin, H.: Cryptanalysis of MD5 compress. In: Eyrocrypt 1996 (1996)
6. Dobbertin, H.: The Status of MD5 After a Recent Attack. In: CryptoBytes The technical newsletter of RSA Laboratories, a division of RSA Data Security, Inc., SUMMER 1996, vol. 2(2) (1996)
7. Franks, J., Hallam-Baker, P., Hostetler, J., Lawrence, S., Leach, P., Luotonen, A., Stewart, L.: HTTP Authentication: Basic and Digest Access Authentication, RFC 2617, June 1999(1999), http://www.ietf.org/rfc/rfc2617.txt
8. Gebhardt, M., Illies, G., Schindler, W.: A note on the practical value of single hash collisions for special file formats. In: Dittmann, J. (ed.) Sicherheit, GI. LNI, vol. 77, pp. 333–344 (2006)
9. Klima, V.: Tunnels in Hash Functions: MD5 Collisions Within a Minute. Cryptology ePrint Archive, Report, /105. (2006), http://eprint.iacr.org/2006/105.pdf
10. Lenstra, A.K., de Weger, B.: On the possibility of constructing meaningful hash collisions for public keys. In: Boyd, C., González Nieto, J.M. (eds.) ACISP 2005. LNCS, vol. 3574, pp. 267–279. Springer, Heidelberg (2005)
11. Leurent, G.: Message Freedom in MD4 and MD5 Collisions: Application to APOP. In: Biryukov, A. (ed.) FSE 2007. LNCS, vol. 4593, pp. 309–328. Springer, Heidelberg (2007)
12. Liang, J., Lai, X.: Improved Collision Attack on Hash Function MD5. Journal of Computer Science and Technology 22(1), 79–87 (2007)
13. Rivest, R.L.: The MD5 Message Digest Algorithm. RFC 1321 (April, 1992), http://www.ietf.org/rfc/rfc1321.txt

14. Myers, J., Rose, M.: Post Office Protocol - Version 3. RFC 1939 (Standard), May 1996. Updated by RFCs 1957, 2449, http://www.ietf.org/rfc/rfc1939.txt
15. Preneel, B., van Oorschot, P.C.: On the Security of Two MAC Algorithms. In: Maurer, U.M. (ed.) EUROCRYPT 1996. LNCS, vol. 1070, pp. 19–32. Springer, Heidelberg (1996)
16. Rosenberg, J., Schulzrinne, H., Camarillo, G., Johnston, A., Peterson, J., Sparks, R., Handley, M., Schooler, E.: SIP: Session Initiation Protocol, RFC 3261, June 2002 (2002), http://www.ietf.org/rfc/rfc3261.txt
17. Sasaki, Y., Naito, Y., Kunihiro, N., Ohta, K.: Improved, collision attack on MD5. Cryptology ePrint Archive, Report 2005/400, http://eprint.iacr.org/2005/400
18. Sasaki, Y., Naito, Y., Kunihiro, N., Ohta, K.: Improved Collision Attacks on MD4 and MD5. IEICE TRANSACTIONS on Fundamentals of Electronics, Communications and Computer Sciences (Japan), E90-A(1), 36–47 (2007) (The initial result was announced as [17])
19. Sasaki, Y., Yamamoto, G., Aoki, K.: Practical Password Recovery on an MD5 Challenge and Response. Cryptology ePrint Archive, Report 2007/101
20. Stevens, M., Lenstra, A., der Weger, B.: Chosen-prefix Collisions for MD5 and Colliding X.509 Certificates for Different Identities. In: Naor, M. (ed.) EUROCRYPT 2007. LNCS, vol. 4515, pp. 1–12. Springer, Heidelberg (2007)
21. Wang, X., Yu, H.: How to Break MD5 and Other Hash Functions. In: Cramer, R.J.F. (ed.) EUROCRYPT 2005. LNCS, vol. 3494, pp. 19–35. Springer, Heidelberg (2005)

A Other Attack Targets: SIP and Digest Authentication

We show that the APOP attack is applicable to SIP [16], which is an application-layer protocol for Internet telephone calls, multimedia conferences *e.t.c.*, and Digest Authentication [7], which is hash-based authentication protocols for HTTP Authentication. Authentication part of SIP is based on Digest Authentication.

We found that these two protocols are vulnerable against the APOP attack. Namely, the first 31 characters of user's password is practically recovered. If the attacker (impersonating server) chooses the configuration of parameters which is suitable for the attack, the response computation is as follows.

$$A1 = \mathrm{MD5}(userID\|\text{“ : ”}\|realm\|\text{“ : ”}\|password)$$
$$A2 = \mathrm{MD5}(Method\|\text{“ : ”}\|uri)$$
$$response = \mathrm{MD5}(A1\|\text{“ : ”}\|nonce\|\text{“ : ”}\|A2),$$

where, *userID* and *password* are pre-shared between the user and the server, *realm, uri* and *nonce* are determined by the server as challenge string, and *Method* is a constant value depends on what for the authentication is done.

In these protocols, the secret password is hashed twice, so their securities look stronger than APOP. However, since collision on $A1$ can be observed from *response*s, the attacker can recover the password by generating *realm*s as suitable for the attack.

B   List of Tables

Table 2. Examples of Guess for the 31st Character

$M_0 = M_0'$	m_0=0x3938313c m_1=0x37322d34 m_2=0x332d3635 m_3=0x2e383933 m_4=0x37373936 m_5=0x3433302d m_6=0x38312d35 m_7=0x61704035 m_8=0x6f777373 m_9=0x645f6472 m_{10}=0x63657465 m_{11}=0x5f726f74 m_{12}=0x2e636264 m_{13}=0x6976746d m_{14}=0x632e7765 m_{15}=0x73752e61
$H_1 = H_1'$	0x769b8c7f 0xc41742a1 0x1c8cefcc 0x195d17f4
M_1	m_0=0x986e1da4 m_1=0x83707d06 m_2=0xa86e1ddd m_3=0xe264eedb m_4=0xff68e19f m_5=0x120ea5b3 m_6=0x7437d3e2 m_7=0x600f543d m_8=0x7c63c5ab m_9=0xe9ead9d9 m_{10}=0xa9b5c51e m_{11}=0x<u>c</u>309f623 m_{12}=0xfd534f1e m_{13}=0xad33c7ad m_{14}=0xfd0380c6 m_{15}=0x7745f36a
M_1'	m_0'=0x986e1da4 m_1'=0x83707d06 m_2'=0xa86e1ddd m_3'=0xe264eedb m_4'=0xff68e19f m_5'=0x120ea5b3 m_6'=0x7437d3e2 m_7'=0x600f543d m_8'=0x7c63c5ab m_9'=0xe9ead9d9 m_{10}'=0xa9b5c51e m_{11}'=0x<u>4</u>309f623 m_{12}'=0xfd534f1e m_{13}'=0xad33c7ad m_{14}'=0xfd0380c6 m_{15}'=0x7745f36a
H_2	0x<u>b</u>d7ade50 0x<u>e</u>17a619d 0x<u>8</u>e940937 0x<u>f</u>d4af95f
H_2'	0x<u>3</u>d7ade50 0x<u>6</u>17a619d 0x<u>0</u>e940937 0x<u>7</u>d4af95f

$M_2^{(3)} = M_2'^{(3)}$	m_0=0x6cbebe2c m_1=0x539e4d17 m_2=0x6f342bd1 m_3=0x78e2b4e9 m_4=0xef5d9c25 m_5=0xe02bd34e m_6=0x774d98bd m_7=0x1f7b0622 m_8=0x4342413e m_9=0x47464544 m_{10}=0x4b4a4948 m_{11}=0x4f4e4d4c m_{12}=0x53525150 m_{13}=0x57565554 m_{14}=0x305a5958 m_{15}=0x33333231
$H_3^{(3)} = H_3'^{(3)}$	0x19835d2f 0xa668c75c 0xe100b765 0x3c0acd29
$M_2^{(4)} = M_2'^{(4)}$	m_0=0x8aabec3f m_1=0xf18741d3 m_2=0xe2c27459 m_3=0x046584ad m_4=0x4f20f254 m_5=0xed9a52ad m_6=0x4bab0526 m_7=0xbcfc89e2 m_8=0x4342413e m_9=0x47464544 m_{10}=0x4b4a4948 m_{11}=0x4f4e4d4c m_{12}=0x53525150 m_{13}=0x57565554 m_{14}=0x305a5958 m_{15}=0x34333231
$H_3^{(4)} = H_3'^{(4)}$	0x53fb9ff6 0x33a379ab 0x1fdb63a3 0x58f86842

MD5 uses little-endian, where the first byte of a challenge string becomes the first byte of m_0 for M_0. In the example in Table 2, M_0 starts with '<'(0x3c), and includes an '@'(0x40). M_1, M_1' are the IV Bridge, where a difference exists in MSB of m_{11}. H_2, H_2' are results of the IV Bridge that have the following differences:

$$\Delta H_2 = (\text{0x80000000}, \text{0x80000000}, \text{0x80000000}, \text{0x80000000}).$$

When the 1st to 30th password characters are

'ABCDEFGHIJKLMNOPQRSTUVWXYZ0123'
(ASCII code: 0x41(=A), 0x42(=B), $\cdots$ 0x5a(=Z), 0x30(=0), $\cdots$ 0x33=(3)),

$M_2^{(3)}$ is yielded by guessing that the 31st character is '3'(0x33) and $M_2^{(4)}$ is produced by guessing that it is '4'(0x34).

In both $M_2^{(3)}$ and $M_2^{(4)}$, m_0 to m_7 are randomly chosen. The first byte of m_8 is fixed to be '>'(0x3e), which we need in the end of the challenge string. The following 30 bytes are fixed as the password. Finally, the last byte is fixed to each guess.

In Table 2, we underline the values that have difference (M_1, M_1', H_2, H_2').

Table 3. Differential Path for the IV Bridge

Step	Shift		Δb_i	
i	s_i	Δm_{i-1}	Numerical difference	Difference in each bit
12	22	$*2^{31}$	2^{21}	$\Delta[-21, -22, \cdots, 31]$
13	7		2^6	$\Delta[-6, -7, -8, 9]$
			2^{21}	$\Delta[21]$
			-2^{31}	$\Delta[-31]$
14	12		2^{31}	$\Delta[31]$
15	17		-2^6	$\Delta[-6]$
			-2^{10}	$\Delta[10, 11, \cdots, -20]$
			2^{31}	$\Delta[31]$
16	22		2^{31}	$\Delta[31]$
17	5		2^{26}	$\Delta[26]$
			-2^{31}	$\Delta[-31]$
18	9		2^8	$\Delta[8]$
			2^{20}	$\Delta[20]$
			2^{24}	$\Delta[-24, -25, \cdots, 28]$
			2^{31}	$\Delta[31]$
19	14	$*2^{31}$	2^{31}	$\Delta[31]$
20	20		2^{31}	$\Delta[31]$
21	5		-2^1	$\Delta[1, 2, \cdots, -8]$
			2^{31}	$\Delta[31]$
22	9		2^{13}	$\Delta[-13, -14, \cdots, 16]$
			2^{29}	$\Delta[29]$
			2^{31}	$\Delta[31]$
23	14		-2^{31}	$\Delta[-31]$
24	20		-2^{31}	$\Delta[-31]$
25	5		-2^6	$\Delta[6, 7, -8]$
			-2^{31}	$\Delta[-31]$
26	9		2^{22}	$\Delta[-22, 23]$
			-2^{31}	$\Delta[-31]$
27	14		-2^{31}	$\Delta[-31]$
28	20		2^{11}	$\Delta[11]$
			-2^{31}	$\Delta[-31]$
29	5		-2^{31}	$\Delta[-31]$
30	9			
31	14			
32	20		$*2^{31}$	$\Delta[*31]$
33	4		$*2^{31}$	$\Delta[*31]$
34	11		$*2^{31}$	$\Delta[*31]$
35	16	$*2^{31}$	$*2^{31}$	$\Delta[*31]$
36	23		$*2^{31}$	$\Delta[*31]$
$\cdots$			$*2^{31}$	$\Delta[*31]$
61	6		$*2^{31}$	$\Delta[*31]$
62	10	$*2^{31}$	$*2^{31}$	$\Delta[*31]$
63	15		$*2^{31}$	$\Delta[*31]$
64	21		$*2^{31}$	$\Delta[*31]$

The symbol '$\Delta[i]$' means that the value of the chaining variable in bit position i changes from 0 to 1. The symbol '$\Delta[-i]$' means that it changes from 1 to 0 instead.

Table 4. Sufficient Conditions and Extra Conditions for the IV Bridge

Chaining variables	Conditions on bits			
	31 - 24	23 - 16	15 - 8	7 - 0
$b_1 - b_8$	- - - - - - - -	- - - - - - - -	- - - - - - - -	- - - - - - - -
b_9	- - - - - - - -	- - - $\bar{0}\bar{0}\bar{0}\bar{0}\bar{0}$	$\bar{0}\bar{0}\bar{0}\bar{0}\bar{0}\bar{0}\bar{0}\bar{0}$	$\bar{0}\bar{0}\bar{0}\bar{0}\bar{0}\bar{0}\bar{0}\bar{0}$
b_{10}	1 - - - - - - 1	- - $\bar{0}\bar{0}\bar{0}\bar{0}\bar{0}\bar{0}$	$\bar{0}\bar{0}\bar{0}\bar{0}\bar{0}\bar{0}\bar{0}\bar{0}$	$\bar{0}\bar{0}\bar{0}\bar{0}\bar{0}\bar{0}\bar{0}\bar{0}$
b_{11}	0 a a a a a a 0	a a 0 1 $\bar{1}\bar{1}\bar{1}\bar{1}$	$\bar{1}\bar{1}\bar{1}\bar{1}\bar{1}\bar{1}\bar{1}\bar{1}$	$\bar{1}\bar{1}\bar{1}\bar{1}\bar{1}\bar{1}\bar{1}\bar{1}$
b_{12}	0 1 1 1 1 1 1 1	1 1 1 0 - - - -	- - - - - - 0 a	a a - - - - - -
b_{13}	1 0 0 0 0 1 0 0	0 0 0 0 1 1 1 0	1 1 1 1 1 1 0 1	1 1 - - - - - -
b_{14}	0 1 1 1 1 1 0 1	1 $\neq$ - 1 a a a 1	a a a a a a 0 0	0 0 - - - - - -
b_{15}	0 - - - - 0 - -	0 0 - 0 1 0 - -	- - - - - - 0 1	1 1 1 1 1 1 1 -
b_{16}	0 - - 0 0 1 0 0	- - - 0 0 0 0 -	0 - - - 0 - - 0	- - - - - - - -
b_{17}	1 - - 1 1 0 1 1	- - - 1 1 1 1 a	1 a a a 1 a - 1	- a - - - - - -
b_{18}	0 - - 0 1 1 1 1	- - - 0 - - - -	- - - - - - - 0	- - - - - - - -
b_{19}	0 - - 1 0 1 - -	- - - - - - - -	- - - - - - - 1	0 0 0 1 0 0 0 -
b_{20}	0 - 0 0 1 0 a a	- - - a - - - 0	1 0 0 - - - - 1	1 1 1 1 1 1 1 -
b_{21}	0 - 1 1 - - - -	- - - - - - - 1	1 1 1 - - - - 1	0 0 0 0 0 0 0 -
b_{22}	1 - 0 0 - - - -	- - - - - - - 0	1 1 1 - - - - 0	0 0 - - - - - -
b_{23}	1 - 1 1 - - - -	- - - - - - - -	- - - - - - - a	a 0 a a a a a -
b_{24}	1 - a - - - - -	0 0 - - - - - a	a a a - - - - 0	1 1 - - - - - -
b_{25}	1 - - - - - - -	0 1 - - - - - -	- - - - - - - 1	0 0 - - - - - -
b_{26}	1 - - - - - - -	0 1 - - - - - -	- - - - 0 - - -	- - - - - - - -
b_{27}	1 - - - - - - -	- - - - - - - -	- - - - 1 - - a	a a - - - - - -
b_{28}	1 - - - - - - -	a a - - - - - -	- - - - 0 - - -	- - - - - - - -
b_{29}	1 - - - - - - -	- - - - - - - -	- - - - - - - -	- - - - - - - -
b_{30}	1 - - - - - - -	- - - - - - - -	- - - - a - - -	- - - - - - - -
b_{31}	0 - - - - - - -	- - - - - - - -	- - - - - - - -	- - - - - - - -
b_{32}	0 - - - - - - -	- - - - - - - -	- - - - - - - -	- - - - - - - -
$\cdots$	- - - - - - - -	- - - - - - - -	- - - - - - - -	- - - - - - - -
b_{48}	c - - - - - - -	- - - - - - - -	- - - - - - - -	- - - - - - - -
$\cdots$	c - - - - - - -	- - - - - - - -	- - - - - - - -	- - - - - - - -
b_{60}	c - - - - - - -	- - - - - - - -	- - - - - - - -	- - - - - - - -
b_{61}	d - - - - - - -	- - - - - - - -	- - - - - - - -	- - - - - - - -
b_{62}	c - - - - - - -	- - - - - - - -	- - - - - - - -	- - - - - - - -
b_{63}	c - - - - - - -	- - - - - - - -	- - - - - - - -	- - - - - - - -
cc_0	a - - - - - - -	- - - - - - - -	- - - - - - - -	- - - - - - - -
bb_0	a - - - - - - -	- - - - - - - -	- - - - - - - -	- - - - - - - -

The notations '0', '1', 'a', 'b', 'c', 'd' and '$\neq$' in bit position j of chaining variable b_i stand for the conditions $b_{i,j} = 0$, $b_{i,j} = 1$, $b_{i,j} = b_{i-1,j}$, $b_{i,j} \neq b_{i-1,j}$, $b_{i,j} = b_{i-2,j}$, $b_{i,j} \neq b_{i-2,j}$, and $b_{i,j} \neq b_{i,j-1}$, respectively.

The last two conditions mean that $bb_{0,31} = cc_{0,31} = dd_{0,31}$, which are required by den Boer and Bosselaers's attack.

We emphasize that the IV Bridge does not set any conditions for b_1-b_8.

Conditions without upper bar are sufficient conditions for the differential path. Conditions with upper bar are extra conditions that are introduced for Phase C. We also considered conditions introduced by [12]. In step i, let Σ_i be $a_{i-1} + f(b_{i-1}, c_{i-1}, d_{i-1}) + m_{i-1} + t_{i-1}$. There exists two conditions on Σ: $\Sigma_{30,22} = 0$, $\Sigma_{32,11} = 0$.

C   A Graphical Explanation of Phase B

Chaining variables highlighted by gray are the input for Phase B. Those values are fixed during Phase B. m_5 and m_6 which are used in step 6 and 7 are also input for Phase B. We omit these variables by including them in the constant. Numbers with parenthesis denotes the order of determining the chaining variables. We explain how Phase B in section 5.2 connects steps 3-7.

- Stage 1 of Phase B fixes chaining variables (1) to 0xffffffff. Then, the output of f in step 6 is equal to c_5. This is indicated by the bold line.
- Next, we fix chaining variable c_7 in order to fix the output of f in step 6. Therefore, chaining variables (2) are fixed. c_7 can be fixed to any value, therefore, we repeat the algorithm 2^{32} times by changing c_7 from 0 to 0xffffffff.

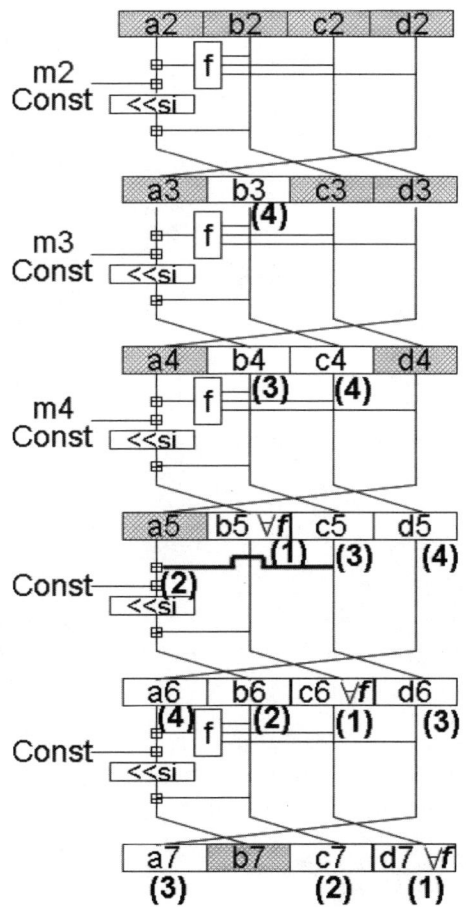

Fig. 1. Graphical Explanation of the Strategy for Phase B

- After that, we can compute the output of f in step 6, which is the same value as that of c_5. Therefore, the chaining variables (3) are fixed.
- All chaining variables input to f in step 7 are fixed, so we can compute its output, this, uniquely determining a_6. Therefore, the chaining variables (4) are fixed, and steps 3-7 will be connected by *Inverse-m* for m_2, m_3 and m_4.

Cryptanalysis of a Hash Function Based on Quasi-cyclic Codes

Pierre-Alain Fouque and Gaëtan Leurent

École Normale Supérieure – Département d'Informatique,
45 rue d'Ulm, 75230 Paris Cedex 05, France
{Pierre-Alain.Fouque,Gaetan.Leurent}@ens.fr

Abstract. At the ECRYPT Hash Workshop 2007, Finiasz, Gaborit, and Sendrier proposed an improved version of a previous provably secure syndrome-based hash function. The main innovation of the new design is the use of a quasi-cyclic code in order to have a shorter description and to lower the memory usage.

In this paper, we look at the security implications of using a quasi-cyclic code. We show that this very rich structure can be used to build a highly efficient attack: with most parameters, our collision attack is faster than the compression function!

Keywords: hash function, provable security, cryptanalysis, quasi-cyclic code, syndrome decoding.

1 Introduction

Following the breakthrough collision attacks by Wang *et al.* against the most widespread hash functions (MD5 in [10], SHA-1 in [9]) the crypto community is trying to design new hash functions. One interesting approach is the construction of *provably secure* hash functions, in which the security of the hash function is proven to rely on some computationally hard problem.

At Mycrypt 2005, Augot, Finiasz, and Sendrier proposed a family of provably secure hash functions based on the syndrome decoding problem called FSB [1]. An improved version of this design (IFSB) was presented at the ECRYPT Hash Workshop 2007 by Finiasz, Gaborit and Sendrier [5]. The new idea introduced in IFSB is to use a quasi-cyclic code instead of a random code. This allows to store a smaller description of the code, and there is a huge speedup when it fits into the CPU cache. This modification was assumed not to lower the security of the construction.

However, this new proposal was broken by Saarinen using a simple linearization technique [7]. The attack does not take advantage of the new elements of the design (*ie.* the quasi-cyclic codes) and can also be used to break the initial function; it is only based on the fact that ratio between two parameters of the hash function is relatively small ($r/w < 2$). Therefore, it does not invalidate the approach of [5]; we can still use a hash function based on a quasi-cyclic code, we just need to choose a larger r.

T. Malkin (Ed.): CT-RSA 2008, LNCS 4964, pp. 19–35, 2008.

This papers studies the IFSB construction, and how the use of a quasi-cyclic code affects its security. We show that if the block size has a small divisor, we can build a very efficient attack which works with a larger r than the linearization attack. We still don't break IFSB for any possible parameters, but most proposed parameters are sensible to this attack.

Our Results. We first point out a strange property of the FSB/IFSB family in Section 3: the mixing of the chaining value with the message is very weak. This has two main consequences: a collision attack is essentially the same as a pseudo-collision attack, and the compression function is not a pseudo random function family (PRF).

In section 4, we introduce our new collision attacks on IFSB based on the structure of quasi-cyclic codes. The main result is an attack using piecewise periodic messages, and a new algorithm to solve a system of cyclic equations. Here is a brief comparison (for pseudo-collisions) of the previous linearization attack and our new cyclic attack (see Table 1 on page 33 for practical figures):

Attack	Conditions	Complexity	Remarks
Linearization	$r \leq 2w$	r^3	r is typically 1024
	if r is bigger	$(4/3)^{r-2w} \cdot r^3$	$\log_2(4/3) \approx 0.415$
Cyclic	$r \leq 4w$	$(n/4w)^3$	$n/4w$ is typically 64
	if r is bigger	$2^{\frac{n(r-4w)}{4wr}} \cdot (n/4w)^3$	$n/4wr$ is typically $1/16$

These attacks are both very efficient when the ratio r/w is below a given threshold, and can be extended to work above the threshold, but the complexity grows exponentially with r. The cyclic attack has a bigger threshold, and when the threshold is exceeded it has a lower factor in the exponent. Note that the linearization attack can break FSB as well as IFSB, whereas our cyclic attack relies on the property of the quasi-cyclic code: it can only break IFSB and also requires r to be a power of 2.

Since these attacks rely on a low r/w ratio, they do not rule out the IFSB construction, but only the parameter proposed in [5]. For instance, one could use the IFSB construction with parameters based on [1], which have a higher r/w. However, the first step of our attack can also be used together with Wagner's attack, so as to remove the dependency in the ration r/w (there is still a requirement that r is a multiple of $n/2w$):

Attack	Complexity	Remarks
Wagner	$r2^{a'} \cdot 2^{r/(a+1)}$	Used as the security parameter
Cyclic + Wagner	$\frac{n}{2w}2^{a'} \cdot 2^{\frac{n}{2w}/(a'+1)}$	a' is a or $a-1$

Since $n/2w$ is typically between $r/2$ and $r/8$, the new attack will have a complexity between the square root and the eighth root of the security parameter. These attacks basically show that the IFSB construction with a quasi-cyclic code is not secure if r, the length of the code has many divisors. The easy choice of a power of two is highly insecure.

In [5], the authors provided some security argument when r is prime and 2 is a primitive root of $\mathbb{Z}/r\mathbb{Z}$. Since this was not a real proof, and they were confident in the security of quasi-cyclic codes even without this argument, some of their parameters do not respect this constraint. Our attacks show that IFSB should only be used with a prime r.

About Provable security. The main motivation for the design of FSB is to have a *proof of security*. In [1], the authors of FSB defined the 2-RNSD problem (2-Regular Null Syndrome Decoding) so that finding a collision in FSB given the matrix $\mathcal{H}$ is equivalent to solving 2-RNSD on the matrix $\mathcal{H}$. They also prove that 2-RNSD is a NP-complete problem, which is a good evidence that there is no polynomial time algorithm to break FSB. However, this does not really prove that finding a collision is hard:

- The fact that there is no polynomial time algorithm to break the function is an *asymptotic* property, but in practice the function is used with a *fixed size*; there might be an algorithm that break the function up to a given size in very little time. Usually, designers look at the best known attack and choose the size of the function so that this attack is unpractical, but there could be a more efficient algorithm (though superpolynomial). Indeed, the first version of FSB did not consider Wagner's generalized birthday attack [4], and the parameters had to be changed.
- More importantly, the fact that a problem is NP-Complete means that there are *some* hard instances, not that *every* instance is hard. For instance, SAT is an NP-Complete problem, but if the specific formula you try to satisfy happen to be in 2-SAT there is a polynomial time algorithm. In the case of FSB, the 2-RNSD problem is NP-Complete but IFSB used instances where $r < 2w$ which are easy [7].
- Moreover, IFSB is an improved version of FSB, but the new components (the final transformation, the change to a quasi-cyclic matrix, and the possibility to use a new constant weight encoder) have no security proof. Indeed, our main result is an algorithm that breaks the 2-RNSD problem when the matrix is quasi-cyclic and r is a power of 2, which is the case for most proposed settings of IFSB.
- Lastly, the security proof considers some specific attack, such as collision, or preimage, but there might be some other undesirable property in the design. In Section 3, we show that the FSB design does not mix properly the chaining value with the message.

2   Design of IFSB and Previous Cryptanalysis

The Fast Syndrome Based (FSB) and Improved Fast Syndrome Based (IFSB) hash functions follow the Merkle-Damgård construction. The compression function is built of two steps:

- A constant weight encoder φ that maps an s-bit word to a n-bit word of Hamming weight w.
- A random matrix $\mathcal{H}$ of size $r \times n$. Typically, r is about one hundred, and n is about one million.

The compression function of FSB takes s bits as input (r bits of chaining variable, and $r - s$ bits of message) and outputs r bits; it is defined by:

$$F(x) = \mathcal{H} \times \varphi(x)$$

$\mathcal{H}$: random $r \times n$ matrix

φ: encodes s bits to n bits with weight w

In the case of IFSB, the matrix $\mathcal{H}$ is quasi-cyclic: $\mathcal{H} = \mathcal{H}_0 || \mathcal{H}_1 || ... \mathcal{H}_{n/r-1}$ and each $\mathcal{H}_i$ is a cyclic $r \times r$ matrix. There is also a final transformation to reduce the size of the digest, but it is not used in our collision attacks.

Notations. We will use the following notations:

- $\mathbf{0}$ and $\mathbf{1}$ are bit-strings of length one, as opposed to 0 and 1.
- if x is a bit-string, x^k is the concatenation of k times x.
- $x^{[i]}$ is the $(i+1)$-th bit of x (i is between 0 and $|x| - 1$).
- $[f(i)]_{i=0}^{p-1}$ is the matrix whose *columns* are the $f(i)$'s.

2.1   Choosing a Constant Weight Encoder

Three constant word encoders are proposed in [5]. The regular encoder is a very simple one, used in most parameters. The optimal encoder and the tradeoff encoder are introduced in order to reduce the size of the matrix: they can use more input bits with the same parameters n and w.

The Regular Encoder. The regular encoder was introduced in [1]; it is the only encoder defined for the FSB family, and the main one for the IFSB family. It is designed to be efficient and is very simple. The message M is split into w words m_i of $\log(n/w)$ bits (all the log in this paper are base 2), and the output word is divided into chunks of n/w bits. Then, each chunk of the output word contains exactly one non-zero bit, with its position defined the corresponding m_i. For an efficient implementation, we will often use $\log(n/w) = 8$.

More precisely, the regular encoder is defined as:

$$\varphi(M) = \varphi(m_0 || m_1 || ...) = \bigoplus_{i=0}^{w-1} \varphi_i(m_i)$$

where φ_i is used to encode one message word into the output chunk i:

$$\varphi_i(x) = 0^{in/w} 0^x 1 0^{n/w-1-x} 0^{n-(in+1)/w}$$

$$\varphi_i(x)^{[k]} = 1 \iff k = in/w + x$$

FSB with the regular encoder just selects one particular column of the matrix $\mathcal{H}$ for each message word:

$$F(M) = \mathcal{H} \times \bigoplus_{i=0}^{w-1} \varphi_i(m_i) = \bigoplus_{i=0}^{w-1} \mathcal{H}_{in/w+m_i}.$$

The Optimal Encoder. The optimal encoder tries to have all the words of weight w in its range, whereas the regular encoders only output regular words. There are $\binom{n}{w}$ such words, as opposed to $(n/w)^w$ regular words; this allows the optimal encoder to use more input bits. The optimal encoder will actually map $\lfloor \log \binom{n}{w} \rfloor$ bits to a subspace of $2^{\lfloor \log \binom{n}{w} \rfloor}$ words. We do not consider the details of the computation, we will only assume that φ and φ^{-1} are efficiently computable.

The Tradeoff Encoder. The main problem of the optimal encoder is that it requires some computations with very large integers. Therefore, the tradeoff encoder uses a combination of the optimal encoder and the regular encoder. Here again, we do not need the details of the construction, we only use the fact that φ and φ^{-1} are efficiently computable.

2.2   Wagner's Generalized Birthday

Wagner's generalized birthday attack [8] is a clever trick to solve the k-sum problem: given some lists $L_1, L_2, .., L_k$ of r-bit values, we want to find $l_1 \in L_1, .., l_k \in L_k$ such that $\bigoplus_{i=1}^{k} l_k = 0$. If each list contains at least $2^{r/k}$ elements there is a good probability that a solution exists, but the best known algorithm is a simple birthday attack in time and memory $\widetilde{\mathcal{O}}(2^{r/2})$. The idea is to build two lists L_A and L_B with all the sums of elements in $L_1, ... L_{k/2}$ and $L_{k/2+1}, ... L_k$ respectively, sort L_A and L_B, and look for a match between the two lists (L_A and L_B contains $2^{r/2}$ elements).

Wagner's algorithm will have a lower complexity, but it requires more elements in the lists. For instance with $k = 4$ we need lists of size $2^{r/3}$ and the algorithm will find one solution using $\widetilde{\mathcal{O}}(2^{r/3})$ time and memory.

The basic operation of the algorithm is the general join $\bowtie_j$: $L \bowtie_j L'$ consists of all elements of $L \times L'$ that agree on their j least significant bits:

$$L \bowtie_j L' = \left\{ (l, l') \in L \times L' \mid (l \oplus l')^{[0..j-1]} = 0^j \right\}.$$

The algorithm for $k = 4$ is described by Figure 1. We first build the list $L_{12} = L_1 \bowtie_{r/3} L_2$ and $L_{34} = L_3 \bowtie_{r/3} L_4$. By the birthday paradox, these lists should contain about $2^{r/3}$ elements. Next, we build $L_{1234} = L_{12} \bowtie_{2r/3} L_{34}$. Since the elements of L_{12} and L_{34} already agree on their $r/3$ lower bits, we are only doing a birthday paradox on the bits $r/3$ to $2r/3$, so we still expect to find $2^{r/3}$ elements. Finally, we expect one of the elements of L_{1234} to be the zero sum. This can be generalized to any k that is a power of two, using a binary tree: if $k = 2^a$, we

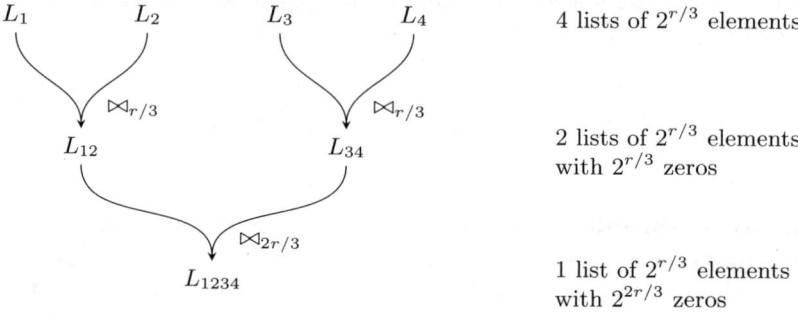

$$\text{Fig. 1. Wagner's algorithm for } k = 4$$

need k lists of $2^{r/(a+1)}$ elements and the time and memory used by the algorithm is $2^a \cdot r2^{r/(a+1)}$.

Finding a collision in FSB with the regular encoder is just an instance of the k-sum problem, so we can use Wagner's algorithm to break it, as shown by Coron and Joux [4]. We look for M and M' such that

$$F(M) \oplus F(M') = \bigoplus_{i=0}^{w-1} \mathcal{H}_{in/w+m_i} \oplus \bigoplus_{i=0}^{w-1} \mathcal{H}_{in/w+m_i'} = 0.$$

This is an instance of the $2w$-sum problem on r bits, with lists of n/w elements: $L_i = \{\mathcal{H}_{in/w+m_i}\}_{m_i=0}^{n/w-1}$ and $L_i' = \{\mathcal{H}_{in/w+m_i'}\}_{m_i'=0}^{n/w-1}$. If $n/w < 2^{r/(\log(w)+2)}$, we cannot apply Wagner's attack with $k = 2w$ as it is, but we can group the lists. For instance, we can build $w/2$ lists of $(n/w)^4$ elements, and this attack is applicable with $k = w/2$ if $(n/w)^4 \geq 2^{r/\log(w)}$. In the end, we will use the largest a that satisfies:

$$\frac{2^a}{a+1} \leq \frac{2w}{r} \log(n/w).$$

This is the main attack to break the k-sum problem, therefore it was used by the authors of FSB to define their security levels.

2.3   Linearization Attack

Wagner noted in [8] that the generalized birthday problem can be solved using linear algebra when $r \leq k$, using a result from Bellare and Micciancio [2, Appendix A]. When we apply this to FSB, we can break it very efficiently when $r \leq 2w$. However, this was overlooked by the designers of ISFB, and almost all the parameters they proposed satisfy $r \leq 2w$. Saarinen rediscovered this attack in [7], and extended it to $r > 2w$ with complexity $(3/4)^{r-2w} \cdot r^3$. He also rediscovered a preimage attack when $r \leq w$ based on the same idea. Here, we will only describe the basic collision attack, using our notations.

Let us choose four distinct elements $a, b, c, d \in \{0..n/w-1\}$, and build a vector u and a matrix Δ: (a^w is the concatenation of w times a)

$$u = (\mathcal{H} \times \varphi(a^w)) \oplus (\mathcal{H} \times \varphi(c^w))$$
$$\Delta_1 = [\mathcal{H} \times (\varphi_i(a) \oplus \varphi_i(b))]_{i=0}^{w}$$
$$\Delta_2 = [\mathcal{H} \times (\varphi_i(c) \oplus \varphi_i(d))]_{i=0}^{w}$$
$$\Delta = \Delta_1 || \Delta_2$$

We solve the equation $\Delta \times x = u$ by linear algebra and we write $x = x_1 || x_2$ such that $\Delta_1 \times x_1 \oplus \Delta_2 \times x_2 = \Delta \times x = u$. Since the matrix Δ has size $r \times 2w$, there is a good probability to find a solution when $r \leq 2w$ (see Appendix A). Then we build the messages M and M' (note that the messages are distinct because a, b, c and d are all distinct):

$$m_i = \begin{cases} a & \text{if } x_1^{[i]} = 0 \\ b & \text{if } x_1^{[i]} = 1 \end{cases} \qquad m_i' = \begin{cases} c & \text{if } x_2^{[i]} = 0 \\ d & \text{if } x_2^{[i]} = 1 \end{cases}$$

Proof. This will give a collision because

$$\mathcal{H} \times \varphi(M) = \mathcal{H} \times \bigoplus \varphi_i(m_i)$$
$$= \mathcal{H} \times \left(\bigoplus_{x_1^{[i]}=0} \varphi_i(a) \oplus \bigoplus_{x_1^{[i]}=1} \varphi_i(b) \right)$$
$$= (\mathcal{H} \times \varphi(a^w)) \oplus \left(\mathcal{H} \times \bigoplus_{x_1^{[i]}=1} (\varphi_i(a) \oplus \varphi_i(b)) \right)$$
$$= (\mathcal{H} \times \varphi(a^w)) \oplus (\Delta_1 \times x_1)$$
$$\mathcal{H} \times \varphi(M') = (\mathcal{H} \times \varphi(c^w)) \oplus (\Delta_2 \times x_2)$$
$$(\mathcal{H} \times \varphi(M)) \oplus (\mathcal{H} \times \varphi(M')) = u \oplus \Delta \times x = 0 \qquad \square$$

3   An IV Weakness

The regular encoder and the tradeoff encoder do not mix the chaining variable input and the message input: we can write $F(c, M) = \mathcal{H} \times \varphi(c||M) = (\mathcal{H}_c \times \varphi_c(c)) \oplus (\mathcal{H}_M \times \varphi_M(M))$, with $\mathcal{H} = \mathcal{H}_c || \mathcal{H}_M$. This means that a collision attack is just the same as a pseudo-collision attack, we only have to work with the smaller matrix $\mathcal{H}_M$. Since we have less degree of freedom, the complexity of the attack might be higher, but it works exactly in the same way.

More importantly, a collision in the compression function is a collision for *any* chaining variable. This is quite unexpected for a hash function, and we believe this should be avoided. In particular, it means that if we replace the IV by a key, we do *not* have a pseudo-random function family. There exists an adversary that knows two messages M_1 and M_2 such that $\mathcal{H}_M \times \varphi(M_1) = \mathcal{H}_M \times \varphi(M_2)$, and we have $F(k, M_1) = F(k, M_2)$ for *any* key k, which gives a distinguisher against a random function. For instance, if FSB is used in the HMAC construction, we can run an existential forgery attack with only one chosen-message MAC, as long as we know one fixed collision. Even if no collisions are known, the security against a PRF-distinguisher is only $2^{n/2}$, instead of the expected 2^n. This also allows to forge signatures for virtually any hash-based signature scheme, or to build cheap multi-collisions.

4   The Cyclic Attack

Our new attack on IFSB relies on the structure of quasi-cyclic codes and uses two new ideas. The first idea is to use a message M such that $\varphi(M)$ is piecewise periodic. This reduces the message space, but when the code is quasi-cyclic, we will see that the hash will become periodic, and we can now work on only one period, instead of the whole hash. This step can be used as a kind of preprecessing for Wagner's attack or for the linearization attack. The second part of the attack is an algorithm to solve the remaining system of cyclic equations, which is more efficient than Wagner's attack or the linearization technique.

4.1   Quasi-cyclic Codes and Rotations

We use $x \lll s$ to denote x rotated by s bits, and we say that x is s-periodic if $x = x \lll s$. Similarly, if x is broken into pieces of k bits $x = x_0||x_1||x_2...$, we define the k-piecewise rotation:

$$x \overset{k}{\lll} s = (x_0 \lll s)||(x_1 \lll s)||(x_2 \lll s)...$$

If $x \overset{k}{\lll} s = x$, we say that x is piecewise periodic. In this paper we will always use $k = r$.

Let us introduce a few definitions and properties of cyclic and quasi-cyclic matrices.

Definition 1. *The matrix $\mathcal{H}$ is cyclic if each row vector is rotated one element to the right relative to the previous row vector:*

$$\mathcal{H} = \begin{bmatrix} \alpha_0 & \alpha_1 & \dots & \alpha_{r-2} & \alpha_{r-1} \\ \alpha_{r-1} & \alpha_0 & \alpha_1 & & \alpha_{r-2} \\ \vdots & \alpha_{r-1} & \alpha_0 & \ddots & \vdots \\ \alpha_2 & & \ddots & \ddots & \alpha_1 \\ \alpha_1 & \alpha_2 & \dots & \alpha_{r-1} & \alpha_0 \end{bmatrix}$$

Property 1. If $\mathcal{H}$ is cyclic, we have:

$$\mathcal{H} \times (x \lll s) = (\mathcal{H} \times x) \lll s$$

Definition 2. $\mathcal{H}$ *is quasi-cyclic if* $\mathcal{H} = (\mathcal{H}_0, \mathcal{H}_1, ...\mathcal{H}_{n/r-1})$, *and each* $\mathcal{H}_i$ *is cyclic.*

$$
\mathcal{H} = \begin{bmatrix} \alpha_0 & \alpha_1 & \cdots & \alpha_{r-2} & \alpha_{r-1} \\ \alpha_{r-1} & \alpha_0 & \alpha_1 & & \alpha_{r-2} \\ \vdots & \alpha_{r-1} & \alpha_0 & \ddots & \vdots \\ \alpha_2 & & \ddots & \ddots & \alpha_1 \\ \alpha_1 & \alpha_2 & \cdots & \alpha_{r-1} & \alpha_0 \end{bmatrix} \begin{bmatrix} \beta_0 & \beta_1 & \cdots & \beta_{r-2} & \beta_{r-1} \\ \beta_{r-1} & \beta_0 & \beta_1 & & \beta_{r-2} \\ \vdots & \beta_{r-1} & \beta_0 & \ddots & \vdots \\ \beta_2 & & \ddots & \ddots & \beta_1 \\ \beta_1 & \beta_2 & \cdots & \beta_{r-1} & \beta_0 \end{bmatrix} \cdots
$$

Property 2. If $\mathcal{H}$ is quasi-cyclic, we have:

$$
\mathcal{H} \times (x \overset{r}{\lll} s) = \sum_{i=0}^{n/r-1} \mathcal{H}_i \times (x_i \lll s)
$$
$$
= (\mathcal{H} \times x) \lll s
$$

Corollary 1. *If* $\mathcal{H}$ *is quasi-cyclic and* x *is piecewise periodic, then* $\mathcal{H} \times x$ *is periodic:*

$$
(\mathcal{H} \times x) \lll s = \mathcal{H} \times (x \overset{r}{\lll} s) = \mathcal{H} \times x.
$$

This simple remark will be the basis of our cyclic attack.

4.2   The Main Attack

The basic idea of our attack is very simple: let us choose M and M' such that $\varphi(M)$ and $\varphi(M')$ are piecewise periodic. Then we know that the output $\mathcal{H} \times \varphi(M)$ and $\mathcal{H} \times \varphi(M')$ are periodic, and we only have to collide on *one* period.

In fact we can even take further advantage of the two messages: let us choose M such that $\varphi(M)$ is piecewise s-periodic, and $M' = \varphi^{-1}(\varphi(M) \overset{r}{\lll} s/2)$. $\varphi(M) \oplus \varphi(M')$ is piecewise $s/2$-periodic, and so is $(\mathcal{H} \times \varphi(M)) \oplus (\mathcal{H} \times \varphi(M')) = \mathcal{H} \times (\varphi(M) \oplus \varphi(M'))$. Our collision search is now a search for M such that the first $s/2$ bits of $\mathcal{H} \times (\varphi(M) \oplus \varphi(M'))$ are zero.

In practice, the smallest period we can achieve with the regular encoder is n/w, so we will require that $n/2w$ is a divisor of r. We divide the encoded word into blocks of size r, and we choose the same m_i for all the chunk of the message that are used in the same block. The choice of a message M so that $\varphi(M)$ is piecewise n/w-periodic is equivalent to choice of n/r values $\mu_i \in \{0..n/w - 1\}$. Each μ_i will be used as the message corresponding to one block of the matrix, so the full message will be $\mu_0^{rw/n} \mu_1^{rw/n} \ldots \mu_{n/r-1}^{rw/n}$ (that is, that is, $m_i = \mu_{\lfloor i/\frac{rw}{n} \rfloor}$). We have

$$
F(M) = \mathcal{H} \times \mu_0^{rw/n} \mu_1^{rw/n} \ldots \mu_{n/r-1}^{rw/n} = \bigoplus_{i=0}^{n/r-1} \mathcal{H} \times \theta_i(\mu_i),
$$

where

$$\theta_i(x) = \varphi_{i\frac{rw}{n}}(x) \oplus \varphi_{i\frac{rw}{n}+1}(x) \oplus \cdots \varphi_{(i+1)\frac{rw}{n}-1}(x)$$

$M' = \varphi^{-1}(\varphi(M) \overset{r}{\lll} s/2)$ can be constructed easily: due to the definition of the regular encoder, we just have to set $\mu_i' = \mu_i + n/2w \pmod{n/w}$. We have now reduced the collision search to the search of $\mu_0, \ldots \mu_{n/r-1}$ such that:

$$F(M) \oplus F(M') = \bigoplus_{i=0}^{n/r-1} h_i(\mu_i) = 0$$

where

$$h_i(x) = \mathcal{H} \times \theta_i(x) \oplus \mathcal{H} \times \theta_i(x + n/2w).$$

We know that the $h_i(x)$'s are $n/2w$ periodic, so we only have to solve an instance of the n/r-sum problem on $n/2w$ bit with lists of $n/2w$ elements: $L_i = \{\bar{h}_i(\mu_i)\}_{\mu_i=0}^{n/2w-1}$ (with $\bar{h}_i(x) = h_i(x)^{[0 \ldots n/2w-1]}$). We can solve it with the same methods as the original one (which was an instance of the $2w$-sum problem on r bits, with lists of n/w elements):

- The linearization attack can be used if $n/r \geq n/2w$, which is equivalent to the condition on the original system: $r \leq 2w$. The complexity will drop from r^3 to $(n/2w)^3$.
- For Wagner's attack, let a_1 be the best a for the original problem and a_2 the best a for the new system. They have to satisfy:

$$\frac{2^a}{a+1} \leq \frac{2w}{r}\log(n/w) \qquad \frac{2^{a'}}{a'+1} \leq \frac{2w}{r}(\log(n/w)-1)$$

In most cases, we will be able the use the same a on the new system, and the complexity of the attack drops from $r2^a \cdot 2^{r/(a+1)}$ (which was used as a security parameter) to $r2^a \cdot 2^{\frac{n}{2w}/(a+1)}$. Since $n/2w$ is usually much smaller than r, this can already break all proposed parameters of FSB, many of them in practical time!

If the we are looking for collision with the same chaining value, instead of pseudo-collision, we will only have $n/r - n/s$ lists instead of n/r.

The next section will introduce a new way to solve this system, which is even more efficient.

4.3   A System of Cyclic Equations

The collision attack on IFSB has now been reduced to a collision attack on much smaller bit-strings. But the small bit-strings still have a strong structure:

we have $\bar{h}_i(x+1) = \bar{h}_i(x) \lll 1$ because $\varphi_i(x+1) = \varphi_i(x) \overset{n/w}{\lll} 1$ and $\mathcal{H}$ is quasi-cyclic. Therefore, each list actually contains every rotations of a single vector: $L_i = \{\bar{h}_i(\mu_i)\}_{\mu_i=0}^{n/2w-1} = \{\bar{h}_i(0) \lll \mu_i\}_{\mu_i=0}^{n/2w-1}$, and we have to solve:

$$\bigoplus_{i=0}^{p-1} H_i \lll \mu_i = 0,$$

where $H_i = \bar{h}_i(0)$ is computed from $\mathcal{H}$, and the μ_i's are the unknown. We have $p = n/r$ for a pseudo-collision attack, and $p = n/r - n/s$ for a collision attack. We will assume that the length of the H_i's is a power of two: $n/w = 2^l$, which is always the case in practice because it make the implementation easier.

First of all, if the sum of all bits of the H_i is non-zero (ie. $\bigoplus_{k,i} H_i^{[k]} = 1$), there is no solution; in this case we will drop one of the H_i with an odd bit-sum by setting $\mu_i' = \mu_i$ instead of $\mu_i' = \mu_i + n/2w$ for this particular i. We now assume that $\bigoplus_{k,i} H_i^{[k]} = 0$.

To describe our algorithm, we will use a set of functions π_k which folds 2^l-bit strings into 2^k-bit strings: $\pi_k(x)$ cuts x into chunks of 2^k bits, and xors them together (π_l is the identity function). We also use $\pi_k^L(x)$ which is the left part of $\pi_k(x)$, and $\pi_k^R(x)$ is the right part (therefore $\pi_{k-1}(x) = \pi_k^L(x) \oplus \pi_k^R(x)$). The algorithm is described by Algorithm 1; it uses linear algebra in a similar way as the attack of section 2.3.

Algorithm 1. Cyclic system solver

1: **for all** μ_i **do**
2:     $\mu_i \leftarrow 0$
3: **for** $1 \le k < l$ **do**
4:     $\Delta \leftarrow [\pi_k(H_i \lll \mu_i)]_{i=0}^{p-1}$
5:     Set u as the solution to $\Delta \times u = \pi_{k+1}^L(\bigoplus_i H_i \lll \mu_i)$
6:     **for all** μ_i **do**
7:         $\mu_i \leftarrow \mu_i + u^{[i]} 2^k$

Proof. The proof of Algorithm 1 uses the fact that after iteration k we have $\pi_{k+1}(\bigoplus_i H_i \lll \mu_i) = 0$. If μ_i are the values at the *beginning* of iteration k, we have $\pi_k(\bigoplus_i H_i \lll \mu_i) = 0$ and:

$$L = \pi_{k+1}^L \left(\bigoplus H_i \lll (\mu_i + u^{[i]} 2^k) \right)$$

$$= \pi_{k+1}^L \left(\bigoplus H_i \lll \mu_i \oplus \bigoplus_{u^{[i]}=1} (H_i \lll \mu_i \oplus H_i \lll (\mu_i + 2^k)) \right)$$

$$= \pi_{k+1}^L \left(\bigoplus H_i \lll \mu_i \right) \oplus \bigoplus_{u^{[i]}=1} \pi_{k+1}^L(H_i \lll \mu_i) \oplus \pi_{k+1}^R(H_i \lll \mu_i)$$

$$= \pi_{k+1}^L \left(\bigoplus H_i \lll \mu_i \right) \oplus \bigoplus_{u^{[i]}=1} \pi_k(H_i \lll \mu_i)$$

$$= \pi_{k+1}^{L}\left(\bigoplus H_i \lll \mu_i\right) \oplus \Delta \times u$$

$$= 0 \quad \text{(By construction of } u\text{)}$$

while

$$R = \pi_{k+1}^{R}\left(\bigoplus H_i \lll (\mu_i + u^{[i]}2^k)\right)$$

$$= \pi_{k+1}^{L}\left(\bigoplus H_i \lll (\mu_i + u^{[i]}2^k)\right) \oplus \pi_k\left(\bigoplus H_i \lll (\mu_i + u^{[i]}2^k)\right)$$

$$= 0 \oplus \pi_k\left(\bigoplus H_i \lll \mu_i\right) \quad \text{(Because } \pi_k(x \lll 2^k) = \pi_k(x))$$

$$= 0$$

Therefore, at the *end* of the iteration, we have $\pi_{k+1}(\bigoplus_i H_i \lll \mu_i) = L||R = 0$ with the new μ_i's. After the last iteration, this reads $\bigoplus_i H_i \lll \mu_i = 0$. $\square$

Complexity Analysis. The complexity of the attack is very low: the only computational intensive step is the linear algebra. Using a simple Gaussian elimination, we can solve the equation $\Delta \times u = c$ where Δ has 2^i rows and p columns in time $p2^{2i}$. The time of the full algorithm is therefore $\sum_{i=1}^{l-1} p2^{2i} = \frac{4^l-4}{3}p$.

The success probability of the algorithm is related to the probability $C(n, p)$ that a random system $\Delta \times u = c$ is consistent, when Δ has n rows and p columns. See Appendix A for an analysis of C. If all the systems are independent, identically distributed, the success probability can be expressed as:

$$P(n, p) = \prod_{i=1}^{\log(n)-1} C(2^i, p).$$

Actually, the systems are not independent and identically distributed because the random bits the H_i's are aligned in a particular manned by the μ_i of the previous step; in particular we have an obvious relation at each step: $\sum \Delta_i = 0$. We will assume that this is the only difference between the algorithm and the resolution of independent systems, which means the probability of success is $P(2^l, p-1)$. However, to solve a cyclic system, we first have to make sure that $\bigoplus_{k,i} H_i^{[k]} = 0$; this means that for one system out of two we have to run the algorithm with $p-1$ bit-strings. In the end, the expected success probability is

$$\Pi(2^l, p) = \frac{P(2^l, p-1) + P(2^l, p-2)}{2}$$

and the expected running time is:

$$\Pi(2^l, p)^{-1} \cdot \frac{4^l - 4}{3}p.$$

To check the hypothesis of independence, we ran the cyclic solver on random systems, and compared the success rate to the theoretical value. Figure 2 shows that we are very close to the actual success probability.

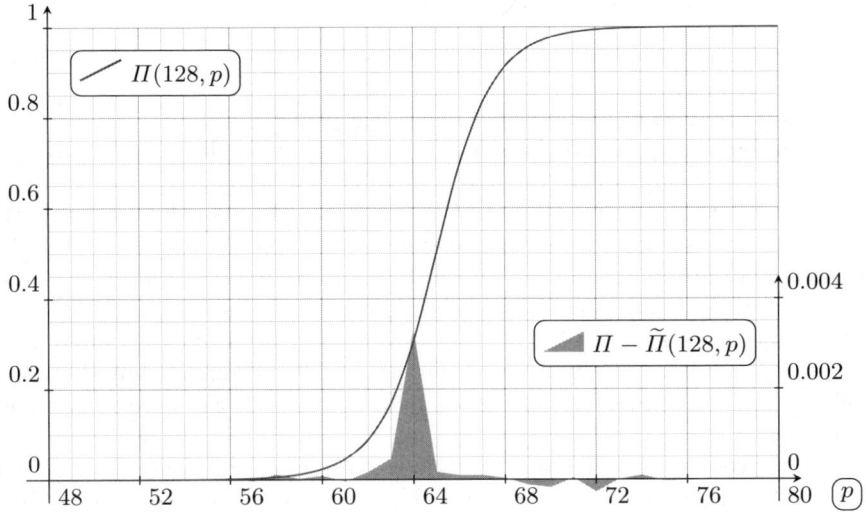

Fig. 2. Theoretical success probability Π versus experimental success probability $\widetilde{\Pi}$. Since the two curves would be on top of each over, we draw the difference between them on a different scale. $\widetilde{\Pi}$ is measured by running the algorithm 2^{20} times with randon systems of the given size.

When we use the cyclic solver to break IFSB, we will have $l = \log(n/2w)$ and $p = n/r$ for pseudo-collision, $p = n/r - n/s$ for collisions. We will analyse the pseudo-collision case in more detail:

1. If $p \geq 2^{l-1}$, ie $r \leq 4w$: we will use p a little larger than 2^{l-1} so that $\Pi(2^l, p)$ is almost 1. In this case, the running time of the full algorithm of the algorithm is essentially:

$$T = \tfrac{4^l - 4}{3} p < 2(n/4w)^3.$$

2. If $p < 2^{l-1}$, ie $r > 4w$, we cannot have a good probability of success and we have to repeat the cyclic solver with a randomized system. When we want to find collisions in a given hash function, we have only one particular system, but we can still apply a random rotation to each word. This will not give an independent system, but since our algorithm only find some special solutions to the cyclic system (the lower bits of the μ_i's are mostly zeros) this randomization could sufficient. Experimentally, with $l = 6$ and $p = 53$ (used for a collision attack against line 4 of Table 1) it works very well.

   In this case the running time of the full algorithm of the algorithm is essentially:

$$T = 2^{p-2^{l-1}} \tfrac{4^l - 4}{3} p < 2^{\frac{n(r-4w)}{4wr}} \cdot 2(n/4w)^3.$$

4.4   Scope of the Attack and Parameters Choice

Our attack is only applicable if r has a small divisor. On the one hand it is very efficient when the parameters are powers of two, but on the other hand it does not improve previous cryptanalysis when r is a prime.

This can be related to previous results about quasi-cyclic codes [3]. If r is a prime such that 2 is primitive modulo n, then the circulant matrix generated by a word of odd weight is invertible, and the code has the same kind of properties than a random code (in particular the syndromes are distributed evenly). This was used as an argument in favor of the IFSB construction in [5]. Note that it does not prove that finding collisions is hard: our attack does not apply, but there might be another way to exploit the quasi-cyclic code. Our attack completes this analysis: when r is carefully chosen there a security argument, but there is an attack for bad r's. Unfortunately, most parameters of IFSB were rather badly chosen.

Table 1 gives an overview of the various parameters of IFSB and the complexity of the linearization attack and the cyclic attack (for pseudo-collisions and collisions). The first part of the table presents the parameters used for performance evaluation in [5]. These parameters use r and n/w are powers of two, which allow our new attack to be used. We can see that it has a lower complexity than the linear attack, especially for the 80 bit security parameters, which have $r > 2w$. The next parts of the table show the parameters recommended by [5]. The recommended parameter set for standard applications has a prime r, which make our attack unusable, but the parameter set for memory constrained environments use a power of two. In the last part of the table, we show the parameters of FSB. They have not been proposed for use with IFSB, but we feel that it would be a natural move to prevent the attacks which need a low r/w ratio. The parameters would have to be tweaked to be used for IFSB, and our complexity evaluation assumes that $n/4w$ ends up being a divisor of r (which is the best case for our attack).

Note that it is still possible to choose parameters of IFSB which would not be broken by known attacks. The parameters of the original FSB [1] have a big r/w ratio which avoids the linearization attack, and if they are adapted to IFSB with a prime r they could also avoid our cyclic attack.

Example 1. To illustrate the attack, let us consider the most interesting setting of the table 1 of [5], designed for 128-bit security. We have:

$$r = 1024 \qquad w = 1024 \qquad s = 8192 \qquad n/w = 256$$

For a collision attack, we can build a cyclic to build a system with $l = n/2w = 128$ and $p = n/r - n/s = 224$. The main operation to solve this system will be linear algebra on a matrix of size 64, which only costs 2^{18} elementary operations (xors on one bit). Since the compression function requires $rw = 2^{20}$ elementary operations, our attack costs less than one call to the compression function!

Table 1. Comparison of the linearization attack and the cyclic attack on the various parameter sets proposed in [5]. The complexity is given in elementary operations; one compression function costs rw elementary operations (typically 2^{20}).

r	w	n	s	n/w	secu	Linear		Cyclic		$n/4w$	n/r	n/s
						psd.	coll	psd.	coll			
512	512	131072	4096	256		2^{27}	2^{27}	2^{19}	2^{19}	64	256	32
512	450	230400	4050	512	64	2^{27}	2^{27}	2^{22}	2^{22}	128	450	32.2
1024	2^{17}	2^{25}	2^{20}	256		2^{30}	2^{30}	2^{19}	2^{19}	64	2^{15}	32
512	170	43520	1360	256	80	2^{100}	-	2^{19}	2^{30}	64	85	32
512	144	73728	1296	512		-	-	2^{22}	2^{63}	128	144	56.9
1024	1024	262144	8192	256		2^{30}	2^{30}	2^{19}	2^{19}	64	256	32
1024	904	462848	8136	512	128	2^{30}	2^{30}	2^{22}	2^{22}	128	452	56.5
1024	816	835584	8160	1024		2^{30}	2^{30}	2^{25}	2^{25}	256	816	102.4
Recommended parameters for standard applications:												
1061	1024	262144	8192	256	128	2^{30}	2^{30}	-	-			
Recommended parameters for memory constrained environments:												
512	320	2560	1280	8	80	2^{27}	2^{61}	2^{11}	2^{11}	2	5	2
Parameters of FSB [1]. Not proposed for IFSB, but could be a way to repair it:												
480	170	43520	1360	256		2^{85}	-	2^{19}	2^{25}	64	90.7	32
400	85	21760	680	256	80	-	-	2^{29}	2^{61}	64	54.4	32
320	42	10752	336	256		-	-	2^{50}	-	64	33.6	32

Example 2. An other interesting set of parameter that we can break are the recommended parameters for memory constrained environments. This function was believed to provide a security of 2^{80}, but we can break it *by hand*! Since we have $n/4w = 2$ the resolution of the cyclic system is almost trivial. The most expensive step of the attack is the construction of the system, which only involves folding the matrix using xors.

4.5   About the Optimal Encoder

The optimal encoder allows to create messages such that $\varphi(M)$ is concentrated on one cyclic block of $\mathcal{H}$. Since (almost) any word of weight w is in the range of φ, if $w \geq r$ we can even choose a piecewise 1-periodic $\varphi(M)$! In this case, we have a very easy pseudo-collision attack:

1. Consider the messages $M_k = \varphi^{-1}\left(0^{kr}1^r0^{n-(k+1)r}1^{w-r}\right)$
2. We have $F(M_k) = \mathcal{H} \times \varphi(M_k) = s_k \oplus t$, where:
   $s_k = \mathcal{H}_k \times 1^r$ is 1-periodic
   $t = \mathcal{H} \times 0^{n-w+r}1^{w-r}$
3. Since s_k is 1-periodic, it can only take two values: 0^r and 1^r: we have at least one collision between M_0, M_1 and M_2.

If the optimal encoder is combined with a quasi-cyclic matrix of non prime length, it is easy to build periodic messages with a very small period. Because of this, we strongly discourage the use of the optimal encoder with quasi-cyclic codes.

Acknowledgement

Part of this work is supported by the Commission of the European Communities through the IST program under contract IST-2002-507932 ECRYPT, and by the French government through the Saphir RNRT project.

References

1. Augot, D., Finiasz, M., Sendrier, N.: A Family of Fast Syndrome Based Cryptographic Hash Functions. In: Dawson, E., Vaudenay, S. (eds.) Mycrypt 2005. LNCS, vol. 3715, pp. 64–83. Springer, Heidelberg (2005)
2. Bellare, M., Micciancio, D.: A New Paradigm for Collision-Free Hashing: Incrementality at Reduced Cost. In: Fumy, W. (ed.) EUROCRYPT 1997. LNCS, vol. 1233, pp. 163–192. Springer, Heidelberg (1997)
3. Chen, C.L., Peterson, W.W., W Jr., E.J.: Some Results on Quasi-Cyclic Codes. Information and Control 15(5), 407–423 (1969)
4. Coron, J.S., Joux, A.: Cryptanalysis of a Provably Secure Cryptographic Hash Function. Cryptology ePrint Archive, Report 2004/013 (2004) http://eprint.iacr.org/
5. Finiasz, M., Gaborit, P., Sendrier, N.: Improved Fast Syndrome Based Cryptographic Hash Functions. In: Rijmen, V. (ed.) ECRYPT Hash Workshop 2007 (2007)
6. Goldman, J., Rota, G.C.: On the foundations of combinatorial theory. IV: Finite vector spaces and Eulerian generating functions. Stud. Appl. Math. 49, 239–258 (1970)
7. Saarinen, M.J.O.: Linearization Attacks Against Syndrome Based Hashes. Cryptology ePrint Archive, Report 2007/295 (2007) http://eprint.iacr.org/
8. Wagner, D.: A Generalized Birthday Problem. In: Yung, M. (ed.) CRYPTO 2002. LNCS, vol. 2442, pp. 288–303. Springer, 2442 (2002)
9. Wang, X., Yin, Y.L., Yu, H.: Finding Collisions in the Full SHA-1. In: Shoup, V. (ed.) CRYPTO 2005. LNCS, vol. 3621, pp. 17–36. Springer, Heidelberg (2005)
10. Wang, X., Yu, H.: How to Break MD5 and Other Hash Functions. In: Cramer, R.J.F. (ed.) EUROCRYPT 2005. LNCS, vol. 3494, pp. 19–35. Springer, Heidelberg (2005)

A   Probability of Solving a Linear System over $\mathbf{F}_2$

In this appendix we study the probability that the equation $\Delta \times x = c$ has at least one solution in x, given a random $n \times p$ binary matrix Δ, and a random vector c. We call this probability $C(n, p)$.

$$
\begin{aligned}
C(n, p) &= \Pr_{\Delta,c} \left[\exists x : \Delta \times x = c \right] \\
&= \Pr_{\Delta,c} \left[c \in \operatorname{Im} \Delta \right] \\
&= \sum_{r=0}^{n} \Pr_{c} \left[c \in \operatorname{Im} \Delta \mid \operatorname{rank}(\Delta) = r \right] \cdot \Pr_{\Delta} \left[\operatorname{rank}(\Delta) = r \right] \\
&= \sum_{r=0}^{n} 2^{r-n} \cdot \Pr_{\Delta} \left[\operatorname{rank}(\Delta) = r \right] \\
&= \sum_{r=0}^{n} 2^{r-n} \cdot 2^{-np} \prod_{i=0}^{r-1} \frac{(2^p - 2^i)(2^n - 2^i)}{2^r - 2^i} \qquad \text{(see [6, Proposition 3])}
\end{aligned}
$$

The case $p > n$. The following lower bound is true for all p, but is mostly useful in the case $p > n$, and very tight when $p \gg n$:

$$
\begin{aligned}
C(n, p) &\geq \Pr_{\Delta} \left[\operatorname{rank}(\Delta) = n \right] = 1 - \Pr_{\Delta} \left[\operatorname{rank}(\Delta) < n \right] \\
&\geq 1 - \sum_{H \text{hyperplan}} \Pr_{\Delta} \left[\operatorname{Im} \Delta \subset H \right] \\
&\geq 1 - 2^n \frac{2^{(n-1)p}}{2^{np}} = 1 - 2^{n-p}
\end{aligned}
$$

It shows that we just have to choose p a little bigger than n to get a very good probability of success.

The case $p \leq n$. When $p \leq n$, we have

$$
C(n, p) \geq 2^{p-n} \cdot \Pr_{\Delta} \left[\operatorname{rank}(\Delta) = p \right] = 2^{p-n} \prod_{i=0}^{p-1} 1 - 2^{i-n}
$$

When $p \ll n$ The quantity $Q(n, p) = \prod_{i=0}^{p-1} 1 - 2^{i-n}$ is very close to one, but we can derive a lower bound it as long as $p \leq n$:

$$
Q(n, p) \geq \prod_{k=1}^{\infty} 1 - 2^{-k} = 0.288788...
$$

This allows us to say that the probability of success of the algorithm when $p < n$ is about 2^{p-n}.

Linear-XOR and Additive Checksums Don't Protect Damgård-Merkle Hashes from Generic Attacks

Praveen Gauravaram[1,*] and John Kelsey[2]

[1] Technical University of Denmark (DTU), Denmark
Queensland University of Technology (QUT), Australia
p.gauravaram@mat.dtu.dk
[2] National Institute of Standards and Technology (NIST), USA
john.kelsey@nist.gov

Abstract. We consider the security of Damgård-Merkle variants which compute linear-XOR or additive checksums over message blocks, intermediate hash values, or both, and process these checksums in computing the final hash value. We show that these Damgård-Merkle variants gain almost no security against generic attacks such as the long-message second preimage attacks of [10, 21] and the herding attack of [9].

1 Introduction

The Damgård-Merkle construction [3, 14] (**DM** construction in the rest of this article) provides a blueprint for building a cryptographic hash function, given a fixed-length input compression function; this blueprint is followed for nearly all widely-used hash functions. However, the past few years have seen two kinds of surprising results on hash functions, that have led to a flurry of research:

1. *Generic attacks* apply to the **DM** construction directly, and make few or no assumptions about the compression function. These attacks involve attacking a t-bit hash function with more than $2^{t/2}$ work, in order to violate some property other than collision resistance. Examples of generic attacks are Joux multicollision [8], long-message second preimage attacks [10, 21] and herding attack [9].
2. *Cryptanalytic attacks* apply to the compression function of the hash function. However, turning an attack on the compression function into an attack on the whole hash function requires properties of the **DM** construction. Examples of cryptanalytic attacks that involve the construction as well as the compression function include multi-block collisions on MD5, SHA-0 and SHA-1 [24,25,26].

These results have stimulated interest in new constructions for hash functions, that prevent the generic attacks, provide some additional protection against cryptanalytic attacks or both. The recent call for submissions for a new hash function standard by NIST [18] has further stimulated interest in alternatives to **DM**.

---

* Author is supported by The Danish Research Council for Technology and Production Sciences grant no. 274-05-0151.

T. Malkin (Ed.): CT-RSA 2008, LNCS 4964, pp. 36–51, 2008.
© Springer-Verlag Berlin Heidelberg 2008

In this paper, we consider a family of variants of **DM**, in which a linear-XOR checksum or additive checksum is computed over the message blocks, intermediate states of the hash function, or both, and is then included in the computation of the final hash value. In a linear-XOR checksum, each checksum bit is the result of XORing together some subset of the bits of the message, intermediate hash states, or both. In an additive checksum, the full checksum is the result of adding together some or all of the message blocks, intermediate hash values, or both, modulo some N. In both cases, the final checksum value is processed as a final, additional block in computing the hash value.

Such **DM** variants can be seen as a special case of a cascade hash. Generic attacks such as the long-message second preimage attack or the herding attack appear at first to be blocked by the existence of this checksum. (For example, see [6] for the analysis of **3C** and MAELSTROM-0 against second preimage and herding attacks.)

Unfortunately, these **DM** variants turn out to provide very little protection against such generic attacks. We develop techniques, based on the multicollision result of Joux [8], which allow us to carry out the generic attacks described above, despite the existence of the checksum. More generally, our techniques permit the construction of a *checksum control sequence*, or CCS, which can be used to control the value of the checksum without altering the rest of the hash computation.

To summarize our results:

1. The generic multicollision, second preimage and herding attacks on **DM** hash functions can be applied to linear-XOR/additive checksum variants of **DM** at very little additional cost, using our techniques.
2. Our techniques are flexible enough to be used in many other situations. Some cryptanalytic attacks on the compression function of a hash, which the linear-XOR/additive checksum appears to stop from becoming attacks on the full hash function, can be carried out on the full hash function at a relatively little additional cost using our techniques. Future generic attacks will almost certainly be able to use our techniques to control checksums at very low cost

1.1 Related Work

In unpublished work, Mironov and Narayan [15] developed a different technique to defeat linear-XOR checksums in hash functions; this technique is less flexible than ours, and does not work for long-message second preimage attacks. However, it is quite powerful, and can be combined with our technique in attacking hash functions with complicated checksums. In [8], Joux provides a technique for finding 2^k collisions for a **DM** hash function for only about k times as much work as is required for a single collision, and uses this technique to attack cascade hashes. The linear-XOR and additive checksum variants of **DM** we consider in this paper can be seen as a special (weak) case of a cascade hash.

Multi-block collisions are an example of a cryptanalytic attack on a compression function, which must deal with the surrounding hash construction.

Lucks [13] and Tuma and Joscak [22] have independently found that if there is a multi-block collision for a hash function with structured differences, concatenation of such a collision will produce a collision on **3C**, a specific hash construction which computes checksum using XOR operation as the mixing function. (**3C** does not prevent Joux multicollision attack over 1-block messages [6, 20].)

Nandi and Stinson [17] have shown the applicability of multicollision attacks to a variant of **DM** in which each message block is processed multiple times; Hoch and Shamir [7] extended the results of [17] showing that generalized sequential hash functions with any fixed repetition of message blocks do not resist multicollision attacks. The MD2 hash function which uses a non-linear checksum was shown to not satisfy preimage and collision resistance properties [11, 16]. Coppersmith [2] has shown a collision attack on a DES based hash function which uses two supplementary checksum blocks computed using XOR and modular addition of the message blocks. Dunkelman and Preneel [4] applied herding attack of [9] to cascade hashes; their technique can be seen as an upper bound on the difficulty of herding **DM** variants with checksums no longer than the hash outputs.

1.2 Impact

The main impact of our result is that new hash function constructions that incorporate linear-XOR/additive checksums as a defense against generic attacks do not provide much additional security. Designers who wish to thwart these attacks need to look elsewhere for defenses. We can apply our techniques to specific hash functions and hashing constructions that have been proposed in the literature or are in practical use. They include **3C**, GOST, MAELSTROM-0 and F-Hash. Both our techniques and the generic attacks which they make possible require the ability to (at least) find many collisions for the underlying compression function of the hash, and so probably represent only an academic threat on most hash functions at present.

1.3 Guide to the Paper

This paper is organised as follows: First, we provide the descriptions of hash functions with linear checksums analysed in this paper. Next, we demonstrate new cryptanlytical techniques to defeat linear-XOR/additive checksums in these designs. We then provide a generic algorithm to carry out second preimage and herding attacks on these designs with an illustration on **3C**. We then demonstrate multi-block collision attacks on these designs. We then compare our cryptanalysis with that of [15]. Finally, we conclude the paper with some open problems.

2 The DM Construction and DM with Linear Checksums

2.1 The DM Construction

The **DM** iterative structure [3, 14] shown in Figure 1 has been a popular framework used in the design of standard hash functions MD5, SHA-1, SHA-224/256 and SHA-384/512.

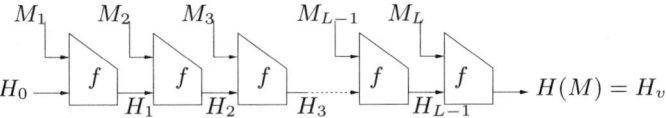

Fig. 1. The Damgård-Merkle construction

The message M, with $|M| \le 2^l - 1$ bits, to be processed using a **DM** hash function H is always padded by appending it with a 1 bit followed by 0 bits until the padded message is l bits short of a full block of b bits. The last l bits are filled in with the binary encoded representation of the length of true message M. This compound message is an integer multiple of b bits and is represented with b-bit data blocks as $M = M_1, M_2, \ldots M_L$. Each block M_i is processed using a fixed-length input compression function f as given by $H_i = f(H_{i-1}, M_i)$ where H_i from $i = 1$ to $L - 1$ are the intermediate states and H_0 is the fixed initial state of H. The final state $H_v = f(H_{L-1}, M_L)$ is the hash value of M.

2.2 Linear-XOR/Additive Checksum Variants of DM

A number of variant constructions have been proposed, that augment the **DM** construction by computing some kind of linear-XOR/additive checksum on the message bits and/or intermediate states, and providing the linear-XOR/additive checksum as a final block for the hash function as shown in Figure 2.

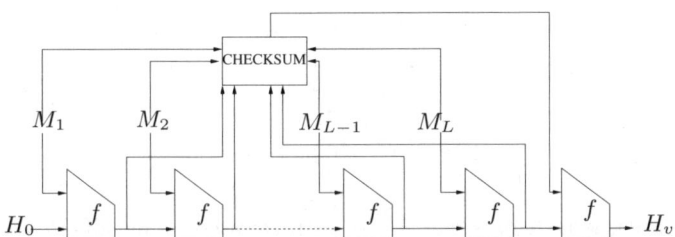

Fig. 2. Hash function structure with a linear-XOR/additive checksum

3C hash function and its variants. The **3C** construction maintains twice the size of the hash value for its intermediate states using iterative and accumulation chains as shown in Figure 3. In its iterative chain, a compression function f with a block size b is iterated in the **DM** mode. In its accumulation chain, the checksum Z is computed by XORing all the intermediate states each of size t bits. The construction assumes that $b > t$. At any iteration i, the checksum value is $\bigoplus_{j=1}^{i} H_j$. The hash value H_v is computed by processing Z padded with 0 bits to make the final data block $\overline{Z}$ using the last compression function.

A 3-chain variant of **3C** called **3CM** is used as a chaining scheme in the MAELSTROM-0 hash function [5]. At every iteration of f in the iterative chain

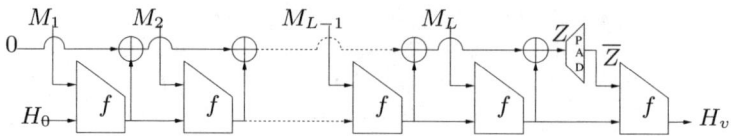

Fig. 3. The **3C**-hash function

of **3CM**, the t-bit value in the third chain is updated using an LFSR. This result is then XORed with the data in the iterative chain at that iteration. All the intermediate states in the iterative chain of **3CM** are XORed in the second chain. Finally, the hash value is obtained by concatenating the data in the second and third chains and processing it using the last f function. F-Hash [12], another variant of **3C**, computes the hash value by XORing part of the output of the compression function at every iteration and then processes it as a checksum block using the last compression function.

GOST hash function. GOST is a 256-bit hash function specified in the Russian standard GOST R 34.11 [19]. The compression function f of GOST is iterated in the **DM** mode and a mod 2^{256} additive checksum is computed by adding all the 256-bit message blocks in an accumulation chain. We generalise our analysis of GOST by assuming that its f function has a block length of b bits and hash value of t bits.

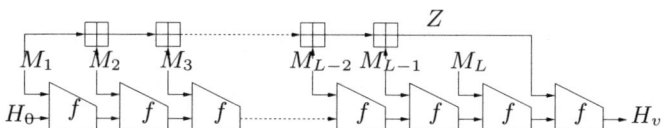

Fig. 4. GOST hash function

An arbitrary length message M to be processed using GOST is split into b-bit blocks $M_1, \ldots, M_{L-1}$. If the last block M_{L-1} is incomplete, it is padded by prepending it with 0 bits to make it a b-bit block. The binary encoded representation of the length of the true message M is processed in a separate block M_L as shown in Figure 4. At any iteration i, the intermediate state in the iterative and accumulation chains is $H_i = f(H_{i-1}, M_i)$ where $1 \leq i \leq L$ and $M_1 + M_2 \ldots + M_i \bmod 2^b$ where $1 \leq i \leq L - 1$. The hash value of M is $H_v = f(Z, H_L)$ where $Z = M_1 + M_2 \ldots + M_{L-1} \bmod 2^b$.

3   New Techniques to Defeat Linear-XOR Checksums

3.1   Extending Joux Multicollisions on DM to Multiple Blocks

Let $C(s, n)$ be a collision finding algorithm on the compression function, where s denotes the state at which the collision attack is applied and n, the number of message blocks present in each of the colliding messages. On a t-bit hash function, a brute force $C(s,n)$ requires about $2^{t/2}$ hash function computations to find a collision with 0.5 probability whereas a cryptanalytic $C(s,n)$ requires less work than that. The Joux multicollision attack [8] finds a sequence of k collisions on a t-bit **DM** hash, to produce a 2^k collision with work only k times the work of a single collision search. For a brute-force collision search, this requires $k \times 2^{t/2}$ evaluations of the compression function. While it is natural to think of constructing such a multicollision from a sequence of single-message-block collisions, it is no more expensive to use the brute-force collision search to find a sequence of multi-message-block collisions.

3.2   Checksum Control Sequences

We define checksum control sequence (CCS) as a data structure which lets us to control the checksum value of the **DM** variant, *without altering the rest of the hash computation*. We construct the CCS by building a Joux multicollision of the correct size using a brute-force collision search. It is important to note that the CCS is not itself a single string which is hashed; instead, it is a data structure which permits us to construct one of a very large number of possible strings, each of which has some effect on the checksum, but leaves the remainder of the hash computation unchanged. That is, the choice of a piece of the message from the CCS affects the checksum chain, but not the iterative chain, of the **DM** variant hash.

For example, a 2^k collision on the underlying **DM** construction of **3C**, in which the sequence of individual collisions is each two message blocks long, is shown in Figure 3. This multicollision gives us a choice of 2^k different sequences of message blocks that might appear at the beginning of this message. When we want a particular k-bit checksum value, we can turn the problem of finding which choices to make from the CCS into the problem of solving a system of k linear equations in k unknowns, which can be done very efficiently using existing tools such as Gaussian elimination [1, Appendix A], [23]. This is shown in Figure 5 for $k = 2$ where we compute the CCS by finding a 2^2 collision using random 2-block messages. Then we have a choice to choose either $H_1^0 \oplus H_2$ or $H_1^1 \oplus H_2$ from the first 2-block collision and either $H_3^0 \oplus H_4$ or $H_3^1 \oplus H_4$ from the second 2-block collision of the CCS to control 2 bits of the checksum without changing the hash value after the CCS.

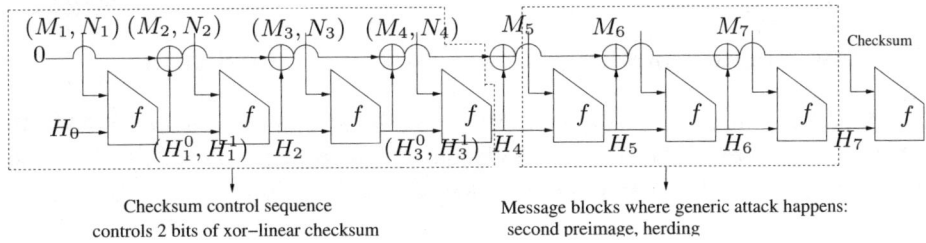

Checksum control sequence
controls 2 bits of xor–linear checksum

Message blocks where generic attack happens:
second preimage, herding

Fig. 5. Using CCS to control 2 bits of the checksum

3.3   Defeating Linear-XOR Checksum in Hash Functions

ALGORITHM: Defeat linear-XOR checksum on 3C
Variables:

1. (e_i^0, e_i^1) : a pair of independent choices of random values after every 2-block collision in the 2^t 2-block collision on **3C** for $i = 1, 2, \ldots, t$.
2. $a = a[1], a[2], \ldots, a[t]$: any t-bit string.
3. $D = D[1], D[2], \ldots, D[t]$: the desired t-bit checksum to be imposed.
4. i, j : temporary variables.

Steps

1. Build a CCS for **3C** by constructing a 2^t 2-block collision on its underlying **DM** using a brute force $C(s, 2)$.
2. Each of the parts of the CCS gives one choice e_i^0 or e_i^1 for $i = 1, 2, \ldots, t$ to determine some random t-bit value that either is or is not XORed into the final checksum value at the end of the CCS. Now $e_i^0 = H_{2i-1}^0 \oplus H_{2i}^0$ and $e_i^1 = H_{2i-1}^1 \oplus H_{2i}^1$ for $i = 1, 2, \ldots, t$.
3. For any t-bit string $a = a[1], a[2], \ldots, a[t]$, let $e^a = e_1^{a[1]}, \ldots, e_t^{a[t]}$.
4. Find a such that $e_1^{a[1]} \oplus e_2^{a[2]} \oplus \ldots \oplus \ldots e_t^{a[t]} = D$. We solve the equation: $\bigoplus_{i=1}^{t} e_i^1 \times a[i] \oplus e_i^0 \times (1 - a[i]) = D$.
5. Each bit position of $e_i^{a[i]}$ gives one equation and turn the above into t equations, one for each bit. Let $\bar{a}[i] = 1 - a[i]$.
6. The resulting system is: $\bigoplus_{i=1}^{t} e_i^1[j] \times a[i] \oplus e_i^0[j] \times \bar{a}[i] = D[j]$ $(j = 1, \ldots, t)$. Here, there are t linear equations in t unknowns that need to be solved for the solution $a[1], a[2], \ldots, a[t]$ which lets us determine the blocks in the CCS that form the prefix giving the checksum D.

Work: It requires $t(2^{t/2}+1)$ evaluations of the compression function to construct the CCS and at most $t^3 + t^2$ bit-XOR operations to solve a system of $t \times t$ equations using Gaussian elimination [1, Appendix A], [23].

Remark 1. Similarly, linear-XOR checksums can be defeated in F-Hash and **3CM**. If a linear-XOR checksum is computed using both the message blocks and intermediate states, linear equations due to XOR of the intermediate states and that of message blocks need to be solved.

4   New Techniques to Defeat Additive Checksums

Consider an additive checksum mod 2^k computed using messages for a **DM** hash function. It is possible to build a checksum control sequence as above, but both its construction and its use require some different techniques.

4.1 Building a CCS with Control of Message Blocks

When the collision finding algorithm is simply brute-force collision search, we can build a CCS for the work required to construct a 2^k Joux multicollision. Using the CCS to control the checksum then requires negligible work.

In this algorithm, we construct a 2^k Joux multicollision, in which each successive collision is two message blocks long. We choose the two-block messages in the collisions in such a way that the additive difference between the pair of two-block messages in each collision is a different power of two. The result is a CCS in which the first collision allows us the power to add 1 to the checksum, the next allows us to add 2, the next 4, and so on until the checksum is entirely controlled[1].

ALGORITHM: Defeat additive checksum on GOST
Steps for Constructing the CCS:

1. Let $h = $ the initial value of the hash function
2. For $i = 0$ to $k - 1$:
   (a) Let A, B be random blocks.
   (b) For $j = 0$ to $2^{t/2} - 1$:
       i. $X[j] = A + j, B - j$
       ii. $X^*[j] = A + j + 2^i, B - j$
       iii. $Y[j] = $ hash of $X[j]$ starting from h
       iv. $Y^*[j] = $ hash of $X^*[j]$ starting from h
3. Search for a collision between list Y and Y^*. Let $u, v = $ values satisfying $Y[u] = Y^*[v]$.
4. $CCS[i] = X[u], X^*[u]$
5. $h = Y[u]$

Steps for Using the CCS:
Using the CCS is very simple; we determine the checksum we would get by choosing $X[0], X[1], X[2], ...$, and then determine what we would need to add to that value to get the desired checksum value. We then use our control over the CCS to add the desired value.

1. Let $T = $ the checksum that is desired.
2. Let $Q = $ the checksum obtained by choosing $X[0], X[1], X[2], ..., X[k-1]$ as the message blocks of the CCS.
3. Let $D = T - Q$.
4. $M = $ an empty message (which will end up with the message blocks chosen from the CCS for this desired checksum).
5. For $i = k - 1$ down to 0:
   (a) If $D > 2^i$ Then:
       i. $M = M||X^*[i]$
       ii. $D = D - 2^i$

---

[1] A variant of this algorithm could be applied to many other checksums based on group operations.

(b) Else:
  i. $M = M||X[i]$

At the end of this process, M contains a sequence of k message blocks which, when put in the place of the CCS, will force the checksum to the desired value.

Work: Constructing the CCS requires k successive brute-force collision searches, each requiring $2^{t/2}$ work. For the specific parameters of the GOST hash, this is 256 successive 2^{129} collision searches, and so requires about 2^{137} work total. (The same CCS could be used for many different messages.) Controlling the checksum with the CCS requires negligible work.

4.2   Building a CCS with Random Message Blocks

If the message blocks are not under our control, or if hash chaining values or other values not under our direct control are used as inputs for the additive checksum, then our attack becomes much less efficient. However, we can still construct a CCS which will be efficient to use, by carrying out an algorithm which is based loosely on Joux's collision attack on cascade hashes.

The idea behind this algorithm is to construct k successive Joux multicollisions, each of $2^{k/2}$ possible message strings. Then, we carry out a collision search on the first $2^{k/2}$-multicollision for a pair of strings that will cause a difference of 1 in the additive checksum, a search on the second $2^{k/2}$-multicollision for a pair that will cause a difference of 2, and so on, until we have the ability to completely control the checksum without affecting the rest of the hash computation.

An algorithm to defeat additive checksum on a t-bit GOST hash function structure H shown in Figure 4 is given below:

ALGORITHM: Defeating checksum in GOST
Variables:

1. i, j, k : integers.
2. $chunk[i]$: a pair of $(b/2) + 1$-block sequences denoted by (e_i^0, e_i^1).
3. H_0 : initial state.
4. H_j^i : the intermediate state on the iterative chain.
5. (M_j^i, N_j^i) : a pair of message blocks each of b bits.
6. T : Table with three columns: a $(b/2) + 1$-collision path, addition modulo 2^b of message blocks in that path and a value of 0 or 1.

Steps:

1. For $i = 1$ to b:
   − For $j = 1$ to $(b/2) + 1$:
      • Find M_j^i and N_j^i such that $f(H_{j-1}^i, M_j^i) = f(H_{j-1}^i, N_j^i) = H_j^i$ where $H_0^1 = H_0$. That is, build a $(b/2) + 1$-block multicollision where each block yields a collision on the iterative chain and there are $2^{(b/2)+1}$ different $(b/2) + 1$-block sequences of blocks all hashing to the same intermediate state $H_{(b/2)+1}^i$ on the iterative chain.

  – Find a pair of paths from the different $(b/2) + 1$-block sequences whose
    additive checksum differs by 2^{i-1} as follows:
    - $T = $ empty table.
    - for $j = 1$ to $2^{(b/2)+1}$
      * $C_j^i \equiv \sum_{k=1}^{(b/2)+1} X_k^i \bmod 2^b$ where X_k^i can be M_k^i or N_k^i.
      * Add to T: $(C_j^i, 0, X_1^i || X_2^i || \ldots X_{(b/2)+1}^i)$
      * Add to T: $(C_j^i + 2^{i-1}, 1, X_1^i || X_2^i || \ldots X_{(b/2)+1}^i)$.
    - Search T to find colliding paths between the entries with 0 and 1
      in the second column of T. Let these paths of $(b/2) + 1$ sequence of
      blocks be e_i^1 and e_i^0 where $e_i^1 \equiv e_i^0 + 2^{i-1} \bmod 2^b$.
  – $chunk[i] = (e_i^0, e_i^1)$.
2. Construct CCS by concatenating individual chunks each containing a pair
   of $(b/2) + 1$ blocks that hash to the same intermediate state on the iterative
   chain. The CCS is $chunk[1] \; || \; chunk[2] \ldots \; || \; chunk[b]$.
3. The checksum at the end of the 2^b $(b/2) + 1$-block collision can be forced to
   the desired checksum by choosing either of the sequences e_i^0 or e_i^1 from the
   CCS which is practically free to use and adding blocks in each sequence over
   modulo 2^b.

Work: Defeating additive checksum in GOST equals the work to construct b
$2^{(b/2)+1}$ 1-block collisions plus the work to find a chunk in each $2^{(b/2)+1}$ 1-block
collision. It is $b \times ((b/2) + 1) \times 2^{t/2}$ evaluations of f and a time and space of
$b \times 2^{b/2+1}$ for a collision search to find b chunks. For GOST, it is 2^{143} evaluations
of f and a time and space of about 2^{137}.

Similarly, additive checksum mod 2^k for a **DM** hash using intermediate states
can be defeated by constructing a CCS with a 2^k Joux multicollision over 2-block
messages. For a **DM** hash with additive checksum mod 2^k computed using both
the message blocks and intermediate states, a $2^{(k/2)+1}$ Joux multicollision using
2-block messages is performed to find a pair of messages (resp. intermediate
states) within the multicollision whose additive checksum differs by any desired
value. This can be done by generating all possible $2^{(k/2)+1}$ checksum values
due to messages (resp. intermediate states) from the multicollision, and doing a
modified collision search for a pair of messages (resp. intermediate states) whose
additive difference is the desired value.

5   Generic Attacks

The fundamental approach used to perform the generic attacks on all the hash
functions with linear checksums is similar. Hence, we discuss it here only for **3C**.
Broadly, it consists of the following steps:

1. Construct a CCS and combine it with whatever other structures such as
   expandable message, diamond structure (or vice versa for some attacks) for
   the generic attack to work.
2. Perform the generic attack, ignoring its impact on the linear checksum.

3. Use the CCS to control the linear checksum, forcing it to a value that permits the generic attack to work on the full hash function.

To find a 2^k-2-block collision on **3C**, first find a 2^k-2-block collision on the iterative chain of **3C** and construct CCS from this end. By defeating each possible 2^k checksum value to a fixed checksum, we can get a 2^k-collision for **3C**. Constructing and using the CCS does not imply random gibberish in the messages produced; using Yuval's trick [27], a brute-force search for the multicollision used in the CCS can produce collision pairs in which each possible message is a plausible-looking one. This is possible when the CCSs to defeat the checksums are constructed from individual collisions as in (Dear Fred/Freddie,)(Enclosed please find/I have sent you) (a check for \$100.00/a little something) and so on, where we can choose either side of the slash for the next part of the sentence. In that case, any choice for the CCS used to defeat the checksum will be a meaningful message.

5.1   Long-Message Second Preimage Attack on 3C

Long message second preimage attack on a t-bit **3C** hash function H:

ALGORITHM: LongMessageAttack(M_{target}) on H
Find the second preimage for a message of $2^d + d + 2t + 1$ blocks.
Variables:

1. M_{target} : the target long message.
2. M_{link} : linking block connecting the intermediate state at the end of the *expandable message* to an intermediate state of the target message.
3. H_{exp} : the intermediate state at the end of the *expandable message*.
4. H_t : the intermediate hash value at the end of the CCS.
5. M_{sec} : the second preimage for H of the same length as M_{target}.
6. M_{pref} : the checksum control prefix obtained from the CCS.

Steps:

1. Compute the intermediate hash values for M_{target} using H:
   - H_0 and h_0 are the initial states of the iterative and accumulation chains respectively.
   - M_i is the ith message block of M_{target}.
   - $H_i = f(H_{i-1}, M_i)$ and $h_i = H_i \oplus h_{i-1}$ are the ith intermediate states on the iterative and accumulation chains respectively.
   - The intermediate states on the iterative and accumulation chains are organised in some searchable structure for the attack, such as hash table. The hash values $H_1, \ldots, H_d$ and those obtained in the processing of t 2-block messages are excluded from the hash table.
2. Build a CCS for H by constructing a 2^t 2-block collision starting from H_0. Let H_t be the multicollision value and h_t be the corresponding checksum value which is random.

3. Construct a $(d, d + 2^d - 1)$-*expandable message* M_{exp} with H_t as the starting state using either of the *expandable message* construction methods [10]. Append M_{exp} to the CCS and process it to obtain H_{exp}.

4. Find M_{link} such that $f(H_{exp}, M_{link})$ collides with one of the intermediate states on the iterative chain stored in the hash table while processing M_{target}. Let this matching value of the target message be H_u and the corresponding state in the accumulation chain be h_u where $d + 2t + 1 \leq u \leq 2^d + d + 2t + 1$.

5. Use the CCS built in step 2 to find the checksum control prefix M_{pref} which adjusts the state in the accumulation chain at that point to the desired value h_u of M_{target}. This is equivalent to adjusting the checksum value at the end of the CCS.

6. Expand the *expandable message* to a message M^* of $u - 1$ blocks long.

7. Return the second preimage $M_{sec} = M_{pref}||M^*||M_{link}||M_{u+1} \cdots$ $M_{2^d+d+1+2t}$ of the same length as M_{target} such that $H(M_{sec}) = H(M_{target})$.

Work: The work to find a second preimage on **3C** equals the work to construct the CCS plus the work to solve a system of $t \times t$ linear equations plus the work to do the second preimage attack on **DM**. Note that constructing and using the CCS is very fast compared to the rest of the attack.

Illustration: Using generic-expandable message algorithm [10], the work to find a second preimage for **3C-SHA-256** for a target message of $2^{54} + 54 + 512 + 1$ blocks is $2^{136} + 54 \times 2^{129} + 2^{203}$ SHA-256 compression function evaluations and $2^{24} + 2^{16}$ bit-XOR operations assuming abundant memory.

5.2   Herding Attack on 3C

The herding attack on a t-bit **3C** hash function H is outlined below:

1. Construct a 2^d hash value wide diamond structure for H and output the hash value H_v as the chosen target which is computed using any of the possible 2^{d-1} checksum values or some value chosen arbitrarily. Let h_c be that checksum value.

2. Build a CCS for H using a 2^t collision over 2-block messages. Let H_t be the intermediate state due to this multicollision on H.

3. When challenged with the prefix message P, process P using H_t. Let $H(H_t, P) = H_p$.

4. Find a linking message block M_{link} such that $H(H_p, M_{link})$ collides with one of the 2^d outermost intermediate states on the iterative chain in the diamond structure. If it is matched against all of the $2^{d+1} - 2$ intermediate states in the diamond structure then a $(1, d + 1)$-*expandable message* must be produced at the end of the diamond structure to ensure that the final herded message is always a fixed length.

5. Use the CCS computed in step 2 to force the checksum of the herded message P to h_c. Let M_{pref} be the checksum control prefix.

6. Finally, output the message $M = M_{pref}||P||M_{link}||M_d$ where M_d are the message blocks in the diamond structure that connect $H(H_p, M_{link})$ to the chosen target H_v. Now $H_v = H(M)$.

Work: The work to herd **3C** equals the work to build the CCS plus the work to solve the system of equations plus the work to herd **DM** [9]. This equals about $t \times 2^{t/2} + 2^{t/2+d/2+2} + d \times 2^{t/2+1} + 2^{t-d-1}$ evaluations of f and $t^3 + t^2$ bit-XOR operations assuming that all of the $2^{d+1} - 2$ intermediate states are used for searching in the diamond structure. Note that the work to build and use the CCS is negligible compared to the rest of the attack.

Illustration: Herding **3C-SHA-256** with $d = 84$ requires $2^{136} + 2^{172} + 84 \times 2^{129} + 2^{171}$ evaluations of SHA-256 compression function and $2^{24} + 2^{16}$ bit-XOR operations.

6   On Carrying Out Generic Attacks Using Collision Attacks

We note that it is difficult to construct the CCSs using cryptanalytic $C(s, n)$ such as the ones built on MD5 and SHA-1 [25, 26] in order to defeat linear checksums to carry out generic attacks. For example, consider two 2-block colliding messages of format $(M_{2.i-1}, M_{2.i})$,$(N_{2.i-1}, N_{2.i})$ for $i = 1, \ldots, t$ on the underlying **MD** of **3C** based on near collisions due to the first blocks in each pair of the messages. Usually, the XOR differences of the nearly collided intermediate states are either fixed or very tightly constrained as in the collision attacks on MD5 and SHA-1 [25, 26]. It is difficult to construct a CCS due to the inability to control these fixed or constrained bits. Similarly, it is also difficult to build the CCSs using colliding blocks of format $(M_{2.i-1}, M_{2.i})$,$(N_{2.i-1}, M_{2.i})$. It is not possible to control the checksum due to 2-block collisions of the format $(M_{2.i-1}, M_{2.i})$, $(M_{2.i-1}, N_{2.i})$ [24] as this format produces a zero XOR difference in the checksum after every 2-block collision.

Though we cannot perform generic attacks on this class of hash functions using structured collisions, we can find multi-block collisions by concatenating two structured collisions. Consider a collision finding algorithm $C(s, 1)$ with $s = H_0$ for the GOST hash function H. A call to $C(s, 1)$ results in a pair of b-bit message blocks (M_1, N_1) such that $M_1 \equiv N_1 + \Delta \mod 2^b$ and $f(H_0, M_1) = f(H_0, N_1) = H_1$. Now call $C(s, 1)$ with $s = H_1$ which results in a pair of blocks (M_2, N_2) such that $N_2 \equiv M_2 + \Delta \mod 2^b$ and $f(H_1, M_2) = f(H_1, N_2) = H_2$. That is, $H(H_0, M_1 || M_2) = H(H_0, N_1 || N_2)$. Consider $M_1 + M_2 \mod 2^b = \Delta + N_1 + N_2 - \Delta \mod 2^b = N_1 + N_2 \mod 2^b$, a collision in the chain which computes additive checksum.

7   Comparison of Our Techniques with That of [15]

Independent to our work, Mironov and Narayanan [15] have found an alternative technique to defeat linear-XOR checksum computed using message blocks. We call this design GOST-x. While our approach to defeat the XOR checksum in GOST-x requires finding a 2^b collision using b random 1-block messages (M_i, N_i) for $i = 1$ to b, their technique considers repetition of the same message block twice for a collision. In contrast to the methods presented in this paper for solving system of linear equations for the whole message, their approach solves the system of linear equations once after processing every few message blocks. We note that this constrained choice of messages would result in a zero checksum at the end of the 2^b multicollision on this structure and thwarts the attempts to perform the second preimage attack on GOST-x. The reason is that the attacker loses the ability to control the checksum after finding the linking message block from the end of the expandable message which matches some intermediate state obtained in the long target message.

However, we note that their technique with a twist can be used to perform the herding attack on GOST-x. In this variant, the attacker chooses the messages for the diamond structure that all have the same effect on the linear-XOR checksum. These messages would result in a zero checksum at every level in the diamond structure. Once the attacker is forced with a prefix, processing the prefix gives a zero checksum to start with and then solving a system of equations will find a set of possible linking messages that will all combine with the prefix to give a zero checksum value. When the approach of [15] is applied to defeat checksums in **3C**, **3CM** and F-Hash, the 2^t 2-block collision finding algorithm used to construct the CCS must output the same pair of message blocks on the either side of the collision whenever it is called. This constraint is not there in our technique, and the approach of [15] is not quite as powerful. However, it could be quite capable of defeating linear-XOR checksums in many generic attacks. Because it is so different from our technique, some variant of this technique might be useful in cryptanalytic attacks for which our techniques do not work.

8 Concluding Remarks

Our research leaves a number of questions open. Among these, the most interesting is, *how much security can be added by adding a checksum to **DM** hashes?* Our work provides a lower bound; for linear-XOR and additive checksums, very little security is added. Joux's results on cascade hashes [8] and more recent results of [4] provide an upper bound, since a checksum of this kind can be seen as a kind of cascade hash.

The other open question is on the properties that would ensure that a checksum would thwart generic attacks, and thus be no weaker than a cascade hash with a strong second hash function. The inability to construct a CCS for the checksum with less work than the generic attack is necessary but apparently not sufficient to achieve this goal, since we cannot rule out the possibility of other attacks on checksums of this kind, even without a CCS. The final open question is on how our techniques might be combined with cryptanalytic attacks on compression functions. It appears to be possible to combine the construction and use of a CCS with some kinds of cryptanalytic attacks, but this depends on fine details of the cryptanalysis and the checksum used.

Acknowledgments. We thank Gary Carter, Ed Dawson, Morris Dworkin, Jonathan Hoch, Barbara Guttman, Lars Knudsen, William Millan, Ilya Mironov, Heather Pearce, Adi Shamir, Tom Shrimpton, Martijn Stam, Jiri Tuma and David Wagner for comments on our work.

References

1. Bellare, M., Micciancio, D.: A new paradigm for collision-free hashing: Incrementality at reduced cost. In: Fumy, W. (ed.) EUROCRYPT 1997. LNCS, vol. 1233, pp. 163–192. Springer, Heidelberg (1997)
2. Coppersmith, D.: Two Broken Hash Functions. Technical Report IBM Research Report RC-18397, IBM Research Center (October 1992)
3. Damgård, I.: A Design Principle for Hash Functions. In: Brassard, G. (ed.) CRYPTO 1989. LNCS, vol. 435, pp. 416–427. Springer, Heidelberg (1990)
4. Dunkelman, O., Preneel, B.: Generalizing the herding attack to concatenated hashing schemes. In: ECRYPT hash function workshop (2007)
5. Filho, D.G., Barreto, P., Rijmen, V.: The MAELSTROM-0 Hash Function. In: 6th Brazilian Symposium on Information and Computer System Security (2006)
6. Gauravaram, P.: Cryptographic Hash Functions: Cryptanalysis, Design and Applications. PhD thesis, Information Security Institute, QUT (June 2007)
7. Hoch, J., Shamir, A.: Breaking the ICE: Finding Multicollisions in Iterated Concatenated and Expanded (ICE) Hash Functions. In: Robshaw, M.J.B. (ed.) FSE 2006. LNCS, vol. 4047, pp. 179–194. Springer, Heidelberg (2006)
8. Joux, A.: Multicollisions in Iterated Hash Functions. Application to Cascaded Constructions. In: Franklin, M. (ed.) CRYPTO 2004. LNCS, vol. 3152, pp. 306–316. Springer, Heidelberg (2004)
9. Kelsey, J., Kohno, T.: Herding Hash Functions and the Nostradamus Attack. In: Vaudenay, S. (ed.) EUROCRYPT 2006. LNCS, vol. 4004, pp. 183–200. Springer, Heidelberg (2006)
10. Kelsey, J., Schneier, B.: Second Preimages on n-bit Hash Functions for Much Less than 2ñ Work. In: Cramer, R.J.F. (ed.) EUROCRYPT 2005. LNCS, vol. 3494, pp. 474–490. Springer, Heidelberg (2005)
11. Knudsen, L., Mathiassen, J.: Preimage and Collision attacks on MD2. In: Gilbert, H., Handschuh, H. (eds.) FSE 2005. LNCS, vol. 3557, pp. 255–267. Springer, Heidelberg (2005)
12. Lei, D.: F-HASH: Securing Hash Functions Using Feistel Chaining. Cryptology ePrint Archive, Report 2005/430 (2005)
13. Lucks, S.: Hash Function Modes of Operation. In: ICE-EM RNSA 2006 Workshop at QUT, Australia (June, 2006)
14. Merkle, R.: One way Hash Functions and DES. In: Brassard, G. (ed.) CRYPTO 1989. LNCS, vol. 435, pp. 428–446. Springer, Heidelberg (1990)
15. Mironov, I., Narayanan, A.: Personal communication (August 2006)
16. Muller, M.: The MD2 Hash Function Is Not One-Way. In: Lee, P.J. (ed.) ASIACRYPT 2004. LNCS, vol. 3329, pp. 214–229. Springer, Heidelberg (2004)
17. Nandi, M., Stinson, D.: Multicollision attacks on some generalized sequential hash functions. Cryptology ePrint Archive, Report 2006/055 (2006)
18. NIST. Cryptographic Hash Algorithm Competition (November, 2007), http://www.csrc.nist.gov/groups/ST/hash/sha-3/index.html
19. Government Committee of the Russia for Standards. GOST R 34.11-94 (1994)

20. Gauravaram, P., Millan, W., Dawson, E., Viswanathan, K.: Constructing Secure Hash Functions by Enhancing Merkle-Damgård Construction. In: Batten, L.M., Safavi-Naini, R. (eds.) ACISP 2006. LNCS, vol. 4058, pp. 407–420. Springer, Heidelberg (2006)
21. Dean, R.D.: Formal Aspects of Mobile Code Security. PhD thesis, Princeton University (1999)
22. Tuma, J., Joscak, D.: Multi-block Collisions in Hash Functions based on 3C and 3C+ Enhancements of the Merkle-Damgård Construction. In: Rhee, M.S., Lee, B. (eds.) ICISC 2006. LNCS, vol. 4296, pp. 257–266. Springer, Heidelberg (2006)
23. Wagner, D.: A Generalized Birthday Problem. In: Yung, M. (ed.) CRYPTO 2002. LNCS, vol. 2442, pp. 288–303. Springer, Heidelberg (2002)
24. Wang, X., Yin, Y.L., Yu, H.: Efficient collision search attacks on SHA-0. In: Shoup, V. (ed.) CRYPTO 2005. LNCS, vol. 3621, pp. 1–16. Springer, Heidelberg (2005)
25. Wang, X., Yin, Y.L., Yu, H.: Finding collisions in the full SHA-1. In: Shoup, V. (ed.) CRYPTO 2005. LNCS, vol. 3621, pp. 17–36. Springer, Heidelberg (2005)
26. Wang, X., Yu, H.: How to Break MD5 and Other Hash Functions. In: Cramer, R.J.F. (ed.) EUROCRYPT 2005. LNCS, vol. 3494, pp. 19–35. Springer, Heidelberg (2005)
27. Yuval, G.: How to swindle Rabin. Cryptologia 3(3), 187–189 (1979)

Efficient Fully-Simulatable Oblivious Transfer

Andrew Y. Lindell

Aladdin Knowledge Systems and Bar-Ilan University, ISRAEL
andrew.lindell@aladdin.com, lindell@cs.biu.ac.il

Abstract. Oblivious transfer, first introduced by Rabin, is one of the basic building blocks of cryptographic protocols. In an oblivious transfer (or more exactly, in its 1-out-of-2 variant), one party known as the sender has a pair of messages and the other party known as the receiver obtains one of them. Somewhat paradoxically, the receiver obtains exactly one of the messages (and learns nothing of the other), and the sender does not know which of the messages the receiver obtained. Due to its importance as a building block for secure protocols, the efficiency of oblivious transfer protocols has been extensively studied. However, to date, there are almost no known oblivious transfer protocols that are secure in the presence of *malicious adversaries* under the *real/ideal model simulation paradigm* (without using general zero-knowledge proofs). Thus, *efficient protocols* that reach this level of security are of great interest. In this paper we present efficient oblivious transfer protocols that are secure according to the ideal/real model simulation paradigm. We achieve constructions under the DDH, Nth residuosity and quadratic residuosity assumptions, as well as under the assumption that homomorphic encryption exists.

1 Introduction

In an oblivious transfer, a sender with a pair of strings m_0, m_1 interacts with a receiver so that at the end the receiver learns exactly one of the strings, and the sender learns nothing [24,11]. This is a somewhat paradoxical situation because the receiver can only learn one string (thus the sender cannot send both) whereas the sender cannot know which string the receiver learned (and so the receiver cannot tell the sender which string to send). Surprisingly, it is possible to achieve oblivious transfer under a wide variety of assumptions and adversary models [11,15,19,23,1,17].

Oblivious transfer is one of the most basic and widely used protocol primitives in cryptography. It stands at the center of the fundamental results on secure two-party and multiparty computation showing that any efficient functionality can be securely computed [25,15]. In fact, it has even been shown that oblivious transfer is *complete*, meaning that it is possible to securely compute any efficient function once given a box that computes oblivious transfer [18]. Thus, oblivious transfer has great importance to the theory of cryptography. In addition to this, oblivious transfer has been widely used to construct efficient protocols for problems of interest (e.g., it is central to almost all of the work on privacy-preserving data mining).

T. Malkin (Ed.): CT-RSA 2008, LNCS 4964, pp. 52–70, 2008.

Due to its general importance, the task of constructing efficient oblivious transfer protocols has attracted much interest. In the semi-honest model (where adversaries follow the protocol specification but try to learn more than allowed by examining the protocol transcript), it is possible to construct efficient oblivious transfer from (enhanced) trapdoor permutations [11] and homomorphic encryption [19,1]. However, the situation is significantly more problematic in the malicious model where adversaries may arbitrarily deviate from the protocol specification. One possibility is to use the protocol compiler of Goldreich, Micali and Wigderson [15] to transform oblivious transfer protocols for semi-honest adversaries into protocols that are also secure in the presence of malicious adversaries. However, the result would be a highly inefficient protocol. The difficulties in obtaining secure oblivious transfer in this model seem to be due to the strict security requirements of *simulation-based definitions* that follow the ideal/real model paradigm.[1] Thus, until recently, the only known oblivious transfer protocols that were secure under this definition, and thus were *fully simulatable*, were protocols that were obtained by applying the compiler of [15]. In contrast, highly-efficient oblivious transfer protocols that guarantee *privacy* (but not simulatability) in the presence of malicious adversaries have been constructed. These protocols guarantee that even a malicious sender cannot learn which string the receiver learned, and that a malicious receiver can learn only one of the sender's input strings. Highly efficient protocols have been constructed for this setting under the DDH and N-residuosity assumptions and using homomorphic encryption [19,23,1,17].

This current state of affairs is highly unsatisfactory. The reason for this is that oblivious transfer is often used as a building block in other protocols. However, oblivious transfer protocols that only provide privacy are difficult – if not impossible – to use as building blocks. Thus, the vast number of protocols that assume (fully simulatable) oblivious transfer do not have truly efficient instantiations today. For just one example, this is true of the protocol of [20] that in turn is used in the protocol of [2] for securely computing the median. The result is that [2] has no efficient instantiation, even though it *is* efficient when ignoring the cost of the oblivious transfers. We conclude that the absence of efficient fully-simulatable oblivious transfer acts as a bottleneck in numerous other protocols.

Our results. In this paper, we construct oblivious transfer protocols that are secure (i.e., fully-simulatable) in the presence of malicious adversaries. Our constructions build on those of [23,1,17] and use cut-and-choose techniques. It is folklore that the protocols of [23,1,17] can be modified to yield full simulatability by adding proofs of knowledge. To some extent, this is what we do. However, a direct application of proofs of knowledge does not work. This is because the known efficient protocols are all information-theoretically secure in the presence of a malicious receiver. This means that only one of the sender's inputs is defined

---

[1] According to this paradigm, a real execution of a protocol is compared to an ideal execution in which a trusted third party receives the parties' inputs and sends them their outputs.

by the protocol transcript and thus a standard proof of knowledge cannot be applied. (Of course, it is possible to have the sender prove that it behaved honestly according to some committed input but this will already not be efficient.) Our protocols yield full simulatability and we provide a full proof of security.

As we show, our protocols are in the order of ℓ times the complexity of the protocols of [23,1,17], where ℓ is such the simulation fails with probability $2^{-\ell+2}$. Thus, ℓ can be taken to be relatively small (say, in the order of 30 or 40). This is a considerable overhead. However, our protocols are still by far the most efficient known without resorting to a random oracle.

Related work. There has been much work on efficient oblivious transfer in a wide range of settings. However, very little has been done regarding fully-simulatable oblivious transfer that is also efficient (without using random oracles). Despite this, recently there has been some progress in this area. In [6], fully simulatable constructions are presented. However, these rely on strong and relatively non-standard assumptions (q-power DDH and q-strong Diffie-Hellman). Following this, protocols were presented that rely on the Decisional Bilinear Diffie-Hellman assumption [16]. Our protocols differ from those of [6] and [16] in the following ways:

1. *Assumptions:* We present protocols that can be constructed assuming that DDH is hard, that there exist homomorphic encryption schemes, and more. Thus, we rely on far more standard and long-standing hardness assumptions.
2. *Complexity:* Regarding the number of exponentiations, it appears that our protocols are of a similar complexity to [6,16]. However, as pointed out in [10], bilinear curves are considerably more expensive than regular Elliptic curves. Thus, the standard decisional Diffie-Hellman assumption is much more efficient to use (curves that provide pairing need keys that are similar in size to RSA, in contrast to regular curves that can be much smaller).
3. *The problem solved:* We solve the basic 1-out-of-2 oblivious transfer problem, although our protocols can easily be extended to solve the *static k-out-of-n* oblivious transfer problem (where static means that the receiver must choose which k elements it wishes to receive at the onset). In contrast, [6] and [16] both solve the considerably harder problem of *adaptive k-out-of-n* oblivious transfer where the receiver chooses the elements to receive one and a time, and can base its choice on the elements it has already received.

In conclusion, if adaptive k-out-of-n oblivious transfer is needed, then [6,16] are the best solutions available. However, if (static) oblivious transfer suffices, then our protocols are considerably more efficient and are based on far more standard assumptions.

2   Definitions

In this section we present the definition of security for oblivious transfer, that is based on the general simulation-based definitions for secure computation; see [14,21,5,7]. We refer the reader to [12, Chapter 7] for full definitions, and provide only a brief overview here. Since we only consider oblivious transfer in this paper, our definitions are tailored to the secure computation of this specific function only.

Preliminaries. We denote by $s \in_R S$ the process of randomly choosing an element s from a set S. A function $\mu(\cdot)$ is **negligible** in n, or just **negligible**, if for every positive polynomial $p(\cdot)$ and all sufficiently large n's it holds that $\mu(n) < 1/p(n)$. A **probability ensemble** $X = \{X(n,a)\}_{a \in \{0,1\}^*; n \in \mathbb{N}}$ is an infinite sequence of random variables indexed by a and $n \in \mathbb{N}$. (The value a will represent the parties' inputs and n the security parameter.) Two distribution ensembles $X = \{X(n,a)\}_{n \in \mathbb{N}}$ and $Y = \{Y(n,a)\}_{n \in \mathbb{N}}$ are said to be **computationally indistinguishable**, denoted $X \overset{c}{\equiv} Y$, if for every non-uniform polynomial-time algorithm D there exists a negligible function $\mu(\cdot)$ such that for every $a \in \{0,1\}^*$,

$$|\Pr[D(X(n,a),a) = 1] - \Pr[D(Y(n,a),a) = 1]| \leq \mu(n)$$

All parties are assumed to run in time that is polynomial in the security parameter. (Formally, each party has a security parameter tape upon which that value 1^n is written. Then the party is polynomial in the input on this tape.)

Oblivious transfer. The oblivious transfer functionality is formally defined as a function f with two inputs and one output. The first input is a pair (m_0, m_1) and the second input is a bit σ. The output is the string m_σ. Party P_1, also known as the sender, inputs (m_0, m_1) and receives no output. In contrast, party P_2, also known as the receiver, inputs the bit σ and receives m_σ for output. Formally, we write $f((m_0, m_1), \sigma) = (\lambda, m_\sigma)$ where λ denotes the empty string. Stated in words, in the oblivious transfer functionality party P_1 receives no output, whereas party P_2 receives m_σ (and learns nothing about $m_{1-\sigma}$).

Adversarial behavior. Loosely speaking, the aim of a secure two-party protocol is to protect an honest party against dishonest behavior by the other party. In this paper, we consider *malicious adversaries* who may arbitrarily deviate from the specified protocol. Furthermore, we consider the *static corruption model*, where one of the parties is adversarial and the other is honest, and this is fixed before the execution begins.

Security of protocols. The security of a protocol is analyzed by comparing what an adversary can do in the protocol to what it can do in an ideal scenario that is secure by definition. This is formalized by considering an *ideal* computation involving an incorruptible *trusted third party* to whom the parties send their inputs. The trusted party computes the functionality on the inputs and returns to each party its respective output. Denote by f the oblivious transfer functionality and let $\overline{M} = (M_1, M_2)$ be a pair of non-uniform probabilistic *expected* polynomial-time machines (representing parties in the ideal model). Such a pair is **admissible** if for at least one $i \in \{1, 2\}$ we have that M_i is honest (i.e., follows the honest party instructions in the above-described ideal execution). Then, the joint execution of f under $\overline{M}$ in the ideal model (on input $((m_0, m_1), \sigma)$), denoted $\text{IDEAL}_{f, \overline{M}}((m_0, m_1), \sigma)$, is defined as the output pair of M_1 and M_2 from the above ideal execution.

We next consider the real model in which a real two-party protocol is executed and there exists no trusted third party. In this case, a malicious party may follow an arbitrary feasible strategy; that is, any strategy implementable by non-uniform probabilistic polynomial-time machines. Let π be a two-party protocol. Furthermore, let $\overline{M} = (M_1, M_2)$ be a pair of non-uniform probabilistic polynomial-time machines (representing parties in the real model). Such a pair is admissible if for at least one $i \in \{1, 2\}$ we have that M_i is honest (i.e., follows the strategy specified by π). Then, the joint execution of π under $\overline{M}$ in the real model (on input $((m_0, m_1), \sigma))$, denoted $\text{REAL}_{\pi, \overline{M}}((m_0, m_1), \sigma)$, is defined as the output pair of M_1 and M_2 resulting from the protocol interaction.

Having defined the ideal and real models, we can now define security of protocols. Loosely speaking, the definition asserts that a secure two-party protocol (in the real model) emulates the ideal model (in which a trusted party exists). This is formulated by saying that admissible pairs in the ideal model are able to simulate admissible pairs in an execution of a secure real-model protocol.

Definition 1. *Let f denote the oblivious transfer protocol and let π be a two-party protocol. Protocol π is said to be a* secure oblivious transfer protocol *if for every pair of admissible non-uniform probabilistic polynomial-time machines $\overline{A} = (A_1, A_2)$ for the real model, there exists a pair of admissible non-uniform probabilistic expected polynomial-time machines $\overline{B} = (B_1, B_2)$ for the ideal model, such that for every $m_0, m_1 \in \{0, 1\}^*$ of the same length and every $\sigma \in \{0, 1\}$,*

$$\left\{ \text{IDEAL}_{f, \overline{B}}(n, (m_0, m_1), \sigma) \right\} \stackrel{c}{\equiv} \left\{ \text{REAL}_{\pi, \overline{A}}(n, (m_0, m_1), \sigma) \right\}$$

Note that we allow the ideal adversary/simulator to run in expected (rather than strict) polynomial-time. This is essential for achieving constant-round protocols; see [4].

3   Oblivious Transfer under the DDH Assumption

In this section we present an oblivious transfer protocol that is secure in the presence of malicious adversaries, under the DDH assumption. The protocol is a variant of the two-round protocol of [23] with some important changes. Before proceeding, we recall the protocol of [23]. Basically, this protocol works by the receiver generating a tuple (g^a, g^b, g^c, g^d) with the following property: if the receiver's input equals 0 then $c = ab$ and d is random, and if the receiver's input equals 1 then $d = ab$ and c is random. The sender receives this tuple and carries out a manipulation that randomizes the tuple so that if $c = ab$ then the result of the manipulation on (g^a, g^b, g^c) is still a DDH tuple and the result of the manipulation on (g^a, g^b, g^d) yields a completely random tuple (if $d = ab$ then the same holds in reverse). The sender then derives a secret key from the manipulation of each of (g^a, g^b, g^c) and (g^a, g^b, g^d), and sends information that enables the receiver to derive the same secret key from the DDH tuple, whereas the key from the non-DDH tuple remains completely random. In addition, the sender encrypts its first message under the key derived from (g^a, g^b, g^c) and its second message under the key derived from (g^a, g^b, g^d). The receiver is able to decrypt the message derived from the DDH tuple but has no information about the other key and so cannot learn anything about the other message. We remark that the sender checks that $g^c \neq g^d$. This ensures that only one of (g^a, g^b, g^c) and (g^a, g^b, g^d) is a DDH tuple.

The secret key that is derived from the non-DDH tuple above is information-theoretically hidden from the receiver. This causes a problem when attempting to construct a simulator for the protocol because the simulator must learn *both* of the sender's inputs in order to send them to the trusted party (and for whatever first message the simulator sends, it can only learn one of the sender's inputs). We remark that if rewinding is used to obtain both messages then this causes a problem because the sender can make its input depend on the first message from the receiver. We therefore change the protocol of [23] so that instead of sending (g^a, g^b, g^c, g^d) where at most one of c or d equals $a \cdot b$, the receiver sends two tuples: one of the tuples is a DDH type and the other is *not*. The parties then interact to ensure that indeed only one of the tuples is of the DDH type. As we will see, this ensures that the receiver obtains only one message. The "interaction" used to prove this is of the simplest cut-and-choose type.

The protocol below uses two commitment schemes for the purpose of coin tossing: a perfectly hiding commitment scheme denoted Com_h, and a perfectly binding commitment scheme, denoted Com_b. We remark that such commitment schemes exist under the Discrete Log assumption, and thus also under the DDH assumption. We assume that the input values m_0, m_1 of the sender are in the group $\mathcal{G}$ that we are working with for the DDH assumption. If they cannot be mapped to $\mathcal{G}$ (e.g., they are too long), then the oblivious transfer can be used to exchange secret keys k_0 and k_1 that are used to encrypt m_0 and m_1, respectively.

Protocol 1

- **Auxiliary input:** *The parties have the description of a group $\mathcal{G}$ of order q, and a generator g for the group. In addition, they have a statistical error parameter ℓ.*

- **Input:** *The sender has a pair of group elements (m_0, m_1) and the receiver has a bit σ.*

- **The protocol:**
  1. For $i = 1, \ldots, \ell$, the receiver P_2 chooses a random bit $\sigma_i \in_R \{0, 1\}$ and random values $a_i^0, b_i^0, c_i^0, a_i^1, b_i^1, c_i^1 \in_R \{1, \ldots, q\}$ under the constraint that $c_i^{\sigma_i} = a_i^{\sigma_i} \cdot b_i^{\sigma_i}$ and $c_i^{1-\sigma_i} \neq a_i^{1-\sigma_i} \cdot b_i^{1-\sigma_i}$. Then, P_2 computes the tuples $\gamma_i^0 = (g^{a_i^0}, g^{b_i^0}, g^{c_i^0})$ and $\gamma_i^1 = (g^{a_i^1}, g^{b_i^1}, g^{c_i^1})$. Note that $\gamma_i^{\sigma_i}$ is a DDH tuple and $\gamma_i^{1-\sigma_i}$ is not.
  P_2 sends all of the pairs $\langle (\gamma_1^0, \gamma_1^1), \ldots, (\gamma_\ell^0, \gamma_\ell^1) \rangle$ to the sender P_1.

  2. Coin tossing:
     (a) P_1 chooses a random $s \in_R \{0, 1\}^\ell$ and sends $\mathsf{Com}_h(s)$ to P_2.

     (b) P_2 chooses a random $s' \in_R \{0, 1\}^\ell$ and sends $\mathsf{Com}_b(s')$ to P_1.

(c) P_1 and P_2 send decommitments to $\mathsf{Com}_h(s)$ and $\mathsf{Com}_b(s')$, respectively, and set $r = s \oplus s'$. Denote $r = r_1, \ldots, r_\ell$.

3. For every i for which $r_i = 1$, party P_2 sends $a_i^0, b_i^0, c_i^0, a_i^1, b_i^1, c_i^1$ to P_1. In addition, for every j for which $r_j = 0$, party P_2 sends a "reordering" of γ_j^0 and γ_j^1 so that all of the γ_j^σ tuples are DDH tuples and all of the $\gamma_j^{1-\sigma}$ tuples are not. This reordering is a bit such that if it equals 0 then the tuples are left as is, and if it equals 1 then γ_j^0 and γ_j^1 are interchanged.

4. P_1 checks that for every i for which $r_i = 1$ it received the appropriate values and that they define γ_i^0 and γ_i^1. Furthermore, it checks that exactly one of γ_i^0 and γ_i^1 is a DDH tuple as defined above and the other is not. If any of the checks fail, P_1 halts and outputs $\bot$. Otherwise it continues as follows:

(a) Denote $\gamma_j^0 = (x_j^0, y_j^0, z_j^0)$ and $\gamma_j^1 = (x_j^1, y_j^1, z_j^1)$. Then, for every j for which $r_j = 0$, party P_1 chooses random $u_i^0, u_i^1, v_i^0, v_i^1 \in_R \{1, \ldots, q\}$ and computes the following four values:

$$w_j^0 = \left(x_j^0\right)^{u_i^0} \cdot g^{v_i^0} \qquad k_j^0 = \left(z_j^0\right)^{u_i^0} \cdot \left(y_j^0\right)^{v_i^0}$$
$$w_j^1 = \left(x_j^1\right)^{u_i^1} \cdot g^{v_i^1} \qquad k_j^1 = \left(z_j^1\right)^{u_i^1} \cdot \left(y_j^1\right)^{v_i^1}$$

(b) Let $j_1, \ldots, j_t$ be the indices j for which $r_j = 0$. Then, P_1 "encrypts" m_0 under all of the keys k_j^0, and m_1 under all of the keys k_j^1, as follows:

$$c_0 = \left(\prod_{i=1}^{t} k_{j_i}^0\right) \cdot m_0 \qquad c_1 = \left(\prod_{i=1}^{t} k_{j_i}^1\right) \cdot m_1$$

P_1 sends P_2 all of the w_j^0, w_j^1 values, as well as the pair (c_0, c_1).

5. For every j for which $r_j = 0$, party P_2 computes $k_j^\sigma = (w_j^\sigma)^{b_j^0}$. Then, P_2 outputs $m_\sigma = c_\sigma \cdot \left(\prod_{i=1}^{t} k_{j_i}^\sigma\right)^{-1}$.

Before proceeding to the proof, we show that the protocol "works", meaning that when P_1 and P_2 are honest, the output is correctly obtained. We present this to "explain" the computations that take place in the protocol, although these are exactly as in the protocol of [23]. First, notice that

$$\left(w_j^\sigma\right)^{b_j^\sigma} = \left(x_j^\sigma\right)^{u_j^\sigma \cdot b_j^\sigma} \cdot \left(g^{v_j^\sigma}\right)^{b_j^\sigma} = \left(g^{a_j^\sigma \cdot b_j^\sigma}\right)^{u_j^\sigma} \cdot \left(g^{b_j^\sigma}\right)^{v_j^\sigma}$$

By the fact that γ_j^σ is a DDH tuple we have that $g^{a_j^\sigma \cdot b_j^\sigma} = z_j^\sigma$ and so

$$\left(w_j^\sigma\right)^{b_j^\sigma} = \left(z_j^\sigma\right)^{u_j^\sigma} \cdot \left(y_j^\sigma\right)^{v_j^\sigma} = k_j^\sigma$$

Thus P_2 correctly computes each key k_j^σ for j such that $r_j = 0$. Given all of these keys, it immediately follows that P_2 can decrypt c_σ, obtaining m_σ. We now proceed to prove the security of the protocol.

Theorem 1. *Assume that the decisional Diffie-Hellman problem is hard in $\mathcal{G}$ with generator g, that Com_h is a perfectly-hiding commitment scheme, and that Com_b is a perfectly-binding commitment scheme. Then, Protocol 1 securely computes the oblivious transfer functionality in the presence of malicious adversaries.*

Proof: We separately prove the security of the protocol for the case that no parties are corrupted, P_1 is corrupted, and P_2 is corrupted. In the case that both P_1 and P_2 are honest, we have already seen that P_2 obtains exactly m_σ. Thus, security holds. We now proceed to the other cases.

P_1 is corrupted. Let $\mathcal{A}_1$ be a non-uniform probabilistic polynomial-time real adversary that controls P_1. We construct a non-uniform probabilistic expected polynomial-time ideal-model adversary/simulator $\mathcal{S}_1$. The basic idea behind how $\mathcal{S}_1$ works is that it uses rewinding in order to ensure that all of the "checked" tuples are valid (i.e., one is a DDH tuple and the other is not), whereas all of the "unchecked" tuples have the property that they are *both* of the DDH type. Now, since the protocol is such that a receiver can obtain a key k_j^σ as long as γ_j^σ was a DDH tuple, it follows that $\mathcal{S}_1$ can obtain all of the k_j^0 and k_j^1 keys. This enables it to decrypt both c_0 and c_1 and obtain both messages input by $\mathcal{A}_1$ into the protocol. $\mathcal{S}_1$ then sends these inputs to the trusted party, and the honest party P_2 in the ideal model will receive the same message that it would have received in a real execution with $\mathcal{A}_1$ (or more accurately, a message that is computationally indistinguishable from that message).

We now describe $\mathcal{S}_1$ formally. Upon input 1^n and (m_0, m_1), the machine $\mathcal{S}_1$ invokes $\mathcal{A}_1$ upon the same input and works as follows:

1. $\mathcal{S}_1$ chooses a random $r \in_R \{0,1\}^\ell$ and generates tuples $\gamma_1^0, \gamma_1^1, \ldots, \gamma_\ell^0, \gamma_\ell^1$ with the following property:
   (a) For every i for which $r_i = 1$, $\mathcal{S}_1$ constructs γ_i^0 and γ_i^1 like an honest P_2 (i.e., one of them being a DDH tuple and the other not, in random order).
   (b) For every j for which $r_j = 0$, $\mathcal{S}_1$ constructs γ_j^0 and γ_j^1 to *both* be DDH tuples.
   $\mathcal{S}_1$ hands the tuples to $\mathcal{A}_1$.
2. *Simulation of the coin tossing:* $\mathcal{S}_1$ simulates the coin tossing so that the result is r, as follows:
   (a) $\mathcal{S}_1$ receives a commitment c_h from $\mathcal{A}_1$.
   (b) $\mathcal{S}_1$ chooses a random $s' \in_R \{0,1\}^\ell$ and hands $c_b = \mathsf{Com}_b(s')$ to $\mathcal{A}_1$.
   (c) If $\mathcal{A}_1$ does not send a valid decommitment to c_h, then $\mathcal{S}_1$ simulates P_2 aborting and sends $\perp$ to the trusted party. Then $\mathcal{S}_1$ outputs whatever $\mathcal{A}_1$ outputs and halts.
   Otherwise, let s be the decommitted value. $\mathcal{S}_1$ proceeds as follows:
   i. $\mathcal{S}_1$ sets $s' = r \oplus s$, rewinds $\mathcal{A}_1$, and hands it $\mathsf{Com}_b(s')$.
   ii. If $\mathcal{A}_1$ decommits to s, then $\mathcal{S}_1$ proceeds to the next step. If $\mathcal{A}_1$ decommits to a value $\tilde{s} \neq s$, then $\mathcal{S}_1$ outputs fail. Otherwise, if it does not decommit to any value, $\mathcal{S}_1$ returns to the previous step and tries again until $\mathcal{A}_1$ does decommit to s. (We stress that in

every attempt, S_1 hands A_1 a commitment to the same value s'. However, the randomness used to generate the commitment $\mathsf{Com}_b(s')$ is independent each time.)[2]

3. Upon receiving a valid decommitment to s from A_1, simulator S_1 decommits to A_1, revealing s'. (Note that $r = s \oplus s'$.)

4. For every i for which $r_i = 1$, simulator S_1 hands A_1 the values $a_i^0, b_i^0, c_i^0, a_i^1, b_i^1$, c_i^1 used to generate γ_i^0 and γ_i^1. In addition, S_1 hands A_1 a random reordering of the pairs.

5. If A_1 does not reply with a valid message, then S_1 sends $\bot$ to the trusted party, outputs whatever A_1 outputs and halts. Otherwise, it receives a series of pairs (w_j^0, w_j^1) for every j for which $r_j = 0$, as well as ciphertexts c_0 and c_1. S_1 then follows the instructions of P_2 for deriving the keys. However, unlike an honest P_2, it computes $k_j^0 = (w_j^0)^{b_j^0}$ and $k_j^1 = (w_j^1)^{b_j^1}$ and uses the keys it obtains to decrypt *both* c_0 and c_1. (Note that for each such j, both γ_j^0 and γ_j^1 are DDH tuples; thus this makes sense.)

Let m_0 and m_1 be the messages obtained by decrypting. S_1 sends the pair to the trusted party as the first party's input, outputs whatever A_1 outputs and halts.

We now prove that the joint output distribution of S_1 and an honest P_2 in an ideal execution is computationally indistinguishable from the output distribution of A_1 and an honest P_2 in a real execution. First, note that the view of A_1 in the simulation with S_1 is indistinguishable from its view in a real execution. The only difference in its view is due to the fact that the tuples γ_j^0 and γ_j^1 for which $r_j = 0$ are both of the DDH type. The only other difference is due to the coin tossing (and the rewinding). However, by the binding property of the commitment sent by A_1 and the fact that P_2 generates its commitment after receiving A_1's, we have that the outcome of the coin tossing in a real execution is statistically close to uniform (where the only difference is due to the negligible probability that A_1 will break the computational binding property of the commitment scheme.) In the simulation by S_1, the outcome is always uniformly distributed, assuming that S_1 does not output fail. Since S_1 outputs fail when A_1 breaks the computational binding of the commitment scheme, this occurs with at most negligible probability (a rigorous analysis of this is given in [13]). We therefore have that, apart from the negligible difference due to the coin tossing, the only difference is due to the generation of the tuples. Intuitively, indistinguishability therefore follows from the DDH assumption. More formally, this is proven by constructing a machine D that distinguishes many copies of DDH tuples from many copies of non-DDH tuples. D receives a series of tuples and runs in exactly the same way as S_1 except that it constructs the γ_j^0 and γ_j^1 tuples (for $r_j = 0$) so that one is a DDH tuple and the other is from its input, in random order. Furthermore, it provides the reordering so that all of the DDH tuples it generates are associated with σ and all of the ones it receives

---

[2] This strategy by S_1 is actually over-simplified and does not guarantee that it runs in expected polynomial-time. This technicality will be discussed below, and we will show how S_1 can be "fixed" so that its expected running-time is polynomial.

externally are associated with $1 - \sigma$. (For the sake of this mental experiment, we assume that D is given the input σ of P_2.) It follows that if D receives a series of DDH tuples, then the view of $\mathcal{A}_1$ is exactly the same as in the simulation with $\mathcal{S}_1$ (because all the tuples are of the Diffie-Hellman type). In contrast, if D receives a series of non-DDH tuples, then the view of $\mathcal{A}_1$ is exactly the same as in a real execution (because only the tuples associated with σ are of the Diffie-Hellman type). This suffices for showing that the output of $\mathcal{A}_1$ in a real execution is indistinguishable from the output of $\mathcal{S}_1$ in an ideal execution (recall that $\mathcal{S}_1$ outputs whatever $\mathcal{A}_1$ outputs). However, we have to show this for the joint distribution of the output of $\mathcal{A}_1$ (or $\mathcal{S}_1$) and the honest P_2. In order to see this, recall that the output of P_2 is m_σ where σ is the honest P_2's input. Now, assume that there exists a polynomial-time distinguisher D' that distinguishes between the REAL and IDEAL distributions with non-negligible probability. We construct a distinguisher D as above that distinguishes DDH from non-DDH tuples. The machine D receives the input σ of P_2 and a series of tuples that are either DDH or non-DDH tuples. D then works exactly as above (i.e., constructing the γ_j^0 and γ_j^1 tuples so that in the reordering step, all the γ_j^σ tuples are those it generated itself and all the $\gamma_j^{1-\sigma}$ tuples are those it received as input). Since D generated all of the γ_j^σ tuples, it is able to "decrypt" c_σ and obtain m_σ. Machine D therefore does this, and invokes D' on the output of $\mathcal{A}_1$ *and* the message m_σ (which is the output that an honest P_2 would receive). Finally D outputs whatever D' does. It is clear that if D receives non-DDH tuples, then the output distribution generated is exactly like that of a real execution between $\mathcal{A}_1$ and P_2. In contrast, if it receives DDH tuples, then the output distribution is exactly like of an ideal execution with $\mathcal{S}_1$. (A subtle point here is that the distribution over the γ tuples generated by D who knows σ is identical to the distribution generated by $\mathcal{S}_1$ who does not know σ. The reason for this is that when all the tuples are of the DDH type, their ordering makes no difference.) We conclude that D solves the DDH problem with non-negligible probability, in contradiction to the DDH assumption. Thus, the REAL and IDEAL output distributions must be computationally indistinguishable, as required.

It remains to prove that $\mathcal{S}_1$ runs in expected polynomial-time. Unfortunately, this is not true! In order to see this, denote by p the probability that $\mathcal{A}_1$ decommits correctly to s when it receives a commitment to a random s'. Next, denote by q the probability that $\mathcal{A}_1$ decommits correctly when it receives a commitment to $s' = s \oplus r$. (Note that this is not random because r is implicit in the way that $\mathcal{S}_1$ generated the tuples. That is, if $r_i = 1$ then γ_i^0 and γ_i^1 are honestly generated, and otherwise they are both of the DDH type.) Now, by the hiding property of the commitment scheme Com_b, the difference between p and q can be at most negligible. Furthermore, the expected running-time of $\mathcal{S}_1$ in the rewinding stage equals p/q times some fixed polynomial factor. In order to see this, observe that $\mathcal{S}_1$ enters the rewinding stage with probability p, and concludes after an expected $1/q$ number of rewindings. It thus remains to bound p/q. (We remark that $\mathcal{S}_1$'s running time in the rest of the simulation is a fixed polynomial and so we ignore this from now on). Unfortunately, even though p and q are at most negligibly far from each other, as we have discussed, the value p/q may not necessarily be polynomial. For example, if $p = 2^{-n}$ and $q = 2^{-n} + 2^{-n/2}$ then $p/q \approx 2^{n/2}$. Thus, the expected running-time of $\mathcal{S}_1$ is not necessarily polynomial. Fortunately, this can be solved using the techniques of [13] who solved an identical problem. Loosely speaking, the technique of [13] works by first estimating p and then ensuring that the number of rewinding attempts does not exceed a fixed polynomial times the estimation of p. It is shown that this yields a simulator that is guaranteed to run in expected polynomial time. Furthermore, the output of the simulator is only negligibly far from the original (simplified) strategy described above. Thus, these techniques can be applied here and the simulator appropriately changed, with the result being that the output is only negligibly different from before, as required.

P_2 is corrupted. As before, we let $\mathcal{A}_2$ be any non-uniform probabilistic polynomial-time adversary controlling P_2 and we construct a non-uniform probabilistic expected polynomial-time simulator $\mathcal{S}_2$. The simulator $\mathcal{S}_2$ extracts the bit σ used by $\mathcal{A}_2$ by rewinding it and obtaining the reordering of tuples that it had previously opened. Formally, upon input 1^n and σ, the simulator $\mathcal{S}_2$ invokes $\mathcal{A}_2$ upon the same input and works as follows:

1. $\mathcal{S}_2$ receives a series of tuples $\gamma_1^0, \gamma_1^1, \ldots, \gamma_\ell^0, \gamma_\ell^1$ from $\mathcal{A}_2$.
2. $\mathcal{S}_2$ hands $\mathcal{A}_2$ a commitment $c_h = \mathsf{Com}_h(s)$ to a random $s \in_R \{0,1\}^\ell$, receives back c_b, decommits to c_h and receives $\mathcal{A}_2$'s decommitment to c_b. $\mathcal{S}_2$ then receives all of the $a_i^0, b_i^0, c_i^0, a_i^1, b_i^1, c_i^1$ values from $\mathcal{A}_2$, for i where $r_i = 1$, and the reorderings for j where $r_j = 0$. If the values sent by $\mathcal{A}_2$ are not valid (as checked by P_1 in the protocol) or $\mathcal{A}_2$ did not send valid decommitments, $\mathcal{S}_2$ sends $\bot$ to the trusted party, outputs whatever $\mathcal{A}_2$ outputs, and halts. Otherwise, it continues to the next step.
3. $\mathcal{S}_2$ rewinds $\mathcal{A}_2$ back to the beginning of the coin-tossing, hands $\mathcal{A}_2$ a commitment $\tilde{c}_h = \mathsf{Com}_h(\tilde{s})$ to a fresh random $\tilde{s} \in_R \{0,1\}^\ell$, receives back some $\tilde{c}_b$, decommits to $\tilde{c}_h$ and receives $\mathcal{A}_2$'s decommitment to $\tilde{c}_b$. In addition, $\mathcal{S}_2$ receives the $a_i^0, b_i^0, c_i^0, a_i^1, b_i^1, c_i^1$ values and reorderings.

   If any of the values are not valid, $\mathcal{S}_2$ repeats this step using fresh randomness each time, until all values are valid.
4. Following this, $\mathcal{S}_2$ rewinds $\mathcal{A}_2$ to the beginning and resends the exact messages of the first coin tossing (resulting in exactly the same transcript as before).
5. Denote by r the result of the first coin tossing (Step 2 above), and $\tilde{r}$ the result of the second coin tossing (Step 3 above). If $r = \tilde{r}$ then $\mathcal{S}_2$ outputs fail and halts. Otherwise, $\mathcal{S}_2$ searches for a value t such that $r_t = 0$ and $\tilde{r}_t = 1$. (Note that by the definition of the simulation, exactly one of γ_t^0 and γ_t^1 is a DDH tuple. Otherwise, the values would not be considered valid.) If no such t exists (i.e., for every t such that $r_t \neq \tilde{r}_t$ it holds that $r_t = 1$ and $\tilde{r}_t = 0$), then $\mathcal{S}_2$ begins the simulation from scratch with the exception that it must find r and $\tilde{r}$ for which all values are valid (i.e., if for r the values sent by $\mathcal{A}_2$ are not valid it does not terminate the simulation but rather rewinds until it finds an r for which the responses of $\mathcal{A}_2$ are all valid).

If S_2 does not start again, we have that it has $a_t^0, b_t^0, c_t^0, a_t^1, b_t^1, c_t^1$ and can determine which of γ_t^0 and γ_t^1 is a DDH tuple. Furthermore, since $\tilde{r}_t = 1$, the reordering that S_2 receives from A_2 after the coin tossing indicates whether the DDH tuple is associated with 0 or with 1. S_2 sets $\sigma = 0$ if after the reordering γ_t^0 is of the DDH type, and sets $\sigma = 1$ if after the reordering γ_t^1 is of the DDH type. (Note that exactly one of the tuples is of the DDH type because this is checked in the second coin tossing.)

6. S_2 sends σ to the trusted party and receives back a string $m = m_\sigma$. Simulator S_2 then computes the last message from P_1 to P_2 honestly, while encrypting m_σ under the keys k_j^σ (and encrypting any arbitrary string of the same length under the keys $k_{1-\sigma}^j$). S_2 hands A_2 these messages and outputs whatever A_2 outputs and halts.

We now prove that the output distribution of A_2 in a real execution with an honest P_1 (with input (m_0, m_1)) is computationally indistinguishable from the output distribution of S_2 in an ideal execution with an honest P_1 (with the same input (m_0, m_1)). We begin by showing that S_2 outputs fail with probability at most $2^{-\ell}$, ignoring for now the probability that $r = \tilde{r}$ in later rewindings (which may occur if S_2 has to start again from scratch). Recall that this event occurs if everything is "valid" after the first coin tossing (where the result is r), and the result of the second coin-tossing after which everything is valid is $\tilde{r} = r$.[3] First, observe that the *distributions* of the strings r and $\tilde{r}$ are identical. This is because S_2 runs the coin tossing in the same way each time (using fresh random coins), and accepts $\tilde{r}$ when all is valid, exactly as what happened with r. Next, note that the distribution over the result of the coin tossing – without conditioning over A_2 sending valid decommitments – is uniform. This holds because the commitment that S_2 hands to A_2 is perfectly hiding and the commitment returned by A_2 to S_2 is perfectly binding. Let R be a random variable that denotes the result of the first coin tossing between A_2 and S_2 in the simulation, and let valid be the event that A_2 replies with valid decommitments and values after the first coin tossing. Finally, for a given $r \in \{0,1\}^\ell$, let obtain_r denote the event that the result of one of the coin tossing attempts in the second stage equals r. (Note that this does not mean that $\tilde{r} = r$ because $\tilde{r}$ is the result that is finally accepted after A_2 sends valid values. However, the decision of A_2 to send valid values may also depend on the randomness used to generate $\mathsf{Com}_h(s)$. Thus, $\tilde{r}$ may not equal r, even though r is obtained in one of the coin tossing attempts in the second stage.) Clearly, fail can only occur if r is obtained at least once as the result of a coin tossing attempt in the second stage (because fail can only occur if $\tilde{r} = r$). We therefore have the following:

$$\Pr[\mathsf{fail}] \leq \sum_{r \in \{0,1\}^\ell} \Pr[R = r \ \& \ \mathsf{valid}] \cdot \Pr[\mathsf{obtain}_r] \tag{1}$$

---

[3] It is very easy to prove that the probability that S_2 outputs fail is at most $2^{-\ell/2}$. However, in order to keep ℓ to a low value, we present a more subtle analysis that demonstrates that S_2 outputs fail with probability at most $2^{-\ell}$.

Before analyzing this probability, we compute $\Pr[\mathsf{obtain}_r]$ for a fixed r. Let p denote the probability (over $\mathcal{A}_2$ and $\mathcal{S}_2$'s coin tosses) that $\mathcal{A}_2$ sends valid values after the coin tossing. It follows that the expected number of trials by $\mathcal{S}_2$ in the second coin tossing is $1/p$. Letting X_r be a Boolean random variable that equals 1 if and only if the result of the second coin tossing attempt equals the fixed r, we have that $E[X_r] = 2^{-\ell}$. By Wald's equation (e.g., see [22, Page 300]), it follows that the expected number of times that r is obtained as the result of a coin tossing attempt in the second stage by $\mathcal{S}_2$ is $1/p \cdot 2^{-\ell}$. Using Markov's inequality, we have that the probability that r is obtained at least once as the result of a coin tossing attempt in the second stage is at most $1/p \cdot 2^{-\ell}$. That is:

$$\Pr[\mathsf{obtain}_r] \leq \frac{1}{p \cdot 2^{\ell}}$$

We are now ready to return to Eq. (1). Denote by p_r the probability that $\mathcal{A}_2$ sends valid values conditioned on the outcome of the coin tossing being r. It follows that

$$p = \sum_{r \in \{0,1\}^{\ell}} \Pr[R = r] \cdot p_r = \sum_{r \in \{0,1\}^{\ell}} \frac{p_r}{2^{\ell}}$$

Furthermore,

$$\Pr[R = r \;\&\; \mathsf{valid}] = \Pr[\mathsf{valid} \mid R = r] \cdot \Pr[R = r] = p_r \cdot \frac{1}{2^{\ell}}$$

Combining the above, we have:

$$\Pr[\mathsf{fail}] \leq \sum_{r \in \{0,1\}^{\ell}} \Pr[R = r \;\&\; \mathsf{valid}] \cdot \Pr[\mathsf{obtain}_r]$$

$$\leq \sum_{r \in \{0,1\}^{\ell}} \frac{p_r}{2^{\ell}} \cdot \frac{1}{p \cdot 2^{\ell}}$$

$$= \frac{1}{p \cdot 2^{\ell}} \cdot \sum_{r \in \{0,1\}^{\ell}} \frac{p_r}{2^{\ell}}$$

$$= \frac{1}{p \cdot 2^{\ell}} \cdot p = \frac{1}{2^{\ell}}$$

We conclude that $\mathcal{S}_2$ outputs fail with probability at most $2^{-\ell}$, as required. Recall that this analysis doesn't take into account the probability that $\mathcal{S}_2$ starts the simulation from scratch. Rather, it just shows that $\mathcal{S}_2$ outputs fail in any simulation attempt (between starts from scratch) with probability at most $2^{-\ell}$. Below, we will show that the probability that $\mathcal{S}_2$ starts from scratch is at most $1/2$. Denote by fail_i the probability that $\mathcal{S}_2$ outputs fail in the ith attempt, given that there is such an attempt. Likewise, denote by repeat_i the probability that $\mathcal{S}_2$ has an ith attempt. We have shown that for every i, $\Pr[\mathsf{fail}_i] = 2^{-\ell}$, and below we show that every repeat happens with probability $1/2$ and so for every i, $\Pr[\mathsf{repeat}_i] = 2^{i-1}$ ($\mathsf{repeat}_1 = 1$ because we always have one attempt). We therefore have:

$$\Pr[\mathsf{fail}] = \sum_{i=1}^{\infty} \Pr[\mathsf{fail}_i] \cdot \Pr[\mathsf{repeat}_i] = \frac{1}{2^{\ell}} \sum_{i=1}^{\infty} \frac{1}{2^{i-1}} = \frac{1}{2^{\ell}} \cdot 2 = \frac{1}{2^{\ell-1}}$$

Given the above, we proceed to show indistinguishability of the ideal and real distributions. Notice that in the case that $\mathcal{S}$ does not output fail, the final transcript as viewed by $\mathcal{A}_2$ consists of the first coin tossing (that is distributed exactly as in a real execution) and the last message from $\mathcal{S}_2$ to $\mathcal{A}_2$. This last message is not generated honestly, in that c_σ is indeed an encryption of m_σ, but $c_{1-\sigma}$ is an encryption of an arbitrary value (and not necessarily of $m_{1-\sigma}$). However, as shown in [23], for any tuple $\gamma_j^{1-\sigma}$ that is *not* a DDH tuple, the value $k_j^{1-\sigma}$ is uniformly distributed in $\mathcal{G}$ (even given $w_j^{1-\sigma}$ as received by $\mathcal{A}_2$). This implies that $c_{1-\sigma}$ is uniformly distributed, independent of the value $m_{1-\sigma}$. Thus, $\mathcal{A}_2$'s view in the execution with $\mathcal{S}_2$ is statistically close to its view in a real execution with P_1 (the only difference being if $\mathcal{S}_2$ outputs fail). This completes the proof regarding indistinguishability.

It remains to prove that $\mathcal{S}_2$ runs in expected polynomial-time. We begin by analyzing the rewinding by $\mathcal{S}_2$ in the coin tossing phase (clearly, the running-time of $\mathcal{S}_2$ outside of the rewinding is strictly polynomial, and so it suffices to bound the expected number of rewinding attempts). Denote by p the probability that $\mathcal{A}_2$ completes the coin tossing phase and provides valid values to $\mathcal{S}_2$. The important point to note here is that each rewinding attempt is successful with probability exactly p (there is no difference between the distribution over the first and second coin tossing attempts, in contrast to the simulation where P_1 is corrupted). Thus, with probability p there are rewinding attempts, and in such a case there are an expected $1/p$ such attempts. This yields an expected number of rewindings of 1. We now analyze the number of times that $\mathcal{S}_2$ is expected to have to begin from scratch (due to there being no t for which $r_t = 0$ and $\tilde{r}_t = 1$). The main observation here is that for any pair r and $\tilde{r}$ which forces $\mathcal{S}_2$ to begin from scratch, interchanging r and $\tilde{r}$ would result in a pair for which $\mathcal{S}_2$ would be able to continue. Now, since r and $\tilde{r}$ are derived through independent executions of the coin tossing phase, the probability that they are in one order *equals* the probability that they are in the opposite order. Thus, the probability that $\mathcal{S}_2$ needs to start from scratch equals at most $1/2$. This implies that the expected number of times that $\mathcal{S}_2$ needs to start from scratch is at most two. We remark that when $\mathcal{S}_2$ starts from scratch, the expected number of times it needs to rewind in order to obtain each of r and $\tilde{r}$ is $1/p$. Thus, overall the expected number of rewinding attempts is $p \cdot \mathcal{O}(1)/p = \mathcal{O}(1)$. We conclude that the overall expected running time of $\mathcal{S}_2$ is polynomial, as required. ∎

Efficiency. The complexity of the protocol is in the order of ℓ times the basic protocol of [23]. Thus, the efficiency depends strongly on the value of ℓ that is taken. It is important to notice that the simulation succeeds except with probability $\approx 2^{-\ell+1}$ (as long as the cryptographic primitives are not "broken"). To be more exact, one should take ℓ and n so that the probability of "breaking" the cryptographic primitives (the commitments for the coin tossing or the security of encryption) is at most $2^{-\ell+1}$. In such a case, our analysis in the proof shows that the ideal and real executions can be distinguished with probability at most $2^{-\ell+2}$. This means that ℓ can be chosen to be relatively small, depending on the level of security desired. Specifically, with $\ell = 30$ the probability of successful undetected cheating is $2^{-28} \approx 3.7 \times 10^{-9}$ which is already very very small. Thus, it is reasonable to say that the complexity of the protocol is between 30 and 40 times of that of [23]. This is a non-trivial price; however, this is far more efficient than known solutions. We also remark that a similar idea can be used to achieve security in the model of covert adversaries of [3]. For deterrent factor $\epsilon = 1/2$ one can use $\ell = 2$ and have the sender choose r singlehandedly with one bit of r equalling 0 and the other equalling 1. This yields very high efficiency, together with simulatability (albeit in the weaker model of covert adversaries).

4   Oblivious Transfer Using Smooth Hashing

The protocol of [23] was generalized by [17] via the notion of smooth projective hashing of [8]. This enables the construction of oblivious transfer protocols that are analogous to [23] under the Nth residuosity and quadratic residuosity assumptions. Protocol 1 can be extended directly in the same way, yielding oblivious transfer protocols that are secure against malicious adversaries, under the Nth residuosity and quadratic residuosity assumptions. We remark that as in the protocol of [17], the instantiation of the protocol under the Nth residuosity assumption is highly efficient, whereas the instantiation under the quadratic residuosity assumption enables the exchange of a single bit only (but is based on a longer-standing hardness assumption). We remark, however, that using Elliptic curves, the solution based on the DDH assumption is by far the most efficient.

5   Oblivious Transfer from Homomorphic Encryption

In this section, we present a protocol based on the protocol of [1] that uses homomorphic encryption. We assume an additive homomorphic encryption scheme (G, E, D), where $G(1^n)$ outputs a key-pair of length n, E is the encryption algorithm and D the decryption algorithm. Note that additive homomorphic operations imply multiplication by a scalar as well. The ideas behind this protocol are similar to above, and our presentation is therefore rather brief.

Protocol 2

- **Input:** *The sender has a pair of strings* (m_0, m_1) *of known length and the receiver has a bit* σ. *Both parties have a security parameter* n *determining the length of the keys for the encryption scheme, and a separate statistical security parameter* ℓ.

- **The protocol:**
  1. Receiver's message:
     (a) *The receiver* P_2 *chooses a key-pair* $(pk, sk) \leftarrow G(1^n)$ *from a homomorphic encryption scheme* (G, E, D).[4]

---

[4] We assume that it is possible to verify that a public-key pk is in the range of the key generation algorithm G. If this is not the case, then a zero-knowledge proof of this fact must be added.

(b) *For $i = 1, \ldots, \ell$, party P_2 chooses a random bit $b_i \in_R \{0, 1\}$ and defines*

$$c_i^{b_i} = E_{pk}(0; r_i^{b_i}) \quad \text{and} \quad c_i^{1-b_i} = E_{pk}(1; r_i^{1-b_i}) .$$

where r_i^0 and r_i^1 are random strings, and $E_{pk}(x; r)$ denotes an encryption of message x using random coins r.

(c) *P_2 sends $pk, \langle c_1^0, c_1^1, \ldots, c_\ell^0, c_\ell^1 \rangle$ to P_1.*

2. Coin tossing:
   (a) *P_1 chooses a random $\tilde{s} \in_R \{0, 1\}^\ell$ and sends $\mathsf{Com}_h(\tilde{s})$ to P_2.*

   (b) *P_2 chooses a random $\hat{s} \in_R \{0, 1\}^\ell$ and sends $\mathsf{Com}_b(\hat{s})$ to P_1.*

   (c) *P_1 and P_2 send decommitments to $\mathsf{Com}_h(\tilde{s})$ and $\mathsf{Com}_b(\hat{s})$, respectively, and set $s = \tilde{s} \oplus \hat{s}$. Denote $s = s_1, \ldots, s_\ell$. Furthermore let S_1 be the set of all i for which $s_i = 1$, and let S_0 be the set of all j for which $s_j = 0$. (Note that S_1, S_0 are a partition of $\{1, \ldots, \ell\}$.)*

3. Receiver's message:
   (a) *For every $i \in S_1$, party P_2 sends the randomness r_i^0, r_i^1 used to encrypt c_i^0 and c_i^1.*

   (b) *In addition, for every $j \in S_0$, party P_2 sends a bit β_j so that if $\sigma = 0$ then $\beta_j = b_j$, and if $\sigma = 1$ then $\beta_j = 1 - b_j$.*

4. Sender's message:
   (a) *For every $i \in S_1$, party P_1 verifies that either $c_i^0 = E_{pk}(0; r_i^0)$ and $c_i^1 = E_{pk}(1; r_i^1)$, or $c_i^0 = E_{pk}(1; r_i^0)$ and $c_i^1 = E_{pk}(0; r_i^1)$. That is, P_1 verifies that in every pair, one ciphertext is an encryption of 0 and the other is an encryption of 1. If this does not hold for every such i, party P_1 halts. If it does hold, it proceeds to the next step.*

   (b) *For every $j \in S_0$, party P_1 defines c_j and c_j' as follows:*
      i. *If $\beta_i = 0$ then $c_j = c_j^0$ and $c_j' = c_j^1$.*

      ii. *If $\beta_i = 1$ then $c_j = c_j^1$ and $c_j' = c_j^0$.*
      This implies that if $\sigma = 0$ then $c_j = E_{pk}(0)$ and $c_j' = E_{pk}(1)$, and if $\sigma = 1$ then $c_j = E_{pk}(1)$ and $c_j' = E_{pk}(0)$.[5]

   (c) *For every $j \in S_0$, party P_1 chooses random ρ_j, ρ_j', uniformly distributed in the group defined by the encryption scheme. Then, P_1 uses the homomorphic properties of the encryption scheme to compute:*

$$c_0 = \left(\sum_{j \in S_1} \rho_j \cdot c_j \right) + E_{pk}(m_0) \quad \text{and} \quad c_1 = \left(\sum_{j \in S_1} \rho_j' \cdot c_j' \right) + E_{pk}(m_1)$$

---

[5] In order to see this, note that if $\sigma = 0$ then $\beta_j = b_j$. Thus, if $\beta_j = b_j = 0$ we have that $c_j = c_j^0 = E_{pk}(0)$ and $c_j' = c_j^1 = E_{pk}(1)$. In contrast, if $\beta_j = b_j = 1$ then $c_j = c_j^1 = E_{pk}(0)$ and $c_j' = c_j^0 = E_{pk}(1)$. That is, in all cases of $\sigma = 0$ it holds that $c_j = E_{pk}(0)$ and $c_j' = E_{pk}(1)$. Analogously, if $\sigma = 1$ the reverse holds.

where addition above denotes the homomorphic addition of cipher-texts and multiplication denotes multiplication by a scalar (again using the homomorphic properties).

(d) P_1 sends (c_0, c_1) to P_2.

5. Receiver computes output: P_2 outputs $D_{sk}(c_\sigma)$ and halts.

Before discussing security, we demonstrate correctness:

1. Case $\sigma = 0$: In this case, as described in Footnote 5, it holds that for every j, $c_j = E_{pk}(0)$ and $c'_j = E_{pk}(1)$. Noting that the multiplication of 0 by a scalar equals 0, we have:

$$c_0 = \left(\sum_{j \in S_1} \rho_j \cdot c_j \right) + E_{pk}(m_0) = E_{pk}(0) + E_{pk}(m_0) = E_{pk}(m_0).$$

   Thus, when P_2 decrypts c_0 it receives m_0, as required.
2. Case $\sigma = 1$: In this case, it holds that for every j, $c_j = E_{pk}(1)$ and $c'_j = E_{pk}(0)$. Thus, similarly to before,

$$c_1 = \cdot \left(\sum_j \rho'_j \cdot c'_j \right) + E_{pk}(m_1) = E_{pk}(0) + E_{pk}(m_1) = E_{pk}(m_1),$$

   and so when P_2 decrypts c_1, it receives m_1, as required.

We have the following theorem:

Theorem 2. *Assume that (G, E, D) is a secure homomorphic encryption scheme, Com_h is a perfectly-hiding commitment scheme and Com_b is a perfectly-biding commitment scheme. Then, Protocol 2 securely computes the oblivious transfer functionality in the presence of malicious adversaries.*

Proof (sketch):   In the case that P_2 is corrupted, the simulator works by rewinding the corrupted P_2 over the coin tossing phase in order to obtain two different openings and reorderings. In this way, the simulator can easily derive the value of P_2's input σ (σ is taken to be 0 if all the c_j ciphertexts for which it obtained both reorderings and openings are encryptions of 0, and is taken to be 1 otherwise). It sends σ to the trusted party and receives back $m = m_\sigma$. Finally, the simulator generates c_σ as the honest party P_1 would (using m), and generates $c_{1-\sigma}$ as an encryption to a random string. Beyond a negligible fail probability in obtaining the two openings mentioned, the only difference with respect to a corrupted P_2's view is the way $c_{1-\sigma}$ is generated. However, notice that:

$$c_{1-\sigma} = \left(\sum_{j \in S_1} \hat{\rho}_j \cdot \hat{c}_j \right) + E_{pk}(m_{1-\sigma})$$

where $\hat{\rho}_j = \rho_j$ and $\hat{c}_j = c_j$, or $\hat{\rho}_j = \rho'_j$ and $\hat{c}_j = c'_j$, depending on the value of σ. Now, if at least one value $\hat{c}_j$ for $j \in S_1$ is an encryption of 1, then the ciphertext $c_{1-\sigma}$ is an encryption of a uniformly distributed value (in the group defined by the homomorphic encryption scheme). This is due to the fact that $\hat{c}_j$ is multiplied by $\hat{\rho}_j$ which is uniformly distributed. Now, by the cut-and-choose technique employed, the probability that for all $j \in S_1$ it holds that $\hat{c}_j \neq E_{pk}(1)$ is negligible. This is due to the fact that this can only hold if for *many* ciphertext pairs c_i^0, c_i^1 sent by P_2 in its first message, the pair is *not* correctly generated (i.e., it is not the case that one is an encryption of 0 and the other an encryption of 1). However, if this is the case, then P_1 will abort except with negligible probability, because S_0 will almost certainly contain one of these pairs (and the sets S_0 and S_1 are chosen as a random partition based on the value s output from the coin tossing).

In the case that P_1 is corrupted, the simulator manipulates the coin tossing so that in the unopened pairs of encryptions, all of the ciphertexts encrypt 0. This implies that both $\left(\sum_{j \in S_1} \rho_j \cdot c_j \right) = E_{pk}(0)$ and $\left(\sum_{j \in S_1} \rho'_j \cdot c'_j \right) = E_{pk}(0)$, in turn implying that $c_0 = E_{pk}(m_0)$ and $c_1 = E_{pk}(m_1)$. Thus, the simulator obtains both m_0 and m_1 and sends them to the trusted party. This completes the proof sketch. A full proof follows from the proof of security for Protocol 1. ∎

Acknowledgements

We would like to thank Nigel Smart for helpful discussions and Benny Pinkas for pointing out an error in a previous version.

References

1. Aiello, W., Ishai, Y., Reingold, O.: Priced Oblivious Transfer: How to Sell Digital Goods. In: Pfitzmann, B. (ed.) EUROCRYPT 2001. LNCS, vol. 2045, pp. 119–135. Springer, Heidelberg (2001)
2. Aggarwal, G., Mishra, N., Pinkas, B.: Secure Computation of the kth-Ranked Element. In: Cachin, C., Camenisch, J.L. (eds.) EUROCRYPT 2004. LNCS, vol. 3027, pp. 40–55. Springer, Heidelberg (2004)
3. Aumann, Y., Lindell, Y.: Security Against Covert Adversaries: Efficient Protocols for Realistic Adversaries. In: Vadhan, S.P. (ed.) TCC 2007. LNCS, vol. 4392, pp. 137–156. Springer, Heidelberg (2007)
4. Barak, B., Lindell, Y.: Strict Polynomial-Time in Simulation and Extraction. SIAM Journal on Computing 33(4), 783–818 (2004)
5. Beaver, D.: Foundations of Secure Interactive Computing. In: Feigenbaum, J. (ed.) CRYPTO 1991. LNCS, vol. 576, pp. 377–391. Springer, Heidelberg (1992)
6. Camenisch, J., Neven, G., Shelat, A.: Simulatable Adaptive Oblivious Transfer. In: Naor, M. (ed.) EUROCRYPT 2007. LNCS, vol. 4515, pp. 573–590. Springer, Heidelberg (2007)
7. Canetti, R.: Security and Composition of Multiparty Cryptographic Protocols. Journal of Cryptology 13(1), 143–202 (2000)
8. Cramer, R., Shoup, V.: Universal Hash Proofs and a Paradigm for Adaptive Chosen Ciphertext Secure Public-Key Encryption. In: Knudsen, L.R. (ed.) EUROCRYPT 2002. LNCS, vol. 2332, pp. 45–64. Springer, Heidelberg (2002)

 9. Dodis, Y., Gennaro, R., Håstad, J., Krawczyk, H., Rabin, T.: Randomness Extraction and Key Derivation Using the CBC, Cascade and HMAC Modes. In: Franklin, M. (ed.) CRYPTO 2004. LNCS, vol. 3152, pp. 494–510. Springer, Heidelberg (2004)
10. Galbraith, S.D., Paterson, K.G., Smart, N.P.: Pairings for Cryptographers. Cryptology ePrint Archive Report 2006/165 (2006)
11. Even, S., Goldreich, O., Lempel, A.: A Randomized Protocol for Signing Contracts. Communications of the ACM 28(6), 637–647 (1985)
12. Goldreich, O.: Foundations of Cryptography: Basic Applications, vol. 2. Cambridge University Press, Cambridge (2004)
13. Goldreich, O., Kahan, A.: How To Construct Constant-Round Zero-Knowledge Proof Systems for NP. Journal of Cryptology 9(3), 167–190 (1996)
14. Goldwasser, S., Levin, L.: Computation of General Functions in Presence of Immoral Majority. In: Menezes, A., Vanstone, S.A. (eds.) CRYPTO 1990. LNCS, vol. 537, pp. 77–93. Springer, Heidelberg (1991)
15. Goldreich, O., Micali, S., Wigderson, A.: How to Play any Mental Game – A Completeness Theorem for Protocols with Honest Majority. In: 19th STOC, pp. 218–229 (1987) For details see [12]
16. Green, M., Hohenberger, S.: Blind Identity-Based Encryption and Simulatable Oblivious Transfer. In: Kurosawa, K. (ed.) ASIACRYPT 2007. LNCS, vol. 4833, pp. 265–282. Springer, Heidelberg (2007)
17. Kalai, Y.T.: Smooth Projective Hashing and Two-Message Oblivious Transfer. In: Cramer, R.J.F. (ed.) EUROCRYPT 2005. LNCS, vol. 3494, pp. 78–95. Springer, Heidelberg (2005)
18. Kilian, J.: Founding Cryptograph on Oblivious Transfer. In: 20th STOC, pp. 20–31 (1988)
19. Kushilevitz, E., Ostrovsky, R.: Replication is NOT Needed: SINGLE Database, Computationally-Private Information Retrieval. In: 38th FOCS, pp. 364–373 (1997)
20. Lindell, Y., Pinkas, B.: An Efficient Protocol for Secure Two-Party Computation in the Presence of Malicious Adversaries. In: Naor, M. (ed.) EUROCRYPT 2007. LNCS, vol. 4515, pp. 52–78. Springer, Heidelberg (2007)
21. Micali, S., Rogaway, P.: Secure Computation. In: Feigenbaum, J. (ed.) CRYPTO 1991. LNCS, vol. 576, pp. 392–404. Springer, Heidelberg (1992)
22. Mitzenmacher, M., Upfal, E.: Probability and Computing. Cambridge University Press, Cambridge (2005)
23. Naor, M., Pinkas, B.: Efficient Oblivious Transfer Protocols. In: 12th SODA, pp. 448–457 (2001)
24. Rabin, M.: How to Exchange Secrets by Oblivious Transfer. Tech. Memo TR-81, Aiken Computation Laboratory, Harvard U (1981)
25. Yao, A.: How to Generate and Exchange Secrets. In: 27th FOCS, pp. 162–167 (1986)

Separation Results on the "One-More" Computational Problems

Emmanuel Bresson[1], Jean Monnerat[2,*], and Damien Vergnaud[3]

[1] DCSSI Crypto Lab, Paris, France
[2] Department of Computer Science & Engineering, University of California San Diego, USA
[3] École Normale Supérieure – C.N.R.S. – I.N.R.I.A.
45 rue d'Ulm, 75230 Paris CEDEX 05, France

Abstract. In 2001, Bellare, Namprempre, Pointcheval and Semanko introduced the notion of *"one-more" computational problems*. Since their introduction, these problems have found numerous applications in cryptography. For instance, Bellare *et al.* showed how they lead to a proof of security for Chaum's RSA-based blind signature scheme in the random oracle model.

In this paper, we provide separation results for the computational hierarchy of a large class of algebraic "one-more" computational problems (*e.g.* the one-more discrete logarithm problem, the one-more RSA problem and the one-more static Computational Diffie-Hellman problem in a bilinear setting). We also give some cryptographic implications of these results and, in particular, we prove that it is very unlikely, that one will ever be able to prove the unforgeability of Chaum's RSA-based blind signature scheme under the sole RSA assumption.

Keywords: "One-more" problems, Black-box reductions, Random self-reducible problems, Algebraic algorithms.

1 Introduction

BACKGROUND. In cryptography, a one-way function f is a function that can be computed by some algorithm in polynomial time (with respect to the input size) but such that no probabilistic polynomial-time algorithm can compute a preimage of $f(x)$ with a non-negligible probability, when x is chosen uniformly at random in the domain of f. At the very beginning of the century, it has been observed that there seems little hope of proving the security of many cryptographic constructions based only on the "standard" one-wayness assumption of the used primitive. The security of some schemes seems to rely on different, and probably stronger, properties of the underlying one-way function. Cryptographers have therefore suggested that one should formulate explicit new computational problems to prove the security of these protocols. For instance, Okamoto and Pointcheval [14] introduced in 2001 a novel class of computational problems, the *gap problems*, which find a nice and rich practical instantiation with the Diffie-Hellman problems. They used the gap Diffie-Hellman problem for solving a more than 10-year old open security problem: the unforgeability of Chaum-van Antwerpen undeniable signature scheme [11].

---

* Supported by a fellowship of the Swiss National Science Foundation, PBEL2–116915.

T. Malkin (Ed.): CT-RSA 2008, LNCS 4964, pp. 71–87, 2008.

In 2001, Bellare, Namprempre, Pointcheval and Semanko [2] introduced the notion of *one-more one-way function*. A function is one-more one-way if it can be computed by some algorithm in polynomial time (in the input size) but for which there exists no probabilistic polynomial-time algorithm $\mathcal{A}$ with non-negligible probability to win the following game:

- $\mathcal{A}$ gets the description of f as input and has access to two oracles;
- an *inversion* oracle that given y in f's codomain returns x in f's domain such that $f(x) = y$;
- a *challenge* oracle that, each time it is invoked (it takes no inputs), returns a random challenge point from f's codomain;
- $\mathcal{A}$ wins the game if it succeeds in inverting all n points output by the challenge oracle using strictly less than n queries to the inversion oracle.

Bellare *et al.* showed how these problems lead to a proof of security for Chaum's RSA-based blind signature scheme [10] in the random oracle model.

The approach consisting in introducing new computational problems to study the security of cryptosystems is not completely satisfactory since the proof of security often relies on an extremely strong assumption which is hard to validate. Nevertheless, it is better to have such a security argument than nothing since as mentioned in [2]: *"These problems can then be studied, to see how they relate to other problems and to what extent we can believe in them as assumptions."* The purpose of this paper is to study the hierarchy of the computational difficulty of the "one-more" problems of Bellare *et al.* and its cryptographic implications. In particular, we prove that it is very unlikely, that one will ever be able to prove the unforgeability of Chaum's RSA-based blind signature scheme under the sole RSA assumption.

RELATED WORK. Since the one-more-inversion problems were introduced in [2], they have found numerous other applications in cryptography.

- Bellare and Palacio [4] proved in 2002 that Guillou-Quisquater and Schnorr identification schemes [12,17] are secure against impersonation under active (and concurrent) attack under the assumption that the *one-more* RSA *problem* and the *one-more discrete logarithm problem* are intractable (respectively).
- Bellare and Neven [3] proved the security of an RSA based transitive signature scheme suggested by Micali and Rivest in 2002 [13] under the assumption of the hardness of the one-more RSA problem.
- Bellare and Sandhu had used the same problem to prove the security of some two-party RSA-based signature protocols [5].
- In [6], Boldyreva proposed a new blind signature scheme – based on Boneh-Lynn-Shacham signature [7] – which is very similar to the RSA blind signature protocol. She introduced a new computational problem: the *one-more static Computational Diffie-Hellman problem* (see also [9]) and proved the security (in the random oracle model) of her scheme assuming the intractability of this problem.
- Paillier and Vergnaud [15] provided evidence that the security of Schnorr signatures [17] cannot be equivalent to the discrete log problem in the standard model. They proposed a method of converting a reduction of the unforgeability of this signature scheme to the discrete logarithm problem into an algorithm solving the

one-more discrete log problem. Their technique applies whenever the reduction belongs to a certain "natural" class of reductions that they refer to as *algebraic reductions*.

CONTRIBUTIONS OF THE PAPER. Following the approach from [15], we give arguments showing that, for any integer $n > 1$, solving the one-more problem with access to the inversion oracle up to n times cannot be reduced to the resolution of this problem with access to this oracle limited to $n + 1$ queries. Our results apply to the class of black-box reductions and are extended in the case of the one-more discrete logarithm problems to a class of algebraic black-box reductions.

These separation results apply to many computational problems used in the cryptographic literature, like the one-more RSA problem and the one-more static Diffie-Hellman problem in a bilinear setting. Due to the equivalence of the unforgeability of Chaum and Boldyreva blind signatures [10,6] and the intractability of the one-more RSA problem and the one-more static Diffie-Hellman problem in a bilinear setting, our results imply that it is very unlikely, that one will ever be able to prove the unforgeability of these schemes under the sole assumption of the one-wayness of their respective underlying primitive.

We stress that our work sheds more light on the computational complexity of these problems but does not explicitly lead to actual way to solve them. Finally, we mention that Brown [8] *independently* found similar separation results[1].

2   Preliminaries

NOTATIONS. Taking an element x uniformly at random from a set X will be denoted $x \leftarrow_U X$. Assigning a value a to a variable x is denoted by $x \leftarrow a$. Algorithms are modeled by probabilistic Turing machines and are usually considered polynomial-time. The term "efficient" will refer to polynomial-time. We write $\mathcal{A}(\star; \varpi)$ the output of algorithm $\mathcal{A}$ when running on input $\star$ and using random ϖ. With $\mathcal{A}(\star)$, we mean the random variable resulting from $\mathcal{A}(\star; \varpi)$ by choosing ϖ uniformly at random. For any algorithm $\mathcal{A}$, $T(\mathcal{A})$ denotes its running time. An algorithm $\mathcal{A}$ with a black-box oracle access to an algorithm $\mathcal{B}$ is denoted $\mathcal{A}^{\mathcal{B}}$.

BLACK-BOX REDUCTIONS. An algorithm $\mathcal{R}$ is said to be a black-box reduction from a problem P_2 to a problem P_1 if for any algorithm $\mathcal{A}$ solving P_1, algorithm $\mathcal{R}^{\mathcal{A}}$ solves P_2 thanks to a black-box access to $\mathcal{A}$. Below, we provide more details about our black-box model. Namely, we describe what we mean by a "black-box access" and give a characterization of the classes of algorithms $\mathcal{A}$ we will consider. In other words, we specify which algorithms are transformed by $\mathcal{R}$ and how $\mathcal{R}$ can interact with them.

BLACK-BOX ACCESS. A black-box access essentially means that $\mathcal{R}$ is allowed to use $\mathcal{A}$ as a subroutine without taking advantage of its internal structure (code). $\mathcal{R}$ can only provide the inputs to $\mathcal{A}$ and observe the resulting outputs. If $\mathcal{A}$ has access to an oracle,

---

[1] His paper appeared on the IACR eprint, while our paper was already under submission. His work is based on the very same core idea but does not explicitly handle the case where reductions make use of rewinding techniques.

the corresponding queries must be answered by $\mathcal{R}$. In other words, the reduction should simulate $\mathcal{A}$'s environment through its input-output interface.

When $\mathcal{A}$ is probabilistic, a new black-box access by $\mathcal{R}$ results in a new execution of $\mathcal{A}$ with fresh random coins. In this paper, we do not consider the random tape of $\mathcal{A}$ to be seen by $\mathcal{R}$. This is in accordance with the work by Barak [1] saying that the knowledge of such randomness can hardly help a black-box reduction.

As usually in the literature, we allow $\mathcal{R}$ to rewind $\mathcal{A}$ with a previously used random tape. Our approach is formalized by restricting the reduction $\mathcal{R}$ to sequentially execute some of the following operations when interacting with $\mathcal{A}$:

- **Launch.** Any previously launched execution of $\mathcal{A}$ is aborted. $\mathcal{R}$ launches a new execution of $\mathcal{A}$ with a fresh random tape ϖ on an input of its choice.
- **Rewind.** Any previously launched execution of $\mathcal{A}$ is aborted. $\mathcal{R}$ restarts $\mathcal{A}$ with a previously used random tape and an input of its choice.
- **Stop.** $\mathcal{R}$ definitely stops the interaction with $\mathcal{A}$.

We assume that all executions with fresh random tapes are uniquely identified so that $\mathcal{R}$ can make some "**Rewind**" without explicitly knowing these random tapes. Note that a call of any above procedure is counted as a single time unit in the complexity of $\mathcal{R}$.

For some results, we will need to consider a weaker model obtained by relaxing the "**Rewind**" queries made by $\mathcal{R}$. Instead, we only tolerate a kind of weak rewinding of $\mathcal{A}$ with the same random tape and its corresponding input. So, in this weaker model, we replace the "**Rewind**" queries by the following one:

- **Relaunch.** Any previously launched execution of $\mathcal{A}$ is aborted. $\mathcal{R}$ restarts $\mathcal{A}$ with a previously used random tape *and* the corresponding input.

Hence, rewinding techniques which restart $\mathcal{A}$ on the same random tape and a *different* input are not allowed in this model. As a consequence, reductions involving *"forking-Lemma"*-like [16] techniques are not considered. We however point out that a "**Relaunch**" query can be useful to $\mathcal{R}$ when $\mathcal{A}$ has access to some oracles. Namely, $\mathcal{R}$ may differently simulate the oracle outputs from an execution to another one in order to gain some information to solve its challlenge.

CLASSES OF ALGORITHMS. For any τ and ε non-negative functions defined on $\mathbf{N}$ and a computational problem P with associated security parameter $k \in \mathbf{N}$, an algorithm $\mathcal{A}$ is said to be an (ε, τ)-P solver if it succeeds in solving P (fed with k) with probability at least $\varepsilon(k)$ and at most time complexity $\tau(k)$ for any $k \in \mathbf{N}$, where the probability is over random tapes of all involved algorithms. We denote by $\mathcal{CL}(P, \varepsilon, \tau)$ the class of such probabilistic algorithms. We say that $\mathcal{R}$ is an $(\varepsilon_1, \tau_1, \varepsilon_2, \tau_r)$-reduction from P_2 to P_1 if it transforms any algorithm in $\mathcal{CL}(P_1, \varepsilon_1, \tau_1)$ into a P_2-solver with success probability greater or equal to ε_2 and the running time of $\mathcal{R}$ is less or equal to τ_r. Usually, black-box reductions transform any adversary with a given success probability without any consideration of the time complexity of this one. In this case, we have $\tau_1(k) = +\infty$ for any $k \in \mathbf{N}$ and use the term of $(\varepsilon_1, \varepsilon_2, \tau_r)$-reduction from P_2 to P_1 reduction. We call such reductions "classical" while those transforming only bounded adversaries are called "sophisticated" reductions. As far as we know, we are not aware of the existence in the literature of "sophisticated" reductions which are not implicitly classical, i.e., which do not succeed in transforming adversaries with a greater complexity than the given τ_1.

BLACK-BOX SEPARATIONS. A black-box reduction $\mathcal{R}$ from P_2 to P_1 can just be seen as an oracle Turing machine solving the problem P_2. Thus it can be transformed through a so-called *meta-reduction* to solve another problem (say, P_3). When the latter is assumed to be hard, an efficient meta-reduction rules out the existence of $\mathcal{R}$, hence proving a separation between problems P_1 and P_2. In other words, it proves that P_2 is strictly harder than P_1, conditioned by the hardness of P_3.

More formally, the construction is as follows. We start from the reduction $\mathcal{R}$ from P_2 to P_1. Our goal is to specify an algorithm $\mathcal{M}$ that solves the problem P_3, having a black-box access to $\mathcal{R}$. The algorithm $\mathcal{M}$ needs to simulate the environment of $\mathcal{R}$, i.e., all its oracles, and in particular, the correct behavior of the P_1-solver. In the present work, such *correct behavior* (viewed from $\mathcal{R}$) is formalized by assuming that the P_1-solver belongs to a certain class $\mathcal{CL}(P_1, \varepsilon, \tau)$ for some given ε and τ. However, in the classical case, the reduction is (formally) able to transform any P_1-solver[2] whatever its running time is ($\tau = +\infty$).

In this work, we will focus on "classical" reductions but also show that our results hold for "sophisticated" reductions in the weaker model, where "**Rewind**" queries are replaced by the "**Relaunch**" ones. In this case, we confine the P_1 to a given finite τ, thus to a smaller class $\mathcal{CL}(P_1, \varepsilon, \tau)$. Hence, by restricting the class of solvers that the reduction $\mathcal{R}$ is able to deal with, we enlarge the class of such reductions. In fact, the smaller τ is, the bigger the number of possible reductions do exist. Thus, excluding the existence of such reductions using a meta-reduction leads to a separation result which is at least as strong as the case where $\tau = +\infty$.

We may wonder whether this is *strictly* stronger to do so. Usually the reduction should not care about the complexity of the P_1-solver which is treated as a black-box and for which an access is counted as a single unit anyway. However, when the P_1-problem is specified with some oracle access to a solver (as for the "one-more" problems), one must be more careful. The main reason is that there may be a correlation between the "external behavior" of a P_1-solver and its complexity. What we call the "external behavior" of an algorithm corresponds to the distribution of both the intermediate outputs (queries) sent to the oracles and the final output. As a result, a reduction may be able to only transform the P_1-solvers with a specific "external behavior", guaranteed by the bound τ. For instance, a one-more discrete logarithm solver which would never query the discrete logarithm oracle must be a discrete logarithm solver. Assuming that no discrete logarithm solver exists with a time less than τ, a sophisticated reduction knows that it does not need to transform such an n-DL solver.

When devising the meta-reduction $\mathcal{M}$, this difficulty is induced in the simulation of the P_1-solver. To be more explicit, the meta-reduction needs to simulate a P_1-solver with the correct "external behavior", since this one may need to comply to a specific form when τ is finite. In our results, the class $\mathcal{CL}(P_1, \varepsilon, \tau)$ will be chosen so that $\mathcal{M}$ will simulate the correct behavior of a solver of this class.

---

[2] Of course, for a cryptographic result to make sense in practice, the solver is restricted to polynomial-time.

3   Random Self-reducible Problems

3.1   Definitions

Let $P = (PG, IG)$ be a generic computational problem, where PG is a parameter generator and IG is an instance generator[3]. Given a security parameter $k \in \mathbf{N}$, the probabilistic
polynomial-time algorithm $PG(1^k)$ generates some parameters param. These parame-
ters notably characterize an instance set $\mathbf{I}$ and a solution set $\mathbf{S}$. The instance generator
IG takes param as input and outputs an instance $\mathtt{ins} \in \mathbf{I}$.

We assume that there exists an efficient verification algorithm $V(\mathtt{param}, \mathtt{ins}, \mathtt{sol})$
that, for any $(\mathtt{ins}, \mathtt{sol}) \in \mathbf{I} \times \mathbf{S}$, outputs 1 if $\mathtt{sol}$ is a solution to $\mathtt{ins}$ (with respect
to param) and 0 otherwise. For an algorithm $\mathcal{A}$, we consider the following experiment:

> Experiment $\mathbf{Exp}_{\mathcal{A}}^{P}(k)$.
> $\quad$ param $\leftarrow PG(1^k)$
> $\quad$ ins $\leftarrow IG(\mathtt{param})$
> $\quad$ sol $\leftarrow \mathcal{A}(\mathtt{param}, \mathtt{ins})$
> $\quad$ Output $V(\mathtt{param}, \mathtt{ins}, \mathtt{sol})$

The success probability of $\mathcal{A}$ is $\mathsf{Succ}_{\mathcal{A}}^{P}(k) = \Pr[\mathbf{Exp}_{\mathcal{A}}^{P}(k) = 1]$, where the probability
is taken over the random coins of all algorithms PG, IG and $\mathcal{A}$.

We introduce the "one-more" variants of the problem P. Let n be a non-negative
integer. We denote by $\mathcal{O}_P$ an oracle which perfectly solves the problem P, that is, on
any input $\mathtt{ins} \in \mathbf{I}$, the oracle outputs some $\mathtt{sol}$ such that $V(\mathtt{param}, \mathtt{ins}, \mathtt{sol}) = 1$.
The one-more n-P problem is defined by the following experiment for an algorithm $\mathcal{A}$.

> Experiment $\mathbf{Exp}_{\mathcal{A}}^{n\text{-}P}(k)$.
> $\quad$ param $\leftarrow PG(1^k)$
> $\quad$ For $i = 0, \ldots, n$, generate $\mathtt{ins}_i \leftarrow IG(\mathtt{param}; \varpi_i)$ with fresh
> $\quad$ random tapes $\varpi_0, \ldots, \varpi_n$.
> $\quad$ $(\mathtt{sol}_0, \ldots, \mathtt{sol}_n) \leftarrow \mathcal{A}^{\mathcal{O}_P}(\mathtt{ins}_0, \ldots, \mathtt{ins}_n)$
> $\quad$ If $V(\mathtt{param}, \mathtt{ins}_i, \mathtt{sol}_i) = 1$ for $i = 0, \ldots, n$ and $\mathcal{A}$ made at
> $\quad$ most n queries to $\mathcal{O}_P$ output 1 else output 0

The success probability of $\mathcal{A}$ is $\mathsf{Succ}_{\mathcal{A}}^{n\text{-}P}(k) = \Pr[\mathbf{Exp}_{\mathcal{A}}^{n\text{-}P}(k) = 1]$, where the probability is taken over the random coins of all involved algorithms. For any functions
$\varepsilon, \tau : \mathbf{N} \to \mathbf{R}$, an algorithm $\mathcal{A} \in \mathcal{CL}(n\text{-}P, \varepsilon, \tau)$ is called an (ε, τ)-n-P solver.

Proposition 1 (Reduction from $(n+1)$-P to n-P). *Let n, m be two integers such that*
$n < m$. *Then the m-P problem cannot be harder than the n-P problem.*

We omit the proof since it is elementary. We now give a definition for a computational
problem P to be random self-reducible.

Definition 2 (Random self-reducibility). *A problem P defined as above is said to be
random self-reducible if there exists an efficient blinding algorithm* B *and an efficient
un-blinding algorithm* UB *such that for any $k \in \mathbf{N}$, any string* param *generated by* PG,
and any element $\mathtt{ins} \in \mathbf{I}$:

---

[3] We separate PG and IG for exposition convenience.

1. B(param, ins; ϖ) *is a uniformly distributed element* ins$_{bl}$ *in* **I**, *w.r.t. the random choice of* ϖ;

2. for any random tape ϖ, *any blinded instance* ins$_{bl}$ *generated by* B *from instance* ins *using random tape* ϖ, *the algorithm* UB *satisfies*

$$V(param, ins_{bl}, sol_{bl}) = 1 \implies V(param, ins, UB(param, sol_{bl}; \varpi)) = 1.$$

In what follows, we denote by Ω the set of random tapes ϖ used by B (and UB) and the time complexity of algorithms B, UB, V by $\tau_{BL}, \tau_{UB}, \tau_{VER}$ respectively. We remark that our definition is similar to that of Okamoto-Pointcheval [14] except that we do not require that UB outputs a uniform element in **S**. Our definition is also like that given by Tompa-Woll [18] with the relaxation of a similar condition. Note that both the discrete logarithm and the RSA inversion problems satisfy this definition.

3.2 Black-Box Separation

Definition 3 (Parameter-invariant reductions). *Let* n *and* n' *be some non-negative integers. A black-box reduction* $\mathcal{R}$ *from* n-P *to* n'-P *is said to be* parameter-invariant *if* $\mathcal{R}$ *only feeds the* n'-P-*solver with challenges containing the same string* param *that was in the challenge given to* $\mathcal{R}$.

We first assume that the reduction executes the $(n+1)$-P-solver at most one time and never rewinds.

Lemma 4 (Separation, basic case). *Let* n *be a non-negative integer and* $\varepsilon, \varepsilon', \tau_r$ *be some positive functions defined on* **N**. *We set* $\tau_{TOT} := \tau_{BL} + \tau_{UB} + \tau_{VER}$. *There exists a meta-reduction* $\mathcal{M}$ *such that, for any parameter-invariant black-box* $(\varepsilon, \varepsilon', \tau_r)$-*reduction* $\mathcal{R}$ *from* n-P *to* $(n+1)$-P *which makes at most only one* "**Launch**" *(and no* "**Rewind**"*) query to the* $(n+1)$-P-*solver,* $\mathcal{M}^{\mathcal{R}}$ *is an* $(\varepsilon', \tau_r + (n+1) \cdot \tau_{TOT})$-$n$-P *solver.*

Proof. First, remark that for any $\epsilon > 0$ there exists a $(n+1)$-P-solver that succeeds with probability ϵ. This is because P is verifiable so the exhaustive search is possible, and that we do not consider the execution time at that point. Moreover, we can assume that this solver always makes $n + 1$ uniformly distributed queries to $\mathcal{O}_P$. We denote this (naive) algorithm by $\mathcal{A}_1$. It receives as input an instance of $(n + 1)$-P, picks $n + 1$ uniform random P-instances ins$_1^*, \ldots,$ ins$_{n+1}^* \in_U$ **I**, and submits them sequentially to $\mathcal{O}_P$. For each query, $\mathcal{A}_1$ checks the answer of $\mathcal{O}_P$ using the verification algorithm V and if one answers is wrong, $\mathcal{A}_1$ outputs $\perp$ and aborts. Otherwise, it does an exhaustive search and outputs the correct answer of the $(n + 1)$-P challenge with probability ϵ.

One may wonder why it is necessary to check the validity of the $\mathcal{O}_P$ oracle, since we have assumed it is a perfect oracle. Indeed, in the real $\mathbf{Exp}_{\mathcal{A}}^{(n+1)\text{-P}}(k)$ experiment, such a verification is always successful ($\mathcal{A}_1$ never outputs $\perp$). However these checks will be crucial in the simulated experiment in which $\mathcal{O}_P$ is simulated by $\mathcal{R}$, which can possibly cheats. We also emphasize that the queries ins$_i^*$ are always the same if the random tape of $\mathcal{A}_1$ and param are the same.

DESCRIPTION OF THE META-REDUCTION. We now explain how to build a meta-reduction $\mathcal{M}$ that solves the n-P problem. First, $\mathcal{M}$ is given access to a $\mathcal{O}_P$-oracle (with at most n queries allowed) and receives an n-P challenge $(\text{param}, \text{ins}_0, \ldots, \text{ins}_n)$. Then it will use $\mathcal{R}$ to solve it, by simulating adversary $\mathcal{A}_1$ (which is assumed to make uniform queries). To this goal, $\mathcal{M}$ simply feeds $\mathcal{R}$ with the given challenge. Remind that $\mathcal{R}$ has oracle access to $\mathcal{O}_P$ and $\mathcal{A}_1$ (the so-called "**Launch**" query), so $\mathcal{M}$ must be able to answer both types of queries. The $\mathcal{O}_P$ queries are answered in a trivial way; namely, $\mathcal{M}$ simply forwards the queries to its oracle $\mathcal{O}_P$. Below, we describe how $\mathcal{M}$ processes the (unique) "**Launch**" query made by $\mathcal{R}$, by simulating the execution of $\mathcal{A}_1$.

> **Launch**$(\text{param}, \text{ins}_0', \ldots, \text{ins}_{n+1}')$
> **for** $i = 0, 1, \ldots, n$ **do**
> $\varpi_i \leftarrow_U \Omega$
> $\text{bl-ins}_i \leftarrow \text{B}(\text{param}, \text{ins}_i; \varpi_i)$
>   Submit bl-ins_i to $\mathcal{O}_P$ (simulated by $\mathcal{R}$) and receive bl-sol_i
> $\text{sol}_i \leftarrow \text{UB}(\text{param}, \text{bl-sol}_i; \varpi_i)$
>   **if** $0 \leftarrow \text{V}(\text{param}, \text{ins}_i, \text{sol}_i)$ **then**
>   **Return** $\perp$
> **Abort** the interaction with $\mathcal{R}$

If all the $\mathcal{O}_P$-queries are correctly answered by $\mathcal{R}$, the meta-reduction aborts the interaction with $\mathcal{R}$ after the $(n+1)$-th $\mathcal{O}_P$-query, and simply returns $(\text{sol}_0, \ldots, \text{sol}_n)$ as the solution of the n-P problem given as input. On the other hand, if the simulation of the oracle appears to be wrong, the "**Launch**" query is answered with $\perp$ — this correctly simulates the behavior of $\mathcal{A}_1$ described at the beginning of the proof. In that case, the reduction eventually outputs a tuple that $\mathcal{M}$ just relays as its own answer. If $\mathcal{R}$ does not ask any "**Launch**" query or decides to prematurely "**Stop**" the interaction with $\mathcal{A}_1$, then $\mathcal{M}$ also outputs the same answer as $\mathcal{R}$.

In simulating $\mathcal{A}_1$, $\mathcal{M}$ needs to run the algorithms B, UB, and V once per query to $\mathcal{O}_P$. Since $\mathcal{M}$ launches $\mathcal{R}$ one time, we get $T(\mathcal{M}^{\mathcal{R}}) \leq \tau_r + (n+1) \cdot \tau_{\text{TOT}}$.

PROBABILITY ANALYSIS. Let $\text{Succ}_{\mathcal{R}}$ and $\text{Succ}_{\mathcal{M}}$ be the events that "$\mathcal{R}$ succeeds" and "$\mathcal{M}$ succeeds" respectively. Let us denote Good the event, where $\mathcal{R}$ correctly answers to all $\mathcal{O}_P$ queries made by $\mathcal{A}_1$. In particular, event $\neg$Good includes the executions in which $\mathcal{R}$ does not make any "**Launch**" query or prematurely stops the interaction with $\mathcal{A}_1$. From the above description, we can easily see that, if Good occurs, $\mathcal{M}$ always recovers the correct solutions sol_i for $i = 0, \ldots, n$. On the other hand, if $\neg$Good occurs, $\mathcal{M}$ outputs whatever $\mathcal{R}$ outputs, and thus we have $\Pr[\text{Succ}_{\mathcal{M}}|\neg\text{Good}] = \Pr[\text{Succ}_{\mathcal{R}}|\neg\text{Good}]$. Then we obtain

$$\begin{aligned} \Pr[\text{Succ}_{\mathcal{M}}] &= \Pr[\text{Succ}_{\mathcal{M}}|\text{Good}] \cdot \Pr[\text{Good}] + \Pr[\text{Succ}_{\mathcal{M}}|\neg\text{Good}] \cdot \Pr[\neg\text{Good}] \\ &= \Pr[\text{Good}] + \Pr[\text{Succ}_{\mathcal{R}}|\neg\text{Good}] \cdot \Pr[\neg\text{Good}] \\ &\geq \Pr[\text{Succ}_{\mathcal{R}} \wedge \text{Good}] + \Pr[\text{Succ}_{\mathcal{R}} \wedge \neg\text{Good}] = \Pr[\text{Succ}_{\mathcal{R}}] \geq \varepsilon', \end{aligned}$$

which concludes the proof. $\square$

Theorem 5 (Separation, general case). *Let* $n, \varepsilon, \varepsilon', \tau_r, \tau_{\text{TOT}}$ *be as in Lemma 4. There exists a meta-reduction* $\mathcal{M}$ *such that, for any parameter-invariant black-box* $(\varepsilon, \varepsilon', \tau_r)$-*reduction* $\mathcal{R}$ *from* n-P *to* $(n+1)$-P *which makes at most* ℓ *queries* "**Launch**" *or* "**Rewind**" *to the* $(n+1)$-P-*solver,* $\mathcal{M}^{\mathcal{R}}$ *is an* $(\varepsilon', \tau_r + (n+1)\ell \cdot \tau_{\text{TOT}})$-$n$-P-*solver.*

Proof. The proof works like in Lemma 4 except that $\mathcal{M}$ needs to deal with possibly many runs and/or rewindings of the $(n+1)$-P-solver $\mathcal{A}_1$. We summarize how we deal with the queries of type "**Launch**" and "**Rewind**".

Launch. For each such query, $\mathcal{M}$ needs to simulate a new execution of $\mathcal{A}_1$ with fresh random coins. The simulation works like for Lemma 4. The main difference is that, when a wrong simulation of $\mathcal{O}_P$ leads to an abortion of $\mathcal{A}_1$, the reduction $\mathcal{R}$ is allowed to continue with a new query "**Launch**" or "**Rewind**". The same holds if $\mathcal{R}$ aborts the current execution by itself by just making a new query. This does not affect the simulation of $\mathcal{A}_1$ by $\mathcal{M}$, which simply waits for a correct simulation of $\mathcal{O}_P$ by $\mathcal{R}$ on the $n+1$ queries[4].

Rewind. The simulation of such an execution is done similarly as for "**Launch**" except that the randomness is not fresh. This implicitly means that $\mathcal{M}$ keeps track of the history of previous $\mathcal{A}_1$'s executions in a non-ambiguous way. Recall that the $\mathcal{O}_P$ queries made by $\mathcal{A}_1$ does not depend on the received n-P instance, but only its random tape and param, which is unchanged by definition. Thus, if $\mathcal{R}$ rewinds several times, $\mathcal{M}$ can simulate correctly: the same queries to the $\mathcal{O}_P$ oracle are repeated from a previous execution, and a wrong answer is still detected, leading to an abortion. In that case, $\mathcal{R}$ continues with another query "**Launch**" or "**Rewind**", or makes its final "**Stop**". As above, $\mathcal{M}$ stops the execution if it receives all answers of the queries sent to $\mathcal{O}_P$ and wins the game[5]. If this does not occur, $\mathcal{M}$ simply answers the $\mathcal{R}$ output to its challenger.

A probability analysis as in Lemma 4 shows that $\mathcal{M}$ succeeds with probability at least ε'. The running time of $\mathcal{M}^{\mathcal{R}}$ is easily upper-bounded by $\tau_r + (n+1)\ell \cdot \tau_{\mathrm{TOT}}$.   □

Remark 6. We observe that the reduction $\mathcal{R}$ solves the n-P problem using an $(n+1)$-P solver as a subroutine, while our meta-reduction $\mathcal{M}$ solves the same problem without such subroutine (but with access to $\mathcal{R}$). Thus, if the n-P problem is hard, the existence of an efficient $\mathcal{M}$ shows that there cannot exist an efficient black-box algorithm $\mathcal{R}$ from the n-P problem to the $(n+1)$-P problem. On the other hand, if the n-P problem is easy, we know that the $(n+1)$-P problem is easy as well, and so basing the security of cryptographic schemes on these problems (m-P for $m \geq n$) would be hopeless.

Remark 7. Note that if the problem P is not efficiently verifiable, the above result does not apply anymore. For instance, this does not work with the one-more (static) Computational Diffie-Hellman, except in a group with an easy decisional Diffie-Hellman problem (DDH). Namely, if $\mathcal{R}$ simulates the oracle by answering random elements, the adversary $\mathcal{A}_1$ cannot easily detect that the simulation is not correct unless this one can solve DDH. However, in the context of pairing cryptography the bilinear pairing allows to efficiently solve DDH. In this case, our results apply to the one-more CDH.

---

[4] Note that we can optimize by simply waiting until $\mathcal{M}$ could get $n+1$ correct answers of fresh queries made to $\mathcal{O}_P$, even if these ones were not made during one single execution of $\mathcal{A}_1$. For this, $\mathcal{M}$ should not re-use a solved ins_i by sending a new blinded ins_i value to $\mathcal{O}_P$ in a subsequent fresh (new random tape) execution of $\mathcal{A}_1$.

[5] Though the queries are the same, $\mathcal{R}$ may simulate $\mathcal{O}_P$ in a different way. In particular, it may succeed in simulating $\mathcal{O}_P$, even if it failed to do so in a previous execution.

3.3   The Case of Sophisticated Reductions

Here, we investigate our separation results in the context of "sophisticated" reductions, i.e., those which are supposed to only transform a class of algorithms with a bounded time complexity. In what follows, we are able to exhibit such a separation under the assumption that the reduction does not rewind the adversary with the same random and a different input.

Theorem 8 (Sophisticated reductions). *Let $n, \varepsilon, \varepsilon', \tau_r, \tau_{\text{TOT}}$ be as in Lemma 4. Consider $\tau_0 : \mathbf{N} \to \mathbf{R}$ an arbitrary time-complexity upper bound of some existing $(n+1)$-P solver succeeding with probability greater or equal to ε. In other words, we assume*

$$\mathcal{CL}((n + 1)\text{-P}, \varepsilon, \tau_0) \neq \emptyset.$$

Let τ such that $\tau(k) \geq \tau_0(k) + (n + 1) \cdot \tau_{\text{TOT}}(k)$ for any $k \in \mathbf{N}$. There exists a meta-reduction $\mathcal{M}$ such that, for any "sophisticated" parameter-invariant black-box $(\varepsilon, \tau, \varepsilon', \tau_r)$-reduction $\mathcal{R}$ from n-P to $(n + 1)$-P making at most ℓ queries "Launch" or "Relaunch" to the $(n + 1)$-P-solver, $\mathcal{M}^{\mathcal{R}}$ is an $(\varepsilon', \tau_r + (n+1)\ell \cdot \tau_{\text{TOT}})$-n-P-solver.

Proof. The proof is similar to that of Theorem 5 except that the existence of a certain $(n + 1)$-P-solver $\mathcal{A}_1$ belonging to $\mathcal{CL}((n + 1)\text{-P}, \varepsilon, \tau)$ needs to be shown ($\mathcal{A}_1$ is no more a naive algorithm). It is constructed as follows.

By definition of τ_0, there exists an algorithm $\mathcal{A}_0$ belonging to the class $\mathcal{CL}((n+1)\text{-}P, \varepsilon, \tau_0)$. The algorithm $\mathcal{A}_1$ receives as input an instance of the $(n + 1)$-P-problem, starts $\mathcal{A}_0$ with this input, processes the $\mathcal{A}_0$'s queries as explained hereafter, and finally outputs whatever $\mathcal{A}_0$ outputs. For each $\mathcal{O}_P$ query ins, $\mathcal{A}_1$ picks a uniformly distributed random tape ϖ and computes $\text{ins}_{\text{bl}} \leftarrow B(\text{param}, \text{ins}; \varpi)$. It then queries ins_{bl} to $\mathcal{O}_P$ and gets the answer sol_{bl}. It checks whether $V(\text{param}, \text{ins}_{\text{bl}}, \text{sol}_{\text{bl}}) \to 1$: if it is the case, it forwards $\text{sol} \leftarrow \text{UB}(\text{param}, \text{sol}_{\text{bl}}; \varpi)$ as the answer to $\mathcal{A}_0$, otherwise it terminates and outputs $\bot$. If $\mathcal{A}_0$ asks less than $n + 1$ queries to $\mathcal{O}_P$, $\mathcal{A}_1$ asks as many uniformly distributed random queries as necessary.

This algorithm $\mathcal{A}_1$ has the same behavior as in Theorem 5 (it always makes $(n + 1)$ uniform queries), except that for a given randomness the queries to $\mathcal{O}_P$ may depend on the input. The rest of the proof works like for Theorem 5. In particular, $\mathcal{M}$ simulates $\mathcal{A}_1$ in the very same way and handles the "**Relaunch**" queries made by $\mathcal{R}$ as the "**Rewind**" ones in the proof of Theorem 5.   □

Remark 9. The difficulty of extending the above proof to "**Rewind**" queries comes from our inability to correctly simulate $\mathcal{A}_1$ after a rewinding of $\mathcal{R}$ with a *different* input. Since $\mathcal{A}_1$ must be restarted with the same random tape, we cannot produce uniform $\mathcal{O}_P$-queries anymore: the blinding on a different input with the same randomness would produce different blinded queries, while in Theorem 5 the queries should not change in case of a "**Rewind**".

4   One-More Discrete Logarithm Problems

4.1   Definitions

Let $k \in \mathbf{N}$ be a security parameter and Gen be an efficient algorithm taking k (or 1^k) as input and which outputs the description of a cyclic group $\mathbf{G}$ of prime order (written multiplicatively), a generator g of $\mathbf{G}$, and the k-bit prime group order $q = \#\mathbf{G}$. We assume that elementary group operations in $\mathbf{G}$ can be done efficiently, i.e., $g_1 g_2$ and g_1^{-1} can be efficiently computed for any $g_1, g_2 \in \mathbf{G}$, and we denote by τ_{exp} the time required for computing an exponentiation g^x, where $x \in [1, q]$. We also consider a perfect discrete logarithm oracle DL_g, i.e., an oracle which on any queried element always answers its discrete logarithm with respect to g. For a non-negative integer n, the n-DL problem (one-more discrete logarithm) consists in extracting the discrete logarithms of $n + 1$ elements of $\mathbf{G}$ with respect to g using at most n queries to the oracle DL_g. More formally, for an algorithm $\mathcal{A}$, we consider the following experiment [2]:

Experiment $\mathbf{Exp}_{\mathsf{Gen}, \mathcal{A}}^{n\text{-DL}}(k)$.

    $(\mathbf{G}, q, g) \leftarrow \mathsf{Gen}(1^k)$

    $(t_0, t_1, \ldots, t_n) \leftarrow_U \mathbf{Z}_q^{n+1}$; $y_i \leftarrow g^{t_i}$ for $i = 0, \ldots, n$

    $(t_0', \ldots, t_n') \leftarrow \mathcal{A}^{\mathsf{DL}_g}(\mathbf{G}, g, q, y_0, \ldots, y_n)$

    Return 1 if the following conditions hold else return 0

      $- t_i' \equiv t_i \pmod{q}$ for all $i = 0, \ldots, n$

      $- \mathsf{DL}_g$ has been queried at most n times

We define the success probability of $\mathcal{A}$ in the above experiment as

$$\mathsf{Succ}_{\mathsf{Gen}, \mathcal{A}}^{n\text{-DL}}(k) = \Pr[\mathbf{Exp}_{\mathsf{Gen}, \mathcal{A}}^{n\text{-DL}}(k) = 1],$$

where the probability is taken over the t_i's and the random tapes of Gen and $\mathcal{A}$.

For any functions $\varepsilon, \tau : \mathbf{N} \to \mathbf{R}$, we denote by $\mathcal{DL}(n, \varepsilon, \tau)$ the set $\mathcal{CL}(n\text{-DL}, \varepsilon, \tau)$. An algorithm $\mathcal{A}$ of this class is said to be an (ε, τ)-n-DL solver.

4.2   Algebraic Separations

First, we note that Theorems 5 and 8 apply to the discrete logarithm problems if we assume that the reduction is base-invariant, i.e., it always feeds the $(n+1)$-DL solver with the same group $\mathbf{G}$ and generator g given in the n-DL experiment. In what follows, we show that we can extend these separation results to some non-invariant base (but same group) reductions under the assumption that these reductions are algebraic. We restrict to classical black-box reductions with rewinding and follow the spirit of Theorem 5. A treatment of "sophisticated" reductions with a rewinding relaxation (as in Theorem 8) can be done in the very same manner so that we omit such a presentation.

ALGEBRAIC ALGORITHMS. We use the concept of algebraic algorithms introduced by Paillier and Vergnaud [15]. Roughly, an algorithm $\mathcal{R}$ is *algebraic* with respect to a cyclic group $\mathbf{G}$ (of order q) if any element of $\mathbf{G}$ output by the algorithm at any step can be described as an explicitly known "multiplicative combination" of its $\mathbf{G}$ inputs. More precisely, there should exist an algorithm Extract which, given the random tape ϖ of $\mathcal{R}$, its inputs $(s, g_1, \ldots, g_\ell) \in \{0, 1\}^* \times \mathbf{G}^\ell$, and its code $\mathsf{co}(\mathcal{R})$, enables to retrieve, for any $y \in \mathbf{G}$ output by $\mathcal{R}$, the coefficients $a_1, \ldots, a_\ell \in \mathbf{Z}_q$ such that

$$y = g_1^{a_1} \cdots g_\ell^{a_\ell}.$$

Moreover, it is required that the procedure Extract runs in polynomial time with respect to $|\mathsf{co}(\mathcal{R})|$ (the size of the code of $\mathcal{R}$) and $\tau = T(\mathcal{R})$. We denote the time complexity of one Extract execution by τ_{EXT}. Though an algebraic algorithm may not be treated as a black-box, we will use the notation $\mathcal{M}^{\mathcal{R}}$ to express the algorithm obtained by an algorithm $\mathcal{M}$ which uses $\mathcal{R}$ as a subroutine and possibly makes calls to Extract.

Definition 10 (Group-invariant reductions). *Let n and n' be two non-negative integers. A reduction $\mathcal{R}$ from n-DL to n'-DL is said to be* group-invariant *if $\mathcal{R}$ exclusively feeds the n'-DL solver with challenges containing the same group $\mathbf{G}$ which was given by* Gen *in the n-DL experiment.*

Theorem 11 (Separation for algebraic reductions). *Let $n, \varepsilon, \varepsilon', \tau_r$ be as in Lemma 4. There exists a meta-reduction $\mathcal{M}$ (non black-box) such that, for any algebraic group-invariant black-box $(\varepsilon, \varepsilon', \tau_r)$-reduction $\mathcal{R}$ from n-DL to $(n + 1)$-DL which makes at most ℓ "**Launch**" or "**Rewind**" queries to the underlying $(n + 1)$-DL-solver, $\mathcal{M}^{\mathcal{R}}$ is an $(\varepsilon', \tau_r + 2(n + 1)\ell \cdot \tau_{\text{EXP}} + \ell \cdot \tau_{\text{EXT}})$-$n$-DL solver.*

Proof. This proof is similar to that of Theorem 5 except that $\mathcal{M}$ needs to simulate the $(n + 1)$-DL-solver $\mathcal{A}_1$ in a different way.

DESCRIPTION OF $\mathcal{M}$. At the beginning, $\mathcal{M}$ is challenged with $(\mathbf{G}, g_1, q, y_0, \dots, y_n)$ and forwards this challenge to $\mathcal{R}$. Then $\mathcal{M}$ has to deal with the queries made by $\mathcal{R}$: "**Launch**", "**Rewind**" and queries to the DL_{g_1} oracle. That latter is simulated in a straightforward way by the meta-reduction, since its own oracle is relative to base g_1 as well and the number of queries asked by $\mathcal{R}$ is less than n. For the "**Launch**" and "**Rewind**" queries, we have to show how $\mathcal{M}$ simulates the $(n + 1)$-DL-solver $\mathcal{A}_1$. We assume that at least one execution of $\mathcal{A}_1$ will terminate correctly: $\mathcal{A}_1$ asks $n+1$ discrete logarithm queries and receives correct answers. We denote this event by Good.

The subtlety is as follows. On a "**Rewind**"-query, $\mathcal{R}$ can specify another base (and the DL-queries made by $\mathcal{A}_1$ will be answered by $\mathcal{R}$ relatively to this base). However, $\mathcal{A}_1$ being started with new inputs but unchanged random tape must ask the same DL-queries (they only depend on the random tape). We now show that this is not a problem, as long as one execution goes correctly (event Good). For convenience of notation, we denote g_1 as y_{-1} and for any $i \in [-1, n]$ we note $\alpha_i = \log_{g_2} y_i$. For completeness we explicitly specify the random tape ϖ of the reduction $\mathcal{R}$ (keep in mind that this randomness is provided by $\mathcal{M}$ which has *non black-box* access to $\mathcal{R}$).

A "**Launch**"-query is processed as follows:

> **Launch$(\mathbf{G}, g_2, q, z_0, \dots, z_{n+1})$**
> $(b_{-1}, b_0, \dots, b_n) \leftarrow \text{Extract}(g_2, \varpi, \text{co}(\mathcal{R}))$ // we have: $g_2 = \prod_{j=-1}^{n} y_j^{b_j}$
>   // up to a permutation, we can assume $b_{-1} \neq 0$
> **for** $i = 0$ to n **do**
> $\quad r_i \leftarrow_U \mathbf{Z}_q$ // $\mathcal{A}_1$ asks at most $n + 1$ queries
> $\quad$ Submit $g_2^{r_i} y_i$ to DL_{g_2} (simulated by $\mathcal{R}$) and receive answer θ_i
> $\quad \alpha_i \leftarrow \theta_i - r_i$ // clearly $\alpha_i = \log_{g_2} y_i = x_i$
> $\alpha_{-1} \leftarrow b_{-1}^{-1}(1 - \sum_{j=0}^{n} b_j \alpha_j)$ // we have $\alpha_{-1} = \log_{g_2} g_1 = x_{-1} \neq 0$
> **for** $i = 0$ to n **do**
> $\quad c_i \leftarrow \alpha_i / \alpha_{-1} \bmod q$ // $c_i = \log_{g_1} y_i$
> **Abort** the interaction with $\mathcal{R}$

From above, it is clear that if a "**Launch**"-query goes successfully ($n+1$ DL-queries that are answered correctly), $\mathcal{M}$ is able to recover all $c_i = \log_{g_1} y_i$ for $i \in [0, n]$.

On the other hand, if the first $\mathcal{A}_1$'s execution that goes successfully[6] is a "**Rewind**"-query, then $\mathcal{M}$ does as follows. Let us denote by g_2' the (new) generator provided by $\mathcal{R}$ as an input of this query, $\mathcal{M}$ still constructs the DL-queries as $g_2^{r_i} y_i$ (and not $g_2'^{r_i} y_i$). However the answers are relative to base g_2' and must be exploited differently. We first note that we have $n + 3$ equations: one for the $\mathsf{Extract}(g_2, \dots)$-query made in the underlying "**Launch**"-query, one for the $\mathsf{Extract}(g_2', \cdots)$ in the successful "**Rewind**", and $n + 1$ equations $\delta\theta_i = r_i + x_i$, for $i = 0, \dots, n$ with $\delta = \log_{g_2} g_2'$. The obtained matrix relative to the linear system with $n + 3$ unknowns $(x_{-1}, x_0, \dots, x_n, \delta)$ is:

$$
\begin{pmatrix}
b_{-1} & b_0 & b_1 & \cdots & b_n & 0 \\
b_{-1}' & b_0' & b_1' & \cdots & b_n' & -1 \\
1 & & & & & -\theta_0 \\
 & 1 & & & & -\theta_1 \\
 & & \ddots & & & \vdots \\
 & & & & 1 & -\theta_n
\end{pmatrix}
\quad
\begin{array}{l}
\texttt{// from } g_2 = \prod_{-1}^n y_i^{b_i} \\[4pt]
\texttt{// from } g_2' = \prod_{-1}^n y_i^{b_i'} \\[14pt]
\texttt{// from } g_2'^{\theta_i} = g_2^{r_i} y_i = g_2^{r_i + x_i}
\end{array}
$$

Up to the sign, the determinant of this matrix is easily seen to be

$$
\pm\Delta = b_{-1}\left(-1 \cdot 1 + \sum_0^n b_i' \theta_i \right) - b_{-1}'\left(\sum_0^n b_i \theta_i \right)
$$

From the two "Extract" equations it is easy to see that for any $i \in [-1, n]$ we have $b_i' = \delta b_i$ with overwhelming probability (if it was not the case, linearly combining the two equations would lead to some x_i by expliciting $(b_i' - \delta b_i) x_i = \cdots)$. It follows immediately that $\Delta = \pm b_{-1} \neq 0$.

As a conclusion, as soon as event Good occurs, the meta-reduction can solve the system and obtain all the $\log_{g_2} y_i$ as well as $\log_{g_2} g_1$, and thus can solve its challenge.

Otherwise (if $\mathcal{R}$ always halts $\mathcal{A}_1$ before it asks $n + 1$ DL-queries or if $\mathcal{R}$ always answers incorrectly), then $\mathcal{M}$ outputs whatever $\mathcal{R}$ outputs. Thus:

$$
\Pr[\mathsf{Succ}_\mathcal{M}] = \underbrace{\Pr[\mathsf{Succ}_\mathcal{M} | \mathsf{Good}]}_{\text{we show } = 1} \Pr[\mathsf{Good}] + \underbrace{\Pr[\mathsf{Succ}_\mathcal{M} | \neg\mathsf{Good}]}_{=\Pr[\mathsf{Succ}_\mathcal{R} | \neg\mathsf{Good}]} \Pr[\neg\mathsf{Good}] \geq \epsilon'
$$

(as in Lemma 4). The running time is easily checked. $\square$

Definition 12 (Regular reductions). *Let $\mathcal{R}$ be a black-box reduction from a problem P_2 to a problem P_1. We denote by* suc *the success event for $\mathcal{R}$ in solving P_2 and* Exec *the event that $\mathcal{R}$ launches at least one complete execution of the P_1-solver and correctly simulates its environment. The reduction $\mathcal{R}$ is said to be* regular *if the event* Succ *always implies the event* Exec.

---

[6] We recall that a "successful" execution is defined by the fact that $\mathcal{A}_1$ receives correct answers, not that $\mathcal{A}_1$ terminates. In fact it never terminates, since $\mathcal{M}$ aborts $\mathcal{A}_1$ as soon as it has enough information to conclude.

This definition captures the fact that the reduction really exploits the access given to the P_1-solver. This assumption seems quite natural, since the reduction would simply be a P_2-solver otherwise. The following result shows that the separation holds under the hardness of the DL problem (rather than the one-more DL) if we assume $\mathcal{R}$ to be regular. Moreover we get an improvement on the "extra" time of $\mathcal{M}$ w.r.t. $\mathcal{R}$, which drops from $2(n+1)\ell \cdot \tau_{\text{EXP}} + \ell \cdot \tau_{\text{EXT}}$ down to $2\ell \cdot \tau_{\text{EXP}} + (n+\ell) \cdot \tau_{\text{EXT}}$.

Theorem 13 (Separation for regular reductions). *Let $n, \varepsilon, \varepsilon', \tau_r$ be as in Lemma 4. There exists a (non-black-box) meta-reduction $\mathcal{M}$ such that, for any regular algebraic group-invariant black-box $(\varepsilon, \varepsilon', \tau_r)$-reduction $\mathcal{R}$ from n-DL to $(n+1)$-DL which makes at most ℓ queries of type "**Launch**" or "**Rewind**" to the $(n+1)$-DL-solver, $\mathcal{M}^{\mathcal{R}}$ is an $(\varepsilon', \tau_r + 2\ell \cdot \tau_{\text{EXP}} + (n+\ell) \cdot \tau_{\text{EXT}})$-DL-solver.*

Proof. The proof differs from that of Theorem 11 in the way $\mathcal{M}$ simulates the oracle DL_{g_1} and feeds the reduction.

DESCRIPTION OF $\mathcal{M}$. At the beginning, $\mathcal{M}$ is challenged with $(\mathbf{G}, g_1, q, y)$. The meta-reduction picks a tuple $(r_0, \ldots r_n) \leftarrow_U \mathbf{Z}_q^n$ and computes $w_i \leftarrow g_1^{r_i}$ for $i = 0, \ldots, n$. Then, $\mathcal{M}$ feeds $\mathcal{R}$ with $(\mathbf{G}, g_1, q, w_0, \ldots, w_n)$ and has to deal with the queries made by $\mathcal{R}$. The DL_{g_1} oracle is simulated using the algebraicity of $\mathcal{R}$ as follows:

> **Query** $\text{DL}_{g_1}(u)$
> $(a, a_0, \ldots, a_n) \leftarrow \text{Extract}(u, \varpi, \text{co}(\mathcal{R}))$ // we have: $u = g_1^a \cdot \prod_{i=0}^n w_i^{a_i}$
> **Return** $a + \sum_{i=0}^n a_i r_i \bmod q$

We now describe how $\mathcal{M}$ simulates $\mathcal{A}_1$ (on "**Launch**" queries).

> **Launch**$(\mathbf{G}, g_2, q, z_0, \ldots, z_{n+1})$
> $(b, c_0, \ldots, c_n) \leftarrow \text{Extract}(g_2, \varpi, \text{co}(\mathcal{R}))$ // we have: $g_2 = g_1^b \cdot \prod_{i=0}^n w_i^{c_i}$
> $\alpha \leftarrow b + \sum_{i=0}^n r_i c_i$ // $\alpha = \log_{g_1}(g_2)$
> $r \leftarrow_U \mathbf{Z}_q$
> Submit $g_2^r \cdot y$ to DL_{g_2} (simulated by $\mathcal{R}$) and receive $r + \beta$ // $\beta = \log_{g_2}(y)$
> $d \leftarrow \alpha \cdot \beta$
> **Abort** the interaction with $\mathcal{R}$

As in previous proof, the interaction is aborted if $\mathcal{R}$ answers incorrectly. By assumption on $\mathcal{R}$, there always exists a successful interaction ($\mathcal{R}$ answering correctly). If this is a "**Launch**"-query, we can easily see from the above that $\mathcal{M}$ successfully outputs $d = \log_{g_1} y$. If this is a "**Rewind**$(g_2', \cdots)$"-query, then we have three unknowns: α, β and δ (the values of $\log_{g_1} g_2$, $\log_{g_2} y$ and $\log_{g_2} g_2'$, respectively) and three equations:

$$\begin{cases} \alpha = b + \sum_i r_i c_i & \text{// from Extract}(g_2, \cdots) \\ \delta \cdot \alpha = b' + \sum_i r_i c_i' & \text{// from Extract}(g_2', \cdots) \\ \delta \cdot \theta = r + \beta & \text{// answer } \theta \text{ to } \text{DL}_{g_2'}(g_2^r y) \end{cases}$$

This is clearly solvable. Thus after one sucessful execution of $\mathcal{A}_1$, $\mathcal{M}$ is always able to compute its solution $\log_{g_1} y = \alpha \cdot \beta$. The "regular" notion ensures that $\mathcal{M}$ has a success probability greater or equal to ε'. $\qquad\square$

Remark 14. Note that Theorem 13 is valid if we only assume that $\mathcal{R}$ correctly answered (at least) a single DL_g query made by $\mathcal{A}_1$ when $\mathcal{R}$ succeeds in solving its n-DL challenge. This condition is a relaxation of the "regular" notion.

5 Some Applications to the Public-Key Cryptography

Here, we derive some cryptographic consequences from the above separation results (mainly Theorem 5). All follow from the following reasoning: if a cryptographic primitive is equivalent to an n-P problem with $n > 1$, then our results show that the security of this primitive cannot rely on the hardness of solving P (aka 0-P) by using classical black-box reductions[7]. Below, we consider cryptographic algorithms which have been proven secure under a "one-more problem" and summarize equivalence results.

5.1 Chaum's Blind Signature

Chaum's RSA-based blind signature [10] originally motivates the introduction of "one-more problems" (see [2]). In this scheme, the public key is (N, e) and the signer's private key is d (with, $ed = 1 \mod \phi(N)$ and the factorization of N is unknown). The signature of a message M is $x = \mathsf{RSA}_{N,e}^{-1}(H(M)) = H(M)^d \mod N$, where $H : \{0,1\}^* \to \mathbf{Z}_N$ is a hash function. The blind signature protocol allows a user to get the signature of a message without revealing it to the signer. To do so, the user picks $r \leftarrow_U \mathbf{Z}_N^*$ and sends $\bar{M} = r^e \cdot H(M) \mod N$ to the signer; the signer computes $\bar{x} = \mathsf{RSA}_{N,e}^{-1}(\bar{M}) = \bar{M}^d \mod N$ and returns $\bar{x}$ to the user, who extracts $x = \bar{x} \cdot r^{-1} \mod N$. In their paper, Bellare *et al.* [2] defined the notion of one-more RSA problems[8] and prove there exists a reduction from n-RSA to the one-more unforgeability of the blind signature in the random oracle model. Briefly speaking, the one-more unforgeability means that no efficient algorithm can produce $n+1$ valid message-signature pairs, after at most n interactions with the signer (remind that in such interaction, the signer does not see the actual message, he just extracts e-th modular roots).

The other direction is fairly simple. One key point is that the forger sees the randomness used in the signature; its "signing" oracle is actually an "RSA-inversion" oracle. We will not go into the details here, just give the rough idea. Assume we have an algorithm $\mathcal{A}$ that solves the n-RSA problem. Building a one-more forger against the Chaum's blind signature scheme is easy: just launch algorithm $\mathcal{A}$, answer its queries using the *"signing"* oracle (an e-th root extractor), and use its output to produce a forgery.

Now assume that the scheme can be proven secure under the standard RSA assumption, using a classical black-box reduction $\mathcal{R}$. Then, from a one-more RSA solver $\mathcal{A}$, we can construct a forger as above. And applying $\mathcal{R}$ to this forger would lead to an efficient RSA-inverter: in other words, we would have inverted RSA starting from algorithm $\mathcal{A}$. But this would contradict our Theorem 5 above. Thus, the unforgeability of Chaum blind signatures cannot be (black-box) based on the standard RSA problem.

---

[7] We can derive a similar conclusion for sophisticated reductions which do not use "forking-Lemma" like techniques. Since we are not aware of such reductions in the literature, we do not expand on this.

[8] As noted in [2], these problems can be hard only if factoring does not reduce to RSA inversion.

5.2   Blind BLS Signature

In 2003, Boldyreva [6] proposed variants of the BLS signature [7], whose security holds in GDH groups [14] (groups in which CDH is hard, but deciding if a 4-tuple is a Diffie-Hellman one can be efficiently decided). The blind signature described in [6] was proven secure (in the random oracle model) under the one-more CDH problem. It considers a cyclic group $\mathbf{G}$ of prime order q generated by g and a bilinear pairing $e : \mathbf{G} \times \mathbf{G} \rightarrow \mathbf{G}'$ to a group $\mathbf{G}'$ of order q. The secret key is an element $x \leftarrow_U \mathbf{Z}_q$ and public key is $y = g^x$. The BLS signature σ of a message M is given by $H(M)^x$, where $H : \{0,1\}^* \rightarrow \mathbf{G}$ is a hash function (modeled by a random oracle). The verification consists in checking whether $e(H(M), y) = e(\sigma, g)$ holds, i.e., we check that $(g, y, H(M), \sigma)$ is a correct DDH-tuple. The blinding signing procedure consists in picking $r \in \mathbf{Z}_q$ and sending $\bar{M} = H(M) \cdot g^r$ to the signer who computes $\bar{\sigma} = \bar{M}^x$. The signature is finally obtained by computing $\sigma = \bar{\sigma} \cdot y^{-r}$.

The security model is the same as for Chaum's blind signature except that the forger has access to an oracle $(\cdot)^x$ which computes the scalar exponentiation in $\mathbf{G}$ on the query with factor x. The one-more CDH problem consists in receiving $n+1$ random elements $h_0, \ldots, h_n \leftarrow_U \mathbf{G}$ and in returning $y_0, \ldots, y_n$ such that $y_i = h_i^x$, while asking at most n queries to the oracle $(\cdot)^x$.

One can show that n-CDH is equivalent to the one-more unforgeability of the blind BLS. Namely, one feeds the n-CDH solver with $h_i := H(m_i)$ for $i = 0, \ldots, n$ with any chosen m_i's and returns the same output as the solver's one. In addition, the oracle $(\cdot)^x$ of the n-CDH-solver is trivially simulated using the same oracle as in the unforgeability game. The other direction was proved by Boldyreva[9]. As for Chaum's blind signature, one deduces from Theorem 5 that one cannot get a black-box reduction from CDH problem to the unforgeability of blind BLS.

6   Conclusion

We presented rigorous arguments that a "one-more" problem n-P maybe not as hard as the corresponding 0-P when P is self-random reducible and efficiently verifiable. This class of problems include RSA inversion problem, computational Diffie-Hellman problem in the pairing context, and discrete logarithm problem. As main cryptographic consequences, we showed that the security of some blind signatures may hardly rely on standard assumption such as the RSA inversion problem or computational Diffie-Hellman problem. Furthermore, we showed that an equivalence result between the security of a primitive and the hardness of an n-P problem rules out the existence of a black-box reduction from 0-P to the security notion. Finally, our results also show that relying the security of a cryptographic primitive on a "one-more" problem n-P clearly

---

[9] To be more precise, she proved the security of this signature under a variant called one-more chosen-target CDH. A one-more chosen-target problem is like the variant presented in this article except that the solver receives m instances (with $m > n + 1$) and solves $n + 1$ instance of his choice with n oracle accesses. This variant is closer to the unforgeability notion, since a forger can make more than n hash evaluations. Bellare *et al.* showed that both variants are equivalent in the case of RSA and the discrete logarithm. We can apply the same technique to show that this also holds for CDH.

does not give any guarantee that the security of the primitive can be relied on the corresponding 0-P problem, i.e., to a standard computational problem.

Acknowledgments. We would like to thank Mihir Bellare, Daniel Brown, and Daniele Micciancio for very interesting discussions and valuable comments on this work.

References

1. Barak, B.: How to Go Beyond the Black-Box Simulation Barrier. In: FOCS 2001, pp. 106–115 (2001)
2. Bellare, M., Namprempre, C., Pointcheval, D., Semanko, M.: The 1-More-RSA-Inversion Problems and the Security of Chaum's Blind Signature Scheme. J. Crypto 16, 185–215
3. Bellare, M., Neven, G.: Transitive Signatures: New Proofs and Schemes. IEEE IT 51(6), 2133–2151
4. Bellare, M., Palacio, A.: GQ and Schnorr Identification Schemes: Proofs of Security against Impersonation under Active and Concurrent Attacks. In: Yung, M. (ed.) CRYPTO 2002. LNCS, vol. 2442, pp. 162–177. Springer, Heidelberg (2002)
5. Bellare, M., Sandhu, R.: The Security of Practical Two-Party RSA Signature Schemes, http://eprint.iacr.org/2001/060
6. Boldyreva, A.: Threshold Signatures, Multisignatures and Blind Signatures Based on the Gap-Diffie-Hellman-Group Signature Scheme. In: Desmedt, Y.G. (ed.) PKC 2003. LNCS, vol. 2567, pp. 31–46. Springer, Heidelberg (2002)
7. Boneh, D., Lynn, B., Shacham, H.: Short Signatures from the Weil Pairing. J. Crypto 17(4), 297–319
8. Brown, D.: Irreducibility to the One-More Evaluation Problems: More May Be Less, http://eprint.iacr.org/2007/435
9. Brown, D., Gallant, R.: The Static Diffie-Hellman Problem, http://eprint.iacr.org/2004/306
10. Chaum, D.: Blind Signatures for Untraceable Payments. In: CRYPTO 1982, pp. 199–203 (1982)
11. Chaum, D., van Antwerpen, H.: Undeniable Signatures. In: Brassard, G. (ed.) CRYPTO 1989. LNCS, vol. 435, pp. 212–216. Springer, Heidelberg (1990)
12. Guillou, L.C., Quisquater, J.-J.: A Practical Zero-Knowledge Protocol Fitted to Security Microprocessor Minimizing Both Transmission and Memory. In: Günther, C.G. (ed.) EUROCRYPT 1988. LNCS, vol. 330, pp. 123–128. Springer, Heidelberg (1988)
13. Micali, S., Rivest, R.L.: Transitive Signature Schemes. In: CT–RSA 2002, pp. 236–243 (2002)
14. Okamoto, T., Pointcheval, D.: The Gap-Problems: A New Class of Problems for the Security of Cryptographic Schemes. In: PKC 2001, pp. 104–118 (2001)
15. Paillier, P., Vergnaud, D.: Discrete-Log-Based Signatures May Not Be Equivalent to Discrete Log. In: Roy, B. (ed.) ASIACRYPT 2005. LNCS, vol. 3788, pp. 1–20. Springer, Heidelberg (2005)
16. Pointcheval, D., Stern, J.: Security Arguments for Digital Signatures and Blind Signatures. J. Crypto (3), 361–396
17. Schnorr, C.-P.: Efficient Signature Generation by Smart Cards. J. Crypto (3), 161–174
18. Tompa, M., Woll, H.: Random Self-Reducibility and Zero Knowledge Interactive Proofs of Possession of Information. In: FOCS 1987, pp. 472–482 (1987)

An Efficient Protocol for
Fair Secure Two-Party Computation

Mehmet S. Kiraz and Berry Schoenmakers

Dept. of Mathematics and Computer Science, TU Eindhoven
P.O. Box 513, 5600 MB Eindhoven, The Netherlands
m.kiraz@tue.nl, berry@win.tue.nl

Abstract. In the 1980s, Yao presented a very efficient constant-round secure two-party computation protocol withstanding semi-honest adversaries, which is based on so-called garbled circuits. Later, several protocols based on garbled circuits covering malicious adversaries have been proposed. Only a few papers, however, discuss the fundamental property of fairness for two-party computation. So far the protocol by Pinkas (Eurocrypt 2003) is the only one which deals with fairness for Yao's garbled circuit approach.

In this paper, we improve upon Pinkas' protocol by presenting a more efficient variant, which includes several modifications including one that fixes a subtle security problem with the computation of the so-called majority circuit. We prove the security of our protocol according to the real/ideal simulation paradigm, as Lindell and Pinkas recently did for the malicious case (Eurocrypt 2007).

1 Introduction

In secure two-party computation there are two parties who are interested in evaluating a public function $f(x, y) = (f_1(x, y), f_2(x, y))$ where x and y are their respective private inputs, and the first party wants to know the value $f_1(x, y)$ and the other party wants to know $f_2(x, y)$ without disclosing more information about their inputs than what is implied by the outputs. There might be only common output $(f_1(x, y) = f_2(x, y))$, or one party receiving no output (e.g., $f_2(x, y) = \perp$).

In his seminal paper, Yao [14] presented a protocol for secure two-party computation in the semi-honest model where the adversary follows the protocol specifications but stores all intermediate values which may be analyzed later to learn additional information. He used a tool called a *garbled circuit*, an encrypted form of a Boolean circuit that implements the function $f(x, y)$. Roughly speaking, in Yao's protocol, the garbled circuit is constructed by one party (Bob), and it is evaluated by the other party (Alice). Recently, several papers appeared, extending Yao's protocol to the malicious case by using *cut-and-choose* techniques [7,3,13,5,6]. However, these protocols do not ensure fairness. Informally speaking, a protocol is fair if either both parties learn their (private) outputs, or none of them learns anything. So, a fair protocol ensures that a malicious party

T. Malkin (Ed.): CT-RSA 2008, LNCS 4964, pp. 88–105, 2008.

cannot gain an advantage by aborting the protocol before the other (honest) party gets its output. Pinkas [11] presented the first fair and secure two-party protocol based on Yao's garbled circuits, which is the starting point of this work.

One of the main ideas of [11] is that the evaluation of the garbled circuit, as performed by Alice, does not result in garbled values (one per output wire) but—roughly speaking—in *commitments* to these garbled values. Basically, Alice will hold commitments for Bob's output wires, and, v.v., Bob will hold commitments for Alice's output wires. The important point here is that both parties are convinced of the correctness of the values contained in the commitments held by the other party. For this we need some special protocol techniques. In [11], blind signatures are used as a building block to achieve correctness of these commitments. However, blind signatures make Pinkas' protocol rather complex and inefficient. In our protocol, we avoid the use of blind signatures, resulting in a savings of a factor k, where k is a security parameter.

Once the correctness of the commitments for the output values is guaranteed, both parties will gradually open their commitments (bit by bit). For this gradual release phase we will use the protocol of [4] implementing their "commit-prove-fair-open" functionality, which is actually proved correct according to the real/ideal simulation paradigm. We will use their results in a black-box way.

1.1 Our Contributions

There have been several recent advances for Yao's garbled circuit approach [3,5,13,6]. In this paper, we revisit fairness while borrowing several improvements and ideas from these recent papers. We thus improve upon the protocol by Pinkas [11] which is the only paper considering fairness for Yao's garbled circuit approach.

Pinkas presents an intricate method which involves a modification of the truth tables for the garbled circuits. A crucial part of his protocol is to let the parties convince each other of the correctness of the committed values for the output wires, as explained above. The difficulty is to show the correctness of Alice's commitments for Bob's output wires. Concretely, blind signatures are used in [11] as a building block for the verification of these commitments (where Bob is the signer, Alice is the receiver of the blind signatures, and a further cut-and-choose subprotocol is used to ensure that Bob only signs correctly formed (blinded) commitments). Instead, we use the well-known OR-proofs [2], to let Alice show that she committed correctly to the garbled output values that she obtained for each of Bob's output wires. Alice needs to render these OR-proofs only for the so-called majority circuit. As a consequence, whereas Pinkas' protocol uses $2\ell m\kappa$ blind signatures, our protocol only needs ℓ OR-proofs (but also ℓm homomorphic commitments), where ℓ is the number of output wires for Bob, m is the number of circuits used for cut-and-choose, and κ is another security parameter (used for cut-and-choose as well). Overall, this leads to an improvement by a factor of κ in computational complexity, and also a significant improvement in communication complexity.

The above application of OR-proofs also critically relies on a slight modification of the circuit representing f, where a new input wire (for Bob) is introduced for every output wire of Bob. Bob will use random values for these new input wires, to blind the values of his output wires. Nevertheless, Alice will still be able to determine a majority circuit. This modification was suggested to us by Pinkas to resolve a subtle problem for the protocol of [11], which we communicated to him [10]; the problem is that a corrupted Alice may learn Bob's private inputs.

The security of our protocol is analyzed according to the real/ideal simulation paradigm, following the proof by Lindell and Pinkas [6] for the malicious case. However, we note that the failure probability for their simulator is quite large in the case that Bob is corrupted; namely $2^{1-m/17}$, where m is the number of circuits used, for a certain type of failure—in addition to other types of failures. This requires a large value for m, even for a low level of security. By two further extensions to the circuit representing f (the first of which actually suffices for [6]), we are able to reduce the failure probability to $2^{-m/4}$ (which we believe is optimal for this type of cut-and-choose protocols relying on majorities).

Roadmap

In Section 2, we describe the problem which covers fairness, the importance of the majority circuit computation and the problem mentioned above for the protocol by Pinkas. In Section 3, we present our improved fair and secure two-party protocol, in Section 4 we analyze its performance and compare with the protocol by Pinkas. Finally, in Section 5 we analyze the security of our protocol, where we also describe how to modify the circuit for f in order to get a more efficient simulation.

2   Main Protocol Ideas

In this section we briefly review some issues for Pinkas' protocol, and we present the main ideas behind our new protocol. We refer to [11] for a list of references on fairness.

Gradual Release. Let us first mention that for the gradual release phase, which concludes our protocol, we will use the protocol for the "commit-prove-fair-open" functionality of [4]. Note that Pinkas [11, Appendix A] uses a special protocol for this, based on *timed-commitments*. We will use the "commit-prove-fair-open" functionality in a black-box manner. The only thing we need to take care of before reaching the gradual release phase is that Alice and Bob hold commitments to the correct values for each other's output wires.

The intuition behind these gradual release protocols (which do not rely on a trusted party) is that Alice and Bob open the commitments bit-by-bit in such a way that the other party can directly check the correctness of each bit so released. Also, a party prematurely aborting the protocol cannot perform the search for the remaining bits in parallel, and [4] even shows how this can be done in a simulatable way.

Majority Circuits. We consider protocols which let Alice decide (at random) for each of m garbled circuits constructed by Bob, whether such a circuit must be opened by Bob (for checking by Alice), or whether the circuit will be evaluated by Alice. Exactly $m/2$ circuits will be checked by Alice, and she will evaluate the remaining half. Among the evaluated circuits there may still be a few incorrect circuits (not computing function f), but with high probability the majority of the evaluated circuits is correct.

We actually need to deal with two types of majority circuits. One is related to Bob's output wires (which we will indicate by index r throughout the paper; to ensure that Bob cannot get any information on Alice's inputs), and one is related to Alice's output wires (to ensure that Alice is guaranteed to get correct output values). These majority circuits can be characterized as follows. First, the *majority value* for an output wire is the bit value that occurs most frequently for that wire when considered over all $m/2$ evaluated circuits (ties can be broken arbitrarily). Further, an output wire is called a *majority wire* if its value is equal to the majority value. And, finally, a circuit is called a *majority circuit for Alice* resp. *Bob* if all of Alice's resp. Bob's output wires are majority wires.

Next, we explain why Alice should to send the output values only for a majority circuit, rather than sending Bob the output values for all evaluated circuits. For example, Bob could have constructed $m - 1$ garbled circuits that compute the function $f(x, y)$ and a single circuit which simply outputs Alice's input values. Since Alice evaluates $m/2$ out of m circuits, the probability would be $1/2$ that this single circuit is selected by Alice. Therefore, Bob would receive the outputs of the incorrect circuit from Alice with probability $1/2$ at the end of the evaluation.

Note that the computation of majority circuit for Bob can be avoided altogether for a protocol tolerating malicious adversaries but not achieving fairness. Namely, Lindell and Pinkas in [6, Section 2.2] show that it suffices to consider the case that only Alice receives private output. It is not clear, however, whether this protocol can be extended to cover fairness as well.

Problem with Pinkas' Computation of Majority Circuit for Bob.

Omitting details, the protocol by Pinkas reaches a state in which for each evaluated garbled circuit GC_j and for each output wire i of Bob, Bob knows a random bit k_{ij} and Alice knows the value $b_{ij} \oplus k_{ij}$, where b_{ij} is the actual output bit for wire i in circuit GC_j. Alice and Bob then use these values to determine a majority circuit r for Bob. Pinkas proposes that Alice can be trusted to construct a garbled circuit for this task, as Alice needs this majority circuit to prevent Bob from cheating. But this way, nothing prevents Alice from constructing an arbitrary circuit which reveals some of Bob's input values and hence some of the $k_{i,j}$ values. Then Alice learns Bob's actual output bits, which should not be possible. Of course, this problem can be solved by running any two-party protocol which is secure against malicious parties (e.g., [6]). However, in this paper, we will not require any additional protocol for computing a majority circuit for Bob. We present a simple modification to the circuit and in this way we show that a majority circuit can be computed without considerable additional cost.

Modified Circuit Randomizing Bob's Output Wires. Alice needs to be able to determine a majority circuit for Bob, but at the same time she should not learn Bob's actual output values. Let C_f denote a circuit computing function $f(x, y)$. We extend circuit C_f to a *randomized circuit* RC_f, as follows [10]. See Figure 1. Hence, for each output wire W_i of Bob, a new input wire W_i' is added as well as a new output wire W_i'', such that $W_i'' = W_i \oplus W_i'$.

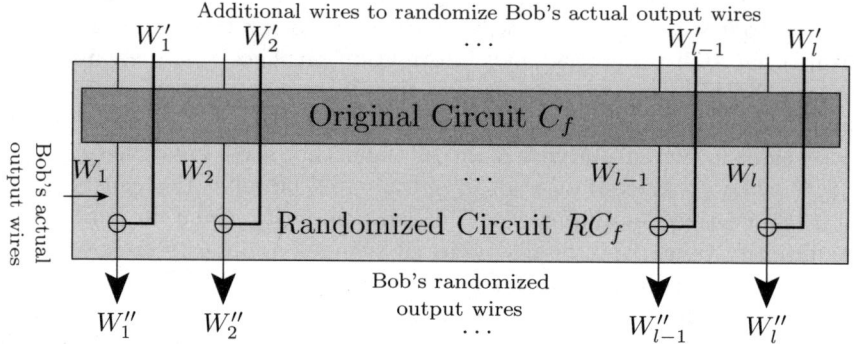

Fig. 1. The randomized circuit RC_f made from the original circuit C_f

Correctness of Garbled Input Values. Bob is not only providing Alice with m garbled circuits, but also with garbled values for each input wire for each circuit to be evaluated. It must be ensured that these garbled values are *correct* (i.e., correspond to the actual input value and fit with the garbled truth tables of the garbled circuits).

The correctness of Alice's garbled inputs will basically follow by definition of 1-out-of-2 oblivious transfer (OT) as in [11]. However, as pointed out in [5], one cannot use plain OT for the malicious case; rather a stronger form of OT such as committed OT, or the potentially weaker form of *committing OT* should be used. Here, committing OT is like plain OT with the added requirement upon completion of the protocol, the sender is committed to both of the input values it used during the protocol. The additional commitments output by the committing OT protocol will be opened by Bob (sender in OT) for the circuits which are checked by Alice.

The correctness of Bob's garbled inputs is not as straightforward to handle. Pinkas [11] originally used OR-proofs [2], whereas later papers [3,13,6]) aimed at using less expensive techniques relying on symmetric operations only (used in combination with cut-and-choose). In this paper, we use the *equality-checker scheme* of Franklin and Mohassel [3] for proving correctness of Bob's inputs.

The equality-checker scheme roughly runs as follows. For each pair of garbled circuits GC_j and $GC_{j'}$ with $1 \leq j < j' \leq m$, and for each input wire i for Bob, and for each $b \in \{0, 1\}$, Bob commits to the garbled values $w_{i,j,b}$ and $w_{i,j',b}$. Alice will check the consistency of these commitments for each pair of garbled circuits that are both opened by Bob. This implies with overwhelming probability that Bob's inputs to any two circuits GC_j and $GC_{j'}$ are equal(see [3] for more details).

Correctness of Committed Outputs. In order that the two parties can safely enter the gradual release phase, one of the main problems that needs to be solved is that both parties are convinced of the correctness of the values contained in the commitments held by the other party. We treat this problem differently for Alice and Bob.

Bob's commitments to Alice's outputs will be guaranteed to be correct by cut-and-choose, exactly as in [11]. For Alice's commitments to Bob's outputs, however, we will use a different approach than in [11], which used blind signatures for this purpose. In our protocol, Alice will first obtain the garbled values for Bob's outputs for all evaluated circuits, and she commits to all of these values. At that point, Bob can safely reveal both garbled values for each output wire (of the randomized circuit, as described above). Additional commitments from Bob will ensure that these garbled values are correct. Finally, Alice proves that she committed to one of these garbled values, from which Bob deduces that Alice indeed committed to correct values for Bob's output wires.

Concretely, we let Alice produce an OR-proof as follows. Suppose Alice committed to a garbled value $w_{i,j}$ for output wire i of Bob in circuit GC_j, and that she received garbled values $w_{i,j,0}$ and $w_{i,j,1}$ from Bob. Using homomorphic commitments, such as Pedersen commitments [9], Alice can prove that either $w_{i,j} = w_{i,j,0}$ or $w_{i,j} = w_{i,j,1}$ without revealing which is the case, applying [2] to the Chaum-Pedersen protocol for proving equality of discrete logarithms [1]. We will use the Fiat-Shamir heuristic to make these proofs non-interactive (and provably secure in the random oracle model).

3   A Fair Secure Two-Party Protocol

The object of the protocol is to evaluate a function of the form $f(x, y) = (f_1(x, y), f_2(x, y))$ securely, where Alice holds input x and gets output $f_1(x, y)$ and Bob holds input y and gets output $f_2(x, y)$. For simplicity, we assume that these inputs and outputs are all of equal length, i.e., $f : \{0, 1\}^\ell \times \{0, 1\}^\ell \to \{0, 1\}^\ell \times \{0, 1\}^\ell$, for some integer value ℓ.

Let RC_f denote the randomized boolean circuit for function f, see Figure 1. We use I_A resp. O_A to denote the set of Alice's input resp. output wires, and I_B resp. O_B to denote the set of Bob's input resp. output wires. Furthermore, we use I'_B to denote the additional input wires for Bob, used in the construction of RC_f from C_f. Note that $|I_A| = |I_B| = |I'_B| = |O_A| = |O_B| = \ell$. Accordingly, we write $x = \langle x_i \in \{0, 1\} : i \in I_A \rangle$ for Alice's input, $y = \langle y_i \in \{0, 1\} : i \in I_B \rangle$ for Bob's input, and $z = \langle z_i \in \{0, 1\} : i \in I'_B \rangle$ for Bob's random input to RC_f. Further, $|RC_f|$ denotes the number of gates in the circuit RC_f, and we $\mathcal{W}$ denote the set of all wires in the circuit RC_f. Hence, $I_A \cup I_B \cup I'_B \cup O_A \cup O_B \subseteq \mathcal{W}$.

In Phase 3 of the protocol, Bob will generate m garbled versions of the circuit RC_f, where m is a security parameter. We will denote these garbled circuits by

GC_j, for $j = 1, \ldots, m$. A garbled circuit GC_j for RC_f is completely determined by the pair of garbled values $(w_{i,j,0}, w_{i,j,1})$ assigned by Bob to each wire $i \in \mathcal{W}$. Here, $w_{i,j,b} \in \{0, 1\}^k$ corresponds to bit value $b \in 0, 1$, where k is another security parameter, denoting the length of the garbled values.

We abstract away most of the details of how garbled circuits are generated and evaluated. For our purposes it suffices to view a garbled circuit (as to be evaluated by Alice) as a concatenation of permuted-encrypted-garbled-4-tuples, one for each (binary) gate in RC_f, and of permuted ordered pairs, one for each output wire of Alice: $GC_j = \langle\langle \text{PEG-4-tuple}_{n,j} : 1 \le n \le |RC_f|\rangle, \langle POP_{i,j} : i \in O_A\rangle\rangle$. The permuted ordered pairs $POP_{i,j}$ are generated at random by Bob, using the garbled values $w_{i,j,0}$ and $w_{i,j,1}$ assigned to wire $i \in O_A$ in circuit j: $POP_{i,j} = w_{i,j,\sigma_{i,j}}, w_{i,j,1-\sigma_{i,j}})$, where $\sigma_{i,j} \in_R \{0, 1\}$.

In the protocol we use two types of commitments, namely homomorphic ('asymmetric') commitments, e.g., Pedersen commitments [9], and other ('symmetric') commitments, e.g., constructed from pseudorandom generators [8]. We let $\text{commit}_P(m; r)$ denote a symmetric commitment to a message m using randomness r generated by party P, and we use $\text{commit}_P^h(m; r)$ to denote homomorphic commitments.

The protocol consists of 10 phases (see also Appendix A for a protocol diagram).

Phase 1 [Generation of garbled input values]. Bob generates garbled values $w_{i,j,b} \in_R \{0, 1\}^k$, for $i \in I_A$, $j = 1, \ldots, m$, and $b \in \{0, 1\}$.

Phase 2 [Committing OT]. Committing OT is run in order for Alice to learn her garbled input values. Bob is the sender with private input $w_{i,j,0}, w_{i,j,1}$ for $i \in I_A$ and $j = 1, \ldots, m$ which were generated in Phase 1, and Alice is the receiver with private input $x_i \in \{0, 1\}$. At the end of committing OT Alice receives w_{i,j,x_i} and Bob gets no information about which of his inputs is chosen. Also, the common output is $A_{i,j,b}^{OT} = \langle \text{commit}_B(w_{i,j,b}; \alpha_{i,j,b}) \rangle$ for $i \in I_A$, $j = 1, \ldots, m$, $b \in \{0, 1\}$. These commitments will be checked by Alice later on, in order to prevent cheating by Bob; in particular to avoid the protocol issue addressed in [5].

Phase 3 [Construction]. In this phase Bob does the following:

- He prepares the garbled circuits GC_j for $j = 1, \ldots, m$ such that the garbled values $w_{i,j,0}, w_{i,j,1}$ for $i \in I_A$ and $j = 1, \ldots, m$ which are generated in Phase 1 are used for the corresponding wires.

- He also generates the commitments $B_{i,j,j',b} = \text{commit}_B(w_{i,j,b}, w_{i,j',b}; \beta_{i,j,j',b})$ for $i \in I_B \cup I_B'$ and j, j' such that $1 \le j < j' \le m, b \in \{0, 1\}$ for the equality-checker scheme. $B_{i,j,j',b}$'s are committed to Bob's garbled input values and they are generated to ensure that Bob's input is consistent for all the circuits (see Section 2).

- He also computes the commitments $C_{i,j} = \text{commit}_B(\sigma_{i,j}; \gamma_{i,j})$ for $i \in O_A$ and $j = 1, \ldots, m$ where $\sigma_{i,j} \in_R \{0, 1\}$. These committed values are used to permute Alice's output values and the correctness will be proved by the cut-and-choose technique, by opening half of them in the opening phase.

- Finally in this phase, the commitments $D_{i,j} = \text{commit}_B(w_{i,j,0}, w_{i,j,1}; \delta_{i,j})$ for $i \in O_B$ and $j = 1, \ldots, m$ are computed. The $D_{i,j}$'s are committed to Bob's garbled output values and they are generated so that Alice can determine a correct majority circuit.

He sends the circuits and the commitments generated above to Alice. Each pair of commitments $(B_{i,j,j',0}, B_{i,j,j',1})$ is sent in random order, in order that Alice does not learn Bob's inputs when Bob opens one commitment for each of these pairs later on in the evaluation phase.

Phase 4 [Challenge]. Alice and Bob run the challenge phase (a coin-tossing protocol) in order to choose a random bit string $\ell = \ell_1 \ldots \ell_m \in_R \{0,1\}^m$ that defines which garbled circuits and which commitments will be opened.

Phase 5 [Opening & Checking].

- Alice asks Bob to open the circuits GC_j for j such that $\ell_j = 1$ which are called *check-circuits*. Similarly, the circuits GC_j for j such that $\ell_j = 0$ will be called *evaluation-circuits*. She also asks Bob to open the corresponding commitments for j such that $\ell_j = 1$. Bob sends the following for opening:
- Bob sends the opening set $\widetilde{GC}_j = \langle\, w_{i,j,b} : i \in W \,\rangle$, for j such that $\ell_j = 1$, $b \in \{0,1\}$ to open the check-circuits.
- He also sends $\widetilde{A}_{j,b}^{OT} = \langle\, \alpha_{i,j,b} : i \in I_A \,\rangle$ for j such that $\ell_j = 1$, $b \in \{0,1\}$ in order to open the corresponding commitments $A_{i,j,b}^{OT}$.
- He also sends the opening set $\widetilde{B}_{j,j',b} = \langle\, \beta_{i,j,j',b} : i \in I_B \cup I_B' \,\rangle$ for j, j' such that $\ell_j = \ell_{j'} = 1$, $1 \le j < j' \le m$, $b \in \{0,1\}$ to open the corresponding commitments $B_{i,j,j',b}$.
- The opening set $\widetilde{C}_j = \langle\, \sigma_{i,j}, \gamma_{i,j} : i \in O_A \,\rangle$ for j such that $\ell_j = 1$ is also sent in this phase to open the corresponding commitments $C_{i,j}$. $-$The opening set $\widetilde{D}_j = \langle\, \delta_{i,j} : i \in O_B \,\rangle$ for j such that $\ell_j = 1$ is also sent to open the corresponding commitments $D_{i,j}$.
- The opening set $\widetilde{B}_{j,j'}^{input} = \langle\, w_{i,j,y_i}, w_{i,j',y_i}, \; \beta_{i,j,j',y_i}, \; w_{i',j,z_{i'}}, w_{i',j',z_{i'}}, \beta_{i',j,j',z_{i'}} : i \in I_B, \; i' \in I_B' \,\rangle$ for j, j' such that $\ell_j = \ell_{j'} = 0$, $1 \le j < j' \le m$. This set contains the garbled values of Bob's input wires for the evaluation-circuits, and sent to Alice which is a part of the equality-checker scheme.

Alice verifies the circuits and the commitments. Note that the consistency check of Bob's input is done now by the equality-checker scheme with the commitment set $\widetilde{GC}_j$ (contains all garbled values) for j such that $\ell_j = 1$ and $b \in 0,1$ and the set $\widetilde{B}_{j,j',b}$ for j, j' such that $\ell_j = \ell_{j'} = 1, 1 \le j < j' \le m$ and $b \in \{0,1\}$. Note that the opening sets $\widetilde{A}_{j,b}^{OT}$, $\widetilde{B}_{j,j',b}$ and $\widetilde{D}_j$ contain only randomness since the corresponding garbled values comes already from the set $\widetilde{GC}_j$. If any of the verifications fail Alice aborts the protocol.

Phase 6 [Evaluation]. Alice does the following in the evaluation phase:

- She first evaluates the circuits GC_j for $\ell_j = 0$ and computes garbled output values.
- She then commits to Bob's output values as $E_{i,j} = commit_A^h(w_{i,j}; \zeta_{i,j})$ for $i \in O_B$ and j such that $\ell_j = 0$ and sends them to Bob. Note that the commitments $E_{i,j}$ are generated to assure Bob that the committed values in $E_{i,j}$ are circuit values. If, for example, Alice commits to values different from garbled output values then she will be detected in OR-proofs in

Phase 9. The crucial property we need here is that these commitments are homomorphic in order to be able to use in OR-proofs.

Phase 7 [Opening of Bob's ordered output]. After Bob receives the commitments $E_{i,j}$ for $i \in O_B$ and j such that $\ell_j = 0$ he opens the commitments $D_{i,j}$ by sending the opening set $\widetilde{D_j} = \langle\, w_{i,j,0}, w_{i,j,1}, \delta_{i,j} : i \in O_B \,\rangle$ for j such that $\ell_j = 0$. Note that the commitments $D_{i,j}$ can be opened since Bob's outputs are randomized (see Figure 1); hence Alice can only see which outputs match (and determine a majority circuit), but she does not learn Bob's output $f_2(x, y)$.

Phase 8 [Decision of majority circuit]. Alice determines a majority circuit GC_r for some r such that $\ell_r = 0$. Note that she can determine a correct majority circuit GC_r without further interaction with Bob since the additional input values that were used to randomize Bob's output wires are all identical for the same wire in each circuit GC_j for j such that $\ell_j = 0$.

Phase 9 [Verification of Alice's commitments]. They run OR-proofs where Alice is the prover, Bob is the verifier. Alice proves that the committed value inside $E_{i,r}$ is either equal to $w_{i,r,0}$ or $w_{i,r,1}$. Alice cannot cheat here, otherwise she has to guess a garbled value, which has chances at most negligibly in k.

Phase 10 [Gradual release]. They then run the protocol for the gradual release to open their respective commitments, namely, $C_{i,j}$'s and $E_{i,j}$'s. At the end of the gradual release:

- Alice learns all $\sigma_{i,j}$ for $i \in O_A$ and $j \in U$ and applies it to $POP_{i,j}$ to learn her actual outputs for j circuits. She takes the majority of j circuits which will result in $f_1(x, y)$.
- Bob also matches the garbled output values that are received from Alice and the additional wire values in terms of bits (he knows the garbled values and the corresponding bits). Bob finally computes his output $f_2(x, y)$ by XORing his randomized output wire for the circuit GC_r with the corresponding additional wires.

4   Performance Analysis

We analyze the overall communication and computational complexity of our protocol, and compare with Pinkas' protocol by ignoring the constructions that are used in both protocols. We assume that Pinkas' protocol also uses the equality-checker for consistency of Bob's input. We also assume that Pinkas' protocol uses committing OT to fix the protocol issue [5]. Note that by the modification presented in Figure 1 we need ℓ additional XOR gate for each output wire of Bob which has only negligible affect to the overall complexity.

As we said before, the problem of Pinkas' protocol with majority circuit computation can be fixed by running any two-party protocol considering malicious adversaries. For example, if the protocol in [6] is used then the communication complexity of majority circuit computation is $O(\ell m^2 \log m)$. We note that there is no need to use such a protocol in our case.

We next consider the parts related to fairness. Note that the way we generate Bob's commitments to Alice's outputs is the same as in Pinkas' protocol (namely, there are $O(\ell m)$ commitments to permutations $\sigma_{i,j}$'s). However, for Alice's commitments to Bob's outputs is much different: Pinkas' protocol has $O(\ell m \kappa)$ commitments which are generated by Alice in order to be blindly signed by Bob, where κ is another security parameter (which are actually timed-commitments for the gradual opening). In our protocol, there are $O(\ell m)$ homomorphic commitments, and ℓ OR-proofs (namely, one proof for each wire). In the gradual phase, we use the protocol of [4] in order to ensure fairness, namely new commitments (together with the proof of correctness) will be generated before applying gradual opening. Note that one can also use the protocol presented by Schoenmakers and Tuyls [12] which is a weaker version of [4] but relatively more efficient.

The major difference between our approach and the construction by Pinkas [11] is in the removal of the blind signatures and of the majority circuit computation. This leads to an improvement by a factor of κ for the computational complexity. The reason is that for every output wire of Bob 2κ blind signatures are needed in Pinkas' protocol while in ours only one proof of knowledge is needed together with a simple modification to the circuit.

5 Security Analysis

In our security analysis we want to take advantage of the frameworks established by [6] and [4] for the real/ideal simulation paradigm, resp., for the malicious case in secure two-party computation (and Yao's protocol in particular) and for the case of fair protocols. To do so we will actually focus on analyzing a variant of our protocol, in which Phase 10 is replaced by Phase 10':

Phase 10' [**Trivial opening.**] Alice opens the commitments $E_{i,r}$ for $i \in O_B$ and Bob opens the commitments $C_{i,j}$ for $i \in O_A$ and j such that $\ell_j = 0$.

This adapted protocol is not fair, but it withstand malicious adversaries. We will argue so by showing how to simulate it, following [6]. From this we conclude that the commitments upon entering Phase 10 in the protocol are correct, as a consequence of which the framework of [4] applies and the simulatability of our protocol follows.

Before we give a simulation we present the following two additional modifications over the circuit RC_f, in order to have an efficient simulation.

Modification 1. We first modify Bob's input wires of the circuit RC_f in the following way: for each input wire of Bob (say W_B), we add an AND gate and an OR gate as shown in Figure 2 in such a way that the AND gate has one new input wire for Alice (say W_A) and the original input wire from Bob (W_B). This composition of gates always reproduces the value of the wire W_B independently of the value of W_A $((W_A \wedge W_B) \vee W_B = W_B)$.

This modification is applied so that the simulator is able to learn the input of the corrupted Bob. Roughly speaking, if the simulator knows the garbled circuit, two garbled values for each input wire of Alice (together with their corresponding bit value) and a garbled value for each input wire of Bob (where he does not know the corresponding bit value) then it is possible to compute the bit value of the garbled value for each input wire of Bob.

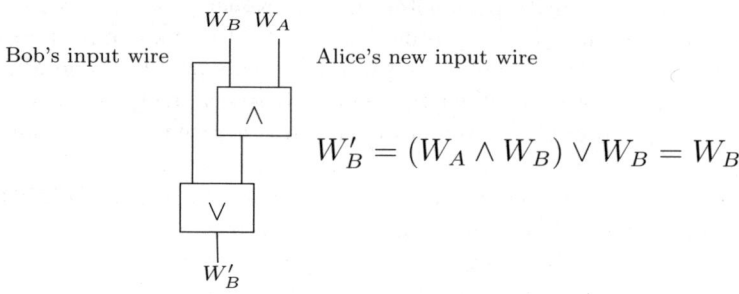

Fig. 2. Additional gates for each of Bob's input wires

We now proceed with the details. Let $w_{A,i,0}$ and $w_{A,i,1}$ denote Alice's garbled input values (for 0 and 1) for $i \in I_A$, and $w_{B,i,b}$ Bob's garbled input value for $i \in I_B \cup I'_B$ and for some $b \in \{0, 1\}$. If the garbled values $w_{A,i,0}, w_{A,i,1}, w_{B,i,b}$ and the garbled circuit are given then by evaluating the garbled AND gate with the garbled inputs $(w_{A,i,0}, w_{B,i,b})$ and $(w_{A,i,1}, w_{B,i,b})$ it is possible to decide the bit value of $w_{B,i,b}$ (i.e., learn the bit value b). Namely, if after these two evaluations the same garbled value is obtained, it means that Bob's garbled input corresponds to the bit 0; otherwise, Bob's input bit is 1. We note that this deduction process does not work for an arbitrary Boolean gate, and this is the reason why we modified the circuit in such a way that the input gates are AND gates. For example, evaluation of an XOR gate (has Alice's input wire and Bob's input wire) using $(w_{A,i,0}, w_{B,i,b})$ and $(w_{A,i,1}, w_{B,i,b})$ would always result in two different garbled values from which one cannot conclude the input bit of the corrupted Bob.

Modification 2. We next modify Bob's output wires of the circuit RC_f in the following way: for each output wire of Bob, we add the construction presented in Modification 1 and add two XOR gates as shown in Figure 3 in such a way that new input wire for Alice is added. This composition of gates always reproduces the original output bit of Bob in the garbled circuit independently of the value of Alice's additional input (i.e., the bit value of the wires W_B, W'_B and W''_B in Figure 3 is the same regardless of Alice's input.). The simulator has to learn the garbled output values of the corrupted Bob together with their corresponding bit (we describe this during the simulation), and by this modification the simulator would be able to learn them.

If the garbled values $w_{A,i,0}, w_{A,i,1}, w_{B,i,b}$ and the garbled circuit are given then by evaluating the modification in Figure 3 it is possible to compute the garbled values for each output wire of Bob together with their corresponding bit.

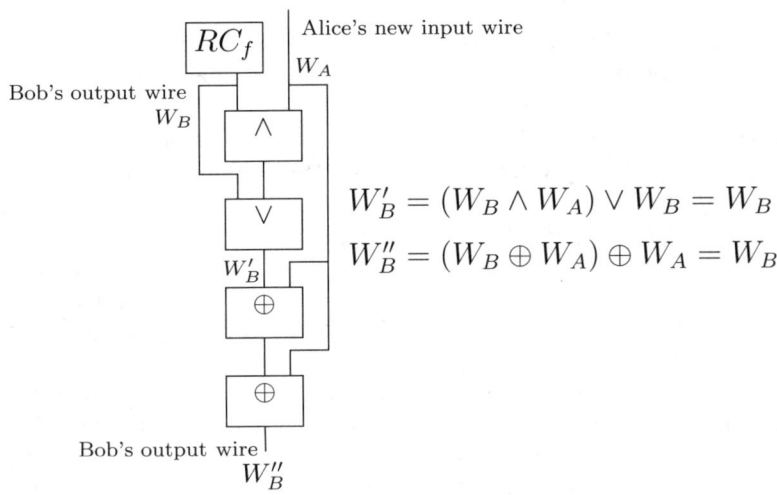

$$W'_B = (W_B \wedge W_A) \vee W_B = W_B$$

$$W''_B = (W_B \oplus W_A) \oplus W_A = W_B$$

Fig. 3. Additional gates for each of Bob's output wires

More precisely, as described above, by evaluating the garbled AND gate with the garbled inputs $(w_{A,i,0}, w_{B,i,b})$ and $(w_{A,i,1}, w_{B,i,b})$ it is possible to compute the output bit value of Bob's garbled output value $w_{B,i,b}$ for the wire W_B in Figure 3 (i.e., learn the bit value b). We here note that the bit value of W'_B is the same as the bit value of W_B, so the bit value of W'_B is also known.

We next show that it is possible to compute both garbled output values (0 and 1) for the output wires of XOR gates from the second construction (for the wires W'_B and W''_B in Figure 3), together with their corresponding bit. Let $\hat{w}$ be the evaluated garbled output value of the OR gate for wire W'_B. By evaluating the garbled XOR gate with the garbled inputs $(w_{A,i,0}, \hat{w})$ and $(w_{A,i,1}, \hat{w})$ where the bit value of $\hat{w}$ is known one can learn both garbled values for the output wires of XOR gates (W'_B and W''_B), and their corresponding bit. Namely, these two evaluations always result in two different garbled values from which it is easy to learn the corresponding bit.

We stress that our protocol is applied to this final modified circuit together with the above modifications, and in this way we show that we have an efficient simulator.

We are now ready to simulate the protocol (the one with the trivial opening) assuming that either Bob or Alice is corrupted.

Case 1- Assume Bob is corrupted

Let R_B be an adversary corrupting Bob; we construct a simulator S_B as follows. Since we assume that the committing OT protocol is secure, we analyze the security of the protocol in the hybrid model with a trusted party computing the committing OT functionality.

The simulator.

1. The simulator S_B chooses a fixed input $x' = 0$ for Alice and uses it only in the beginning of the protocol (namely, to run the OT phase) but it is not used later on.

2. S_B invokes R_B and obtains the garbled input values $w_{i,j,0}$ and $w_{i,j,1}$ for $i \in I_A$ and $j \in \{1, \ldots, m\}$ which are R_B's inputs from the committing OT protocol (in the hybrid model).

3. S_B receives all of the garbled circuits and the commitments from R_B.

4. S_B then runs the challenge phase to generate the random challenge values.

5. Now the input of R_B will be extracted as follows. The simulator S_B receives all of the required decommitments from R_B based on the challenge values, including the garbled values that correspond to Bob's input. Let $w_{i,j}$ be Bob's garbled input value for $i \in I_B \cup I'_B$ and j such that $\ell_j = 0$. S_B verifies that all the commitments are correct as Alice would do in Phase 5. If any of the checks fail, S_B sends an abort message to R_B, sends $\perp$ to the trusted party and halts, outputting whatever R_B outputs. If none of the checks fail, S_B obtains $m/2$ input for Bob for $m/2$ circuits because of Modification 1. More precisely, the simulator knows $w_{i,j,0}$, $w_{i,j,1}$ for $i \in I_A$ and $w_{i,j}$ for $i \in I_B \cup I'_B$ for j such that $\ell_j = 0$, and by Modification 1 the simulator can learn the input bit of Bob for each $w_{i,j}$ for $i \in I_B \cup I'_B$ for j such that $\ell_j = 0$. (In the real case, this does not happen since Alice learns only one garbled input value from OT for each her input wire.) If no input value appears more than $m/4$ times, then S_B outputs fail[1]. We show below that fail also does not occur with high probability. Otherwise, S_B sets y to be the value that appears more than $m/4$ times and sends it to the trusted party. Trusted party replies with $f_2(x, y)$ to S_B.

6. Now the simulator knows $f_2(x, y)$ but it has to convert this value into the corresponding garbled values. The simulator S_B first computes the evaluation-circuits as in the real protocol and obtains one garbled output value per wire. The complementary values will appear as well which are in general not the correct ones since the simulator computes the garbled circuit in the case that $x' = 0$. However, Modification 2 has been applied in order to learn both garbled output values of Bob, and the corresponding bits. As we described above, the simulator learns the output bit of $w_{i,j}$ for $i \in I_B \cup I'_B$ and j such that $\ell_j = 0$ from the AND gate in Figure 3 (for the wire W_B). This bit value is the same as the bit value for the wire W'_B in Figure 3. Then, by decrypting the XOR gates the simulator learns both garbled values, and their corresponding bits. In the real case, this does not happen since Alice learns only one garbled input value from OT for each her input wire. Hence, since the simulator knows the private output of the corrupted party and corresponding garbled output values it then computes the commitments $\text{commit}_A(w_{i,j}; \zeta_{i,j})$ for $i \in O_B$ and j such that $\ell_j = 0$ as Alice does in the real protocol in Phase 6 and sends to R_B.

7. The commitments $\text{commit}_B(w_{i,j,0}, w_{i,j,1}; \delta_{i,j})$ for $i \in O_B$ and j such that $\ell_j = 0$ are opened by R_B for the evaluation-circuits as in the real protocol.

---

[1] The majority of inputs are computed in order to have a correct output by the cut-and-choose technique.

8. The simulator then determines the majority circuit since it knows the garbled output values and the corresponding bits as in the real protocol.

9. Since the simulator knows the values in the commitments $E_{i,r}$ it can produce the proof.

10'. Alice opens the commitments $E_{i,r}$ for $i \in O_B$ and Bob opens the commitments $C_{i,j}$ for $i \in O_A$ and j such that $\ell_j = 0$.

Analysis. We claim that the view of R_B in the simulation with S_B is statistically close to its view in a hybrid execution of the protocol with a trusted party computing the committing OT protocol. (Note that our protocol is not statistically secure since the simulation is in the hybrid model for committing OT functionality, and it depends on the implementation of OT subprotocol, the commitment schemes and the OR-proofs used).

First of all, we show that if Alice aborts the protocol depending a cheating behavior by Bob, then Bob does not get any information about Alices input. This is only possible either at Phase 5 (at the opening & checking phase) or at Phase 7 (at the opening of $D_{i,j}$'s) while checking the correctness of the circuits and the commitments. In this case, the decision to abort is based on Bob's construction of the circuits as well as commitments (including commitments from the OT phase), and on the random inputs of the parties, and is independent of Alice's input. Thus, Bob does not get any information if Alice aborts the protocol. Thus, we know that the difference between Alice receives "abort" in an ideal execution with S_B and that Alice outputs "abort" in a real execution with R_B is negligible. From here on, we therefore consider the case that Alice does not abort the protocol.

We now prove that the circuits and the commitments are correct with overwhelming probability. First of all, we note that the additional modifications does not compromise the security of the garbled circuit since, by definition of garbled circuit, having one garbled value for each input wire for a gate results in always one garbled output value, which ensures privacy. If Alice does not abort then with probability $2^{-m/4}$ at most $m/4$ of the circuits are bad (including the commitments). Also, we know that the equality-checker scheme [3] assures with high probability a majority of the evaluation-circuits obtain the same input and OT assures with high probability that the values received from OT are correct garbled values, and therefore **fail** does not occur with negligible probability. The simulator S_B can then decide on a majority circuit, prove that the commitments are committed to the garbled values of output wires of R_B and open the commitments for only this circuit.

We next show that if S_B does not output any fail message, the simulated view of R_B is identically distributed to its view in an execution of the protocol. Actually, they are identical since S_B just runs the honest Alice's instructions when interacting with corrupted Bob. Since S_B uses independent random coins in the challenge phase and follows Alice's instructions each time, the above process results in a distribution that is identical to the view of R_B in a real execution with Alice. As we mentioned before the protocol is not statistically secure since the simulation is considered in the hybrid model for committing OT functionality, and it depends on the implementation of OT subprotocol, the commitment schemes and the OR-proofs.

Case 2- Assume Alice is corrupted

The security analysis when Alice is corrupted is very similar to the proof of [6]. During the protocol Alice sees the circuits and the commitments and they run a secure committing OT where she gets only the garbled values corresponding to her input bits. On a high level, in the simulation, the simulator first extracts the input of Alice from OT functionality in the hybrid model and then sends the input x to the trusted party and learn the output value. Given the output, the simulator constructs the garbled circuits. The simulator constructs the garbled circuits where some of them correctly computes $f(x, y)$ and some of them compute a constant function which always outputs Alice's real output. Namely, the output of this garbled circuit is always equal to the value which is received from the trusted party. The simulator then chooses the challenge value in such a way that all the check-circuits correctly compute the function $f(x, y)$ while all the other circuits (representation of constant function) are going to be evaluation-circuits which compute the constant function. We refer to [6] for details.

Remark. As we said before, in one part of the proof of [6], the failure probability of the simulator is bounded above by $2^{1-m/17}$. The reason is that the rewinding process is used in the case that Bob is corrupted. We note that Modification 1 is sufficient to have a better bound of [6]. Modification 2 is not necessary for [6] since the way our protocol permits two private outputs is different from theirs. In our case, once Alice evaluates the circuits we know that she can compute only one garbled output value. And, Bob accepts it as output if and only if the value is the same as the circuit garbled value. In our security analysis, Modification 2 lets the simulator learn the corresponding garbled value and in this way, we avoid running the rewinding procedure in [6], which results in more efficient simulation.

Also, note that sending garbled output values or actual bit values to Bob in [6] does not compromise the security of the protocol, however, in our protocol Alice has to send garbled output values but not the actual values (bits). Therefore, we highlight that in our protocol the correctness of outputs comes from checking whether the received values are garbled values of the circuit.

Acknowledgements. We would like to thank Peter van Liesdonk and José Villegas for their comments on the presentation.

References

1. Chaum, D., Pedersen, T.P.: Wallet databases with observers. In: Brickell, E.F. (ed.) CRYPTO 1992. LNCS, vol. 740, pp. 89–105. Springer, Heidelberg (1993)
2. Cramer, R., Damgård, I., Schoenmakers, B.: Proofs of partial knowledge and simplified design of witness hiding protocols. In: Desmedt, Y.G. (ed.) CRYPTO 1994. LNCS, vol. 839, pp. 174–187. Springer, Heidelberg (1994)

3. Franklin, M., Mohassel, P.: Efficiency tradeoffs for malicious two-party computation. In: Yung, M., Dodis, Y., Kiayias, A., Malkin, T.G. (eds.) PKC 2006. LNCS, vol. 3958, pp. 458–473. Springer, Heidelberg (2006)
4. Garay, J.A., MacKenzie, P.D., Prabhakaran, M., Yang, K.: Resource fairness and composability of cryptographic protocols. In: Halevi, S., Rabin, T. (eds.) TCC 2006. LNCS, vol. 3876, pp. 404–428. Springer, Heidelberg (2006)
5. Kiraz, M.S., Schoenmakers, B.: A protocol issue for the malicious case of Yao's garbled circuit construction. In: The 27th Symposium on Information Theory in the Benelux, pp. 283–290 (2006)
6. Lindell, Y., Pinkas, B.: An efficient protocol for secure two-party computation in the presence of malicious adversaries. In: Naor, M. (ed.) EUROCRYPT 2007. LNCS, vol. 4515, pp. 52–78. Springer, Heidelberg (2007)
7. Malkhi, D., Nisan, N., Pinkas, B., Sella, Y.: Fairplay – a secure two-party computation system. In: USENIX Security, pp. 287–302 (2004)
8. Naor, M.: Bit commitment using pseudorandomness. Journal of Cryptology 4, 151–158 (1991)
9. Pedersen, T.: A threshold cryptosystem without trusted party. In: Davies, D.W. (ed.) EUROCRYPT 1991. LNCS, vol. 547, pp. 522–526. Springer, Heidelberg (1991)
10. Pinkas, B.: Personal communication (2005)
11. Pinkas, B.: Fair secure two-party computation. In: Advances in Cryptology–Eurocrypt 2003. LNCS, vol. 2656, pp. 87–105. Springer, Heidelberg (2003)
12. Schoenmakers, B., Tuyls, P.: Practical two-party computation based on the conditional gate. In: Lee, P.J. (ed.) ASIACRYPT 2004. LNCS, vol. 3329, pp. 119–136. Springer, Heidelberg (2004)
13. Woodruff, D.P.: Revisiting the efficiency of malicious two-party computation. In: Naor, M. (ed.) EUROCRYPT 2007. LNCS, vol. 4515, pp. 79–96. Springer, Heidelberg (2007), http://eprint.iacr.org/2006/397
14. Yao, A.: How to generate and exchange secrets. In: 27th IEEE Symposium on Foundations of Computer Science, pp. 162–168 (1986)

A   Protocol Diagram

Common Input: f
Compute: $f(x,y) = (f_1(x,y), f_2(x,y))$

Alice   Bob

Private Input: $x = \langle x_i \in \{0,1\}, i \in I_A \rangle$ **Private Input:** $y = \langle y_i \in \{0,1\}, i \in I_B \rangle$

Phase 1: Generation of garbled input values.

Generate $w_{i,j,b} \in_R \{0,1\}^k, i \in I_A, j = 1, \ldots, m, b \in \{0,1\}$.

Phase 2: Committing OT Run in parallel, for $i \in I_A$.

<u>Receiver</u>   <u>Sender</u>
Private Input: x_i **Private Input:** $\langle w_{i,j,0}, w_{i,j,1} : j = 1, \ldots, m \rangle$
Committing OT subprotocol

$\longleftrightarrow$

Private Output: $\langle w_{i,j,x_i} : j = 1, \ldots, m \rangle$ **Private Output:** $\langle \alpha_{i,j,0}, \alpha_{i,j,1} : j = 1, \ldots, m \rangle$

Common Output:
$A_{i,j,b} = \langle \text{ commit}_B(w_{i,j,b}; \alpha_{i,j,b}) \rangle$ for $j = 1, \ldots, m, b \in \{0,1\}$

Phase 3: Construction.

Compute GC_j for $j = 1, \ldots, m$ s.t. for all $i \in I_A$
$\langle w_{i,j,0}, w_{i,j,1} : j = 1, \ldots, m \rangle$ are used for the
corresponding wires in GC_j.

Compute $B_{i,j,j',b} = \text{commit}_B(w_{i,j,b}, w_{i,j',b}; \beta_{i,j,j',b})$
for $i \in I_B \cup I_B'$ and $1 \leq j < j' \leq m, b \in \{0,1\}$.

Compute $C_{i,j} = \text{commit}_B(\sigma_{i,j}; \gamma_{i,j})$ for $i \in O_A$ and
$j = 1, \ldots, m$ where $\sigma_{i,j} \in_R \{0,1\}$.

Compute $D_{i,j} = \text{commit}_B(w_{i,j,0}, w_{i,j,1}; \delta_{i,j})$ for
$i \in O_B$ and $j = 1, \ldots, m$.

$\langle GC_j : j = 1, \ldots, m \rangle, \langle C_{i,j} : i \in O_A, j = 1, \ldots, m \rangle, \langle D_{i,j} : i \in O_B, j = 1, \ldots, m \rangle,$
$\langle B_{i,j,j',b_{i,j,j'}}, B_{i,j,j',1-b_{i,j,j'}} : i \in I_B \cup I_B', 1 \leq j < j' \leq m, b_{i,j,j'} \in_R \{0,1\} \rangle$

$\longleftarrow$

Phase 4: Challenge.

Secure coin-flipping subprotocol

$\longleftrightarrow$

Common Output: $\ell = \ell_1 || \ell_2 || \ldots || \ell_m$ s.t. $\ell_i \in_R \{0,1\}$

Phase 5: Opening & Checking.

$$\widetilde{GC}_j = \langle w_{i,j,b} : i \in \mathcal{W}, b \in \{0,1\}\rangle, \ell_j = 1.$$
$$\widetilde{A}^{OT}_{j,b} = \langle \alpha_{i,j,b} : i \in I_A\rangle, \ell_j = 1, b \in \{0,1\}.$$
$$\widetilde{B}_{j,j',b} = \langle \beta_{i,j,j',b} : i \in I_B \cup I'_B\rangle, \ell_j = \ell_{j'} = 1, 1 \leq j < j' \leq m,$$
$$b \in \{0,1\}.$$
$$\widetilde{C}_j = \langle \sigma_{i,j}, \gamma_{i,j} : i \in O_A\rangle, \ell_j = 1.$$
$$\widetilde{D}_j = \langle \delta_{i,j} : i \in O_B\rangle \text{ for } \ell_j = 1.$$
$$\widetilde{B}^{input}_{j,j'} = \langle w_{i,j,y_i}, w_{i,j',y_i}, \beta_{i,j,j',y_i}, w_{i',j,z_{i'}}, w_{i',j',z_{i'}},$$
$$\beta_{i',j,j',z_{i'}} : i \in I_B, i' \in I'_B\rangle \text{ for } \ell_j = \ell_{j'} = 0, 1 \leq j < j' \leq m$$

$$\langle \widetilde{GC}_j : \ell_j = 1, b \in \{0,1\}\rangle, \langle \widetilde{B}_{j,j',b} : \ell_j = \ell_{j'} = 1, 1 \leq j < j' \leq m, b \in \{0,1\}\rangle,$$
$$\langle \widetilde{C}_j : \ell_j = 1\rangle, \langle \widetilde{D}_j : \ell_j = 1\rangle, \langle \widetilde{B}^{input}_{j,j'} : \ell_j = \ell_{j'} = 0, 1 \leq j < j' \leq m\rangle$$

$\longleftarrow$

Check GC_j for $\ell_j = 1$ using $\widetilde{GC}_j$.
$A^{OT}_{i,j,b} \overset{?}{=} \text{commit}_B(w_{i,j,b}; \alpha_{i,j,b})$ for $i \in I_A, \ell_j = 1, b \in \{0,1\}$.
$B_{i,j,j',b} \overset{?}{=} \text{commit}_B(w_{i,j,b}, w_{i,j',b}; \beta_{i,j,j',b})$ for $i \in I_B \cup I'_B, \ell_j = \ell_{j'} = 1, 1 \leq j < j' \leq m, b \in \{0,1\}$.
$C_{i,j} \overset{?}{=} \text{commit}_B(\sigma_{i,j}; \gamma_{i,j})$ for $i \in O_A, \ell_j = 1$.
$D_{i,j} \overset{?}{=} \text{commit}_B(w_{i,j,0}, w_{i,j,1}; \delta_{i,j})$ for $i \in O_B, \ell_j = 1$.
$B_{i,j,j',y_i} \overset{?}{=} \text{commit}_B(w_{i,j,y_i}, w_{i,j',y_i}; \beta_{i,j,j',y_i})$ for $i \in I_B, \ell_j = \ell_{j'} = 0, 1 \leq j < j' \leq m$.
$B_{i',j,j',z_{i'}} \overset{?}{=} \text{commit}_B(w_{i',j,z_{i'}}, w_{i',j',z_{i'}}; \beta_{i',j,j',z_{i'}})$ for $i' \in I'_B, \ell_j = \ell_{j'} = 0, 1 \leq j < j' \leq m$.

Phase 6: Evaluation.
Evaluate GC_j for $\ell_j = 0$, using $\widetilde{B}^{input}_{j,j'}$.
Compute $E_{i,j} = \text{commit}^h_P(w_{i,j}; \zeta_{i,j})$ for $i \in O_B$ and $\ell_j = 0$.

$$\langle E_{i,j} : i \in O_B, \ell_j = 0\rangle$$

$\longrightarrow$

Phase 7: Opening of $D_{i,j}$.

$$\widetilde{D}_j = \langle w_{i,j,0}, w_{i,j,1}, \delta_{i,j} : i \in O_B\rangle \text{ for } \ell_j = 0.$$

$$\langle \widetilde{D}_j : \ell_j = 0\rangle$$

$\longleftarrow$

$D_{i,j} \overset{?}{=} \text{commit}_B(w_{i,j,0}, w_{i,j,1}; \delta_{i,j})$ for $i \in O_B, \ell_j = 0$

Phase 8: Decision of majority circuit.
Determine a majority circuit GC_r for some r s.t. $\ell_r = 0$ where
only majority of Bob's output wires are counted.

Phase 9: Verification of Alice's commitments. Run in parallel, for $i \in O_B$, r s.t. $\ell_r = 0$.

Prover		**Verifier**
Private Input: $w_{i,r}$	**Common Input:** $E_{i,r}, w_{i,r,0}, w_{i,r,1}, \epsilon_{i,r}$	**Private Input:** $\perp$
	OR-Proofs subprotocol	
	$\longleftrightarrow$	
Private Output: $\perp$	**Common Output:** Proof of validity	**Private Output:** $\perp$

Phase 10: Gradual Release. Run in parallel, for $i \in O_A, i' \in O_B, \ell_j = 0$.

Private Input: $w_{i',r}, \zeta_{i',r}$	**Common Input:** $C_{i,j}, E_{i',r}$,	**Private Input:** $\sigma_{i,j}, \delta_{i,j}$
	Gradual Release subprotocol	
	$\longleftrightarrow$	
Private Output: $\sigma_{i,j}$		**Private Output:** $w_{i',r}$

Apply $\sigma_{i,j}$ to $POP_{i,j}$ for $i \in O_A$ and match $w_{i,r}$ with $(w_{i,j,0}, w_{i,j,1})$ for $i \in O_A$ and determine a majority circuit for Alice.

Match $w_{i',r}$ with $(w_{i',r,0}, w_{i',r,1})$ for $i' \in O_B$ to find the randomized output bits and compute XOR with the corresponding additional input wires.

Private Output: $f_1(x, y)$

Private Output: $f_2(x, y)$

Efficient Optimistic Fair Exchange Secure in the Multi-user Setting and Chosen-Key Model without Random Oracles

Qiong Huang[1], Guomin Yang[1], Duncan S. Wong[1], and Willy Susilo[2]

[1] Department of Computer Science,
City University of Hong Kong, Hong Kong
[2] School of Computer Science & Software Engineering,
University of Wollongong, Australia

Abstract. Optimistic fair exchange is a kind of protocols to solve the problem of fair exchange between two parties. Almost all the previous work on this topic are provably secure only in the random oracle model. In PKC 2007, Dodis et al. considered optimistic fair exchange in a multi-user setting, and showed that the security of an optimistic fair exchange in a single-user setting may no longer be secure in a multi-user setting. Besides, they also proposed one and reviewed several previous construction paradigms and showed that they are secure in the multi-user setting. However, their proofs are either in the random oracle model, or involving a complex and very inefficient NP-reduction. Furthermore, they only considered schemes in the *certified-key model* in which each user has to show his knowledge of the private key corresponding to his public key.

In this paper, we make the following contributions. First, we consider a relaxed model called *chosen-key model* in the context of optimistic fair exchange, in which the adversary can arbitrarily choose public keys without showing the knowledge of the private keys. We separate the security of optimistic fair exchange in the *chosen-key* model from the *certified-key* model by giving a concrete counterexample. Second, we strengthen the previous *static* security model in the multi-user setting to a more practical one which allows an adversary to choose a key *adaptively*. Third, we propose an *efficient* and *generic* optimistic fair exchange scheme in the multi-user setting and *chosen-key* model. The security of our construction is proven *without random oracles*. We also propose some efficient instantiations.

1 Introduction

Optimistic fair exchange, introduced by Asokan, Schunter and Waidner [1], is a kind of protocols to solve the problems in fairly exchanging items between two parties, say Alice and Bob. In such a protocol, there is an arbitrator who is semi-trusted by Alice and Bob and involves only if one party attempts to cheat the other or simply crashes. Since the introduction, it has attracted many researchers' attention, such as [2,3,11,20,13,16,19,26,25,4,23,12] and so on.

T. Malkin (Ed.): CT-RSA 2008, LNCS 4964, pp. 106–120, 2008.

There are two popular paradigms for building optimistic fair exchange schemes. One is based on *verifiably encrypted signatures* [8], such as [2,3,11], and the other is based on *sequential two-party multisignatures*, such as [20]. Park et al.'s sequential two-party multisignature based optimistic fair exchange [20] was broken and repaired by Dodis and Reyzin [13]. However, Dodis-Reyzin schemes are *setup-driven* [27,28], which require key registration for all users with the arbitrator. In the same year, Micali proposed a fair electronic exchange protocol for contract signing with an invisible trusted party [19], using a CCA2 secure public key encryption scheme with *recoverable randomness* (i.e., the decryption algorithm can extract from the ciphertext both the plaintext and the randomness used for generating the ciphertext) and a signature scheme that is existentially unforgeable under chosen message attacks. The idea is similar to that of the verifiably encrypted signature paradigm. Later, Bao et al. [4] showed that the scheme does not satisfy the fairness requirement. A dishonest Bob can get Alice's full commitment without letting Alice get his obligation. They also provided an improvement to avoid such an attack.

To the best of our knowledge, almost all verifiably encrypted signature schemes and sequential multisignature schemes, even though efficient, are proven secure in the random oracle model only, which is only heuristic. The only schemes which are proven secure without random oracles are the verifiably encrypted signature scheme and the multisignature scheme proposed by Lu et al. [17]. Both schemes are based on Waters' signature scheme [24], and have been proven secure in the *certified-key* model [17] (or the *registered-key* model [5]), in which the adversary is required to certify that the public keys it includes in the signing oracle and in its forgery are properly generated and it knows the corresponding private keys.

Recently, Dodis et al. [12] considered optimistic fair exchange in a multi-user setting. Prior to their work, almost all previous results considered the single-user setting only, in which there are only one signer and one verifier (along with an arbitrator). A more practical setting is the multi-user setting, in which there are many signers and many verifiers (along with an arbitrator), so that a dishonest party can collude with some other parties in an attempt of cheating another party. Though the security of both encryption and signature in the single-user setting is preserved in the multi-user setting, Dodis et al. [12] showed that this is not necessarily true for optimistic fair exchange. They showed a counterexample that is secure in the single user setting but insecure in the multi-user setting. Furthermore, they proposed a formal definition of optimistic fair exchange in the multi-user setting, and presented a generic construction. Their generic construction is *setup-free* (i.e. no key registration is required between users and the arbitrator) and can be built if there exist one-way functions in the random oracle model, or if there exist trapdoor one-way permutations in the standard model. However, all the schemes presented in [12] were proven secure in the *certified-key* model only. If the adversary is allowed to choose public keys arbitrarily without requiring to show its knowledge of the corresponding private keys, these schemes may not be secure.

Our Results: Our contributions are in three-fold. First, we note that optimistic fair exchange schemes secure in the certified-key model may not be secure in the chosen-key model [18]. We separate these two models by presenting a counterexample. Namely, we present a scheme which is secure in the certified-key model but insecure in the chosen-key model. The crux of the problem is to allow the adversary in the chosen-key model to arbitrarily set public keys *without* showing its knowledge of the corresponding private keys (cf. certified-key model). Hence, the model is more realistic and it provides the adversary with more flexibility and power in attacking other honest parties in the system.

Second, we further strengthen the security model in the multi-user setting for optimistic fair exchange first proposed by Dodis et al. [12]. In particular, we notice that in [12], the model capturing the security against the arbitrator is a *static* model which requires the malicious arbitrator to fix its keys before seeing the challenging public key of the signer. We propose to strengthen it to an *adaptive* model which allows the arbitrator to set its keys with reference to the value of the challenging public key of the signer.

Third, we propose an *efficient* and *generic* construction of optimistic fair exchange in the multi-user setting and chosen-key model, and prove the security *without random oracles*. The construction is based on a conventional signature [14,24] and a ring signature [21,24,6,22,10,15], both of which can be constructed efficiently without random oracles. This also contributes a new paradigm for constructing optimistic fair exchange, besides the existing ones: the verifiably encrypted signatures based approach and the sequential two-party multisignature based one. In our generic construction, we further show that the ring signature scheme used in our construction does not need to be with the highest level of existential unforgeability considered in [6], namely *unforgeability with respect to insider corruptions*. Instead, *unforgeability against a static adversary* [10] will suffice. We also propose some efficient instantiations of our generic construction.

Organization: In the next section, we review the definition of optimistic fair exchange, and modify Dodis et al.'s security games to adapt the chosen-key model. In Sec. 3, we give a counterexample to separate the security level between the certified-key model and the chosen-key model. Our generic construction is then proposed and shown secure in the multi-user setting and under the chosen-key model in Sec. 4. Some efficient instantiations are also discussed in the section. Finally, we conclude this paper in Sec. 5.

2   Definitions and Security Model

2.1   Definitions in the Multi-user Setting and Chosen-Key Model

The definition for non-interactive optimistic fair exchange (OFE) follows the one in the multi-user setting given in [12] but having the authenticity assumption on public keys removed. This implies that we do not restrict ourselves to the *certified-key* model [17], but consider the definition under a stronger security model, called the *chosen-key* model [18]. We will give more details shortly

(Sec. 2.2) and make some additional remarks to discuss some subtleties in the definitions. Readers can refer to [12] for the detailed definition.

The *correctness* condition can be defined in a natural way. The *ambiguity* property requires that any "resolved signature" $\mathsf{Res}(m, \mathsf{PSig}(m, SK_{U_i}, APK), ASK, PK_{U_i})$ is *computationally indistinguishable* from an "actual signature" $\mathsf{Sig}(m, SK_{U_i}, APK)$.

2.2   Chosen-Key Model

Note that [12] only considers OFE in the certified-key model [17]. In such a model, it is assumed that the authenticity of public keys of users in the system can be verified and each user should show his knowledge of the corresponding private key in some *public key registration stage* for defending against key substitution attacks. Alternatively, the adversary is required to show that the public keys included in queries to the signing oracle and in its forgery are properly generated.

In this paper, we consider a stronger security model for OFE, the *chosen-key* model, which was originally introduced by Lysyanskaya et al. in the context of aggregate signature [18]. An adversary in a chosen-key model can arbitrarily set public keys *without* showing its knowledge of the corresponding private keys. The only limitations are that the adversary cannot replace the challenge user's public key and all the public keys chosen by the adversary should fall into some public key space (which is defined under some system-wide parameters and known to all parties in the system). Such relaxation gives the adversary more flexibility and power in attacking other (honest) parties in the system. Schemes secure in the certified-key model may not necessarily be secure in the chosen-key model.

For example, let us consider the Security Against Verifiers under the chosen-key model (Sec. 2.3). After receiving a partial signature from the challenge signer, the adversary may ask the arbitrator for resolving it into a full signature with respect to a *different* public key chosen maliciously by the adversary according to the challenge signer's public key and the partial signature received. Based on this attacking approach, in Sec. 3, we describe a concrete OFE scheme as an example for showing that a scheme secure in the certified-key model does not necessarily be secure in the chosen-key model.

2.3   Security Model

The security of optimistic fair exchange consists of three aspects: security against signers, security against verifiers, and security against the arbitrator. The definitions of them in the multi-user setting and chosen-key model are given as follows.

- SECURITY AGAINST SIGNERS: Intuitively, we require that no PPT adversary A should be able to produce a partial signature with non-negligible probability, which looks good to verifiers but cannot be resolved to a full signature by the honest arbitrator. This ensures the fairness for verifiers, that is, if the signer has committed to a message, the verifier will always be able to get the full commitment of the signer. Formally, we consider the following experiment:

$$\mathsf{Setup}^{\mathsf{TTP}}(1^k) \rightarrow (ASK, APK)$$
$$(m, \sigma', PK^*) \leftarrow A^{O_{\mathsf{Res}}}(APK)$$
$$\sigma \leftarrow \mathsf{Res}(m, \sigma', ASK, PK^*)$$
$$\text{success of } A := [\mathsf{PVer}(m, \sigma', PK^*, APK) = \mathsf{accept}$$
$$\wedge \ \mathsf{Ver}(m, \sigma, PK^*, APK) = \mathsf{reject}]$$

where oracle O_{Res} takes as input a *valid*[1] partial signature σ' of user U_i on message m, i.e. (m, σ', PK_{U_i}), and outputs a full signature σ on m under PK_{U_i}. In this experiment, the adversary can arbitrarily choose public keys, and it may not know the corresponding private key of PK^*. The advantage of A in the experiment $\mathrm{Adv}_A(k)$ is defined to be A's success probability.

- SECURITY AGAINST VERIFIERS: This security notion requires that any PPT verifier B should not be able to transform a partial signature into a full signature with non-negligible probability if no help has been obtained from the signer or the arbitrator. This requirement has some similarity to the notion of *opacity* for verifiably encrypted signature [8]. Formally, we consider the following experiment:

$$\mathsf{Setup}^{\mathsf{TTP}}(1^k) \rightarrow (ASK, APK)$$
$$\mathsf{Setup}^{\mathsf{User}}(1^k) \rightarrow (SK, PK)$$
$$(m, \sigma) \leftarrow B^{O_{\mathsf{PSig}}, O_{\mathsf{Res}}}(PK, APK)$$
$$\text{success of } B := [\mathsf{Ver}(m, \sigma, PK, APK) = \mathsf{accept}$$
$$\wedge \ (m, \cdot, PK) \notin Query(B, O_{\mathsf{Res}})]$$

where oracle O_{Res} is described in the previous experiment, the partial signing oracle O_{PSig} takes as input a message m and returns a valid partial signature σ' on m under PK, and $Query(B, O_{\mathsf{Res}})$ is the set of valid queries B issued to the resolution oracle O_{Res}. In the experiment, B can ask the arbitrator for resolving any partial signature with respect to any public key (adaptively chosen by B, probably without the knowledge of the corresponding private key), with the limitation described in the experiment. The advantage of B in the experiment $\mathrm{Adv}_B(k)$ is defined to be B's success probability.

- SECURITY AGAINST THE ARBITRATOR: Intuitively, this security notion requires that any PPT arbitrator C should not be able to generate with non-negligible probability a full signature without explicitly asking the signer for generating one. This ensures the fairness for signers, that is, no one can

---

[1] By 'valid', we mean that σ' is a valid partial signature on m under public key PK_{U_i}, alternatively, the input (m, σ', PK_{U_i}) of O_{Res} satisfies the condition that $\mathsf{PVer}(m, \sigma', PK_{U_i}, APK) = \mathsf{accept}$.

frame the actual signer on a message with a forgery. Formally, we consider the following experiment:

$$\mathsf{Setup}^{\mathsf{User}}(1^k) \rightarrow (SK, PK)$$
$$(ASK^*, APK) \leftarrow C(PK)$$
$$(m, \sigma) \leftarrow C^{O_{\mathsf{PSig}}}(ASK^*, APK, PK)$$
$$\text{success of } C := [\mathsf{Ver}(m, \sigma, PK, APK) = \mathsf{accept}$$
$$\wedge (m, \cdot) \notin Query(C, O_{\mathsf{PSig}})]$$

where the partial signing oracle O_{PSig} is described in the previous experiment, ASK^* is C's state information, which might not be the corresponding private key of APK, and $Query(C, O_{\mathsf{PSig}})$ is the set of queries C issued to the partial signing oracle O_{PSig}. The advantage of C in this experiment $\mathrm{Adv}_C(k)$ is defined to be C's success probability.

Definition 1. *A non-interactive optimistic fair exchange scheme is said to be* secure in the multi-user setting and chosen-key model *if there is no PPT adversary that wins any of the experiments above with non-negligible advantage.*

Remark 1. (Differences From [12]) Though the experiments of Security Against Signers and Security Against Verifiers remain in the same form as those in [12], we put no requirement on that the adversary has to register a public key before using it. In other words, the adversary can freely choose public keys (from the public key space) and use them during the attack, without proving its knowledge of the corresponding private keys. In [12] on the other hand, the authenticity assumption of public keys is made in all the experiments.

On the Security Against the Arbitrator, our corresponding experiment seems to be stronger than the one considered in [12], in which the adversary has to fix APK before learning the challenge signer's public key PK. This *static* form of adversarial key generation seems to be unnecessarily weak. We propose a strengthened one which allows the adversary to *adaptively* set APK based on the value of PK generated using $\mathsf{Setup}^{\mathsf{User}}$. In this way, the security model considered in this paper will be at least as strong as that in [12]'s, if not stronger. This observation is also supported by the counterexample given in Sec. 3.

3 Separating Chosen-Key Model from Certified-Key Model

As reviewed in the introduction, OFE in the single-user setting can normally be built from verifiably encrypted signature or from sequential two-party multisignature. Dodis et al. [12] showed that secure OFE in the multi-user setting can also be built from these primitives, but only the verifiably encrypted signature based ones may support the setup-free feature [27,28]. Also note that in [12], all the security analysis were carried out in the *certified-key* model [17] and therefore, they may not remain secure in the *chosen-key* model [18]. In the following, we give a concrete example for showing that a secure OFE in the certified-key model may no longer be secure in the chosen-key model. The example is based on Lu et al.'s [17] verifiably encrypted signature scheme. Readers can refer to [17] for Lu et al.'s scheme **WVES** .

3.1   A WVES-Based OFE

Observe that Lu et al.'s **WVES** is an OFE in the *single-user* setting and the *certified-key* model , under which, **WVES.Kg** and **WVES.AKg** constitute the OFE registration protocol Setup, and **WVES.Sig**, **WVES. Ver**, **WVES.ESig**, **WVES.EVer** and **WVES.Adj** are corresponding to Sig, Ver, PSig, PVer and Res, respectively. In the single-user setting and certified-key model [13,12], Security Against Signers is due to the correctness of **WVES**. That is, if η is a valid verifiably encrypted signature, the adjudicator can always convert it to an ordinary signature. Security Against Verifiers is due to the opacity property [8] of **WVES**.

The Security Against the Arbitrator does not trivially follow the unforgeability of the verifiably encrypted signature scheme, since in the corresponding experiment, the malicious arbitrator knows more secret information than a public verifier does. To show its security, we build a forger $\mathcal{F}$ of Waters' signature scheme using the malicious arbitrator/adjudicator C. Given the system parameters and a public key $A = e(g, g)^{\alpha}$, $\mathcal{F}$ randomly picks $\beta \leftarrow \mathbb{Z}_p$ and sends the system parameters, A and $(\beta, v := g^{\beta})$ to C^2. The rest of the proof goes essentially the same as that in [17], except that $\mathcal{F}$ uses its signing oracle to simulate the PSig oracle. If C outputs a valid forgery (S_1, S_2), i.e., $\mathbf{Ver}(PK, M, (S_1, S_2)) = \mathsf{accept}$, $\mathcal{F}$ simply outputs $\sigma^* := (S_1, S_2)$ on M as its forgery for Waters' signature scheme. By the validity of (S_1, S_2), we have that σ^* is also a valid forgery with respect to the challenge public key. Besides, the above scheme can easily be shown to be secure in the *multi-user* setting and the certified-key model as well.

3.2   An Attack under Chosen-Key Model

If we retain the multi-user setting but upgrade the model from certified-key model to the chosen-key model, we will see that the **WVES**-based OFE above will no longer be secure.

Let us consider the Security Against Verifiers. In the chosen-key model, the adversary (i.e. the verifier in the experiment) can first ask the challenge signer for a partial signature on some message under the challenge public key PK. Then, the adversary makes up a new public key PK' according to the partial signature and PK, and queries the challenger for resolving the partial signature with respect to PK' rather than to PK. The adversary finally tries to find out the full signature under PK from the resolved signature. In the *chosen-key* model, since the adversary can arbitrarily pick public keys without showing its

---

[2] Alternatively, C picks its key pair and shows its knowledge of ASK. This is due to the restriction of certified-key model. Readers can refer to [12] for detailed discussions about this.

knowledge of the corresponding private keys, such an attack approach is possible. Below is the detail of the actual attack against the **WVES**-based OFE.

(In)Security Against Verifiers: Upon receiving the challenge signer's public key $PK = e(g, g)^\alpha$ from the challenger, the adversary B queries O_{PSig} for a partial signature $\sigma' = (K_1, K_2, K_3)$ on message M. Then B generates another public key $PK' := PK \cdot e(g, g)^b$ where $b \leftarrow \mathbb{Z}_p$, and queries O_{Res} for resolving a partial signature in the form $\sigma'' = (K_1 \cdot g^b, K_2, K_3)$ under the public key PK'. Note that σ'' is a valid partial signature on M under PK'. Upon receiving the resolved signature $\sigma = (S_1, S_2)$, B outputs the full signature under the challenge public key PK as $\tilde{\sigma} = (S_1/g^b, S_2)$ and wins the game.

Therefore, **WVES**-based OFE is insecure in the multi-user setting under the *chosen-key* model. We should also emphasize that this does not contradict with the results given in [17] as their schemes were originally designed for security in the *certified-key* model only.

4 An Efficient and Generic Construction without Random Oracles

In this section, we propose an OFE proven secure in the multi-user setting and the chosen-key model, that is, under the adversarial model formalized in Sec. 2.3. Our construction is based on two primitives: conventional signature [14] and ring signature [21]. Since there exist signature schemes and ring signature schemes proven secure without random oracles, it is possible for us to construct a secure OFE without random oracle also. Refers can refer to [14] for the security definition of conventional signatures. In the following, we first briefly review the definition of ring signature.

(Ring Signature:) The notion of ring signature was introduced by Rivest et al. in Asiacrypt 2001 [21] and has later been widely studied [6,10,22,15].

The security of a ring signature scheme includes two parts, *anonymity* (or *ambiguity*) and *unforgeability*. The strongest computational complexity based security notions of them are *anonymity against attribution attacks/full key exposure* and *unforgeability with respect to insider corruption*, respectively [6,10]. In our construction of OFE (to be shown later), we actually do not require a ring signature scheme to equip with such a strong level of anonymity and unforgeability. Instead, *unforgeability under an adaptive attack, against a static adversary* [10] will suffice. It is defined as follows.

$$(sk_i, pk_i) \leftarrow \mathsf{RS.KG}(1^k), \text{ for } i = 1, \cdots, \ell$$

$$R := \{pk_i\}_{i=1}^{\ell}$$

$$(R, m, \sigma) \leftarrow A^{O_{\mathsf{RS.Sig}}}(R)$$

success of $A := [\mathsf{RS.Ver}(m, \sigma, R) = \mathsf{accept} \wedge (\cdot, m, R) \notin Query(A, O_{\mathsf{RS.Sig}})]$

where A is a PPT adversary, $O_{\mathsf{RS.Sig}}$ is the ring signing algorithm which takes as input an index i, a message m, a list of public keys S such that $S \cap R \neq \emptyset$

and $pk_i \in R$, and outputs a ring signature σ on m under the ring S using the signing key sk_i, and $Query(A, O_{\mathsf{RS.Sig}})$ is the set of ring signing queries (of the form (i, m, S)) issued by A. The advantage of A in the experiment is defined to be its success probability. A ring signature scheme is said to be *(existentially) unforgeable under an adaptive attack, against a static adversary* (where 'static' means that the adversary should not corrupt any honest user and its forgery should be with respect to the prescribed ring R,) if there is no PPT adversary which wins the experiment with non-negligible advantage. It's readily seen that the above unforgeability is weaker than the *unforgeability with respect to insider corruption* considered in [6]. For our purpose, the number ℓ of (honestly generated) public keys is 2 and the size of the ring S in a signing query issued by A is also 2 (i.e., $\ell = 2$ and $|S| = 2$).

4.1 The Construction

Let $\mathsf{SIG} = (\mathsf{KG}, \mathsf{Sig}, \mathsf{Ver})$ be a conventional signature scheme and $\mathsf{RS} = (\mathsf{KG}, \mathsf{Sig}, \mathsf{Ver})$ a ring signature scheme. Our construction idea is as follows. The partial signature will be a conventional signature generated using SIG, and the full signature is the partial signature in conjunction with a ring signature generated under RS. The 'ring' members of the ring signature are the signer and the arbitrator. To resolve a partial signature, the arbitrator simply produces a ring signature. One of the main reasons of employing a ring signature scheme in our construction is that the unforgeability game of ring signature (that is, unforgeability under an adaptive attack, against a static adversary, as stated above) fits well in the chosen-key model for OFE. That is, the adversary can ask for a ring signature with respect to a ring which includes public keys not being certified. Below are the details of our generic construction denoted by OFE.

- $\mathsf{Setup^{TTP}}$: The arbitrator runs $(ask, apk) \leftarrow \mathsf{RS.KG}(1^k)$ and sets $(ASK, APK) := (ask, apk)$.
- $\mathsf{Setup^{User}}$: Each user U_i runs $(\hat{sk}_i, \hat{pk}_i) \leftarrow \mathsf{SIG.KG}(1^k)$ and $(\bar{sk}_i, \bar{pk}_i) \leftarrow \mathsf{RS.KG}(1^k)$. U_i then sets $(SK_{U_i}, PK_{U_i}) := ((\hat{sk}_i, \bar{sk}_i), (\hat{pk}_i, \bar{pk}_i))$.
- Sig: On input a message m, the signer U_i first produces a conventional signature σ' as the partial signature, i.e. $\sigma' \leftarrow \mathsf{SIG.Sig}(\hat{sk}_i, m)$, and then completes the signing process by generating a ring signature on m and σ', i.e. $\sigma^{\mathsf{RS}} \leftarrow \mathsf{RS.Sig}(\bar{sk}_i, m\|\sigma'\|PK_{U_i}, R)$ where $R := \{\bar{pk}_i, apk\}$. The full signature is then set as $\sigma := (\sigma', \sigma^{\mathsf{RS}})$.
- Ver: On input a message m and a signature σ purportedly produced by U_i, where $\sigma = (\sigma', \sigma^{\mathsf{RS}})$, the verifier checks the validity of σ' and σ^{RS} by running $\mathsf{SIG.Ver}(m, \sigma', \hat{pk}_i)$ and $\mathsf{RS.Ver}(m\|\sigma'\|PK_{U_i}, \sigma^{\mathsf{RS}}, R)$ respectively, where $R := \{\bar{pk}_i, apk\}$. If both output accept, it returns accept; otherwise, it returns reject.
- PSig: On input a message m, the signer U_i computes a conventional signature, i.e. $\sigma' \leftarrow \mathsf{SIG.Sig}(\hat{sk}_i, m)$, and returns σ' as the partial signature.
- PVer: On input a message m and a partial signature σ' purportedly produced by U_i, the verifier returns $\mathsf{SIG.Ver}(m, \sigma', \hat{pk}_i)$.

- Res: On input a message m and a partial signature σ' of user U_i, the arbitrator first checks the validity of σ' by running OFE.PVer($m, \sigma', PK_{U_i}, APK$). If σ' is invalid, it rejects the input by outputting $\perp$; otherwise, it computes $\sigma^{\mathsf{RS}} \leftarrow \mathsf{RS.Sig}(ask, m\|\sigma'\|PK_{U_i}, R)$, where $R := \{\bar{pk}_i, apk\}$. The arbitrator returns $\sigma := (\sigma', \sigma^{\mathsf{RS}})$.

As in [12], one cannot view σ' as the full signature of the signer, even though it is itself a valid conventional signature. The signer's full commitment to a message comprises the partial signature σ' generated using SIG, along with a ring signature σ^{RS} produced by the signer or the arbitrator using RS. The *correctness* of the construction simply follows that of SIG and RS, and the *ambiguity* follows the *anonymity* requirement is satisfied due to that of the ring signature RS.

Remark 2.   One may notice that Dodis et al.'s generic OFE construction [12] uses a similar idea to ours. They employ a conventional signature as the partial signature and use an additional OR-signature to complete the generation of the full signature. An OR-signature itself can be viewed as a two-user ring signature. Even though OR-signature can express much richer languages, almost all the constructions of OR-signature follow the Fiat-Shamir heuristic, thus can only be proven secure in the random oracle model, or otherwise, require to have complex NP-reduction and non-interactive witness indistinguishable proofs of knowledge involved, that could be very inefficient. By applying our idea, an efficient and generic OFE scheme without random oracles can be built, as there are already quite a number of efficient conventional signature schemes and ring signature schemes proven secure without random oracles available in the literature.

Intuitively, for our construction above, the Security Against Signers holds unconditionally; the Security Against Verifiers follows the unforgeability property of the ring signature RS, and the Security Against the Arbitrator is guaranteed by the unforgeability of SIG. Thus, we have the following theorem.

Theorem 1. *The generic construction of optimistic fair exchange scheme OFE above is secure in the* multi-user *setting and* chosen-key *model, provided that SIG is a conventional signature scheme that is existentially unforgeable against chosen message attacks and RS is a secure ring signature scheme that is with* basic anonymity *and existential unforgeability under an adaptive attack, against a static adversary.*

Proof. Theorem 1 immediately follows from the following lemmas.   □

Lemma 1. *The optimistic fair exchange scheme OFE above is unconditionally secure against signers.*

Proof. Obviously, for any message m and any valid signature σ' on m under the verification key $\bar{pk}_i$, the arbitrator can always produce a ring signature σ^{RS} on $m\|\sigma'\|PK_{U_i}$ under the ring $R := \{\bar{pk}_i, apk\}$. Therefore, no adversary can win the game.   □

Lemma 2. *The optimistic fair exchange scheme OFE above is secure against verifiers if RS is unforgeable under adaptive attacks against a static adversary.*

Proof. Suppose that B is a PPT adversary which breaks the Security Against Verifiers with probability ϵ_B. We construct a PPT algorithm $\bar{B}$ to break the existential unforgeability of RS with the same probability.

On input a security parameter 1^k and given two public keys pk_0 and pk_1, which are the (honestly generated) challenge public keys as in the unforgeability game of ring signature (See page 113), $\bar{B}$ randomly generates a key pair $(\hat{sk}, \hat{pk})$ of SIG by running $(\hat{sk}, \hat{pk}) \leftarrow$ SIG.KG(1^k), flips a bit $b \leftarrow \{0, 1\}$, and sets $APK := pk_b$ and $PK := (\hat{pk}, pk_{1-b})$. It then runs B on input (APK, PK), and simulates oracle O_{PSig} using the secret key $\hat{sk}$ and oracle O_{Res} using $\bar{B}$'s ring signing oracle. More in detail, to answer an PSig query of m, $\bar{B}$ computes and returns SIG.Sig$(\hat{sk}, m)$ to B. To answer an Res query of (m, σ', PK_{U_i}), if σ' is a valid partial signature on m under PK_{U_i}, $\bar{B}$ queries its ring signing oracle for getting a ring signature σ^{RS} on message $m\|\sigma'\|PK_{U_i}$ under the ring $\{pk_0, pk_1\}$ using the secret key corresponding to pk_b, and then sends $(\sigma', \sigma^{\mathsf{RS}})$ back to B.

At the end of the experiment, when B outputs its forgery $(\tilde{m}, \tilde{\sigma})$, where $\tilde{\sigma} = (\tilde{\sigma}', \tilde{\sigma}^{\mathsf{RS}})$, without loss of generality, we assume that B has already got $\tilde{\sigma}'$ from a query to oracle O_{PSig}. The other case that B produced $\tilde{\sigma}'$ by itself will be covered by the Security Against the Arbitrator, which is to be shown later.

Obviously, the simulation above is perfect, and thus B wins the game with probability ϵ_B. We have that OFE.Ver$(\tilde{m}, \tilde{\sigma}, PK, APK) =$ accept and $(\tilde{m}, \cdot, PK) \notin Query(B, O_{\mathsf{Res}})$. The former also implies that SIG.Ver$(\tilde{m}, \tilde{\sigma}', \hat{pk}) =$ accept and RS.Ver$(\tilde{m}\|\tilde{\sigma}'\|PK, \sigma^{\mathsf{RS}}, (pk_0, pk_1)) =$ accept hold. Since $(\tilde{m}, \cdot, PK) \notin Query(B, O_{\mathsf{Res}})$, $\bar{B}$ has never issued a query to its ring signing oracle on input $\tilde{m}\|\tilde{\sigma}'\|PK$. Therefore, $\tilde{\sigma}^{\mathsf{RS}}$ is a valid ring signature on the new message $\tilde{m}\|\tilde{\sigma}'\|PK$ under the ring $\{pk_0, pk_1\}$. We then let $\bar{B}$ output $(\tilde{m}\|\tilde{\sigma}'\|PK, \tilde{\sigma}^{\mathsf{RS}})$ and $\bar{B}$ wins its own game with probability ϵ_B. □

Lemma 3. *The optimistic fair exchange scheme* OFE *above is secure against the arbitrator if* SIG *is unforgeable under chosen-message attacks.*

Proof. Suppose that C is a PPT adversary which breaks the Security Against the Arbitrator with probability ϵ_C. We build a PPT algorithm $\bar{C}$ to break the unforgeability of the conventional signature scheme SIG with the same probability.

Given the challenge verification key pk of SIG (along with a signing oracle O_{sk}), $\bar{C}$ runs RS.KG(1^k) to get a key pair $(\bar{sk}, \bar{pk})$ and feeds $PK := (pk, \bar{pk})$ as input to C, which then returns an arbitrator public key APK and begins to issue queries to O_{PSig}. This oracle can perfectly be simulated by $\bar{C}$ using O_{sk}. Namely, on input a message m, $\bar{C}$ forwards it to O_{sk} and relays the oracle's answer to C as a valid partial signature. Finally, C outputs its forgery $(\tilde{m}, \tilde{\sigma})$ where $\tilde{\sigma} = (\tilde{\sigma}', \tilde{\sigma}^{\mathsf{RS}})$, such that OFE.Ver$(\tilde{m}, \tilde{\sigma}, PK, APK) =$ accept and $(\tilde{m}, \cdot) \notin Query(C, O_{\mathsf{PSig}})$. We then have that $\tilde{\sigma}'$ is a valid signature on $\tilde{m}$, and $\tilde{m}$ has never been issued by $\bar{C}$ to its signing oracle. We simply let $\bar{C}$ output $(\tilde{m}, \tilde{\sigma}')$. Obviously $(\tilde{m}, \tilde{\sigma}')$ is a valid forgery for SIG, and $\bar{C}$ wins the unforgeability game with advantage ϵ_C. □

4.2   Instantiations

There are quite a number of efficient conventional signature schemes and ring signature schemes without random oracles available in the literature, like [24,7], [22,15,10] and many others. Using these schemes and applying our generic construction, we can get many concrete and efficient OFE schemes proven secure without random oracles in the multi-user setting and chosen-key model. For example, we can use Waters' signature scheme [24] as SIG and Shacham-Waters' ring signature scheme [22] as RS. Note that in such an instantiation, Waters' signature scheme may work in a group of composite order [22] rather than in a group of prime order [24], so that SIG and RS can share the same set of system parameters. Besides, it is necessary to mention that there is a global setup process before any execution of the scheme. The requirement of having such a setup process stems from that of Shacham-Waters' ring signature scheme. For this instantiation, the ambiguity of the scheme is based on sub-group decision assumption [9,22], while the security against verifiers and security against the arbitrator are based on computational Diffie-Hellman assumption. The OFE.Sig algorithm of the resulting scheme requires no pairing operation, and the OFE.Ver algorithm requires four pairings. A main disadvantage of this instantiation is that the size of system parameters is large. It is determined by the output length of the underlying hash function used in Waters' signature scheme [24,22].

Alternatively, we may consider another instantiation, which enjoys much shorter system parameters but suffers from stronger underlying assumptions, i.e. strong Diffie-Hellman assumption [7,15]. In this instantiation, we employ Boneh-Boyen's weakly secure signature scheme [7] plus a one-time signature scheme as SIG[3], and Groth's ring signature scheme (in the common reference string model) [15] as RS. The reason that we use Boneh-Boyen's weakly secure signature scheme plus a one-time signature scheme as SIG is the same as the one behind the combination of Waters signature and Shacham-Waters ring signature. (SIG and RS share system parameters.) Note that for RS, we do not need to use the signature compression technique as in [15] since the ring in our case merely consists of two users. The Sig algorithm of the resulting scheme does not require any pairing operation either, while the Ver algorithm requires nine pairings.

In these two instantiations, each user has two key pairs, one for the conventional signature and the other one for ring signature, just as in the generic construction (Sec. 4.1). To make the instantiations more practical and efficient, people may wish to combine the two key pairs into one. Boyen's ring signature [10] (or, say, his mesh signature) is a good candidate for this purpose. In Boyen's ring signature scheme, the adversary can make not only ring signature queries, but also atomic (or conventional) signature queries. Boyen's scheme works in the common reference string model. The anonymity holds unconditionally, and the unforgeability is guaranteed by the *Poly Strong Diffie-Hellman* assumption introduced by Boyen [10], which is a stronger variant of the Strong Diffie-Hellman

---

[3] It is easy to see that a weakly secure signature scheme plus a one-time signature scheme lead to a signature scheme that is unforgeable against chosen message attacks. We skip the detailed proof here.

(SDH) assumption. In the resulting OFE scheme, the signer Alice and the arbitrator Charlie form a ring. We view an atomic signature of Alice as her partial signature, and the combination of the atomic signature and a ring signature as Alice's full commitment. We can see that, similar to the generic construction, the security against signers of this optimized instantiation also holds unconditionally. The security against verifiers will hold due to the unforgeability of Boyen's (two-user) ring signature scheme, and the security against the arbitrator follows the unforgeability of the (single-user) ring signature scheme. Any forgery of Alice's atomic signature σ' on a message m, where $\sigma' = (S, t) = (g^{\frac{1}{a+bm+ct}}, t)$ and (a, b, c) is Alice's secret key, can be trivially transformed into a forgery of the ring signature scheme under the ring consisting of Alice only, i.e. we set $s_0 := 0$ and randomly select t' from its domain, then the forgery is $(S_0, S_1, t_0, t_1) := (1, S, t', t)$. The validity of the forgery is readily seen. Though this instantiation relies on a stronger assumption, it enjoys higher efficiency and fewer system parameters. It also requires fewer pairing operations for OFE.Ver than that of the second instantiation, and has fewer system parameters than that of the first instantiation. The OFE.Sig does not require any pairing operation, and OFE.Ver requires only four pairings. Each user including the arbitrator needs to manage only one key pair (unlike the first two instantiations in which each user has two key pairs), and the public key consists of only three points on the elliptic curve (if we employ the *symmetric* group setting, i.e. $\mathbf{e} : \mathbb{G} \times \mathbb{G} \to \mathbb{G}_t$) [10].

5   Conclusion

In this paper we considered optimistic fair exchange in the multi-user setting and separated the security of optimistic fair exchange in the certified-key model from that in the *chosen-key* model. We proposed the efficient generic construction of optimistic fair exchange in the multi-user setting and chosen-key model and proved its security without random oracles. Our scheme is built from a conventional signature and a ring signature, both of which can be efficiently constructed without random oracles. We also discussed some efficient instantiations of our generic construction.

Acknowledgements

We'd like to thank the anonymous reviewers for their invaluable comments. The work was supported by grants from CityU (Project Nos. 7001959 and 7002001) and the Research Grants Council of the Hong Kong Special Administrative Region, China (RGC Ref. No. CityU 122107).

References

1. Asokan, N., Schunter, M., Waidner, M.: Optimistic protocols for fair exchange. In: CCS, pp. 7–17. ACM Press, New York (1997)
2. Asokan, N., Shoup, V., Waidner, M.: Optimistic fair exchange of digital signatures (extended abstract). In: Nyberg, K. (ed.) EUROCRYPT 1998. LNCS, vol. 1403, pp. 591–606. Springer, Heidelberg (1998)

3. Asokan, N., Shoup, V., Waidner, M.: Optimistic fair exchange of digital signatures. IEEE Journal on Selected Areas in Communication 18(4), 593–610 (2000)

4. Bao, F., Wang, G., Zhou, J., Zhu, H.: Analysis and improvement of Micali's fair contract signing protocol. In: Wang, H., Pieprzyk, J., Varadharajan, V. (eds.) ACISP 2004. LNCS, vol. 3108, pp. 176–187. Springer, Heidelberg (2004)

5. Barak, B., Canetti, R., Nielsen, J.B., Pass, R.: Universally composable protocols with relaxed set-up assumptions. In: FOCS 2004, pp. 186–195. IEEE Computer Society Press, Los Alamitos (2004)

6. Bender, A., Katz, J., Morselli, R.: Ring signatures: Stronger definitions, and constructions without random oracles. In: Halevi, S., Rabin, T. (eds.) TCC 2006. LNCS, vol. 3876, pp. 60–79. Springer, Heidelberg (2006), http://eprint.iacr.org/

7. Boneh, D., Boyen, X.: Short signatures without random oracles. In: Cachin, C., Camenisch, J.L. (eds.) EUROCRYPT 2004. LNCS, vol. 3027, pp. 56–73. Springer, Heidelberg (2004)

8. Boneh, D., Gentry, C., Lynn, B., Shacham, H.: Aggregate and verifiably encrypted signatures from bilinear maps. In: EUROCRYPT 2003. LNCS, vol. 2656, pp. 416–432. Springer, Heidelberg (2003)

9. Boneh, D., Goh, E.-J., Nissim, K.: Evaluating 2-DNF formulas on ciphertexts. In: Kilian, J. (ed.) TCC 2005. LNCS, vol. 3378, pp. 325–341. Springer, Heidelberg (2005)

10. Boyen, X.: Mesh signatures: How to leak a secret with unwitting and unwilling participants. In: Naor, M. (ed.) EUROCRYPT 2007. LNCS, vol. 4515, pp. 210–227. Springer, Heidelberg (2007)

11. Camenisch, J., Damgård, I.: Verifiable encryption, group encryption, and their applications to separable group signatures and signature sharing schemes. In: Okamoto, T. (ed.) ASIACRYPT 2000. LNCS, vol. 1976, pp. 331–345. Springer, Heidelberg (2000)

12. Dodis, Y., Lee, P.J., Yum, D.H.: Optimistic fair exchange in a multi-user setting. In: Okamoto, T., Wang, X. (eds.) PKC 2007. LNCS, vol. 4450, pp. 118–133. Springer, Heidelberg (2007) Also at Cryptology ePrint Archive, Report 2007/182, http://eprint.iacr.org/

13. Dodis, Y., Reyzin, L.: Breaking and repairing optimistic fair exchange from PODC 2003. In: DRM 2003, pp. 47–54. ACM Press, New York (2003)

14. Goldwasser, S., Micali, S., Rivest, R.: A digital signature scheme secure against adaptive chosen-message attack. SIAM J. Computing 17(2), 281–308 (1988)

15. Groth, J.: Ring signatures of sub-linear size without random oracles. In: Arge, L., Cachin, C., Jurdziński, T., Tarlecki, A. (eds.) ICALP 2007. LNCS, vol. 4596, pp. 423–434. Springer, Heidelberg (2007)

16. Kremer, S.: Formal Analysis of Optimistic Fair Exchange Protocols. PhD thesis, Université Libre de Bruxelles (2003)

17. Lu, S., Ostrovsky, R., Sahai, A., Shacham, H., Waters, B.: Sequential aggregate signatures and multisignatures without random oracles. In: Vaudenay, S. (ed.) EUROCRYPT 2006. LNCS, vol. 4004, pp. 465–485. Springer, Heidelberg (2006)

18. Lysyanskaya, A., Micali, S., Reyzin, L., Shacham, H.: Sequential aggregate signatures from trapdoor permutations. In: Cachin, C., Camenisch, J.L. (eds.) EUROCRYPT 2004. LNCS, vol. 3027, pp. 74–90. Springer, Heidelberg (2004)

19. Micali, S.: Simple and fast optimistic protocols for fair electronic exchange. In: PODC 2003, pp. 12–19. ACM Press, New York (2003)

20. Park, J.M., Chong, E.K.P., Siegel, H.J.: Constructing fair-exchange protocols for e-commerce via distributed computation of RSA signatures. In: PODC 2003, pp. 172–181. ACM Press, New York (2003)
21. Rivest, R., Shamir, A., Tauman, Y.: How to leak a secret. In: Boyd, C. (ed.) ASIACRYPT 2001. LNCS, vol. 2248, pp. 552–565. Springer, Heidelberg (2001)
22. Shacham, H., Waters, B.: Efficient ring signatures without random oracles. In: Okamoto, T., Wang, X. (eds.) PKC 2007. LNCS, vol. 4450, pp. 166–180. Springer, Heidelberg (2007)
23. Wang, G.: An abuse-free fair contract signing protocol based on the RSA signature. In: Proceedings of 14th International Conference on World Wide Web, WWW 2005, pp. 412–421. ACM Press, New York (2005)
24. Waters, B.: Efficient identity-based encryption without random oracles. In: Cramer, R.J.F. (ed.) EUROCRYPT 2005. LNCS, vol. 3494, pp. 114–127. Springer, Heidelberg (2005)
25. Zhang, Z., Zhou, Y., Feng, D.: Efficient and optimistic fair exchanges based on standard RSA with provable security. Cryptology ePrint Archive, Report 2003/178 (2004), http://eprint.iacr.org/
26. Zhu, H.: Constructing optimistic fair exchange protocols from committed signatures. Cryptology ePrint Archive, Report 2005/012 (2003), http://eprint.iacr.org/
27. Zhu, H., Bao, F.: Stand-alone and setup-free verifiably committed signatures. In: Pointcheval, D. (ed.) CT-RSA 2006. LNCS, vol. 3860, pp. 159–173. Springer, Heidelberg (2006)
28. Zhu, H., Susilo, W., Mu, Y.: Multi-party stand-alone and setup-free verifiably committed signatures. In: Okamoto, T., Wang, X. (eds.) PKC 2007. LNCS, vol. 4450, pp. 134–149. Springer, Heidelberg (2007)

Legally-Enforceable Fairness in Secure Two-Party Computation

Andrew Y. Lindell

Aladdin Knowledge Systems and Bar-Ilan University, Israel
andrew.lindell@aladdin.com, lindell@cs.biu.ac.il

Abstract. In the setting of secure multiparty computation, a set of mutually distrustful parties wish to securely compute some joint function of their private inputs. The computation should be carried out in a secure way, meaning that the properties *privacy, correctness, independence of inputs, fairness* and *guaranteed output delivery* should all be preserved. Unfortunately, in the case of no honest majority – and specifically in the important two-party case – it is impossible to achieve fairness and guaranteed output delivery. In this paper, we show how a legal infrastructure that respects digital signatures can be used to enforce fairness in two-party computation. Our protocol has the property that if one party obtains output while the other does not (meaning that fairness is breached), then the party not obtaining output has a digitally signed cheque from the other party. Thus, fairness can be "enforced" in the sense that any breach results in a loss of money by the adversarial party.

1   Introduction

In the setting of secure multiparty computation, a set of parties with private inputs wish to jointly compute some functionality of their inputs. Loosely speaking, the security requirements of such a computation are that nothing is learned from the protocol other than the output (privacy), that the output is distributed according to the prescribed functionality (correctness), that parties cannot make their inputs depend on other parties' inputs (independence of inputs), that the adversary cannot prevent the honest parties from successfully computing the functionality (guaranteed output delivery), and that if one party receives output then so do all (fairness). The generality of secure multiparty computation has made it a very important and useful tool for proving the feasibility of carrying out a variety of tasks. Indeed, in a number of different settings, it has been shown that *any* two-party or multiparty function can be securely computed [21,14,13,3,8]. Stated more simply, any distributed task that a set of parties wish to compute, can be computed in a secure way. This implies feasibility for a multitude of tasks, including those as simple as coin-tossing and agreement, and as complex as electronic voting, electronic auctions, electronic cash schemes, anonymous transactions, and privacy-preserving data mining.

T. Malkin (Ed.): CT-RSA 2008, LNCS 4964, pp. 121–137, 2008.

Fairness. Unfortunately, the above description is misleading and inaccurate. Indeed, it is possible to securely compute any functionality, where security implies all of the properties described above. However, this is only true if there exists an *honest majority* amongst the participating parties. In the case of no honest majority, and specifically in the important two-party case, it is *impossible* to achieve fairness and guaranteed output delivery [10] in general (although some non-trivial functions can be securely computed; see [12]). A number of different approaches have been taken to achieve some sort of fairness despite this impossibility:

1. *Gradual release:* In this approach, the output is not revealed all at once. Rather, it is released gradually with the property that if an abort occurs, then the adversary has not learned much more about the output than the honest parties. The drawback of this approach is that it is inherently expensive (requiring many rounds), and if the adversary is more powerful than the honest parties, fairness may still be breached. See, for example, the early works of [2,15] and more recent works [20,11] (a good survey of the many works in this area can be found in [20]).

2. *The optimistic model:* In this approach, a trusted server is used with the following property. If all the parties behave honestly, then there is no need for the server (it is not contacted). However, if fairness is breached, then the server may be contacted in which case fairness is restored. This approach can be highly efficient. However, its drawbacks are the need for new infrastructure (in the form of such a server) and the fact that the server must be *trusted* to not collude with the adversary. See [18,19,1,4] for some works in this area.

In this paper, we present an approach to achieving fairness, that is inspired by the work of [9] and has some similarities with the optimistic model. The authors of [9] consider the question of fair exchange of signatures. They make a highly interesting observation that in order for a signature to be *enforced*, it needs to be presented at a court of law. In this case, the other party will actually see the signature. They therefore run a computation with the following property: the first party to receive output obtains something called a "keystone", and the second party then receives its signature (i.e., the signature it should receive as output). The keystone by itself gives nothing and so if the first party aborts after receiving it, no damage has been done (fairness has not been breached). In contrast, after receiving its signature, the second party may abort and the first party is left only with a useless keystone. Nevertheless, the interesting property here is that given the keystone *and* the second party's signature (i.e., the signature that the second party received as output), it is possible to construct the signature that the first party should receive as output. Thus, if the second party wishes to enforce its signed contract in a court of law, it can only do so by essentially revealing the signature that the first party should receive, thereby restoring fairness. In [9], the above notion is formalized and called a concurrent signature scheme. In addition, [9] present an efficient construction of such a scheme that is secure under the discrete logarithm assumption in the random oracle model.

Our results. In this paper, we extend the idea of [9] to general secure two-party computation. One byproduct is that we show that the problem of concurrent signatures can easily be cast as a standard secure two-party computation problem, and therefore random oracles are not necessary. We stress, however, that in contrast to the construction of [9], our protocol is *not* efficient. Thus, it should be viewed as a "feasibility proof" that concurrent signatures can be constructed under general assumptions and without random oracles.

The basic idea of our approach is as follows. We construct a protocol with the property that either both parties receive output (and so fairness is preserved) or one party receives output while the other receives a digitally-signed cheque from the other party that it can take to a court of law or a bank. This cheque can contain any sum of money, as agreed by the parties. The protocol further has the property that the only way that a party can evade paying the sum in the cheque is to reveal the other party's output, thereby restoring fairness.

It is instructive to compare our approach to the *optimistic model*. On the one hand, both solutions use a trusted party that is only contacted in the case of attempted cheating. However, the optimistic model guarantees fairness always, whereas our approach allows an adversary to breach fairness as long as it is willing to pay the cheque (although if the sum in the cheque is set appropriately, such an event is unlikely to ever occur). In addition, even if the adversary is not willing to pay the cheque, it can prevent the other party from receiving its output until the court or bank processes the cheque, at which point it can provide the other party with its output and evade payment. Thus, our approach provides a somewhat weaker security guarantee. On the other hand, in the optimistic model a dedicated server must be set up and trusted. The advantage of our model is that it uses existing infrastructure (like courts and banks) that *are* trusted. However, our approach does assume that digital signature law and digital cheques are respected, and so may not always be applicable. In summary, we believe that our approach provides an interesting alternative to the optimistic one.

Organization. As a warm-up to see how our construction works, we first present a simple solution to the problem of concurrent signatures in Section 2 that is based on general protocols for secure computation. Then, in Section 3 we present the definitions that we need, as well as a formal definition of *legally-enforceable fairness*. Finally, in Section 4 we show how every two-party functionality can be securely computed with legally-enforceable fairness.

2   Concurrent Signatures

As a warm-up, we present a protocol for concurrent signatures that is based on general secure two-party computation. We rely on the intuitive description in the Introduction of what concurrent signatures are. A formal definition can be found in [9]. We also use a general protocol for secure computation, where party P_1 always receives output first (e.g., the protocols of [14,13] has this property); see definitions in Section 3 below.

We assume a public-key infrastructure for digital signatures. In particular, P_1 has a pair of signing/verification keys denoted (sk_1, vk_1) and P_2 has an analogous pair (sk_2, vk_2). Furthermore, each party knows the other's public verification key. Without loss of generality, we assume that a signature includes the message being signed upon. The aim of the protocol is for P_1 to receive a message m_2 that is signed by P_2, and for P_2 to receive a message m_1 that is signed by P_1. The protocol appears below in Figure 1.

Concurrent Signatures

The parties use a secure two-party protocol to compute the following functionality:

- **Inputs:**
    1. Party P_1 inputs its pair of keys (sk_1, vk_1), party P_2's verification-key vk_2, and the messages m_1 and m_2 to be signed upon
    2. Party P_2 inputs its pair of keys (sk_2, vk_2), party P_1's verification-key vk_1, and the messages m_1 and m_2 to be signed upon
- **Outputs:**
    1. If the keys do not match (i.e., P_2 inputs vk_1' that is different to the vk_1 input by P_1 or vice versa), or they are not valid (i.e., sk_1 is not the signing key associated with vk_1),[1] or $m_1 \neq m_2$, then the functionality outputs $\bot$ to both parties. Otherwise:
    2. Party P_1 receives $\sigma_2 = \mathsf{Sign}_{sk_2}(m_2, \sigma_1)$, where σ_1 is defined next.
    3. Party P_2 receives $\sigma_1 = \mathsf{Sign}_{sk_1}(m_1)$; recall that by our convention a signature contains the message and so σ_1 contains m_1.

Fig. 1. A protocol for concurrent signatures without random oracles

We now informally describe why our protocol achieves concurrent signatures. First, observe that if both parties received output in the secure protocol, then they both have mutual signatures on the appropriate messages m_1 and m_2. However, if only P_1 receives output (because it receives output first), then P_2 does not receive the signature it should receive on the message m_1. Nevertheless, the signature σ_2 obtained by P_1 *contains* the signature σ_1 that P_2 should receive. Therefore, if P_1 wishes to enforce its signature by taking P_2 to court, it will necessarily reveal σ_1 – the signature that P_2 should receive – thereby restoring fairness. Relying on the constructions of secure protocols by [14,13], for example, we have the following theorem:

Theorem 1. *Assuming the existence of enhanced trapdoor permutations, there exist protocols for concurrent signatures, as defined in* [9].

We remark that the constructions of [9] rely on specific assumptions and assume a random oracle, whereas our construction relies only on general assumptions and is in the standard model. However, the constructions of [9] are highly efficient, while ours are not.

---

[1] We assume that such validity can be verified given the signing and verification key-pair. This is without loss of generality.

We also remark that P_1 does not receive a "pure" signature on m_2. Rather, it receives a signature on m_2 together with σ_1. This may have ramifications in some applications. For example, consider the case that P_1 wishes to show a third party a signature on m_2 without revealing σ_1 or m_1 (since σ_2 contains σ_1 which in turn contains m_1, this information is revealed). If necessary, it is possible to prevent this by defining that P_1 receives a signature on m_2 together with an *encryption* of σ_1 under a key belonging to P_2. This will have the same effect regarding fairness, but now σ_2 does not reveal anything about σ_1 or m_1 to a third party, as desired.

3   Definitions

3.1   Standard Definitions

We use the standard definition of two-party computation for the case of no honest majority, where no fairness is guaranteed. In particular, this means that the adversary always receives output first, and can then decide if the honest party also receives output. We remark that the definition in [13, Section 7] only allows a corrupted P_1 to receive output without the honest party receiving output. The variant that we use is obtained via a straightforward modification to the ideal model; see [16]. We refer the reader to [13, Section 7] for full definitions of security for secure two-party computation, and present a very brief description here only.

Preliminaries. A function $\mu(\cdot)$ is negligible in n, or just negligible, if for every positive polynomial $p(\cdot)$ and all sufficiently large n's it holds that $\mu(n) < 1/p(n)$. A probability ensemble $X = \{X(a,n)\}_{a \in \{0,1\}^*; n \in \mathbb{N}}$ is an infinite sequence of random variables indexed by a and $n \in \mathbb{N}$. (The value a will represent the parties' inputs and n the security parameter.) Two distribution ensembles $X = \{X(a,n)\}_{n \in \mathbb{N}}$ and $Y = \{Y(a,n)\}_{n \in \mathbb{N}}$ are said to be computationally indistinguishable, denoted $X \stackrel{c}{\equiv} Y$, if for every non-uniform polynomial-time algorithm D there exists a negligible function $\mu(\cdot)$ such that for every $a \in \{0,1\}^*$,

$$|\Pr[D(X(a,n)) = 1] - \Pr[D(Y(a,n)) = 1]| \leq \mu(n)$$

All parties are assumed to run in time that is polynomial in the security parameter. (Formally, each party has a security parameter tape upon which the value 1^n is written. Then the party is polynomial in the input on this tape.)

Secure two-party computation. A two-party protocol problem is cast by specifying a random process that maps sets of inputs to sets of outputs (one for each party). This process is called a functionality and is denoted $f : \{0,1\}^* \times \{0,1\}^* \rightarrow \{0,1\}^* \times \{0,1\}^*$, where party P_1 is supposed to receive the first output and party P_2 the second output. For simplicity, we assume that $f_1 = f_2$ and thus that both parties receive the same output. When considering protocols for securely computing any functionality, this is without loss of generality (encryption can be used so that P_1 cannot read P_2's portion of the output and vice versa).

Security is formalized by comparing a real protocol execution to an ideal model setting where a trusted party is used to carry out the computation. In this ideal model, the parties send their inputs to the trusted party who first sends the output to the adversary. (The adversary controls one of the parties and can instruct it to behave arbitrarily) After the adversary receives the output it either sends continue to the trusted party instructing it to also send the output to the honest party, or halt in which case the trusted party sends $\perp$ to the honest party. The honest party outputs whatever it received from the trusted party and the adversary outputs whatever it wishes. We stress that the communication between the parties and the trusted party is ideally secure. The pair of outputs of the honest party and an adversary $\mathcal{A}$ in an ideal execution where the trusted party computes f is denoted $\text{IDEAL}_{f,\mathcal{A}(w)}(x_1, x_2, n)$, where x_1, x_2 are the respective inputs of P_1 and P_2, w is an auxiliary input received by $\mathcal{A}$, and n is the security parameter.

In contrast, in the real model, a real protocol π is run between the parties without any trusted help. Once again, an adversary $\mathcal{A}$ controls one of the parties and can instruct it to behave arbitrarily. At the end of the execution, the honest party outputs the output specified by the protocol π and the adversary outputs whatever it wishes. The pair of outputs of the honest party and an adversary $\mathcal{A}$ in an real execution of a protocol π is denoted $\text{REAL}_{\pi,\mathcal{A}(w)}(x_1, x_2, n)$, where x_1, x_2, w and n are as above.

Finally, we present the notion of a "hybrid model" where the parties run a protocol π as well as having access to a trusted party. In this paper, we will use this to model external authorities that exist in the real world. For example, we will consider a certificate authority (for a public-key infrastructure) and a bank. In this model, the protocol π contains both standard messages that are sent between the parties as well as ideal messages that are sent between the parties and the trusted party. The pair of outputs of the honest party and an adversary $\mathcal{A}$ in a hybrid execution of a protocol π with a trusted party computing a functionality g is denoted $\text{HYBRID}^g_{\pi,\mathcal{A}(w)}(x_1, x_2, n)$, where x_1, x_2, w and n are as above.

Given the above, we can now define the security of a protocol π (we present this for the hybrid model because our protocols are in this model and by taking g to be a function with no output we have the real model as well).

Definition 2. *Let π be a probabilistic polynomial-time protocol and let f be a probabilistic polynomial-time two-party functionality. We say that π securely computes f with abort in the g-hybrid model if for every non-uniform probabilistic polynomial-time adversary $\mathcal{A}$ attacking π there exists a non-uniform probabilistic polynomial-time adversary $\mathcal{S}$ for the ideal model so that for every $x_1, x_2, w \in \{0,1\}^*$,*

$$\left\{ \text{IDEAL}_{f,\mathcal{S}(w)}(x_1, x_2, n) \right\}_{n \in \mathbb{N}} \stackrel{c}{\equiv} \left\{ \text{HYBRID}^g_{\pi,\mathcal{A}(w)}(x_1, x_2, n) \right\}_{n \in \mathbb{N}}$$

This is called "security with abort" because when fairness is not guaranteed, the adversary is allowed to abort early (after it received output but before the honest party receives output).

Immediate message receipt. Any protocol that aims to achieve any notion of fairness must deal with the question of when to declare that one of the parties has *not* sent a message. This can be dealt with by introducing time into the model and allowing only a certain delay (as would be the case in practice), or can be achieved by assuming synchronous computation (i.e., the protocol proceeds in rounds and in each round all parties send and receive messages). For the sake of simplicity, we assume the latter. Furthermore, we assume that the receipt of messages is *immediate*, meaning that after a party receives a message, its next-message is sent straight away. In particular, this means that if it does not send such a message, it will not send it later.

Secure signature schemes. A signature scheme consists of three probabilistic polynomial-time algorithms (Gen, Sign, Vrfy) such that for every n, every (vk, sk) in the range of $\mathsf{Gen}(1^n)$, and every $m \in \{0,1\}^*$, $\mathsf{Vrfy}(vk, m, \mathsf{Sign}(sk, m)) = 1$. A signature scheme is existentially secure against chosen message attacks if any non-uniform probabilistic polynomial-time adversary given access to a signing oracle can generate a forgery with at most negligible probability. (A forgery is a valid signature on *any* message that the adversary did not query to its signing oracle.) See [13] for full definitions. By convention, we assume that a signature $\sigma = \mathsf{Sign}(sk, m)$ *contains the message* m. Thus, in order to verify a signature, it suffices to have the verification key vk and σ. We also assume that the validity of a signing pair can be verified given vk and sk (this is without loss of generality; in particular, the random coins used to run Gen can be included as part of sk).

The certificate-authority functionality. We assume a public-key infrastructure for our protocols, and formalize this via the certificate-authority functionality of [6]. This functionality provides simple register and retrieve instructions and is denoted $\mathcal{F}_{CA}$; see Figure 2.

Functionality $\mathcal{F}_{CA}$

1. Upon receiving the first message (Register, sid, v) from a party P, send (Register, sid, v) to the adversary. Upon receiving back ok from the adversary, check that $sid = P$ and that this is the first request from P. If yes, then record the pair (P, v); otherwise, ignore the message.
2. Upon receiving a message (Retrieve, sid) from a party P', send (Retrieve, sid, P') to the adversary, and wait for an ok from the adversary. Then, if there is a recorded pair (sid, v) output (Retrieve, sid, v) to P'. Else output (Retrieve, $sid, \perp$) to P'.

Fig. 2. The ideal certification authority functionality $\mathcal{F}_{CA}$

3.2   A Simple Bank Functionality

Our notion of legally-enforceable fairness assumes the existence of an external authority that can force parties to carry out some action. In today's society such an authority exists in the form of a court of law (or a bank that respects digital cheques), and we assume the existence of digital signature law that can be used to enforce payment when one party holds a *cheque* that has been digitally signed by another party. We use the following notation:

> **Cheques:** A cheque for \$$\alpha$ for a party P_j from a party P_i is denoted $chq = \mathsf{cheque}(cid, P_i \rightarrow P_j, \alpha, z)$, where cid is a unique identifier and z is an auxiliary-information field (like a "notes" field on a regular paper cheque). A cheque is only valid when signed and so chq contains the information cid, $P_i \rightarrow P_j$, α and z in some standardized form, all signed with party P_i's signing key. Thus, denoting P_i's signing key-pair as (vk_i, sk_i), we have that a cheque is the signature $chq = \mathsf{Sign}_{sk_i}(cid, P_i \rightarrow P_j, \alpha, z)$. Recall that by our convention, a signature also contains the message and so the signature is all that is needed.

We assume that a digitally signed message constitutes a legally binding cheque that is respected by banks and by courts of law.

We define a functionality that represents a "bank". It is not supposed to be a full-fledged abstraction of the banking system. Rather, it is a minimal functionality that is fulfilled by the real banking system in use today. The bank that we define is such that upon receiving a cheque $\mathsf{cheque}(cid, P_i \rightarrow P_j, \alpha, z)$, the bank transfers \$$\alpha$ from P_i's account to P_j's account, and sends P_i a copy of the cheque. (This is analogous to the scan of a cheque that is sent by many banks to customers.) For the sake simplicity, we initialize the accounts of all parties to \$0 and allow the current balance of an account to be any value, positive or negative; we denote the current balance of a party P_i by $\mathsf{balance}_i$. The bank functionality is defined as follows:

Functionality $\mathcal{F}_{\mathsf{bank}}$

The functionality $\mathcal{F}_{\mathsf{bank}}$ runs with a certificate authority $\mathcal{F}_{CA}$ and parties $P_1, \ldots, P_n$. It initializes values $\mathsf{balance}_i = 0$ for all i and a set $\mathsf{used} = \phi$, and works as follows:

> Upon receiving a message $chq = \mathsf{cheque}(cid, P_i \rightarrow P_j, \alpha, z)$ from a party P_j, the functionality first checks that chq is a valid cheque from P_i to P_j (it does this using P_i's verification-key vk_i as retrieved from $\mathcal{F}_{CA}$), and that $(cid, i, j) \notin \mathsf{used}$. If not, it ignores the message. If yes, it does the following:
> 1. Set $\mathsf{balance}_i = \mathsf{balance}_i - \alpha$
> 2. Set $\mathsf{balance}_j = \mathsf{balance}_j + \alpha$
> 3. Add (cid, i, j) to the set used
> 4. Send chq to party P_i

Fig. 3. The ideal bank functionality $\mathcal{F}_{\mathsf{bank}}$

$\mathcal{F}_{\text{bank}}$ records tuples of the form (cid, i, j) in the set used in order to ensure that the same cheque is not cashed twice.

3.3   Legally-Enforceable Fairness

We are now ready to formally define what it means for a protocol to securely compute a functionality f with legally-enforceable fairness. The basic idea is that an adversary has three choices regarding the output of the protocol:

1. The adversary can abort the protocol before anyone learns anything (preserving fairness).
2. The protocol can conclude with both parties receiving output (preserving fairness).
3. The protocol can conclude with the adversary receiving output while the honest party does not. However, in case this happens, the honest party receives a cheque for \$$\alpha$ from the adversary that it can cash at the bank.

We formalize this by defining a functionality $\mathcal{F}_f^\alpha$ that incorporates the bank functionality and computes f as above; see Figure 4.

Functionality $\mathcal{F}_f^\alpha$

The functionality $\mathcal{F}_f^\alpha$ runs with parties P_1 and P_2, with initial balances balance_1 and balance_2, and an adversary $\mathcal{A}$. Let P_i denote the corrupted party and P_j the honest party $(i, j \in \{1, 2\}, i \neq j)$. Functionality $\mathcal{F}_f^\alpha$ works as follows:

1. $\mathcal{F}_f^\alpha$ receives inputs x_1 and x_2 from parties P_1 and P_2. If either of the inputs equal $\bot$ or are invalid, then $\mathcal{F}_f^\alpha$ sends $\bot$ to both P_1 and P_2 and halts.
2. If $\mathcal{F}_f^\alpha$ receives two valid inputs, it sends $y = f(x_1, x_2)$ to $\mathcal{A}$ and waits for $\mathcal{A}$'s response.
   (a) If $\mathcal{A}$ replies with fair then $\mathcal{F}_f^\alpha$ sends y to P_j
   (b) If $\mathcal{A}$ replies with unfair (or doesn't reply), then $\mathcal{F}_f^\alpha$ sets $\text{balance}_i = \text{balance}_i - \alpha$ and $\text{balance}_j = \text{balance}_j + \alpha$.

Fig. 4. The ideal functionality $\mathcal{F}_f^\alpha$ for computing f with legally-enforceable fairness

Definition 3. *Let π be a protocol and f a two-party functionality. We say that π securely computes f with α-legally-enforceable fairness if π securely computes the functionality $\mathcal{F}_f^\alpha$ according to Definition 2.*

As a sanity check, we show that any protocol that securely computes f with legally-enforceable fairness also securely computes f under the standard notion of security. In order to do this, however, we must make a small modification to the definition in the case of an unfair abort (i.e., where the adversary receives output but the honest party does not), so that the honest party does not necessarily output $\bot$ but may simply remain in a waiting state. (This is essentially equivalent, but the change is necessary on a technical level to make the models match.) Given this modification, we prove the following:

Claim 4. *Let π be a protocol that securely computes a two-party functionality f with α-legally-enforceable fairness. Then, π securely computes f with abort, as in Definition 2.*

Proof: Let $\mathcal{A}$ be a real adversary that attacks protocol π. We wish to construct a simulator $\mathcal{S}$ for the standard notion of security as in Definition 2. In order to do this, we first consider the simulator $\mathcal{S}'$ that is guaranteed to exist by the fact that π securely computes f with legally-enforceable fairness; note that $\mathcal{S}'$ is designed to work with $\mathcal{F}_f^\alpha$. We construct $\mathcal{S}$ from $\mathcal{S}'$ as follows:

1. When $\mathcal{S}'$ sends an input x to the trusted party, $\mathcal{S}$ forwards the same x to its trusted party. Then, when it receives back y, it hands it to $\mathcal{S}'$ and waits for $\mathcal{S}'$'s reply:
2. If $\mathcal{S}'$ replies with fair then $\mathcal{S}$ sends continue to the trusted party (indicating that the honest party should also receive output) and outputs whatever $\mathcal{S}'$ outputs.
3. If $\mathcal{S}'$ replies with unfair or does not reply, then $\mathcal{S}$ sends halt to the trusted party (indicating that the honest party should not receive output) and outputs whatever $\mathcal{S}'$ outputs.

It is clear that the joint output distribution of $\mathcal{S}$ and the honest party is identical to the joint output distribution of $\mathcal{S}'$ and the honest party (which in turn is indistinguishable from the output of $\mathcal{A}$ and the honest party in a real protocol execution). Therefore, the output distributions of $\mathcal{S}$ and $\mathcal{A}$ and the honest party are indistinguishable, as required. ∎

4   A Protocol with Legally-Enforceable Fairness

Before formally presenting the protocol, we informally describe how it works. The protocol has a number of phases, as follows:

1. **Registration phase:** In this phase, the parties register their digital signature keys with the certificate authority $\mathcal{F}_{CA}$.
2. **Main computation phase:** In this phase, the parties run a protocol for secure computation with the property that party P_1 receives a cheque for \$$\alpha$, while P_2 receives nothing. The important property of the cheque received by P_1 is that it itself contains an encrypted cheque for P_2 for \$$\alpha$, much like the signature-inside-a-signature in our solution for concurrent signatures in Section 2. We call this cheque for P_2 a counter-cheque because it enables P_2 to "counter" the cheque that P_1 has received. That is, if P_1 sends its cheque to the bank authority, then upon receiving a copy of this cheque from $\mathcal{F}_{\text{bank}}$, party P_2 can decrypt its own counter-cheque and return the balances to their original values. (Using this terminology, the cheque that P_1 receives contains an encrypted counter-cheque for P_2.) In addition to the above, and crucial to our solution as we will see below, the counter-cheque for P_2 contains the

function output! We stress that the counter-cheque is *encrypted* and so only P_2 can obtain the output that is inside it.

3. **Output exchange phase:** In this phase, P_1 is supposed to send its cheque to P_2, upon which P_2 decrypts it, obtains the counter-cheque and sends it back to P_1. Following this, both parties read the function output from the counter-cheque and output it.

We now informally analyze the above flow in terms of what happens if one of the parties aborts before both have received output. We present our analysis in a step-by-step fashion, showing what happens if an abort occurs at any given step:

1. *Either party aborts in the main computation phase:* In this case no one learns anything and so fairness is preserved.
2. *P_1 aborts by not sending its cheque to P_2:* As in the previous case, here no one learns anything and so fairness is preserved. (Recall that P_1's output from the main computation phase is its cheque and this does not directly contain the output. Rather, the output only appears in the counter-cheque for P_2 that is encrypted so that only P_2 can decrypt it. This means that if P_1 does not send its cheque to P_2 then no one receives output.)
3. *P_2 aborts by not sending its counter-cheque back to P_1:* In this case, P_2 has already received output (because the output is contained in its counter-cheque), while P_1 has not. Thus, if P_2 aborts at this point, without sending P_1 the counter-cheque, fairness will have been breached. This is where the cheques and legal enforcement comes in. In this situation, P_1 has a cheque from P_2 for \$$\alpha$. Thus, the only way for P_2 to avoid paying \$$\alpha$ to P_1 is for it to present its counter-cheque. But, the counter-cheque contains the output and so if P_2 presents the cheque, fairness is restored!

We conclude that either fairness is preserved, or P_1 can force P_2 to pay it \$$\alpha$. We remark that now that cheques are included as part of the model, we must also ensure that it is not possible for a dishonest party to obtain a cheque that cannot be countered by an honest party. Otherwise an adversary could inflict financial damage on an honest party (note that this property is implicit in the definition of $\mathcal{F}_f^{\alpha}$ for legally-enforceable fairness). Now, in our protocol informally described above, the cheque received by P_1 contains a counter-cheque for P_2. Therefore, if a corrupted P_1 sends this cheque to the bank, P_2 will receive a copy containing a counter-cheque that it can be used to restore the bank balances to their original values. Likewise, P_2 only receives its cheque after P_1 has received its own, and so P_1 can always counter any cheque sent by P_2 to the bank.

We are now ready to formally describe the protocol π_{α} for securely computing f with α-legally-enforceable fairness. The protocol is in the $\mathcal{F}_{CA}, \mathcal{F}_{\text{bank}}$-hybrid model, meaning that the parties have access to a trusted certificate authority and a trusted bank (or court of law that enforces digital cheques), and appears below in Figure 5.

A Protocol π_α for Securely Computing f

Inputs: P_1 has x_1 and P_2 has x_2; both parties have α and n.

The protocol:

1. **Phase 0 – registration:** Prior to any execution of the protocol:
   (a) P_1 chooses $(vk_1, sk_1) \leftarrow \mathsf{Gen}(1^n)$ and sends $(\mathsf{Register}, P_1, vk_1)$ to $\mathcal{F}_{CA}$.
   (b) P_2 chooses $(vk_2, sk_2) \leftarrow \mathsf{Gen}(1^n)$ and sends $(\mathsf{Register}, P_2, vk_2)$ to $\mathcal{F}_{CA}$.

2. **Phase 1 – main computation:** The parties use a secure two-party protocol to compute the following functionality:
   Inputs:
   (a) Party P_1 inputs its signing key-pair (vk_1, sk_1), party P_2's public key vk_2 (obtained by sending $(\mathsf{Retrieve}, P_2)$ to $\mathcal{F}_{CA}$), its input x_1, α and n. In addition, P_1 inputs a random string $cid_2 \in_R \{0,1\}^n$.
   (b) Party P_2 inputs its signing key-pair (vk_2, sk_2), party P_1's public key vk_1 (obtained by sending $(\mathsf{Retrieve}, P_1)$ to $\mathcal{F}_{CA}$), its inputs x_2, α and n. In addition, P_2 inputs a random string $cid_1 \in_R \{0,1\}^n$ and a random r of appropriate length (see below).
   Outputs: If the α and n values received from P_1 and P_2 are the same, the keys match (i.e., P_2 inputs $vk_1' = vk_1$ and vice versa), and the key-pairs that are input are valid (i.e., sk_i is associated with vk_i), then the functionality sets $cid = cid_1 \| cid_2$ and defines the outputs as follows:
   (a) Party P_1 receives the cheque $\mathsf{chq}_1 = \mathsf{cheque}(cid, P_2 \to P_1, \alpha, z)$, where $z = r \oplus \mathsf{chq}_2$, $\mathsf{chq}_2 = \mathsf{cheque}(cid, P_1 \to P_2, \alpha, y)$ and $y = f(x_1, x_2)$.
   (b) Party P_2 receives nothing.
   If the checks do not pass, both parties receive $\perp$, in which case they output $\perp$.

3. **Phase 2 – exchange outputs:**
   (a) If P_1 did not receive output from phase 1 it halts and outputs $\perp$. Otherwise, it sends chq_1 to P_2.
   (b) P_2 waits to receive a value chq_1 from P_1. Upon receiving such a chq_1, party P_2 checks that the identifier in chq_1 begins with cid_1 and that the cheque is valid with respect to vk_2. If not, P_2 ignores the message and continues waiting. Otherwise, P_2 computes $\mathsf{chq}_2 = r \oplus z$, sends chq_2 to P_1, and outputs the value y inside chq_2.
   (c) If P_1 did not receive chq_2 from P_2, or if the identifier in chq_2 does not end with cid_2, or if chq_2 is invalid with respect to vk_1, then P_2 sends chq_1 to $\mathcal{F}_{\mathsf{bank}}$ (in order to receive \$$\alpha$). Else, P_1 outputs the value y inside chq_2.

4. **Additional instructions:**
   (a) If P_1 receives a valid chq_2 with an identifier ending with cid_2 from $\mathcal{F}_{\mathsf{bank}}$ (indicating a payment made to P_2), then P_1 sends chq_1 with cid from the same execution to $\mathcal{F}_{\mathsf{bank}}$.
   (b) If P_2 receives a valid cheque chq_1 with an identifier beginning with cid_1 from $\mathcal{F}_{\mathsf{bank}}$ (indicating a payment made to P_1), then P_2 works as follows:
      i. If P_2 received already chq_1 from P_1, then it sends chq_2 with cid to $\mathcal{F}_{\mathsf{bank}}$.
      ii. If P_2 did not receive chq_1 from P_1, it takes chq_1 as the value it is waiting for in Step 3b above and proceeds according to those instructions. In addition it sends chq_2 to $\mathcal{F}_{\mathsf{bank}}$, as derived in Step 3b.

Fig. 5. A protocol for computing f with α-legally-enforceable fairness

Theorem 5. *Assume that the protocol used in phase 1 is secure by Definition 2 and that the signature scheme is existentially unforgeable under chosen-message attacks. Then, Protocol π_α securely computes f with α-legally-enforceable fairness in the $\mathcal{F}_{CA}, \mathcal{F}_{\text{bank}}$-hybrid model.*

Proof: The proof is in the hybrid model, where we assume that $\mathcal{F}_{CA}$, $\mathcal{F}_{\text{bank}}$ and the computation of phase 1 are carried out with the help of a trusted third party (see [5] and [13]). We separately analyze the case that P_1 is corrupted and the case that P_2 is corrupted. Recall that what we need to show is that π_α securely computes the functionality $\mathcal{F}_f^\alpha$.

P_1 is corrupted. Let $\mathcal{A}$ be a (hybrid-model) adversary controlling party P_1. We construct a simulator $\mathcal{S}$ as follows:

1. $\mathcal{S}$ invokes $\mathcal{A}$ upon its input x_1 and auxiliary-input α and n. In addition, $\mathcal{S}$ chooses a key-pair $(vk_2, sk_2) \leftarrow \text{Gen}(1^n)$ for P_2.
2. When $\mathcal{A}$ sends a $(\text{Retrieve}, P_2)$ message intended for $\mathcal{F}_{CA}$, simulator $\mathcal{S}$ replies with $(\text{Retrieve}, P_2, vk_2)$.
3. When $\mathcal{A}$ sends a $(\text{Register}, P_1, vk_1)$ message intended for $\mathcal{F}_{CA}$, simulator $\mathcal{S}$ records vk_1.
4. $\mathcal{S}$ obtains $\mathcal{A}$'s inputs $((vk_1, sk_1), vk_2, x_1, \alpha, 1^n, cid_2)$ for the trusted party computing the functionality in phase 1. If the key vk_2 is not the same key that $\mathcal{S}$ chose for P_2, the key vk_1 in the key-pair is not the same key registered by $\mathcal{A}$, the key-pair (vk_1, sk_1) is not valid, α or n are not the same as above or x_1 is not valid for f (e.g., it is of the wrong length), then $\mathcal{S}$ sends input $\perp$ to the trusted party computing $\mathcal{F}_f^\alpha$ (as P_1's input), hands $\perp$ to $\mathcal{A}$ as its output from phase 1, outputs whatever $\mathcal{A}$ outputs and halts. Otherwise, $\mathcal{S}$ proceeds to the next step.
5. $\mathcal{S}$ chooses $cid_1 \in_R \{0,1\}^n$ and a random string r of appropriate length as in the protocol (this length is known), and sets $cid = cid_1 \| cid_2$. Then, $\mathcal{S}$ computes the cheque $\text{chq}_1 = \text{cheque}(cid, P_2 \rightarrow P_1, \alpha, r)$ (using sk_2 that it chose above) and hands chq_1 to $\mathcal{A}$ as its output from phase 1.
6. $\mathcal{S}$ receives $\mathcal{A}$'s next messages:
   (a) If $\mathcal{A}$ sends some chq_1' intended for P_2, then $\mathcal{S}$ checks that it is a valid cheque with identifier cid. If not, it ignores the message. Otherwise, it works as follows:
      i. If the value z' contained inside does not equal r as chosen by $\mathcal{S}$, then $\mathcal{S}$ outputs fail and halts.
      ii. Otherwise, $\mathcal{S}$ sends x_1 (obtained above from $\mathcal{A}$) to the trusted party computing $\mathcal{F}_f^\alpha$, receives back an output value y, and sends back fair. Then, $\mathcal{S}$ computes $\text{chq}_2 = \text{cheque}(cid, P_1 \rightarrow P_2, \alpha, y)$, using sk_1 obtained from $\mathcal{A}$ above and hands it to $\mathcal{A}$.
   (b) If $\mathcal{A}$ sends some chq_1'' intended for $\mathcal{F}_{\text{bank}}$, then $\mathcal{S}$ checks that it is a valid cheque with identifier cid. If not, it ignores the message. Otherwise, it works as follows:
      i. If the value z'' contained inside chq_1'' does not equal r as chosen by $\mathcal{S}$, then $\mathcal{S}$ outputs fail and halts.

ii. Otherwise:
  A. If $\mathcal{S}$ already received y above, then it simulates $\mathcal{F}_{\text{bank}}$ sending P_1 the cheque chq_2 (as would occur after P_2 sends chq_2 to $\mathcal{F}_{\text{bank}}$).
  B. If not, $\mathcal{S}$ sends x_1 to the trusted party computing $\mathcal{F}_f^\alpha$, receives back the output y, and sends back fair. Then, $\mathcal{S}$ computes $\mathsf{chq}_2 = \mathsf{cheque}(cid, P_1 \rightarrow P_2, \alpha, y)$, using sk_1 obtained from $\mathcal{A}$ above and hands it to $\mathcal{A}$. Finally, $\mathcal{S}$ simulates $\mathcal{F}_{\text{bank}}$ sending P_1 the cheque chq_2 (as would occur after P_2 sends chq_2 to $\mathcal{F}_{\text{bank}}$).

7. If $\mathcal{S}$ did not send x_1 in Step 6 above, then it sends $\perp$ as P_1's input to the trusted party computing $\mathcal{F}_f^\alpha$.
8. $\mathcal{S}$ outputs whatever $\mathcal{A}$ outputs and halts.

This completes the simulation. First, observe that the view of $\mathcal{A}$ in such a simulation with $\mathcal{S}$ is identical to its view in a hybrid execution of Protocol π_α with an honest P_2 (and where a trusted party runs $\mathcal{F}_{CA}$ and phase 1). Furthermore, conditioned on $\mathcal{S}$ not outputting fail, the joint distribution of $\mathcal{A}$ and P_2's output in a hybrid execution is identical to that of $\mathcal{S}$ and P_2's output in an ideal execution. This is due to the fact that if $\mathcal{S}$ does not output fail, then the value output by P_2 in a hybrid execution is either $\perp$ or y, just as in an ideal execution. Furthermore, $\mathcal{S}$ works so that whenever P_2 would receive $\perp$ (resp., y) in a hybrid execution, it receives $\perp$ (resp., y) in an ideal execution (by sending $\perp$ and fair to the trusted party, appropriately). In addition, if $\mathcal{A}$ sends a valid chq_1 to $\mathcal{F}_{\text{bank}}$, then by the instructions of π_α party P_2 would send the appropriate chq_2 to $\mathcal{F}_{\text{bank}}$. This ensures that the account balances of P_1 and P_2 are unchanged, as is the case in the simulation by $\mathcal{S}$. It thus remains to show that $\mathcal{S}$ outputs fail with at most negligible probability. This is proven by a straightforward reduction to the security of the signature scheme with respect to the key vk_2. In order to see that such a reduction is possible, note that the simulation by $\mathcal{S}$ can be run in an identical way with access to a signing oracle that computes signatures with sk_2 (as provided to an adversary in the signature-forging experiment). Furthermore, if $\mathcal{S}$ outputs fail then $\mathcal{A}$ must have generated a signature that was not provided by the oracle to $\mathcal{S}$, meaning that $\mathcal{A}$ successfully forged a signature (contradicting the security of the signature scheme).

We remark that in this corruption case, the unfair command is never used (this is because P_2 receives output first and so it is never the case that $\mathcal{A}$ receives output while P_2 does not).

P_2 is corrupted. Let $\mathcal{A}$ be a (hybrid-model) adversary controlling party P_2. We construct a simulator $\mathcal{S}$ as follows:

1. $\mathcal{S}$ invokes $\mathcal{A}$ upon its input x_2 and auxiliary-input α and n. In addition, $\mathcal{S}$ chooses keys (vk_1, sk_1) for P_1.
2. When $\mathcal{A}$ sends a $(\mathsf{Retrieve}, P_1)$ message intended for $\mathcal{F}_{CA}$, simulator $\mathcal{S}$ replies with $(\mathsf{Retrieve}, P_1, vk_1)$.
3. When $\mathcal{A}$ sends a $(\mathsf{Register}, P_2, vk_2)$ message intended for $\mathcal{F}_{CA}$, simulator $\mathcal{S}$ records vk_2.

4. $\mathcal{S}$ obtains $\mathcal{A}$'s inputs $((vk_2, sk_2), vk_1, x_2, \alpha, 1^n, cid_1, r)$ for the trusted party computing the functionality in phase 1. If the key vk_1 is not the same key that $\mathcal{S}$ chose for P_1, the key vk_2 in the key-pair is not the same as registered by $\mathcal{A}$, the key-pair (vk_2, sk_2) is not valid, α or n are not the same as above, or x_2 is not valid for f, then $\mathcal{S}$ sends the input value $\perp$ to the trusted party computing $\mathcal{F}_f^\alpha$ (as P_2's input), hands $\perp$ to $\mathcal{A}$ as its output from phase 1, outputs whatever $\mathcal{A}$ outputs and halts.

   Otherwise, $\mathcal{S}$ sends x_2 to the trusted party and receives back the output y. $\mathcal{S}$ continues with the simulation as follows:

5. $\mathcal{S}$ sets the identifier $cid = cid_1 \| cid_2$ and computes $z = r \oplus \mathsf{chq}_2$ where $\mathsf{chq}_2 = \mathsf{cheque}(cid, P_1 {\to} P_2, \alpha, y)$. Then, $\mathcal{S}$ computes $\mathsf{chq}_1 = \mathsf{cheque}(cid, P_2 {\to} P_1, \alpha, z)$ and hands chq_1 to $\mathcal{A}$ as the message it receives from P_1 in phase 2.

6. $\mathcal{S}$ receives $\mathcal{A}$'s next messages:

   (a) If $\mathcal{A}$ sends some cheque chq_2' intended for P_1, then $\mathcal{S}$ checks that it is a valid cheque with identifier cid. If not, it ignores the message. Otherwise, it works as follows:

      i. If the value z' contained inside does not equal y, then $\mathcal{S}$ outputs fail and halts.

      ii. If z' does equal y, then $\mathcal{S}$ sends fair to the trusted party computing $\mathcal{F}_f^\alpha$.

   (b) If $\mathcal{A}$ sends some cheque chq_2'' intended for $\mathcal{F}_{\mathrm{bank}}$, then $\mathcal{S}$ checks that it is a valid cheque with identifier cid. If not, it ignores the message. Otherwise, it works as follows:

      i. If the value z'' contained inside does not equal y, then $\mathcal{S}$ outputs fail and halts.

      ii. Otherwise:

         A. If $\mathcal{S}$ already sent fair above, then it simulates $\mathcal{F}_{\mathrm{bank}}$ sending P_2 the cheque chq_1 (as would occur after P_1 sends chq_1 to $\mathcal{F}_{\mathrm{bank}}$).

         B. If not, $\mathcal{S}$ sends fair to the trusted party computing $\mathcal{F}_f^\alpha$ and simulates $\mathcal{F}_{\mathrm{bank}}$ sending P_2 the cheque chq_1.

7. If $\mathcal{S}$ does not send fair in Step 6 above, then it sends unfair to the trusted party computing $\mathcal{F}_f^\alpha$.

8. $\mathcal{S}$ outputs whatever $\mathcal{A}$ outputs and halts.

The analysis for this case is almost identical to the previous one. Namely, there can only be a difference between an execution of π_α with $\mathcal{A}$ and the simulation with $\mathcal{S}$ in the ideal model if $\mathcal{A}$ succeeds in forging a signature. This completes the proof.   ∎

Sequential composition. In general, secure protocols are guaranteed to remain secure when run sequentially many times [5]. However, this is *not* necessarily the case when some joint state is kept between executions [17]. In Protocol π_α, the parties' signing and encryption keys are used in many executions and so sequential composition is not automatically guaranteed. This can be solved in two ways. One solution is to use different keys in each execution. However, since we prefer to assume a standard public-key infrastructure, the user's keys must be fixed throughout. A second solution is to ensure that the signatures in each execution are valid only in that execution by including a unique identifier inside each signature, and having the party who verifies the signature check that the identifier is correct. This solution was used in [7] and shown to achieve the necessary level of security. Observe that in Protocol π_α, the parties include random cheque identifiers cid_1 and cid_2 in order to achieve this exact effect. (Note that if these identifiers were not included, then for example P_1 can send a chq_1 from a previous execution thereby causing the output of P_2 to be that of a previous execution and not the current one.) In order to prove security under sequential composition, the only difference is that S may output fail even when A provides a signature that was provided by the signing oracle. Specifically, this can happen if a cid_i appears twice in two different executions (in such a case, A can provide the cheque from the previous execution, where the output may be different to y). However, since $cid_i \in_R \{0,1\}^n$ and there are only a polynomial number of executions, this can happen with at most negligible probability. Thus, Protocol π_α remains secure also under sequential composition.

Acknowledgements

We would like to thank Jonathan Katz for helpful discussions about how to formalize the notion of legally-enforceable fairness, and for comments on the write-up.

References

1. Asokan, N., Schunter, M., Waidner, M.: Optimistic Protocols for Fair Exchange. In: 4th CCS, pp. 8–17 (1997)
2. Beaver, D., Goldwasser, S.: Multiparty Computation with Faulty Majority. In: 30th FOCS, pp. 468–473 (1989)
3. Ben-Or, M., Goldwasser, S., Wigderson, A.: Completeness Theorems for Non-Cryptographic Fault-Tolerant Distributed Computation. In: 20th STOC, pp. 1–10 (1988)
4. Cachin, C., Camenisch, J.: Optimistic Fair Secure Computation. In: Bellare, M. (ed.) CRYPTO 2000. LNCS, vol. 1880, pp. 93–111. Springer, Heidelberg (2000)
5. Canetti, R.: Security and Composition of Multiparty Cryptographic Protocols. Journal of Cryptology 13(1), 143–202 (2000)
6. Canetti, R.: Universally Composable Signature, Certification, and Authentication. In: 17th IEEE Computer Security Foundations Workshop (CSFW), pp. 219–235 (2004)
7. Canetti, R., Rabin, T.: Universal Composition with Joint State. In: Boneh, D. (ed.) CRYPTO 2003. LNCS, vol. 2729, pp. 265–281. Springer, Heidelberg (2003)
8. Chaum, D., Crépeau, C., Damgard, I.: Multi-party Unconditionally Secure Protocols. In: 20th STOC, pp. 11–19 (1988)
9. Chen, L., Kudla, C., Paterson, K.: Concurrent Signatures. In: Cachin, C., Camenisch, J.L. (eds.) EUROCRYPT 2004. LNCS, vol. 3027, pp. 287–305. Springer, Heidelberg (2004)

10. Cleve, R.: Limits on the Security of Coin Flips when Half the Processors are Faulty. In: 18th STOC, pp. 364–369 (1986)
11. Garay, J., MacKenzie, P., Prabhakaran, M., Yang, K.: Resource Fairness and Composability of Cryptographic Protocols. In: Halevi, S., Rabin, T. (eds.) TCC 2006. LNCS, vol. 3876, pp. 404–428. Springer, Heidelberg (2006)
12. Gordon, S.D., Hazay, C., Katz, J., Lindell, Y.: Complete Fairness in Secure Two-Party Computation (manuscript, 2007)
13. Goldreich, O.: Foundations of Cryptography. Basic Applications, vol. 2. Cambridge University Press, Cambridge (2004)
14. Goldreich, O., Micali, S., Wigderson, A.: How to Play any Mental Game – A Completeness Theorem for Protocols with Honest Majority. In: 19th STOC, pp. 218–229 (1987)
15. Goldwasser, S., Levin, L.: Fair Computation of General Functions in Presence of Immoral Majority. In: Menezes, A., Vanstone, S.A. (eds.) CRYPTO 1990. LNCS, vol. 537, pp. 77–93. Springer, Heidelberg (1991)
16. Goldwasser, S., Lindell, Y.: Secure Computation Without Agreement. Journal of Cryptology 18(3), 247–287 (2005)
17. Lindell, Y., Lysysanskaya, A., Rabin, T.: On the Composition of Authenticated Byzantine Agreement. Journal of the ACM 53(6), 881–917 (2006)
18. Micali, S.: Secure Protocols with Invisible Trusted Parties. Presentation on Multi-Party Secure Protocols, Weizmann Institute of Science, Israel (June 1998)
19. Micali, S.: Simple and Fast Optimistic Protocols for Fair Electronic Exchange. In: 22nd PODC, pp. 12–19 (2003)
20. Pinkas, B.: Fair Secure Two-Party Computation. In: Biham, E. (ed.) EUROCRYPT 2003. LNCS, vol. 2656, pp. 87–105. Springer, Heidelberg (2003)
21. Yao, A.: How to Generate and Exchange Secrets. In: 27th FOCS, pp. 162–167 (1986)

Security of **NMAC** and **HMAC**
Based on Non-malleability

Marc Fischlin[*]

Darmstadt University of Technology, Germany
marc.fischlin@gmail.com
www.fischlin.de

Abstract. We give an alternative security proof for **NMAC** and **HMAC** when deployed as a message authentication code, supplementing the previous result by Bellare (Crypto 2006). We show that (black-box) non-malleability and unpredictability of the compression function suffice in this case, yielding security under different assumptions. This also suggests that some sort of non-malleability is a desirable design goal for hash functions.

1 Introduction

HMAC is one of the most widely deployed cryptographic algorithms today. Proposed by Bellare et al. [3] it is nowadays standardized in several places like ANSI X9.71 and incorporated into SSL, SSH and IPSec. It is used as a universal tool to derive keys, to provide a pseudorandom function or simply to authenticate messages. Roughly, for keys k_{in}, k_{out} algorithm HMAC, and its generalized version NMAC, are defined as

$$\text{HMAC}_{(k_{in}, k_{out})}(M) := H(\text{IV}, k_{out} || H(\text{IV}, k_{in} || M))$$

$$\text{NMAC}_{(k_{in}, k_{out})}(M) := H(k_{out}, H(k_{in}, M))$$

where H is an iterated hash function like MD5 or SHA1, based on some compression function h.

In the original paper of Bellare et al. [3] algorithms HMAC and NMAC have been shown to be a pseudorandom function assuming that the compression function h is pseudorandom *and* collision-resistant. With the emerging attacks on the collision-resistance on popular hash functions like MD5 and SHA1 [11,12] the trustworthiness of HMAC and NMAC was slightly tarnished, but subsequently Bellare [5] proved both algorithms to be pseudorandom under the sole assumption that the compression function is pseudorandom. This result is complemented by several works [8,9,10] showing that weaknesses in collision-resistance can be actually exploited to successfully attack HMAC and NMAC.

---

[*] This work was supported by the Emmy Noether Program Fi 940/2-1 of the German Research Foundation (DFG).

T. Malkin (Ed.): CT-RSA 2008, LNCS 4964, pp. 138–154, 2008.
© Springer-Verlag Berlin Heidelberg 2008

Our results. Here, we present alternative assumptions about the compression function to yield a security proof for HMAC and NMAC *when used as a message authentication code* (MAC), instead of being deployed as a pseudorandom function. We require two orthogonal properties of the compression function which are both implied simultaneously if, for instance, the compression function h is pseudorandom:

- non-malleability: learning images of the iterated compression function (with an unknown key involved) does not lend any additional power to create another hash value under this key.[1]
- unpredictability: it is infeasible to predict the output of the iterated compression function (with an unknown key involved).

Intuitively, unpredictability says that it is impossibile to guess images from scratch (i.e., with no other images available). This is a very weak form of a MAC but does not guarantee the common notion of security under adaptive chosen-message attacks. Adding non-malleability then provides this stronger security notion, as seeing other images does not facilitate the task.

We also show that, if there are pseudorandom functions at all, then there are compression functions which obey these two properties but are not pseudorandom. Hence, our result shows security under weaker prerequisites on the compression function, strengthening the confidence in the security of NMAC and HMAC when deployed as a MAC. Moreover, the result here indicates that non-malleability (or at least some relaxation thereof) is an eligible property for hash functions and their designs.

Related results. We stress that Bellare [5], although using a stronger assumption, also derives a stronger statement, namely, that the pseudorandomness of the compression function carries over. Since HMAC is used for distinct purposes such as key derivation or as a pseudorandom function, this security claim is required for such cases. Our result merely supplements Bellare's more general result and shows that security of MACs is somewhat easier to achieve than pseudorandomness (cf. [1]).

In [5] Bellare also considers weaker requirements for HMAC and NMAC used as a MAC. He introduces the notion of privacy-preserving MACs and shows that this condition suffices to guarantee security of the MAC, together with the fact that the compression function is computationally almost universal which, in turn, follows from the pseudorandomness of the compression function. Although resembling each other, privacy-preserving MACs and our notion of non-malleability are in general incomparable (as we discuss in Section 5.2). Our result can therefore be seen as an alternative security claim based on different assumptions. At the same time, for some specific cases, our notion of non-malleability implies privacy-preservation and thus also helps to characterize this property.

---

[1] Here, depending on the padding of the hash function, we may require a special form of "black-box" non-malleability, implemented via so-called simulatable images.

2   Preliminaries

We start by recalling HMAC and NMAC and then define our two properties, non-malleability and unpredictability, before formalizing security of message authentication codes.

2.1   HMAC and NMAC

Algorithms HMAC and NMAC are built from iterated hash functions based on a compression function h, mapping $\{0,1\}^n \times \{0,1\}^b$ to $\{0,1\}^n$. For such a compression function let the iteration of the compression function $h^*(k, M)$ for input $k \in \{0,1\}^n$ and $M = M[1] \ldots M[n]$, consisting of b-bit blocks $M[i]$, be given by the value z_n, where $z_0 = k$ and $z_{i+1} = h(z_i, M[i+1])$ for $i = 0, 1, \ldots, n-1$. For notational convenience we write $B = \{0,1\}^b$ and $B^{\leq N} = \cup_{i \leq N} B^i$ and $B^+ = \cup_{i \in \mathbb{N}} B^i$ such that $h^* : \{0,1\}^n \times B^+ \to \{0,1\}^n$.

Given the iterated compression function we can define the hash function H as follows. Let $H(k, M) = h^*(k, \mathsf{pad}(M))$ where $\mathsf{pad}(M)$ stands for the message padded to a multiple of b bits. Here $\mathsf{pad}(\cdot)$ is an arbitrary one-to-one function, and we write $\mathrm{Time}_L(\mathsf{pad})$ for the time to compute the padding function for any input of length at most L, and $\mathrm{Extend}_L(\mathsf{pad})$ for the maximal number of bits the padding function adds to each string of at most L bits. We furthermore assume that the padding length only depends on the input length. For example, the standard padding appends a 1-bit to M and then adds the smallest number of 0-bits to obtain a multiple of b bits, such that $\mathrm{Time}_L(\mathsf{pad}) = O(L + b)$ and $\mathrm{Extend}_L(\mathsf{pad}) = b + 1$.

We presume that the hash function's description also contains a fixed, public value IV and we set $H(M) = H(\mathrm{IV}, M)$. Define algorithms HMAC and NMAC now as:

$$\mathsf{HMAC}_{(k_{\mathrm{in}}, k_{\mathrm{out}})}(M) := H(k_{\mathrm{out}} \| H(k_{\mathrm{in}} \| M))$$
$$\mathsf{NMAC}_{(k_{\mathrm{in}}, k_{\mathrm{out}})}(M) := H(k_{\mathrm{out}}, H(k_{\mathrm{in}}, M))$$

where keys $k_{\mathrm{in}}, k_{\mathrm{out}}$ for NMAC consist of n-bits each and are used instead of IV, and keys $k_{\mathrm{in}}, k_{\mathrm{out}} \in \{0,1\}^b$ for HMAC are prepended to the strings.

In practice, HMAC is typically used with dependent keys $k_{\mathrm{in}} = k \oplus \mathsf{ipad}$ and $k_{\mathrm{out}} = k \oplus \mathsf{opad}$ for fixed constants $\mathsf{ipad} = \mathtt{0x3636}\ldots\mathtt{36}$ and $\mathsf{opad} = \mathtt{0x5c5c}\ldots\mathtt{5c}$ and a key k of at most b bits. In either case, one can view HMAC as a special case of NMAC with $k_{\mathrm{in}}^{\mathsf{NMAC}} = h(\mathrm{IV}, k_{\mathrm{in}}^{\mathsf{HMAC}})$ and $k_{\mathrm{out}}^{\mathsf{NMAC}} = h(\mathrm{IV}, k_{\mathrm{out}}^{\mathsf{HMAC}})$.

2.2   Non-malleability and Simulatability

Non-malleability of a cryptographic function refers to the (in)ability to construct an image which is related to previously seen images. This is formalized by considering an experiment in which an adversary $\mathcal{A}$ can first ask to see hash values $y_i = H(x_i)$ of pre-images x_i distributed according to some distribution $\mathcal{X}$. Then $\mathcal{A}$ tries to find a hash value y^* of a related pre-image x^*, where related pre-images are specified through a relation R. The success probability of $\mathcal{A}$ should not be significantly larger than in an experiment of a simulator $\mathcal{S}$ which does not get to learn the images y_i but should still be able to find a related value y^*.

Non-malleability for hash functions has been defined in [2] and we follow their approach but state the property in terms of concrete security. The authors of [2] point out that the most general notion for hash functions and arbitrary distributions and relations is not achievable. Fortunately, here we deal with the easier problem of considering only very special distributions and specific relations.

Below we formalize non-malleability for the *compression function* instead of the hash function to make a claim about the security propagation. In the adversary's experiment we let $\mathcal{A}$ "bias" the distribution of the pre-images x_i via a parameter $p_i \in B^+$ passed to the (stateful) distribution $\mathcal{X}(k, \cdot)$, using a random seed k. Formally, oracle GenSample takes the parameter p_i as input, computes $x_i \leftarrow \mathcal{X}(k, p_i)$ and the image $y_i = h(x_i)$. After having seen some sample images adversary $\mathcal{A}$ outputs an image y^* and a transformation T which maps $x_1, x_2, \ldots$ to x^*. That is, the adversary does not need to know the pre-image x^* when creating y^*, but must only commit to the transformation which determines x^* once $x_1, x_2, \ldots$ become known (in fact, T produces x^* from $k, p_1, p_2, \ldots$ from which $x_1, x_2, \ldots$ can be derived). The simulator $\mathcal{S}$ only gets a restricted oracle GenSample$_0$ which samples the x_i's but does not return the image y_i to $\mathcal{S}$. Still, the simulator should create a valid pair (T, y^*).

Definition 1 (Non-Malleable Compression Function). *A compression function $h : \{0,1\}^n \times \{0,1\}^b \to \{0,1\}^n$ is called $(t_{\mathcal{A}}, t_{\mathcal{S}}, Q, N, \mu)$-non-malleable with respect to distribution $\mathcal{X}$ and relation R if for any algorithm $\mathcal{A}$ with running time $t_{\mathcal{A}}$ there exists a simulator $\mathcal{S}$ with running time $t_{\mathcal{S}}$, such that*

$$\mathrm{Prob}\Big[\boldsymbol{Exp}^{nm\text{-}adv}_{h,\mathcal{A}} = 1 \Big] \leq \mathrm{Prob}\Big[\boldsymbol{Exp}^{nm\text{-}sim}_{h,\mathcal{S}} = 1 \Big] + \mu$$

where

Experiment $\boldsymbol{Exp}^{nm\text{-}adv}_{h,\mathcal{A}}$	**Experiment $\boldsymbol{Exp}^{nm\text{-}sim}_{h,\mathcal{S}}$**
$k \leftarrow \{0,1\}^n$	$k \leftarrow \{0,1\}^n$
$(T, y^*) \leftarrow \mathcal{A}^{\mathsf{GenSample}(k,\cdot)}$	$(T, y^*) \leftarrow \mathcal{S}^{\mathsf{GenSample}_0(k,\cdot)}()$
where GenSample(k, p_i) *computes*	*where* GenSample$_0(k, p_i)$ *picks*
$x_i \leftarrow \mathcal{X}(k, p_i)$	$x_i \leftarrow \mathcal{X}(k, p_i)$
$y_i = h(x_i)$	
and returns y_i	
$x^* \leftarrow T(k, p_1, p_2, \ldots)$	$x^* \leftarrow T(k, p_1, p_2, \ldots)$
Return 1 iff	*Return 1 iff*
$R(T, k, p_1, p_2, \ldots, x^*)$	$R(T, k, p_1, p_2, \ldots, x^*)$
$\wedge (x^*, y^*) \notin \{(x_1, y_1), (x_2, y_2), \ldots\}$	
$\wedge h(x^*) = y^*$	$\wedge h(x^*) = y^*$

Here $\mathcal{A}$ and $\mathcal{S}$ each make at most Q queries to their oracle, each query having at most N blocks.

We consider here a special distribution $\mathcal{X}_{\mathsf{NMAC}}(k, \cdot)$ which, on input $p_i = M_i[1] \ldots M_i[j+1]$ first computes the j-th iteration of the compression function

$z_i = h^*(k, M_i[1] \ldots M_i[j])$ for random key k and then returns the pre-image $x_i = (z_i, M_i[j + 1])$ (from which $y_i = h(x_i)$ is then derived). The relation R_{NMAC} for input $(T, k, p_1, p_2, \ldots, x^*)$ merely checks that the transformation T computes the output $(h^*(k, M[1] \ldots M[j]), M[j+1])$ for some constants $M[1], \ldots, M[j+1] \in B$ hardwired into T, and such that $M[1] \ldots M[j + 1]$ has at most N blocks, and no p_i is a prefix of $M[1] \ldots M[j + 1]$. The prefix-check is necessary to prevent standard extension attacks.

In a refinement of the non-malleability notion we consider π-*simulatable* compression functions which allow to simulate images (given only a fraction $\pi(p_i)$ of the parameter) and which can potentially be used to construct a *black-box* non-malleability simulator. Simulatability essentially says that images can be created without the (complete) pre-image, and thus immediately suggests a strategy for constructing the non-malleability simulator. Below we formalize the notion of simulatability by demanding that no efficient distinguisher can tell apart whether it is communicating with the GenSample oracle or the oracle SimAnswer simulating images:

Definition 2 (π-Simulatability). *A compression function $h : \{0,1\}^n \times \{0,1\}^b \to \{0,1\}^n$ is called $(t_D, t_{\text{Sim}}, Q, N, \sigma)$-$\pi$-simulatable for distribution $\mathcal{X}$ if there is an algorithm SimAnswer running in time t_{Sim} such that for any algorithm $\mathcal{D}$ running in time t_D and making at most Q queries, each of at most N blocks,*

$$\text{Prob}\left[\mathcal{D}^{\text{GenSample}(k,\cdot)} = 1\right] \leq \text{Prob}\left[\mathcal{D}^{\text{SimAnswer}(\pi(\cdot))} = 1\right] + \sigma$$

where oracle GenSample is defined as in Definition 1, and where we assume that $\mathcal{D}$ never queries its oracles about the same value twice. The probabilites are taken over $\mathcal{D}$'s random choices, and $k \leftarrow \{0,1\}^n$ in the first case and the randomness of SimAnswer in the second case.

Given algorithm SimAnswer one can construct a black-box non-malleability simulator as follows. Consider another algorithm Interface which basically provides the interface between the simulator's oracle GenSample_0 and the queries made by the simulated adversary $\mathcal{A}$. Then the non-malleability simulator is of the form $\mathcal{S} = \mathcal{A}^{\text{Interface}(\cdot)}$, where Interface on input p_i forwards this value to oracle GenSample_0 of $\mathcal{S}$ and then computes $y_i \leftarrow \text{SimAnswer}(\pi(p_i))$ and returns it to $\mathcal{A}$. This simulator basically inherits the properties of SimAnswer, namely, runs in time $t_\mathcal{S} = t_\mathcal{A} + Q \cdot (\text{Time}(\text{SimAnswer}) + \text{Time}(\pi))$, makes at most Q queries and is σ-close. This is under one condition: it must be possible to map the difference in the output behavior (T, y^*) of $\mathcal{A}$ when communicating with oracle GenSample or with oracle SimAnswer to a distinguisher with binary output. This will indeed be the case for our application.

2.3   Unpredictability

A trivial example of a (π-simulatable) non-malleable compression function is a constant function: any information available through images is known beforehand and therefore redundant. Of course, such examples do not yield a good

MAC, and in order to avoid such contrived cases we introduce the mild assumption of unpredictability. Basically, a compression function is unpredictable if one cannot determine a parameter $p \in B^*$ (specifying a distribution as in the non-malleability definition), a message block $m \in B$ and its image $z = h(h^*(k,p), m)$, *before* the key k is chosen:

Definition 3 (Unpredictablility). *A compression function* $h : \{0,1\}^n \times \{0,1\}^b \to \{0,1\}^n$ *is* (t, N, ρ)*-unpredictable if for any algorithm P running in time t, the probability that for $(p, m, z) \leftarrow P()$ and $k \leftarrow \{0,1\}^n$ we have $h(h^*(k,p), m) = z$ and $p\|m \in B^{\leq N}$, is at most ρ.*

In the sequel we often view p and m as one message $M = p\|m$ such that the definition says one cannot predict $h^*(k, M) = h(h^*(k,p), m)$. A trivial example of a $(t, N, 2^{-n})$-unpredictable compression function is the identity function (on the key part), $h(k, m) = k$. Another example are pseudorandom compression functions, as we prove formally in Section 5.

The examples of a constant compression function $h(k, m) = 0^n$ and (a modification of) the "identity-on-key-part" function $h(k, m) = k$ also separate the notions of non-malleability and unpredictability. The former function is clearly π-simulatable (for any π) and non-malleable —the simulator can easily simulate the oracle's answers— but not unpredictable. In contrast, a slight modification of the latter function, namely $h(k, m) = k \oplus \text{lsb}_n(m)$ for the n least significant bits $\text{lsb}_n(m)$ of $m \in \{0,1\}^b$, is unpredictable, yet malleable for $\mathcal{X}_{\text{NMAC}}$ and R_{NMAC}. Malleability follows as an adversary is able to recover the key k from a single query about message $m = 0^b$ and can then output the image $k \oplus 1^n$ for message $m^* = 1^b$ (and the corresponding transformation). A simulator, on the other hand, needs to output the image $k \oplus m$ for constant m in clear without having any information about the key k.

We claim that $h(k, m) = k \oplus \text{lsb}_n(m)$ is not π-simulatable for any π with output length strictly less than n. A distinguisher $\mathcal{D}$ merely forwards distinct random values $m_0, m_1 \in \{0,1\}^b$ and checks that the replies y_0, y_1 satisfy $y_0 \oplus y_1 \oplus m_0 \oplus m_1 = 0^n$. For the function h this will be the case with probability 1. For SimAnswer this probability cannot be more than $\frac{1}{2}$, since SimAnswer lacks at least one bit of information about $\text{lsb}_n(m_0 \oplus m_1)$ from $\pi(m_0), \pi(m_1)$.

2.4   Message Authentication Codes

Our goal is to show that non-malleability of the compression function (plus unpredictability) gives a secure message authentication code. Formally, a message authentication code $\mathcal{M} = (\text{KeyGen}, \text{MAC}, \text{Vf})$ consists of three (probabilistic) algorithms, the key generation algorithm returning a key $K \leftarrow \text{KeyGen}()$, the MAC algorithm computing a message authentication code $\tau \leftarrow \text{MAC}(K, M)$ for a message M, and a verification algorithm deciding upon acceptance $a \leftarrow \text{Vf}(K, M, \tau)$ for a message M and a putative MAC τ. MACs generated by the key-holder should always be accepted, i.e., for any $K \leftarrow \text{KeyGen}()$, any message M and any $\tau \leftarrow \text{MAC}(K, M)$ we always have $\text{Vf}(K, M, \tau) = 1$.

Definition 4. *A MAC* $\mathcal{M} = (\mathsf{KeyGen}, \mathsf{MAC}, \mathsf{Vf})$ *is called* (t, Q, L, ϵ)*-unforgeable if for any algorithm* $\mathcal{B}$ *running in time* t*, the probability that for* $K \leftarrow \mathsf{KeyGen}()$*,* $(M^*, \tau^*) \leftarrow \mathcal{B}^{\mathsf{MAC}(K, \cdot)}()$ *making at most* Q *queries* $M_1, M_2, \ldots$ *to* $\mathsf{MAC}(K, \cdot)$*, each query and* M^* *of at most* L *bits, we have* $\mathsf{Vf}(K, M^*, \tau^*) = 1$ *and* $M^* \notin \{M_1, M_2, \ldots\}$*, is at most* ϵ*.*

In the definition above we let the adversary make only a single verification query. Bellare et al. [6] have shown that for HMAC and NMAC this implies security against adversaries which make v arbitrarily interleaved verification queries for fresh messages not queried previously. This comes with a loss of at most a factor v in security (and some minor change in the running time parameter t).

3   Security of **NMAC**

We first show that NMAC is a secure MAC, given that the compression function is non-malleable, simulatable and unpredictable. For simplicity we refer to NMAC as both the MAC algorithm and the scheme (with straightforward key generation and verification algorithm).

We simply state the theorem and present the proof for π-simulatable, where π is a constant function 0; afterwards we discuss that the theorem holds for more general functions. We also remark that we can give a proof based on non-malleability and unpredictability only (i.e., without simulatability) if the padding function is prefix-free. See Appendix A.

Theorem 1. *Let the compression function* $h : \{0,1\}^n \times \{0,1\}^b \rightarrow \{0,1\}^n$ *be* $(t_{\mathcal{A}}, t_{\mathcal{S}}, Q, N, \mu)$*-non-malleable with respect to distribution* $\mathcal{X}_{\mathsf{NMAC}}$ *and relation* R_{NMAC}*, and* $(t_{\mathcal{D}}, t_{\mathsf{Sim}}, Q, \sigma)$*-0-simulatable for* $\mathcal{X}_{\mathsf{NMAC}}$*. Assume further that* h *is* $(t_{\mathcal{S}}, N, \rho)$*-unpredictable. Then* NMAC *is a* (t', Q', L', ϵ')*-unforgeable MAC where*

$$t' = \min\{t_{\mathcal{A}}, t_{\mathcal{S}}, t_{\mathsf{Sim}}\} - (Q+1) \cdot \mathit{Time}_{bN}(\mathsf{pad})$$
$$- \Theta(NQ \log Q \cdot (\mathit{Time}(h) + n + b))$$
$$Q' = Q, \qquad L' = bN - \mathit{Extend}_{bN}(\mathsf{pad}), \qquad \epsilon' \leq 4QN^2 \cdot (\rho + \mu) + \sigma.$$

Proof. We first make two simplifying assumptions about the MAC adversary $\mathcal{B}$. First we assume that $\mathcal{B}$ always pads any message before outputting it or submitting it to the MAC oracle. Since the padding function is one-to-one, the padded forgery attempt is still distinct from all (padded) submissions. Furthermore, the adversary's running time only increases by $(Q+1) \cdot \mathit{Time}_L(\mathsf{pad})$ and the length by at most $\mathit{Extend}_L(\mathsf{pad})$ bits for each message. Secondly, we presume that $\mathcal{B}$ never submits the same message twice. Since the MAC computation is deterministic such queries can be easily answered by keeping track of previous queries and these checks only add the running time $O(bNQ \log Q)$.

For a successful attacker $\mathcal{B}$ on the MAC we distinguish between two cases:

Case NoColl: In the final output the MAC adversary $\mathcal{B}$ returns a (now padded) message M^* such that $h^*(k_{\mathrm{in}}, M^*) \neq h^*(k_{\mathrm{in}}, M_i)$ for all previous queries M_i, $i = 1, 2, \ldots, Q$.

Case Coll: The MAC adverary $\mathcal{B}$ returns a forgery attempt M^* such that $h^*(k_{\text{in}}, M^*) = h^*(k_{\text{in}}, M_i)$ for some i.

Adding (the bounds for) the two probabilities then gives an upper bound on $\mathcal{B}$'s success probability.

Case NoColl. Assume $\mathcal{B}$ succeeds and that the first case happens. Consider the following adversary $\mathcal{A}_{\text{out}}$ against the non-malleability for distribution $\mathcal{X}_{\text{NMAC}}$ and relation R_{NMAC}. Adversary $\mathcal{A}_{\text{out}}$ initially chooses a key $k_{\text{in}} \leftarrow \{0,1\}^n$ at random. It next runs a simulation of $\mathcal{B}$ and answers each query M_i to the MAC oracle by computing locally $z_i = h^*(k_{\text{in}}, M_i)$ and submitting $p_i = \mathsf{pad}(z_i)$ to its oracle GenSample. Adversary $\mathcal{A}_{\text{out}}$ sets $\tau_i = y_i$ for the oracle's answer and returns τ_i to $\mathcal{B}$. When $\mathcal{B}$ eventually produces its output (M^*, τ^*) attacker $\mathcal{A}_{\text{out}}$ computes $z^* = h^*(k_{\text{in}}, M^*)$ and prepares the transformation $T(k, p_1, p_2, \dots)$ which computes $h^*(k, \mathsf{pad}(z^*))$ for fixed value $\mathsf{pad}(z^*)$. $\mathcal{A}_{\text{out}}$ finally outputs (T, y^*) for $y^* = \tau^*$.

It is easy to see that $\mathcal{A}_{\text{out}}$'s success probability in the non-malleability experiment equals the success probability of $\mathcal{B}$ attacking the MAC scheme, given that $h^*(k_{\text{in}}, M^*) \neq h^*(k_{\text{in}}, M_i)$ for all i. The latter implies that $z^* \neq z_1, z_2, \dots$ and therefore $(x^*, y^*) \neq (x_i, y_i)$ for all i in the non-malleability experiment. Furthermore, the padding function is one-to-one and appends only the same amount of bits for each $z^*, z_1, z_2, \dots$ such that no p_i is a prefix of $\mathsf{pad}(z^*)$. Hence, by assumption, there exists a simulator $\mathcal{S}$ making $t_\mathcal{S}$ steps and which is μ-close to $\mathcal{A}_{\text{out}}$'s probability, but which does not get to see any image under $h^*(k_{\text{out}}, \cdot)$. In particular, without any knowledge about k_{out}, the simulator outputs an image y^* and a transformation T involving a constant $c \in B^{\leq N}$ such that $h^*(k_{\text{out}}, c) = y^*$. But since the compression function is $(t_\mathcal{S}, N, \rho)$-unpredictable, the claim for this case follows (in particular, $\epsilon' \leq \mu + \rho$).

Case Coll. Now consider the second case, that $\mathcal{B}$ succeeds and finds a collision in the inner function. In fact, we only need that $\mathcal{B}$ is able to generate such a collision $M^* \neq M_1, M_2, \dots$, possibly not even succeeding in forging a MAC. We then construct an attacker $\mathcal{A}_{\text{in}}$ on the non-malleability of the compression function, again with respect to distribution $\mathcal{X}_{\text{NMAC}}$ and relation R_{NMAC}. The idea is that such a collision can be guessed in advance and can then be used to predict the image for the second value.

But first assume that, instead of communicating with oracle $\mathsf{NMAC}_{(k_{\text{in}}, k_{\text{out}})}$, adversary $\mathcal{B}$ instead receives the answers from oracle $\mathsf{SimAnswer}(\pi(h^*(k_{\text{in}}, \cdot))) = \mathsf{SimAnswer}(0)$ guaranteed by the π-simulatability (where π is constantly 0). We claim that the probability of $\mathcal{B}$ producing a collision for the inner function cannot drop by more than σ when communicating with $\mathsf{SimAnswer}$. Else, one can easily devise a distinguisher $\mathcal{D}$ separating GenSample and SimAnswer.

More formally, distinguisher $\mathcal{D}$ picks k_{in} itself and is given access to an oracle either implementing $h^*(k_{\text{out}}, \cdot)$ or SimAnswer and runs a simulation of $\mathcal{B}$. Each query M_i of $\mathcal{B}$ is answered by first computing $z_i = \mathsf{pad}(h^*(k_{\text{in}}, M_i))$. Then $\mathcal{D}$ checks if this value has appeared before, in which case $\mathcal{D}$ fetches the previous answer and handing it back to $\mathcal{B}$. Else, $\mathcal{D}$ forwards z_i to its oracle and returns the answer to $\mathcal{B}$. When $\mathcal{B}$ eventually outputs a forgery attempt (M^*, τ^*) the distinguisher verifies that $M^* \neq M_1, M_2, \ldots$ (if not, it outputs 0), computes $z^* = \mathsf{pad}(h^*(k_{\mathrm{in}}, M^*))$ and checks that there is a collision between z^* and some z_i. If so, then $\mathcal{D}$ outputs 1, in any other case it returns 0.

The probability of returning 1 when given access to oracle $\mathsf{GenSample}$ is identical to the probability that $\mathcal{B}$ generates a collision on the inner function between M^* and some (distinct) M_i when attacking NMAC. On the other hand, if $\mathcal{D}$ is given access to $\mathsf{SimAnswer}$ then the probability for returning 1 corresponds exactly to the probability that $\mathcal{B}$ produces such a collision when given access to $\mathsf{SimAnswer}$ instead. By assumption this difference cannot be more than σ, taking into account that $\mathcal{D}$ essentially runs in the same time as $\mathcal{B}$ but needs to compute the inner function and check for collisions.

Given $\mathcal{B}$ with access to $\mathsf{SimAnswer}$ we now build adversary $\mathcal{A}_{\mathsf{in}}$ playing against the non-malleability of the inner function. Adversary $\mathcal{A}_{\mathsf{in}}$ is granted access to oracle $\mathsf{GenSample}$. It first picks random indices i_0 between 1 and Q as well as j_0, ℓ_0 between 0 and $N - 1$. It also flips a coin $c \leftarrow \{0, 1\}$. Then it runs a black-box simulation of $\mathcal{B}$ for oracle $\mathsf{SimAnswer}$ (which works independent of the actual content of the queries of $\mathcal{B}$). Only for the i_0-th query $M_{i_0} = M_{i_0}[1] \ldots M_{i_0}[n_{i_0}]$, if $c = 0$, adversary $\mathcal{A}_{\mathsf{in}}$ also forwards $p_{i_0} = M_{i_0}[1] \ldots M_{i_0}[j_0 + 1]$ to its oracle $\mathsf{GenSample}$ to receive a value $z'_{i_0} = h^*(k_{\mathrm{in}}, M_{i_0}[1] \ldots M_{i_0}[j_0 + 1])$.[2] Note that $\mathcal{A}_{\mathsf{in}}$ does not forward this value to $\mathcal{B}$ but rather uses the value generated by $\mathsf{SimAnswer}$. If $c = 1$ then $\mathcal{A}$ does not call its oracle at this point.

When $\mathcal{B}$ finally outputs (M^*, τ^*) and we have $c = 1$ then $\mathcal{A}_{\mathsf{in}}$ has not queried oracle $\mathsf{GenSample}$ so far, and now submits $p^* = M^*[1] \ldots M^*[\ell_0 + 1]$ to receive the value $z' = h^*(k_{\mathrm{in}}, M^*[1] \ldots M^*[\ell_0 + 1])$.[3] Adversary $\mathcal{A}_{\mathsf{in}}$ then returns $y^* = z'$ and the transformation $T(k_{\mathrm{in}}, , p_1, p_2, \ldots)$ which for fixed $j_0, M_{i_0}[1], \ldots, M_{i_0}[j_0 + 1]$ computes the value $x^* = (h^*(k_{\mathrm{in}}, M_{i_0}[1] \ldots M_{i_0}[j_0]), M_{i_0}[j_0 + 1])$. Else, if $c = 0$, then $\mathcal{A}_{\mathsf{in}}$ returns $y^* = z'_{i_0}$ and the transformation $T(k_{\mathrm{in}}, , p_1, p_2, \ldots)$ which for fixed $\ell_0, M^*[1], \ldots, M^*[\ell_0 + 1]$ computes $x^* = (h^*(k_{\mathrm{in}}, M^*[1] \ldots M^*[\ell_0]), M^*[\ell_0 + 1])$.

For the success probability note that, given that $\mathcal{B}$ creates a collision between some message M_i and M^*, adversary $\mathcal{A}_{\mathsf{in}}$ predicts $i_0 = i$ and the right block

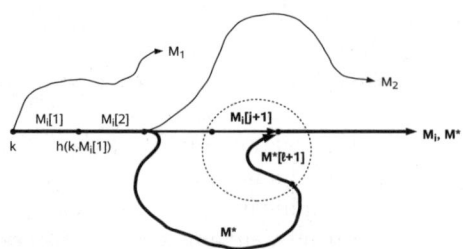

Fig. 1. Proof idea to Theorem 1: collisions among values

---

[2] We assume that $j_0 < n_{i_0}$; else stop immediately with no output.
[3] Again, if ℓ_0 exceeds the number of blocks of M^* we stop with no output.

numbers j_0, ℓ_0 such that $M_{i_0}[1] \ldots M_{i_0}[j_0+1] \neq M^*[1] \ldots M^*[\ell_0+1]$ collide under $h^*(k_{\text{in}}, \cdot)$ but $x_i = h^*(k_{\text{in}}, M_i[1] \ldots M_i[j_0]) \neq x^* = h^*(k_{\text{in}}, M^*[1] \ldots M^*[\ell_0])$, with probability at least $1/QN^2$. See Figure 1. Furthermore, either M_{i_0} is a prefix of M^* or vice versa (or neither one is a prefix of the other one), and we make the "right" choice c with probability at least $1/2$ to submit the message which is not a prefix to oracle GenSample. Hence, $\mathcal{A}_{\text{in}}$'s success probability is only a factor $1/2QN^2$ smaller than the one of $\mathcal{B}$ in this case. The overall running time of $\mathcal{A}_{\text{in}}$ is essentially equal to the one of $\mathcal{B}$, plus some time to prepare the submissions and the final output.

By the non-malleability there must exist a simulator with μ-close success rate to $\mathcal{A}_{\text{in}}$. But this simulator never gets to see any information about k_{in} and must first commit to y^* and $M^* = M^*[1] \ldots M^*[n^*]$ for $n^* \leq N$ (via T), allowing only a success probability ρ by the unpredictability of the compression function. Hence, the probability of $\mathcal{B}$ succeeding for a collision on the inner function is at most $2QN^2(\mu + \rho) + \sigma$.   □

Compared to the proof in [5], showing that NMAC is pseudorandom given that h is pseudorandom, we obtain a different loss factor in the success probability, switching the roles of the number of queries and the length of messages ($\Theta(QN^2)$ vs. $\Theta(Q^2N)$ as in [5]).

In the proof we have assumed that $\pi(\cdot)$ is the constant 0-function. We note that the theorem still holds for more general cases, as long as there is a function Π such that $\pi(h^*(k, M)) = \Pi(M)$ for all k, M. Recall the idea from the second part of the proof that we can simulate images for the outer function without knowing $h^*(k_{\text{in}}, M_i)$. Yet, in order to execute SimAnswer($\pi(h^*(k_{\text{in}}, \cdot))$) we need the information $\pi(h^*(k_{\text{in}}, M_i))$. This information is easy to derive for constant π, but also if we can deduce from M_i (known to us) via Π. Hence, the proof goes through and we obtain:

Corollary 1. *Let the compression function $h : \{0,1\}^n \times \{0,1\}^b \to \{0,1\}^n$ be $(t_{\mathcal{A}}, t_{\mathcal{S}}, Q, N, \mu)$-non-malleable with respect to distribution $\mathcal{X}_{\text{NMAC}}$ and relation R_{NMAC}, and $(t_{\mathcal{D}}, t_{\text{Sim}}, Q, \sigma)$-$\pi$-simulatable for $\mathcal{X}_{\text{NMAC}}$, such that there exists a function Π with $\pi(h^*(k, M)) = \Pi(M)$ for all $k \in \{0,1\}^n$ and all $M \in B^+$. Assume further that h is $(t_{\mathcal{S}}, N, \rho)$-unpredictable. Then NMAC is a (t', Q', L', ϵ')-unforgeable MAC where*

$$t' = \min\{t_{\mathcal{A}}, t_{\mathcal{S}}, t_{\text{Sim}}\}$$
$$\quad - \Theta(NQ \log Q \cdot (Time(h) + n + b + Time_{bN}(\mathsf{pad}) + Time_{bN}(\Pi)))$$
$$Q' = Q, \qquad L' = bN - Extend_{bN}(\mathsf{pad}), \qquad \epsilon' \leq 4QN^2 \cdot (\rho + \mu) + \sigma.$$

As an example for such a function π consider a 0-simulatable compression function which appends some (fixed) subset of its input M to the output, and let π denote the projection onto the corresponding suffix. Then this information is clearly computable from $\Pi(M)$. Somewhat interestingly, the simulatability requirement depends to the padding function of the hash function. Namely, in Appendix A we show that for prefix-free paddings the simulatability requirement vanishes completely.

4 Security of HMAC

As explained in [3], one can map HMAC to NMAC by considering $k_{\text{in}}^{\text{NMAC}} = h(\text{IV}, k_{\text{in}}^{\text{HMAC}})$ and $k_{\text{out}}^{\text{NMAC}} = h(\text{IV}, k_{\text{out}}^{\text{HMAC}})$ and

$$\text{HMAC}(k_{\text{in}}^{\text{HMAC}} \| k_{\text{out}}^{\text{HMAC}}, M) = \text{NMAC}(k_{\text{in}}^{\text{NMAC}} \| k_{\text{out}}^{\text{NMAC}}, M).$$

Put differently, HMAC includes a first step in which the NMAC-keys are computed via $h(\text{IV}, \cdot)$ and the HMAC-keys are used as input block.

To transfer the security claims from NMAC, Bellare [5] defines the "dual" compression function $\overline{h} : B \times \{0,1\}^n \to \{0,1\}^n$ with $\overline{h}(b,a) = h(a,b)$. Then, if h and $\overline{h}$ are pseudorandom, the security of NMAC carries over to HMAC. For the precise claim and a discussion about the validity see [5, Sections 5.1 and 5.4].

We can extend the notions of non-malleability, π-simulatability and unpredictability to these special cases. For this consider distribution $\mathcal{X}_{\text{HMAC}}$ which, for input (k,p) computes $k' = h(\text{IV}, k)$ and samples $x \leftarrow \mathcal{X}_{\text{NMAC}}(k', p)$. Analogously, relation R_{HMAC} merely checks that the transformation T in the first step computes $k' = h(\text{IV}, k)$ and proceeds as R_{NMAC}. Then we demand that h is non-malleable with respect to $\mathcal{X}_{\text{HMAC}}$ and R_{HMAC} and π-simulatable for $\mathcal{X}_{\text{HMAC}}$.

Similarly, we say that h is HMAC-unpredictable if it is infeasible to predict $h(h(\text{IV}, k), M)$; a more formal characeterization is easy to deduce. All assumptions are implied if h and $\overline{h}$ are pseudorandom (where we only need that $\overline{h}$ is pseudorandom with respect to distinguishers that make only a single oracle call). Under the assumptions about non-malleability, π-simulatability for suitable π and unpredictability we conclude:

Theorem 2. *Let the compression function $h : \{0,1\}^n \times \{0,1\}^b \to \{0,1\}^n$ be $(t_{\mathcal{A}}, t_{\mathcal{S}}, Q, N, \mu)$-non-malleable with respect to distribution $\mathcal{X}_{\text{HMAC}}$ and relation R_{HMAC}, and $(t_{\mathcal{D}}, t_{\text{Sim}}, Q, \sigma)$-$\pi$-simulatable for $\mathcal{X}_{\text{HMAC}}$, such that there exists a function Π with $\pi(h^*(k, M)) = \Pi(M)$ for all $k \in \{0,1\}^n$ and all $M \in B^+$. Assume that h is $(t_{\mathcal{S}}, N, \rho)$-HMAC-unpredictable. Then HMAC is a (t', Q', L', ϵ')-unforgeable MAC where*

$$t' = \min\{t_{\mathcal{A}}, t_{\mathcal{S}}, t_{\text{Sim}}\}$$
$$\quad - \Theta(NQ \log Q \cdot (\mathit{Time}(h) + n + b + \mathit{Time}_{bN}(\mathsf{pad}) + \mathit{Time}_{bN}(\Pi)))$$
$$Q' = Q, \qquad L' = bN - \mathit{Extend}_{bN}(\mathsf{pad}), \qquad \epsilon' \le 4QN^2 \cdot (\rho + \mu) + \sigma.$$

For the single-keyed HMAC variant with $k_{\text{in}} = k \oplus \text{ipad}$ and $k_{\text{out}} = k \oplus \text{opad}$ one can in principle adapt the notions of non-malleability, simulatability and unpredictability for this case, too. Both properties are then implied if $\overline{h}$ is pseudorandom under related-key attacks [7] (see also [5, Section 5.3]), and single-keyed HMAC is a secure MAC under these versions of non-malleability, simulatability and unpredictability.

5 Relations Among Security Notions

In this section we show that pseudorandom compression function have our two properties but are stronger than both properties together. We also discuss that there are (simulatable) non-malleable compression functions which are not privacy-preserving according to the notion in [5]. Below we simply write $\$_{N,n}$ for a randomly chosen function from all mappings with domain $B^{\leq N}$ and range $\{0,1\}^n$.

Definition 5. *A function* $f : \{0,1\}^n \times B^+ \to \{0,1\}^n$ *is called* (t, Q, N, δ)-*pseudorandom if, for any algorithm* $\mathcal{D}$ *running in time* t, *we have*

$$\mathrm{Prob}\left[\mathcal{D}^{f(k,\cdot)} = 1\right] - \mathrm{Prob}\left[\mathcal{D}^{\$_{N,n}(\cdot)} = 1\right] \leq \delta$$

where the probability in the first case is over $\mathcal{D}$'s *coin tosses and* $k \leftarrow \{0,1\}^n$, *and in the second case over* $\mathcal{D}$'s *coin tosses and the choice of* $\$_{N,n}$. *In both cases* $\mathcal{D}$ *makes at most* Q *queries to its function oracle, each query of at most* N *blocks.*

Below we sometimes make use of the following fact of Bellare et al. [4] about cascaded iteration of pseudorandom functions. The statement basically says that the cascaded evaluation is also pseudorandom with respect to *prefix-free distinguishers*, i.e., distinguishers such that no oracle query is a prefix of another one:

Lemma 1 ([4]). *If the compression function* $h : \{0,1\}^n \times \{0,1\}^b \to \{0,1\}^n$ *is* $(t, Q, 1, \delta)$-*pseudorandom, then* $h^* : \{0,1\}^n \times B^+ \to \{0,1\}^n$ *is also* (t', Q, N, δ')-*pseudorandom with respect to prefix-free distinguishers, where* $t' = t - \Theta(QNb + N\,Time(h))$ *and* $\delta' = N\delta$.

5.1 Pseudorandom $\Rightarrow$ Non-malleable $\wedge$ Unpredictable $\wedge$ Simulatable

Recall from the security of NMAC that we defined $\mathcal{X}_{\mathsf{NMAC}}$ to be the distribution which, for random $k \in \{0,1\}^n$ and parameter $p = M[1] \ldots M[j+1] \in B^+$ outputs the pre-image $x = (h^*(k, M[1] \ldots M[j]), M[j+1])$. Also, R_{NMAC} is the relation which, on input $T, k, p_1, p_2, \ldots, x^*$ checks that T computes $x^* = (h^*(k, M^*[1] \ldots M^*[j]), M^*[j+1])$ for some fixed $M^*[1] \ldots M^*[j+1]$ of at most N blocks (and checks that no p_i is a prefix of $M^*[1] \ldots M^*[j+1]$).

Proposition 1. *Let* h *be* $(t, Q, 1, \delta)$-*pseudorandom. Then it is* $(t_\mathcal{A}, t_\mathcal{S}, Q, N, \mu)$-*non-malleable with respect to distribution* $\mathcal{X}_{\mathsf{NMAC}}$ *and relation* R_{NMAC}, *where*

$$t_\mathcal{A} = t - \Theta(QNb \log Q \cdot (Time(h) + n)), \quad t_\mathcal{S} = t + O(bNQ \log Q + n), \quad \mu = 2N\delta.$$

Proof. Consider any adversary $\mathcal{A}$ against non-malleability, running in time $t_\mathcal{A}$ and making at most Q queries. We may assume that $\mathcal{A}$ never queries its oracle GenSample about a parameter whose prefix has been submitted before; such values could be computed by $\mathcal{A}$ itself easily and skipping these oracle queries can only increase $\mathcal{A}$'s success probability. This increases the adversary's running time by at most $O(bNQ \log Q \cdot (Time(h) + n))$.

Construct the non-malleability simulator $\mathcal{S}$ running a black-box simulation of $\mathcal{A}$ as follows. Each time the adversary submits a parameter $p_i = M_i[1] \ldots M_i[j+1]$

to its oracle, $\mathcal{S}$ emulates the oracle perfectly except that, instead of using the iterated compression function, $\mathcal{S}$ uses lazy sampling to simulate a truly random function.[4] When the adversary $\mathcal{A}$ eventually outputs (T, y^*) the simulator, too, stops with this output.

For the analysis we first consider $\mathcal{A}$'s behavior if, instead of giving it access to $h^*(k, \cdot)$ we use a truly random function $\$_{N,n}$, also to check that $y^* = \$_{N,n}(M^*)$ for the final output (instead of verifying $y^* = h^*(k, M^*)$). By the pseudorandomness of h (and therefore of h^*) we get that the probability of $\mathcal{A}$ winning in this new experiment is at least $N\delta$-close to the original success probability. This can be easily turned formally into a (prefix-free[5]) distinguisher, simulating $\mathcal{A}$ in a black-box way and calling its function oracle for each message and the final check that T is of the right form and that y^* is an image for M^*.

Next, consider the slightly changed experiment in which we give $\mathcal{A}$ random answers for its message queries as before, but then evaluate correctly $h^*(k, M^*)$ to compare it to y^*. It is easy to see that the success probability of $\mathcal{A}$ in this experiment cannot grow more than $N\delta$, again by the pseudorandomness. That is, it is once more straightforward to construct a prefix-free distinguisher turning this into a formal statement.

But the final experiment is identical to the success probability of the simulator, showing the claim.   □

The proposition above shows that the same remains true for π-simulatable compression functions (for arbitrary π) if we let SimAnswer simply return random strings.

Corollary 2. *Let h be $(t, Q, 1, \delta)$-pseudorandom. Then it is $(t_{\mathcal{A}}, t_{\mathsf{Sim}}, Q, N, \mu)$-$\pi$-simulatable with respect to distribution $\mathcal{X}_{\mathsf{NMAC}}$, where π is arbitrary and*

$$t_{\mathcal{A}} = t - \Theta(QNb \log Q \cdot (Time(h) + n)), \quad t_{\mathsf{Sim}} = t + O(bNQ \log Q + n), \quad \mu = N\delta.$$

Finally, we show that pseudrandomness implies unpredictability:

Proposition 2. *Let h be $(t, 1, 1, \delta)$-pseudorandom. Then it is also (t', N, ρ)-unpredictable, where $t' = t - \Theta(QNb + N \cdot Time(h))$ and $\rho = N(\delta + 2^{-n})$.*

Proof. Since h is pseudorandom it remains pseudorandom if we make a single query of (at most) N blocks. Only the running time drops slightly and the distinguishing advantage increases to $N\delta$. But this implies that any algorithm P trying to predict a function value cannot be better than for a truly random function —which can be predicted with probability at most $N \cdot 2^{-n}$— plus the distinguishing advantage of $N\delta$.   □

---

[4] Meaning that $\mathcal{S}$ picks an independent random string when supposed to evaluate the function on a new value, or repeats a previously given answer for previously evaluated values.

[5] Here we use the fact that no p_i is a prefix of the message encoded in T.

5.2 Non-malleable ∧ Unpredictable ∧ Simulatable ⇏ Pseudorandom

We next prove that non-malleability, simulatability and unpredictability together do not imply pseudorandomness. Since this result only serves as a separation we drop the viewpoint of concrete security and adopt the usual "asymptotic" notion (with regard to parameter n). For example, being pseudorandom means to be $(\text{poly}(n), \text{poly}(n), \text{poly}(n), \delta(n))$-pseudorandom for any polynomial $\text{poly}(n)$ and some negligible function $\delta(n)$ (which may depend on the polynomial). The other security notions can be adopted analogously.

Below we show that pseudorandomness is stronger than non-malleability, unpredictability and π-simulatability for $\pi = 0$. Afterwards we discuss that the claim also holds for a non-trivial functions π like the rightmost-bit function rmb.

Proposition 3. *Assume that there exists a pseudorandom compression function* $h : \{0,1\}^n \times \{0,1\}^b \to \{0,1\}^n$ *and a pseudorandom generator* $G : \{0,1\}^{\lfloor n/2 \rfloor} \to \{0,1\}^n$. *Then there exists a compression function* $h_{\text{sep}} : \{0,1\}^n \times \{0,1\}^b \to \{0,1\}^n$ *which is (a) non-malleable with respect to distribution* $\mathcal{X}_{\text{NMAC}}$ *and relation* R_{NMAC}, *(b) unpredictable, (c) 0-simulatable for* $\mathcal{X}_{\text{NMAC}}$, *but (c) not pseudorandom.*

Proof. First consider the truncated compression function $h_{\text{trunc}}(k, x)$ which splits the key k into $\lfloor n/2 \rfloor$ bits k_L and the remaining $n - \lfloor n/2 \rfloor$ bits k_R, then computes $h(G(k_L), x)$ and outputs only the first $\lfloor n/2 \rfloor$ bits of the result. It follows easily from the pseudorandomness of G and h that this compression function is pseudorandom, too, and therefore non-malleable, unpredictable and 0-simulatable.

Now define $h_{\text{sep}}(k, x) = h_{\text{trunc}}(k, x) || 0^{n - \lfloor n/2 \rfloor}$. With this definition one can compute the iterated output $h_{\text{sep}}^*(k, M)$ for any $M \in B^+$ by evaluating h_{trunc} and appending $n - \lfloor n/2 \rfloor$ bits $\pi(x)$ to the output in each stage (including the final evaluation).

It is clear that h_{sep} is not pseudorandom, because every function evaluation yields only zeros in the right half of output. It is, however, still unpredictable. If it was not unpredictable this would easily contradict unpredictability of h_{trunc} by cutting off the $n - \lfloor n/2 \rfloor$ rightmost bits from the output. Moreover, h_{sep} inherits non-malleability for $\mathcal{X}_{\text{NMAC}}$ and R_{NMAC} from h_{trunc} because one can easily transform attackers and simulators for the two cases (by appending or cutting off zeros in the output). Simulatability can be shown easily, too, as one can output random strings followed by a sequence of zeros.   □

Again, the claim remains true for π-simulatability for some non-trivial functions π. For instance, if $\pi(x)$ returns the rightmost bit rmb(x) of x and we use the compression function h_{sep} in the proof and replace the padding with zeros by $(\text{rmb}(x))^{n - \lfloor n/2 \rfloor}$ —call this function h_{rmb}— then the proposition still holds. In particular, the rmb-simulatability (together with non-malleability and unpredictability) would suffice to prove NMAC to be secure.

We also remark that the example h_{rmb} shows that there are π-simulatable (and non-malleable and unpredictable) compression functions which are *not* a privacy-preserving MAC according to [5]. Such a MAC has the additional property that one cannot tell if either of two messages has been MACed. However, for inputs with distinct rightmost bit this is easy for h_{rmb}, of course. Still, 0-simulatable compression functions are privacy-preserving, giving an alternative characterization of this notion.

Acknowledgments

We thank the anonymous reviewers for valuable comments.

References

1. An, J.H., Bellare, M.: Constructing VIL-MACs from FIL-MACs: Message Authentication under Weakened Assumptions. In: Wiener, M.J. (ed.) CRYPTO 1999. LNCS, vol. 1666, pp. 252–269. Springer, Heidelberg (1999)
2. Boldyreva, A., Cash, D., Fischlin, M., Warinschi, B.: Non-Malleable Hash Functions, Non-Malleable Hash Functions (manuscript, 2007)
3. Bellare, M., Canetti, R., Krawczyk, H.: Keying Hash Functions for Message Authentication. In: Koblitz, N. (ed.) CRYPTO 1996. LNCS, vol. 1109, pp. 1–15. Springer, Heidelberg (1996)
4. Bellare, M., Canetti, R., Krawczyk, H.: Pseudorandom Functions Revisited: The Cascade Construction and Its Concrete Security. In: Proceedings of the Annual Symposium on Foundations of Computer Science FOCS 1996, pp. 514–523. IEEE Computer Society Press, Los Alamitos (1996)
5. Bellare, M.: New Proofs for NMAC and HMAC: Security Without Collision-Resistance. In: Dwork, C. (ed.) CRYPTO 2006. LNCS, vol. 4117, pp. 602–619. Springer, Heidelberg (2006)
6. Bellare, M., Goldreich, O., Mityagin, A.: The Power of Verification Queries in Message Authentication and Authenticated Encryption. Number 2004/309 in Cryptology eprint archive (2004), http://eprint.iacr.org
7. Bellare, M., Kohno, T.: A Theoretical Treatment of Related-Key Attacks: RKA-PRPs, RKA-PRFs, and Applications. In: Biham, E. (ed.) EUROCRYPT 2003. LNCS, vol. 2656, pp. 491–506. Springer, Heidelberg (2003)
8. Contini, S., Yin, Y.L.: Forgery and Partial Key-Recovery Attacks on HMAC and NMAC Using Hash Collisions. In: Lai, X., Chen, K. (eds.) ASIACRYPT 2006. LNCS, vol. 4284, pp. 37–53. Springer, Heidelberg (2006)
9. Kim, J., Biryukov, A., Preneel, B., Hong, S.: On the Security of HMAC and NMAC Based on HAVAL, MD4, MD5, SHA-0 and SHA-1. In: De Prisco, R., Yung, M. (eds.) SCN 2006. LNCS, vol. 4116, pp. 242–256. Springer, Heidelberg (2006)
10. Rechberger, C., Rijmen, V.: On Authentication With HMAC and Non-Random Properties. In: Financial Cryptography FC 2007. LNCS, vol. 4886, pp. 119–133. Springer, Heidelberg (2007)
11. Wang, X., Yu, H.: How to Break MD5 and Other Hash Functions. In: Cramer, R.J.F. (ed.) EUROCRYPT 2005. LNCS, vol. 3494, pp. 19–35. Springer, Heidelberg (2005)
12. Wang, X., Yin, Y.L., Yu, H.: Finding Collisions in the Full SHA-1. In: Shoup, V. (ed.) CRYPTO 2005. LNCS, vol. 3621, pp. 17–36. Springer, Heidelberg (2005)

A   Security of NMAC for Prefix-Free Paddings

Here we show that regular (i.e., not necessarily black-box) non-malleability suffices to show security of NMAC and HMAC, given that the padding function pad() is prefix-free. Formally, the function pad() is called *prefix-free* if for any distinct $M \neq M'$ the value $\mathsf{pad}(M)$ is not equal to any block-wise prefix of $\mathsf{pad}(M')$. An example of a prefix-free padding is the standard padding with the exception that we prepend a block containg the bit size of the original input. Then, any messages with distinct length are clearly not a prefix of each other, and for equal length messages $\mathsf{pad}(M)$ is either longer than any block-wise prefix of $\mathsf{pad}(M')$ or of equal length, in which case they must be distinct since $M \neq M'$.

Theorem 3. *Let the compression function* $h : \{0,1\}^n \times \{0,1\}^b \rightarrow \{0,1\}^n$ *be* $(t_\mathcal{A}, t_\mathcal{S}, Q, N, \mu)$-*non-malleable with respect to distribution* $\mathcal{X}_{\mathsf{NMAC}}$ *and relation* R_{NMAC}. *Assume further that* h *is* $(t_\mathcal{S}, N, \rho)$-*unpredictable and that* pad *is prefix-free. Then* NMAC *is a* (t', Q', L', ϵ')-*unforgeable MAC where*

$$t' = \min\{t_\mathcal{A}, t_\mathcal{S}\} - (Q+1) \cdot Time_{bN}(\mathsf{pad}) - \Theta(NQ \log Q \cdot (Time(h) + n + b))$$

$$Q' = Q, \qquad L' = bN - Extend_{bN}(\mathsf{pad}), \qquad \epsilon' \leq 2QN^2 \cdot (\rho + \mu).$$

Proof. We remark that we again presume that $\mathcal{B}$ pads each message first and that it never queries the MAC oracle about the same message twice. The first part of the proof (the case that there is no collision between the inner value for M^* and any M_i) is identical to the "non-prefix-free" case and is therefore omitted. We next consider the second case that $\mathcal{B}$ causes $h^*(k_{\mathrm{in}}, M_i) = h^*(k_{\mathrm{in}}, M_i)$ for some i. We again construct a non-malleability attacker $\mathcal{A}_{\mathrm{in}}$ against the inner function.

Initially, adversary $\mathcal{A}_{\mathrm{in}}$ picks a random index i_0 between 1 and Q and indices j_0, ℓ_0 between 0 and $N-1$. It next invokes the MAC adversary $\mathcal{B}$ and simulates each answer for query M_i as follows.

- For $i \neq i_0$ adversary $\mathcal{A}_{\mathrm{in}}$ first submits $p_i = M_i$ to its oracle GenSample to get an answer $z_i = y_i$. Then it computes $\tau_i = h^*(k_{\mathrm{out}}, \mathsf{pad}(z_i))$ and returns this value to $\mathcal{B}$.
- For $i = i_0$ and $M_{i_0} = M_{i_0}[1] \ldots M_{i_0}[n_{i_0}]$ it sends $p_{i_0} = M_{i_0}[1] \ldots M_{i_0}[j_0 + 1]$ to the oracle[6] to receive a value $z'_{i_0} = h^*(k_{\mathrm{in}}, M_{i_0}[1] \ldots M_{i_0}[j_0+1])$. Compute $z_{i_0} = h(z'_{i_0}, M_{i_0}[j_0 + 2] \ldots M_{i_0}[n_{i_0}])$ and $\tau_{i_0} = h^*(k_{\mathrm{out}}, \mathsf{pad}(z_{i_0}))$ and return the latter value to $\mathcal{B}$.

When $\mathcal{B}$ stops with output (M^*, τ^*) (for the padded message M^*) adversary $\mathcal{A}_{\mathrm{in}}$ sets its output to $y^* = z'_{i_0}$ for the answer of the previously guessed index i_0. The adversary furthermore defines $T(k_{\mathrm{in}}, , p_1, p_2, \ldots)$ for fixed $\ell_0, M^*[1], \ldots, M^*[\ell_0 + 1]$ to compute $x^* = (h^*(k_{\mathrm{in}}, M^*[1] \ldots M^*[\ell_0]), M^*[\ell_0 + 1])$.[7]

To analyze $\mathcal{A}_{\mathrm{in}}$'s success probability let ℓ denote some index such that, when computing $h^*(k_{\mathrm{in}}, M^*[1] \ldots M^*[\ell + 1])$ iteratively, this value matches any of

---

[6] We assume that $j_0 < n_{i_0}$; else stop immediately with no output.
[7] We assume that $\ell_0 + 1$ is at most the number of blocks in M^*; else stop immediately with no output.

the intermediate values $h(k_{\mathrm{in}}, M_i[1] \ldots M_i[j + 1])$ for some i, j but such that $M^*[1] \ldots M^*[\ell + 1] \neq M_i[1] \ldots M_i[j + 1]$. See again Figure 1 on Page 146. According to the assumption a collision between the value for M^* and an image of some query occurs and such indices must exist.

Furthermore, by the prefix-free padding function pad it follows from $M^* \neq M_1, M_2, \ldots$ that no M_i is equal to a block-wise prefix of M^*. Hence, there must exists indices ℓ_0, i_0, j_0 such that we obtain a collision on the inner function and no M_i for $i \neq i_0$ and neither $M_{i_0}[1] \ldots M_{i_0}[j_0 + 1]$ is a prefix of $M^*[1] \ldots M^*[\ell_0 + 1]$. But then $\mathcal{A}_{\mathrm{in}}$ picks the right indices i_0, j_0, ℓ_0 with probability at least $1/QN^2$. Under this condition, $(x^*, y^*) \neq (x_i, y_i)$ for all i and $\mathcal{A}_{\mathrm{in}}$ wins if $\mathcal{B}$ generates a collision for the inner function.

Overall, $\mathcal{A}_{\mathrm{in}}$ has a success probability which is smaller by a factor $1/QN^2$, and there must exist a simulator with μ-close success rate. But this simulator never gets to see any information about k_{in} and must first commit to y^* and $M^* = M^*[1] \ldots M^*[n^*]$ for $n^* \leq N$ (via T), allowing only a success probability ρ by the unpredictability of the compression function. The overall running time of $\mathcal{A}_{\mathrm{in}}$ is essentially equal to the one of $\mathcal{B}$, plus the time to pad each message, to check for double queries and to evaluate the compression function for each oracle call. □

Aggregate Message Authentication Codes

Jonathan Katz[1] and Andrew Y. Lindell[2]

[1] University of Maryland
jkatz@cs.umd.edu
[2] Aladdin Knowledge Systems and Bar-Ilan University
andrew.lindell@aladdin.com, lindell@cs.biu.ac.il

Abstract. We propose and investigate the notion of *aggregate message authentication codes (MACs)* which have the property that multiple MAC tags, computed by (possibly) different senders on multiple (possibly different) messages, can be aggregated into a shorter tag that can still be verified by a recipient who shares a distinct key with each sender. We suggest aggregate MACs as an appropriate tool for authenticated communication in mobile ad-hoc networks or other settings where resource-constrained devices share distinct keys with a single entity (such as a base station), and communication is an expensive resource.

1 Introduction

Aggregate signatures, introduced by Boneh et al. [5,16], allow t distinct signatures by t (possibly different) signers on t (possibly different) messages to be *aggregated* into a shorter signature that still suffices to convince a verifier that each signer did indeed sign the appropriate message. Since their introduction, various aggregate signature schemes have been proposed [12,11,6,13,4]. To the best of our knowledge, however, no formal attention has yet been dedicated to the *private-key* analogue of aggregate signatures: aggregate message authentication codes (MACs). In this paper, we initiate a formal study of this primitive.

One reason for the relative lack of attention focused on aggregate MACs may be the (incorrect) perception that they are of limited value. Indeed, the applications suggested in [5] — such as compressing certificate chains, or reducing the message size in secure routing protocols — are all specific to the public-key (rather than the shared-key) setting. Nevertheless, we suggest that aggregate MACs *can* be very useful in specific domains. As perhaps the most compelling example, consider the problem of authenticated communication in a mobile ad-hoc network (MANET), where communication is considered a highly "expensive" resource because of its effect on the battery life of the nodes. Here, there is a collection of t nodes $U_1, \ldots, U_t$, each of whom is interested in sending messages to a base station B. We assume that the base station shares in advance a key k_i with each node U_i, and that node U_i authenticates any outgoing message m_i by computing $\mathsf{tag}_i = \mathsf{Mac}_{k_i}(m_i)$.

Most nodes cannot communicate directly with the base station due to the limited range of their wireless devices, and so all communication is instead routed

T. Malkin (Ed.): CT-RSA 2008, LNCS 4964, pp. 155–169, 2008.
© Springer-Verlag Berlin Heidelberg 2008

among the nodes themselves until it reaches the base station. For simplicity in this example, let us assume that nodes are arranged in a (logical) binary tree so that each node U_i at a leaf sends (m_i, tag_i) to its parent, and each internal node U_j forwards to its own parent all the communication from its children in addition to (m_j, tag_j). The root U^* in this example is the only node that is able to communicate directly with the base station, and it forwards to the base station the communication from all nodes in the network along with its own contribution (m^*, tag^*).

The messages themselves may be very short — corresponding, e.g., to temperature readings or even just an indicator bit. For the sake of argument, let us say that messages are 16 bits long. (Replay attacks can be addressed by using a counter shared by the base station and all nodes in the network; this counter would be authenticated by each node along with the message, but would not need to be transmitted and so does not affect the communication complexity in the calculation that follows.) Furthermore, let us assume that the length of a MAC tag is 160 bits (e.g., if HMAC is used), and take $t = 10^4$. The communication from the root node alone to the base station is then $(160 + 16) \cdot t = 1.76 \times 10^6$ bits, while the total communication in the network is (approximately) $(160 + 16) \cdot (2t \log t) \approx 4.6 \times 10^7$ bits.

The above description assumes MACs used in the 'standard' manner, meaning that all MAC tags are transmitted together with the messages. If an aggregate MAC were available, however, then each node U_j would be able to *combine* its own MAC tag with those of its children. Say this aggregation can be performed while maintaining the MAC tag length, even of aggregated tags, at 160 bits. (Our construction will achieve this.) The communication from the root to the base station will now be only $160 + 16t \approx 1.6 \times 10^5$ bits, and the total communication in the network will be improved to roughly $16(2t \log t) + 160t \approx 5.7 \times 10^6$ bits; this is roughly an order of magnitude improvement in each case.

Aggregate MACs could also be used to improve the communication complexity in schemes such as those of [14] or [9] which deal with aggregation of *data*. We do not explore this further here, as we view the use of such techniques as tangential to the main thrust of this paper.

1.1   Our Contributions

Motivated in part by scenarios such as the above, we formally introduce here the notion of aggregate MACs and initiate the first detailed study of this primitive. After giving appropriate definitions, we show a simple and highly efficient construction of aggregate MACs based on a wide variety of existing (standard) MACs. We remark that the existence of *efficient* aggregate MACs is somewhat surprising since algebraic (i.e., number-theoretic) properties of the underlying signature scheme are used to perform aggregation in the setting of aggregate signatures. In contrast, here we would like to avoid number-theoretic constructions and base aggregate MACs on primitives like block ciphers and hash functions that have limited algebraic structure. Summarizing, we prove the following informally-stated theorem:

Theorem (basic construction – informally stated): *If there exists a secure message authentication code, then there exists a secure aggregate message authentication code with complexity as outlined below.*

The complexity of our construction is as follows:

- *Aggregate MAC tag length:* equal to a single tag in the basic MAC scheme
- *Computation of a MAC tag in the aggregate scheme:* the same as for the basic MAC scheme
- *Computation of MAC tag aggregation:* linear in the length of the tags to be aggregated
- *Verification of ℓ aggregated MACs:* equal to the time it takes to verify ℓ MACs in basic scheme

As can be seen from above, the complexity of our aggregate construction is essentially the same as for a regular MAC scheme. This may be somewhat surprising since in the public-key setting of aggregate signatures, it is significantly harder to obtain secure aggregation. Nevertheless, the reason for this will become clear after seeing our construction in Section 3.

Lower bound. Our aggregate scheme works very well when the receiver wishes to verify the authenticity of all the aggregated messages. However, if the receiver wishes to verify only one or a few messages, it must still verify them all. (This is similar to the case of CBC encryption that requires the receiver to decrypt the entire message even if it only wants to read the last block.) In Section 4, we explore a variant of our main construction that offers a trade-off between the length of an aggregate tag and the time required to verify individual messages. We also show a lower bound showing that if constant or logarithmic-time verification of individual messages is desired, then the aggregated tag length must be linear in the total number of messages whose tags are aggregated (and so the trivial approach of concatenating individual tags is optimal up to a multiplicative factor in the security parameter).

Related work. Subsequent to our work on this paper, we became aware of two other recent papers [7,3] that, *inter alia*, use what are essentially aggregate MACs (and, in fact, use essentially the same construction we show in Section 3). The key additional contributions of our work are: (1) we provide a formal definition of the problem and a proof of security for our construction; (2) we suggest extensions of the construction offering the time/length trade-off discussed above; and (3) we show a lower bound on the required tag length when fast verification of individual messages is required.

2   Definitions

Our definitions are based on those given in [5,16] for aggregate signatures. Rather than exploring numerous possible special cases of the definitions, we make our definitions as general as possible (and our construction will achieve these definitions). We begin with a functional definition. The security parameter, which determines the length of the key, will be denoted by n.

Definition 1. *An* aggregate message authentication code *is a tuple of probabilistic polynomial-time algorithms* (Mac, Agg, Vrfy) *such that:*

- **Authentication algorithm** Mac: *upon input a key* $k \in \{0, 1\}^n$ *and a message* $m \in \{0, 1\}^*$, *algorithm* Mac *outputs a tag* tag. *We denote this procedure by* tag $\leftarrow$ Mac$_k(m)$.
- **Aggregation algorithm** Agg: *upon input two sets of message/identifier[1] pairs* $M^1 = \{(m_1^1, \mathsf{id}_1^1), \ldots, (m_{\ell_1}^1, \mathsf{id}_{\ell_1}^1)\}$, $M^2 = \{(m_1^2, \mathsf{id}_1^2), \ldots, (m_{\ell_2}^2, \mathsf{id}_{\ell_2}^2)\}$ *and associated tags* tag^1, tag^2, *algorithm* Agg *outputs a new tag* tag. *We stress that this algorithm is unkeyed.*
- **Verification algorithm** Vrfy: *upon receiving a set of key/identifier pairs* $\{(k_1, \mathsf{id}_1), \ldots, (k_t, \mathsf{id}_t)\}$, *a set of message/identifier pairs* $M = \{(m_1, \mathsf{id}_1'), \ldots, (m_\ell, \mathsf{id}_\ell')\}$, *and a tag* tag, *algorithm* Vrfy *outputs a single bit, with '1' denoting acceptance and '0' denoting rejection. We denote this procedure by* Vrfy$_{(k_1, \mathsf{id}_1), \ldots, (k_n, \mathsf{id}_t)}(M, \mathsf{tag})$. *(In normal usage,* $\mathsf{id}_i' \in \{\mathsf{id}_1, \ldots, \mathsf{id}_t\}$ *for all* i.*)*

The following correctness conditions are required to hold:

- *For all* $k, \mathsf{id}, m \in \{0, 1\}^*$, *it holds that* Vrfy$_{k, \mathsf{id}}(m, \mathsf{Mac}_k(m)) = 1$. *(This is essentially the correctness condition for standard MACs.)*
- *Let* M^1, M^2 *be two sets of message/identifier pairs with[2]* $M^1 \cap M^2 = \emptyset$, *and let* $M = M^1 \cup M^2$. *If:*
  1. Vrfy$_{(k_1, \mathsf{id}_1), \ldots, (k_t, \mathsf{id}_t)}(M^1, \mathsf{tag}^1) = 1$, *and*
  2. Vrfy$_{(k_1, \mathsf{id}_1), \ldots, (k_t, \mathsf{id}_t)}(M^2, \mathsf{tag}^2) = 1$,

  then Vrfy$_{(k_1, \mathsf{id}_1), \ldots, (k_n, \mathsf{id}_n)}(M, \mathsf{Agg}(M^1, M^2, \mathsf{tag}^1, \mathsf{tag}^2)) = 1$.

The second correctness condition states that the aggregation of MAC tags still enables correct verification.

The use of identifiers is merely a technical way to differentiate between different senders: in order to know which secret key to use for verification, the receiver needs to know which message is associated with which sender. (Thus, in the second correctness condition, enforcing $M^1 \cap M^2 = \emptyset$ just means that aggregation is not applied if the same sender authenticated the same message twice.) Note that identifiers are not needed in the setting of aggregate signatures where each sender is associated with a unique public key which, in effect, serves as an identifier. For simplicity in what follows, we write Vrfy$_{k_1, \ldots, k_t}(\cdot, \cdot)$ for the verification algorithm, and we sometimes find it convenient to set $\mathsf{id}_i = i$ (it can be checked that this has no effect on our results).

An aggregate MAC would be used as follows. A receiver R who wants to receive authenticated messages from t senders begins by sharing uniformly random keys $k_1, \ldots, k_t \in \{0, 1\}^n$ with each sender (i.e., key k_i is shared with the sender with identity id_i). When sender id_i wishes to authenticate a message m_i, it simply computes tag$^i \leftarrow \mathsf{Mac}_{k_i}(m_i)$. Given a tag computed in this way, and a second

---

[1] We discuss the role of the identifiers below.

[2] This technical condition ensures that the same message/identifier pair does not appear in both M^1 and M^2.

tag tag^j computed by sender id_j on the message m_j, these two tags can be aggregated by computing the value $\mathsf{tag} \leftarrow \mathsf{Agg}(\{(m_i, \mathsf{id}_i)\}, \{(m_j, \mathsf{id}_j)\}, \mathsf{tag}^i, \mathsf{tag}^j)$. The receiver can then check that sender id_i authenticated m_i, and that sender id_j authenticated m_j, by computing

$$\mathsf{Vrfy}_{k_i, k_j}\left(\{(m_i, \mathsf{id}_i), (m_j, \mathsf{id}_j)\}, \mathsf{tag}\right)$$

and verifying that the output is 1. Note that we do not assume $\mathsf{id}_i \neq \mathsf{id}_j$. (But, as per footnote 2, we do assume $(m_i, \mathsf{id}_i) \neq (m_j, \mathsf{id}_j)$.)

As in the case of aggregate signatures, our definition of security corresponds to existential unforgeability under an adaptive chosen-message attack [8]. Because we are in the shared-key setting, however, there are some technical differences between our definition and the security definition for aggregate signatures. In particular, we consider an adversary who may adaptively corrupt various senders and learn their secret keys, and require security to hold also in such a setting.

Definition 2. *Let $\mathcal{A}$ be a non-uniform probabilistic polynomial-time adversary, and consider the following experiment involving $\mathcal{A}$ and parameterized by a security parameter n:*

- **Key generation:** *Keys $k_1, \ldots, k_t \in \{0, 1\}^n$, for $t = \mathsf{poly}(n)$, are generated.*
- **Attack phase:** *$\mathcal{A}$ may query the following oracles:*
  - **Message authentication oracle Mac:** *On input (i, m), the oracle returns $\mathsf{Mac}_{k_i}(m)$.*
  - **Corruption oracle Corrupt:** *upon input i, the oracle returns k_i.*
- **Output:** *The adversary $\mathcal{A}$ outputs a set of message/identifier pairs $M = \{(m_1, \mathsf{id}_1), \ldots, (m_\ell, \mathsf{id}_\ell)\}$ and a tag tag. (We stress that all the pairs in M are required to be distinct.)*
- **Success determination:** *We say $\mathcal{A}$ succeeds if (1) $\mathsf{Vrfy}_{k_1, \ldots, k_t}(M, \mathsf{tag}) = 1$ and (2) there exists a pair $(m_{i^*}, id_{i^*}) \in M$ such that*
  1. *$\mathcal{A}$ never queried $\mathsf{Corrupt}(\mathsf{id}_{i^*})$, and*
  2. *$\mathcal{A}$ never queried $\mathsf{Mac}(\mathsf{id}_{i^*}, m_{i^*})$.*

We say that the aggregate MAC scheme $(\mathsf{Mac}, \mathsf{Agg}, \mathsf{Vrfy})$ is secure if for all $t = \mathsf{poly}(n)$ and all non-uniform probabilistic polynomial-time adversaries $\mathcal{A}$, the probability that $\mathcal{A}$ succeeds in the above experiment is negligible.

We do not consider verification queries even though, in general, they may give the adversary additional power [1]. This is justified by the fact that our eventual construction satisfies the conditions stated in [1] for which verification queries do *not* give any additional power. (Of course, they prove this only for the case of standard MACs but it is easy to see that their proof carries over to our setting as well.) Note also that we need not allow "aggregate" queries, since the aggregation algorithm Agg is unkeyed.

3   Constructing Aggregate MACs

In this section, we show that aggregate MACs can be constructed from essentially any standard message authentication code. We begin by illustrating the idea using as a building block the simple (standard) message authentication code constructed from a pseudorandom function F with output length n as follows: $\mathsf{Mac}_k(m) = F_k(m)$. In this case, given tags $\mathsf{tag}_1, \ldots, \mathsf{tag}_\ell$ associated with message/identifier pairs (m_i, i), respectively, we can aggregate these tags by simply computing the XOR of all the tag values; i.e.,

$$\mathsf{tag} = \mathsf{tag}_1 \oplus \mathsf{tag}_2 \oplus \cdots \oplus \mathsf{tag}_\ell.$$

(For simplicity, we consider identifiers $1, \ldots, \ell$ above. However, as we will see in the formal description below, these identifiers need not be distinct.) Verification is carried out in the obvious way: given a set of message/identifier pairs $M = \{(m_1, 1), \ldots, (m_\ell, \ell)\}$ and tag, the receiver outputs 1 if and only if

$$\mathsf{tag} = \bigoplus_{i=1}^{\ell} F_{k_i}(m_i).$$

As for the security of this scheme, we may argue informally as follows: say an adversary outputs $\{(m_1, \mathsf{id}'_1), \ldots, (m_\ell, \mathsf{id}'_\ell)\}$ and tag such that there exists an i for which $\mathcal{A}$ did not query either $\mathsf{Corrupt}(\mathsf{id}'_i)$ or $\mathsf{Mac}(\mathsf{id}'_i, m_i)$. Let $i^* = \mathsf{id}'_i$. Then, from the point of view of the adversary, the value $F_{k_{i^*}}(m_i)$ looks random. Since XORing a random(-looking) value with any other (uncorrelated) strings yields a random(-looking) string, we see that the value $\bigoplus_{i=1}^{\ell} F_{k_{\mathsf{id}'_i}}(m_i)$ computed by the receiver also looks random to the adversary, and cannot be guessed by the adversary with probability much better than 2^{-n}. We conclude that tag is a valid forgery with probability only negligibly better than 2^{-n}, and so the adversary does not succeed in outputting a valid forgery except with negligible probability.

Extending the above ideas, we may realize that the proof does not require the individual MAC tag $F_{k_{i^*}}(m_i)$ to be *pseudorandom*, but instead only requires that it be *unpredictable*. But this holds for *any* secure (standard) MAC, by the definition of security for MACs. Thus, as far as security is concerned, the above approach works for *any* underlying MAC. On the other hand, verification in the aggregate MAC requires that verification in the underlying MAC be done by re-computing the MAC tag and checking equality with what is received. (I.e., $\mathsf{Vrfy}_k(m, \mathsf{tag})$ outputs 1 if and only if $\mathsf{Mac}_k(m) = \mathsf{tag}$.) We may assume, without loss of generality, that verification is done this way for any *deterministic* MAC; for randomized MACs (and, in particular, MACs where messages have more than one valid tag for a given key), however, verification *cannot* be done this way. This means that certain randomized MACs (e.g., XOR-MAC [2]) cannot be utilized directly in the above construction, although we remark that any randomized MAC could be "derandomized" using a pseudorandom function. In any case, most commonly-used MACs are deterministic, and thus the restriction is not a serious one.

We now describe our aggregate MAC scheme formally, and rigorously prove its security with respect to Definition 2. Let (Mac, Vrfy) denote a standard message authentication code where Mac is a *deterministic* algorithm. (We will ignore the Vrfy algorithm from now on since, as noted above, we can perform verification by simply re-running Mac.) We have the following construction:

Construction 1. (Aggregate MAC Scheme)
Let Mac *be a deterministic algorithm. We define* (Mac*, Agg*, Vrfy*) *as follows:*

- **Algorithm** Mac*: *upon input* $k \in \{0,1\}^n$ *and* $m \in \{0,1\}^*$, *outputs* $\mathsf{Mac}_k(m)$.
- **Algorithm** Agg*: *upon input two sets* M^1, M^2 *of message/identifier pairs and two tags* $\mathsf{tag}^1, \mathsf{tag}^2$, *the algorithm outputs* $\mathsf{tag} = \mathsf{tag}^1 \oplus \mathsf{tag}^2$.
- **Algorithm** Vrfy*: *upon input a set of keys* $k_1, \ldots, k_t \in \{0,1\}^n$ *and a set* $M = \{(m_1, i_1), \ldots, (m_\ell, i_\ell)\}$ *of message/identifier pairs where* $i_\ell \in \{1, \ldots, t\}$ *for all* ℓ, *algorithm* Vrfy* *computes* $\mathsf{tag}' = \bigoplus_{j=1}^{\ell} \mathsf{Mac}_{k_{i_j}}(m_j)$, *and outputs 1 if and only if* $\mathsf{tag}' = \mathsf{tag}$. *(We stress that the input to* Vrfy* *is taken to be a set, and so all the tuples in* M *are distinct.)*

It is easy to verify correctness of the above scheme. As for security, we have:

Theorem 1. *If* (Mac, Vrfy) *is existentially unforgeable under an adaptive chosen-message attack and* Mac *is deterministic, then* (Mac*, Agg*, Vrfy*) *given in Construction 1 is a secure aggregate message authentication code.*

Proof: Fix a probabilistic polynomial-time adversary $\mathcal{A}$ and some $t = \mathsf{poly}(n)$ as in Definition 2. We construct a probabilistic polynomial-time algorithm $\mathcal{F}$ that interacts with an instance of (Mac, Vrfy) and attempts to produce a valid forgery for a previously-unauthenticated message. $\mathcal{F}$ is given access to an oracle $\mathsf{Mac}_{k^*}(\cdot)$ for an unknown key k^*, and proceeds as follows:

1. It chooses a random $i^* \leftarrow \{1, \ldots, t(n)\}$.
2. For $i = 1$ to $t(n)$:
   (a) If $i \neq i^*$, choose $k_i \leftarrow \{0,1\}^n$.
   (b) If $i = i^*$, do nothing (however, we *implicitly* set $k_{i^*} = k^*$).
3. Run $\mathcal{A}(1^n)$, answering its queries as follows:
   Query Mac(i, m): If $i \neq i^*$ then $\mathcal{F}$ answers the query using the known key k_i. If $i = i^*$ then $\mathcal{F}$ queries its own MAC oracle $\mathsf{Mac}_{k^*}(\cdot)$ and returns the result.
   Query Corrupt(i): If $i \neq i^*$ then give $\mathcal{A}$ the known key k_i. If $i = i^*$ then abort.
4. At some point, $\mathcal{A}$ outputs $M = \{(m_1, \mathsf{id}'_1), \ldots, (m_\ell, \mathsf{id}'_\ell)\}$ and tag. Let j be the first index such that (1) $\mathcal{A}$ never queried Corrupt(id'_j) and (2) $\mathcal{A}$ never queried Mac(id'_j, m_j). (We assume without loss of generality that some such j exists.) If $\mathsf{id}'_j \neq i^*$ then abort; otherwise, proceed as described below.
5. Assuming $\mathsf{id}'_j = i^*$, algorithm $\mathcal{F}$ computes

$$\mathsf{tag}^* = \mathsf{tag} \oplus \left(\bigoplus_{i \neq j} \mathsf{Mac}_{k_{\mathsf{id}'_i}}(m_i) \right),$$

where $\mathcal{F}$ computes $\mathsf{Mac}_{k_{\mathsf{id}_i'}}(m_i)$ using the known key $k_{\mathsf{id}_i'}$ when $\mathsf{id}_i' \neq i^*$, and computes $\mathsf{Mac}_{\mathsf{id}_i'}(m_i)$ by querying its MAC oracle $\mathsf{Mac}_{k^*}(\cdot)$ when $\mathsf{id}_i' = i^*$. Finally, $\mathcal{F}$ outputs (m_j, tag^*).

The proof follows easily from the following observations:

- The probability that $\mathcal{F}$ aborts is exactly $1/t(n)$, which is inverse polynomial. Furthermore, conditioned on not aborting, the simulation that $\mathcal{F}$ provides for $\mathcal{A}$ is perfect.
- If $\mathcal{A}$ succeeds in a given execution (and $\mathcal{F}$ does not abort), then $\mathcal{F}$ outputs a valid forgery. To see this, note that when $\mathcal{A}$ succeeds this means that

$$\bigoplus_{i=1}^{\ell} \mathsf{Mac}_{k_{\mathsf{id}_i'}}(m_i) = \mathsf{tag},$$

where we stress that $\mathsf{Mac}_{k_{\mathsf{id}_i'}}(m_i)$ is a *fixed*, well-defined value by virtue of the fact that Mac is deterministic. Thus, the value tag^* output by $\mathcal{F}$ is equal to the (well-defined) value $\mathsf{Mac}_{k_{i^*}}(m_j) = \mathsf{Mac}_{k^*}(m_j)$. Furthermore, $\mathcal{F}$ has never queried its own MAC oracle with the message m_j since, by assumption, $\mathcal{A}$ never queried $\mathsf{Mac}(i^*, m_j)$ prior to step 5 of $\mathcal{F}$'s execution, above, and $\mathcal{F}$ will not query m_j to its MAC oracle in step 5 since all tuples in the set M must be distinct.

This completes the proof.   ■

Efficiency. Our construction for aggregate MACs is highly efficient. Consider the example of a mobile ad-hoc networks (MANET) as described in the introduction. If the nodes are arranged as a binary (or any other) tree, then each node receives a set of messages together with a single tag from each of its children. In order to forward the messages on, all the node needs to do is to concatenate the lists of messages, compute its own MAC, and XOR all the tags together.

4   An Extension and a Lower Bound

A limitation of the construction given in the previous section is that the receiver must re-compute the (individual) MAC tags on *all* ℓ messages whose tags have been aggregated. This is not a limitation in the MANET example given above. However, in some cases, the receiver may only be interested in verifying the authenticity of a *single* message (or some small subset of the messages). In such cases, the requirement to re-compute the MAC tags of all the messages is undesirable.

In this section, we present a simple idea that offers a trade-off between the length of the aggregate tag and the time required to verify a single message. To achieve authentication of a single message in *constant* time (i.e., independent of the number of aggregated tags ℓ), our approach yields a tag of length $\mathcal{O}(\ell \cdot T)$, where we take T to be the length of the tag in some underlying (standard) MAC.

This is not of much interest because we can achieve a tag of length $\mathcal{O}(\ell \cdot T)$ by just concatenating the tags of a standard MAC (i.e., aggregation equals concatenation). However, our approach yields a tradeoff where the product of the authentication time and tag length is $\mathcal{O}(\ell \cdot T)$. In the previous section, we achieved authentication in time ℓ with a tag of length T. At the other extreme, concatenating MAC tags gives authentication in constant time (i.e., requires verifying a single MAC) but has a tag of length $\ell \cdot T$. Our approach, described below, allows essentially anything in between. In particular, one can achieve authentication in time $\mathcal{O}(\sqrt{\ell})$ with a tag of length $\mathcal{O}(\sqrt{\ell} \cdot T)$.

It is interesting to wonder whether this is optimal. In this direction, we also present a lower bound showing that this approach *is* asymptotically optimal (up to a multiplicative factor of T) when considering verification that takes constant, or at most logarithmic, time. That is, we show that any aggregate MAC scheme that enables authentication in logarithmic time (in ℓ, the number of aggregated MACs) must have a tag of length at least $\Omega(\ell)$.

We stress that in this section, we consider the running time as a function of the number ℓ of messages. Of course, it also takes time to compute and verify a single MAC tags. However, this is a fixed overhead for every value of the security parameter, and so what is really of interest is how many MAC tags need to be computed to verify a *single* message, when the number of aggregated MACs is ℓ.

4.1   The Construction

Before presenting our construction, we first describe the problem in a bit more detail. Recall from Definition 1 that the receiver holds a set of keys $k_1, \ldots, k_t$, and is assumed to receive a set of message/identifier pairs $M = \{(m_1, \mathsf{id}_1), \ldots, (m_\ell, \mathsf{id}_\ell)\}$ and a tag tag. In this section, we assume the receiver does not care to simultaneously verify the authenticity of *all* messages in M (with respect to the identifier associated with each message) as in the previous section, but instead is interested only in verifying authenticity of *one* of the messages m_i (with respect to the associated identifier id_i). Obviously, the only solutions of interest are those that are more efficient than verifying everything.

A fairly straightforward solution is as follows. Fix some parameter ℓ'. Then run multiple instances of the "base aggregation scheme" from the previous section in parallel, but only aggregating at most ℓ' messages/tags using any given instance. (We stress that each sender still holds only one key, the verifier still holds one key per sender, and the Mac^* algorithm is unchanged. All that changes is the way aggregation and verification are performed.) The net result is that a set of message/identifier pairs $M = \{(m_1, \mathsf{id}_1), \ldots, (m_\ell, \mathsf{id}_\ell)\}$ is now authenticated by a sequence of $\ell^* = \lceil \ell/\ell' \rceil$ tags $\mathsf{tag}_1, \ldots, \mathsf{tag}_{\ell^*}$ generated according to the base scheme, where tag_1 authenticates $m_1, \ldots, m_{\ell'}$ (with respect to the appropriate associated identities), tag_2 authenticates $m_{\ell'+1}, \ldots, m_{2\ell'}$, etc. To verify the authenticity of any particular message m_i, the verifier need only re-compute MAC tags for (at most) $\ell' - 1$ other messages.

The tag when ℓ messages are authenticated is now the length of $\lceil \ell/\ell' \rceil$ basic MAC tags (i.e., length $\lceil \ell/\ell' \rceil \cdot T$), and the time for verifying any particular message is improved to $\mathcal{O}(\ell')$ (instead of $\mathcal{O}(\ell)$ as previously). Thus, for example, setting $\ell' = \sqrt{\ell}$ we obtain verification of time $\mathcal{O}(\sqrt{\ell})$ and a tag that is comprised of $\sqrt{\ell}$ basic MAC tags. We remark that the time required to verify *all* the messages is essentially the same as before. Achieving *constant* verification time for any single message using this approach would result in a tag of (total) length linear in the number of messages being authenticated. In particular, when $\ell' = 1$ we obtain an "aggregate" scheme which simply concatenates MAC tags of all the messages being authenticated.

4.2   A Lower Bound

As we have mentioned, when constant verification time is desired (i.e., $\ell' = 1$ in the scheme of the previous section), the result is a MAC tag that consists of ℓ basic MAC tags (i.e., the aggregation works by just concatenating MAC tags). This is rather disappointing and it would be highly desirable to improve this situation. In this section we show that it is impossible to achieve a better result since the above is essentially optimal. Informally speaking, we show that if verification can be carried out in constant time (or even in time $\mathcal{O}(\log \ell)$), then the tag must be at least $\Omega(\ell)$ bits long.

Before proceeding further, we observe this does not contradict the positive result we obtained above. This is because we must have $T = \omega(\log n)$ (otherwise an adversary can guess a valid MAC tag, in the underlying scheme, with non-negligible probability) and because ℓ, the number of aggregated MACs, can be at most polynomial in n (or else it does not make much sense to talk about security of the scheme). Thus, the tag length of our previous construction when $\ell' = \mathcal{O}(\log \ell)$ is

$$T \cdot \ell / \log \ell = \omega(\log n) \cdot \ell / \mathcal{O}(\log n) = \omega(\ell),$$

as required. We now formally state and prove the lower bound:

Claim 1. *Any aggregate MAC scheme in which verification of a single message can be carried out in time $\mathcal{O}(\log \ell)$ (where ℓ denote the total number of messages authenticated by an aggregate tag) has tags whose length is $\Omega(\ell)$.*

Proof: We begin by providing intuition as to why the claim is true. Assume that there exists an aggregate MAC scheme where verification of a single message takes time $\log \ell$, and the tags are of length less than ℓ. The main observation is that if verification takes time $\log \ell$, then the Vrfy algorithm can only read at most $\log \ell$ of the messages whose MACs are aggregated. If each of these messages consists of one bit only, then it is possible to try all possible combinations of the $\log \ell$ bits to see which passes verification (this takes time $2^{\log \ell} = \ell$ which is feasible). A key point here, of course, is that only a correct combination should pass or this could be used to efficiently construct a forgery of the aggregate MAC scheme. This implies that it is possible to reconstruct $\log \ell$ of the (single-bit) messages *given only the MAC tag*. However, this holds for all subsets of $\log \ell$ bits and so *all* of the messages can be reconstructed in polynomial-time given only the MAC tag. But this means that it is possible to reconstruct any ℓ-bit message from a tag of length less than ℓ. Stated differently, it means that an arbitrary ℓ-bit message can be compressed, something that is known to be impossible! Our formal proof follows this intuition with some minor changes, the main one being that we show how to reconstruct one bit at a time rather than blocks of $\log \ell$ bits. Furthermore, we derive our contradiction through lower bounds for probabilistic communication complexity rather than through compression; this is easier because of the negligible probability of error that exists when working with any cryptographic primitive.

We will use the *public random string model* of communication complexity, where two parties share a common random string and the question is how many bits must they communicate in order to correctly compute a function f. Given a protocol Π, the *error* of this protocol is given by $\max_{x,y} \{\Pr[\Pi(x,y) = f(x,y)]\}$ where the probability is taken over the parties' common random string as well as any internal randomness they might use. We let $CC_\epsilon(f)$ denote the minimum number of bits need to compute f, where this minimum is taken over all possible protocols with error at most ϵ. It is known that there exist functions $f : \{0,1\}^\ell \times \{0,1\}^\ell \to \{0,1\}$ for which $CC_\epsilon(f) = \Omega(\ell)$; the inner-product function $IP(x,y) = \sum_{i=1}^\ell x_i y_i \bmod 2$ is one example. See [10,15] for more on communication complexity.

Let $(\mathsf{Mac}^*, \mathsf{Agg}^*, \mathsf{Vrfy}^*)$ be an aggregate MAC scheme in which verification of any message can be carried out in time $\mathcal{O}(\log \ell)$. At the very least, this implies that given any set of messages $M = \{m_1, \ldots, m_\ell\}$ and a single identifier id (using a single identifier just simplifies the proof), verification of a single message m_i (with respect to id) can be carried out by examining only $w = \mathcal{O}(\log \ell)$ other messages in M. (This is due to the fact that it is not possible to read more than $\log \ell$ messages in $\log \ell$ time.) We show that such a scheme implies that the probabilistic communication complexity (in the public random string model) of *every* function $f : \{0,1\}^\ell \times \{0,1\}^\ell \to \{0,1\}$ is essentially the length of a tag for ℓ messages. However, since there exist functions with communication complexity $\Omega(\ell)$ (see the discussion in the previous paragraph) it follows that the tag length must also be $\Omega(\ell)$.

We begin by describing a protocol for computing any function $f : \{0,1\}^\ell \times \{0,1\}^\ell \to \{0,1\}$ with communication complexity that is equal to the tag length plus 1. We stress that this protocol is for the setting of communication complexity and not cryptography. Thus, the parties A and B are fully honest and the only question is how many bits must be sent (there is no requirement on privacy, etc.). Loosely speaking, the protocol we describe works by having A compute an aggregate MAC on her input and then send the tag to B. Party B then reconstructs A's input from the tag, as described in the intuitive discussion above. Finally, given A's full input, B computes the output and sends it to A.

Before formally describing the protocol, we show how to encode a single ℓ-bit input into an aggregate MAC over ℓ messages. Let $x = x_1 \cdots x_\ell$ be A's input. Then, A defines messages $m_1, \ldots, m_\ell$ by $m_i = \langle i \rangle \| x_i$, where $\langle i \rangle$ is the binary encoding of i. The set $\{m_1, \ldots, m_\ell\}$ is a valid encoding of x because it fully defines x (all bits of x are represented, and their positions in x are given by the encoding of j that is included in every m_j). We remark that since only one id is used here, we ignore it from here on. We now describe the protocol:

Protocol 1. (communication complexity protocol for any function f)

- **Inputs:** A has $x \in \{0,1\}^\ell$ and B has $y \in \{0,1\}^\ell$.
- **Public random string:** both parties share a random string $k \in \{0,1\}^n$ for some sufficiently large n.
- **The protocol:**
    1. A's first step:
        (a) Party A encodes its input $x = x_1 \cdots x_\ell$ into a set of ℓ messages $M = \{m_1, \ldots, m_\ell\}$ where $m_i = \langle i \rangle \| x_i$ for every i.
        (b) A computes $\mathsf{tag}_i \leftarrow \mathsf{Mac}_k^*(m_i)$ for all i, and then aggregates all the results into a single tag tag^* by using the algorithm Agg^*.
        (c) A sends tag^* to B.
    2. Upon receiving tag^* from A, party B works as follows for $i = 1, \ldots, \ell$:
        (a) B sets $m_i = \langle i \rangle \| 0$ (this can be viewed as a guess that $x_i = 0$) and attempts to run $\mathsf{Vrfy}_k^*(M, \mathsf{tag}^*)$. However, Vrfy^* expects to receive M and in general may read up to $\log \ell$ other messages in M.[3] Therefore, B proceeds as follows:
        (b) Let $\mathsf{Vrfy}_k^*((j_1, m_{j_1}), \ldots, (j_t, m_{j_t}), \mathsf{tag}^*)$ be the algorithm defined by Vrfy^* after it has read the t messages indexed by $j_1, \ldots, j_t$ with content $m_{j_1}, \ldots, m_{j_t}$; note that m_i is not included in this notation as we assume it is read first. (Essentially, this algorithm is defined by fixing the prefix of its execution until this point.)
        (c) For $t = 0, \ldots, \log \ell$, B works as follows:
            i. If $t = \ell$, then return the output bit of
            $$\mathsf{Vrfy}_k^*((j_1, m_{j_1}), \ldots, (j_t, m_{j_t}), \mathsf{tag}^*)$$
            ii. Else , invoke $\mathsf{Vrfy}_k^*((j_1, m_{j_1}), \ldots, (j_t, m_{j_t}), \mathsf{tag}^*)$ and let j_{t+1} be the next message read by $\mathsf{Vrfy}_k^*((j_1, m_{j_1}), \ldots, (j_t, m_{j_t}), \mathsf{tag}^*)$.
            iii. Recursively invoke $\mathsf{Vrfy}_k^*((j_1, m_{j_1}), \ldots, (j_t, m_{j_t}), (j_{t+1}, 0), \mathsf{tag}^*)$ and $\mathsf{Vrfy}_k^*((j_1, m_{j_1}), \ldots, (j_t, m_{j_t}), (j_{t+1}, 1), \mathsf{tag}^*)$, and return the logical OR of their outputs.
            If the output of Vrfy^* from the above procedure equals 1, then B sets $x_i = 0$. Otherwise, it sets $x_i = 1$.
    3. Given $x = x_1, \ldots, x_\ell$, B computes $f(x, y)$ and returns the result to A.
    4. Both parties output $f(x, y)$.

Note that B's procedure is such that if *any* of the recursive threads returns 1 then B sets $x_i = 0$. However, if this occurs, then this means that there exists a subset of $\log \ell$ messages $m_{j_1}, \ldots, m_{j_{\log \ell}}$ such that Vrfy accepts $m_i = \langle i \rangle \| 0$ relative to this subset. On the other hand, if this does not occur, then Vrfy rejects for all such sets, in which case B sets $x_i = 1$.

---

[3] Without loss of generality we assume that Vrfy first reads m_i and then up to $\log \ell$ other messages.

It is clear that if B reconstructs x correctly then the protocol is correct. It therefore remains to show that B correctly reconstructs x except with negligible probability. (Actually, in the context of communication complexity it suffices to show that this holds except with some constant probability. However, we show something stronger.)

We separately analyze the case that $x_i = 0$ and $x_i = 1$. In the case of $x_i = 0$ we have that A generated tag* with $x_i = 0$ and some setting of the other bits. Therefore, there must exist some subset of $\log \ell$ messages that results in Vrfy accepting (this subset is defined by A's real input x). Thus, when $x_i = 0$, party B *always* sets $x_i = 0$. (This follows from the correctness condition of MACs that states that if a tag is correctly constructed, then Vrfy will always output 1.)

The more challenging case is that of $x_i = 1$. Assume that there exists a message $x = x_1, \dots, x_\ell$ and an i such that with probability p, party B's procedure on an aggregate tag tag* computed from x is such that $x_i = 1$ but B sets $x_i' = 0$. We use this to construct an adversary $\mathcal{A}$ that breaks the MAC scheme (Mac*, Agg*, Vrfy*) with probability p/ℓ.[4] Adversary $\mathcal{A}$ encodes x into a set M exactly as party A does. It then uses its Mac* oracle to compute an aggregate MAC on the set of ℓ messages M; let tag* be the result. Next, $\mathcal{A}$ chooses uniformly distributed bits $b_1, \dots, b_\ell \in \{0, 1\}$ and constructs a new set M' where $m_i' = \langle i \rangle \| 0$ and for all $j \neq i$, $m_j' = \langle j \rangle \| b_j$. Finally, $\mathcal{A}$ outputs the set M' and the tag tag*. (We note that $\mathcal{A}$ uses oracles whereas the party A used the public random string k. However, party A's procedure does not use k in any way except to compute Mac* legitimately and so $\mathcal{A}$ can simulate this using its Mac* oracle.)

We claim that $\mathcal{A}$ succeeds in breaking the MAC scheme with probability p/ℓ. This is due to the following facts:

1. The set M' is such that $\mathcal{A}$ never queried $\text{Mac}(m_i')$ (because $m_i = \langle i \rangle \| 1$ but $m_i' = \langle i \rangle \| 0$).
2. $\mathcal{A}$ did not send any Corrupt queries
3. With probability $2^{-\log \ell} = 1/\ell$ the random bits chosen by $\mathcal{A}$ that are read by Vrfy_k are equal to those that result in B's procedure erring. Therefore,

$$\text{Prob}[\text{Vrfy}_k^*(M', \text{tag}^*)] = \frac{p}{\ell}.$$

We conclude that $\mathcal{A}$ succeeds in its attack with probability p/ℓ, implying that p must be negligible (by the assumption that the scheme is secure). This implies that Protocol 1 (probabilistically) computes f with communication complexity $|\text{tag}^*| + 1$. Since there exist functions f for which the communication complexity is $\Omega(\ell)$, this therefore implies that $|\text{tag}^*| = \Omega(\ell)$ as required.

We conclude by remarking that it is actually only required that the adversary $\mathcal{A}$ run in polynomial-time in order to reach a contradiction regarding the MAC.

---

[4] We are being slightly informal here. What we prove is that for a given n and pair (x, i), a non-uniform adversary will succeed in breaking the MAC scheme with probability that is polynomially related to the probability that B's procedure errs regarding the ith bit. This will then imply that for all sufficiently large n's, the probability $p = p(n)$ must be negligible, as required.

In contrast, B can run in time 2^{ℓ} and this makes no difference (the bounds in communication complexity hold irrespective of the computational complexity of the parties). The crucial point is that $\mathcal{A}$ does *not* run B and so its complexity does not depend on B. Thus B could just try all 2^{ℓ} strings to see if one results in the MAC being accepted. The probability of $\mathcal{A}$ generating a successful forgery remains the same because it is simply based on a random guess.   ∎

In summary, it is not possible to do (much) better than our solution of the previous section when constant- or logarithmic-time verification is required. An interesting question remains as to whether it is possible to do better than the tradeoff achieved by our construction when ℓ' is asymptotically larger than $\log \ell$. It would also be interesting to close the remaining multiplicative factor of T (the tag length of the underlying MAC).

Acknowledgments

The work of the first author was supported by NSF grant #0627306, and by the US Army Research Laboratory and the UK Ministry of Defence under Agreement Number W911NF-06-3-0001. The views and conclusions contained in this document are those of the authors and should not be interpreted as representing the official policies, either expressed or implied, of the US Army Research Laboratory, the US Government, the UK Ministry of Defense, or the UK Government. The US and UK Governments are authorized to reproduce and distribute reprints for Government purposes, notwithstanding any copyright notation herein.

References

1. Bellare, M., Goldreich, O., Mityagin, A.: The Power of Verification Queries in Message Authentication and Authenticated Encryption, http://eprint.iacr.org/2004/309
2. Bellare, M., Guérin, R., Rogaway, P.: XOR MACs: New Methods for Message Authentication Using Finite Pseudorandom Functions. In: Coppersmith, D. (ed.) CRYPTO 1995. LNCS, vol. 963, pp. 15–28. Springer, Heidelberg (1995)
3. Bhaskar, R., Herranz, J., Laguillaumie, F.: Aggregate Designated Verifier Signatures and Application to Secure Routing. Intl. J. Security and Networks 2(3/4), 192–201 (2007)
4. Boldyreva, A., Gentry, C., O'Neill, A., Yum, D.H.: Ordered Multisignatures and Identity-Based Sequential Aggregate Signatures, with Applications to Secure Routing. In: ACM CCCS (2007)
5. Boneh, D., Gentry, C., Lynn, B., Shacham, H.: Aggregate and Verifiably Encrypted Signatures from Bilinear Maps. In: Biham, E. (ed.) EUROCRYPT 2003. LNCS, vol. 2656, pp. 416–432. Springer, Heidelberg (2003)
6. Gentry, C., Ramzan, Z.: Identity-Based Aggregate Signatures. In: Yung, M., Dodis, Y., Kiayias, A., Malkin, T.G. (eds.) PKC 2006. LNCS, vol. 3958, pp. 257–273. Springer, Heidelberg (2006)
7. Chan, H., Perrig, A., Song, D.: Secure Hierarchical In-Network Aggregation in Sensor Networks. In: ACM CCCS, pp. 278–287 (2006)

8. Goldwasser, S., Micali, S., Rivest, R.: A Digital Signature Scheme Secure against Adaptive Chosen-Message Attacks. SIAM J. Computing 17(2), 281–308 (1988)
9. Hu, L., Evans, D.: Secure Aggregation for Wireless Networks. In: Workshop on Security and Assurance in Ad-Hoc Networks, pp. 384–394 (2003)
10. Kushilevitz, E., Nisan, N.: Communication Complexity. Cambridge University Press, Cambridge (1996)
11. Lu, S., Ostrovsky, R., Sahai, A., Shacham, H., Waters, B.: Sequential Aggregate Signatures and Multisignatures Without Random Oracles. In: Vaudenay, S. (ed.) EUROCRYPT 2006. LNCS, vol. 4004, pp. 465–485. Springer, Heidelberg (2006)
12. Lysyanskaya, A., Micali, S., Reyzin, L., Shacham, H.: Sequential Aggregate Signatures from Trapdoor Permutations. In: Cachin, C., Camenisch, J.L. (eds.) EUROCRYPT 2004. LNCS, vol. 3027, pp. 74–90. Springer, Heidelberg (2004)
13. Mu, Y., Susilo, W., Zhu, H.: Compact Sequential Aggregate Signatures. In: 2007 ACM Symposium on Applied Computing (SAC), pp. 249–253 (2007)
14. Przydatek, B., Song, D., Perrig, A.: SIA: Secure Information Aggregation in Sensor Networks. In: SenSys 2003, pp. 255–265 (2003)
15. Raz, R.: Lecture Notes on Circuit Complexity and Communication Complexity. IAS Summer School,
http://www.wisdom.weizmann.ac.il/~ranraz/lecturenotes/index.html
16. Shacham, H.: New Paradigms in Signature Schemes. PhD Thesis, Stanford University (2005)

Boosting AES Performance on a Tiny Processor Core

Stefan Tillich and Christoph Herbst

Graz University of Technology,
Institute for Applied Information Processing and Communications,
Inffeldgasse 16a, A–8010 Graz, Austria
{Stefan.Tillich,Christoph.Herbst}@iaik.tugraz.at

Abstract. Notwithstanding the tremendous increase in performance of desktop computers, more and more computational work is performed on small embedded microprocessors. Particularly, tiny 8-bit microcontrollers are being employed in many different application settings ranging from cars over everyday appliances like doorlock systems or room climate controls to complex distributed setups like wireless sensor networks. In order to provide security for these applications, cryptographic algorithms need to be implemented on these microcontrollers. While efficient implementation is a general optimization goal, tiny embedded systems normally have further demands for low energy consumption, small code size, low RAM usage and possibly also short latency. In this work we propose a small enhancement for 8-bit Advanced Virtual RISC (AVR) cores, which improves the situation for all of these demands for implementations of the Advanced Encryption Standard. Particularly, a single 128-bit block can be encrypted or decrypted in under 1,300 clock cycles. Compared to a fast software implementation, this constitutes an increase of performance by a factor of up to 3.6. The hardware cost for the proposed extensions is limited to about 1.1 kGates.

Keywords: Advanced Encryption Standard, instruction set extensions, 8-bit microcontroller, AVR architecture, hardware-software codesign.

1 Introduction

In recent years, small 8-bit microcontrollers have experienced an increase in popularity due their suitability for exciting new applications in the embedded systems field. A good example is the advent of wireless sensor networks, which require data processing with low energy overhead. In general, the application of such small microcontrollers is conditioned by constraints in energy budget and/or device cost. A common problem encountered by system designers is the relatively low speed and limited memory of 8-bit microcontrollers. Modern architectures like AVR have alleviated the problem to a certain extent, but careful software implementation remains nevertheless a topic of importance.

Providing security to embedded applications demands the use of strong cryptographic algorithms. In this field, symmetric cryptographic primitives can provide users with confidentiality and integrity of data as well as authentication

T. Malkin (Ed.): CT-RSA 2008, LNCS 4964, pp. 170–186, 2008.

services. An important and increasingly popular symmetric algorithm is the Advanced Encryption Standard (AES) algorithm [16], which has been standardized by NIST in 2001 to replace the aging Data Encryption Standard.

Processing of cryptographic algorithms is normally a rather heavy burden on 8-bit microcontrollers and it is generally desirable to keep the overhead for cryptography as low as possible. In the present work we propose to enhance an 8-bit AVR core with some custom instructions (instruction set extensions) in order to speed up AES encryption and decryption. The rest of this paper is organized as follows. In Section 2 we give an overview of previous and related work on instruction set extensions for cryptography. Section 3 provides an overview of the general AVR microcontroller architecture. We present the AES extensions in Section 4. Subsequently, we deal with general implementation issues related to both hardware and software in Section 5. We describe details of our hardware implementation, give performance estimations, and compare the results to related work in Section 6. Conclusions are drawn in Section 7.

2   Previous Work on Instruction Set Extensions

Nahum et al. were the first to suggest to base RISC processor design on the need for supporting a large set of cryptographic software implementations [15]. Jean-François Dhem showed in his Ph.D. thesis the first concrete enhancements for a processor architecture (ARM7M) in the form of long integer modulo support for public-key algorithms [8]. First publications regarding concrete instruction set extensions for secret-key primitives started to appear around 2000 [4,12,19]. A bulk of research has been done in the following years, dealing with cryptography enhancements of general-purpose processors for both public-key and secret-key algorithms. Topics ranged from automatic design space exploration (e.g., [17]) over efficient implementation (e.g., [13]) to resistance against side-channel attacks (e.g., [22]). Closely related to the field of instruction set extensions is the work on dedicated cryptographic processors like the *CryptoManiac* [25] and the *Cryptonite* [3], which are both VLIW architectures.

While earlier work for the secret-key domain tended to focus on a broad support of algorithms, more recent work concentrated on single cryptographic primitives. Due to its increasing importance after standardization, the AES algorithm has received particular attention. Nadehara et al. suggested to map the so-called round lookup (T lookup) of AES into a dedicated functional unit [14]. Bertoni et al. and Tillich et al. suggested independently to allow for a finer granularity of operations, separating the S-box lookup from the ShiftRows and MixColumns transformation [2,21].

So far, almost all architectural extensions for cryptography have been proposed for processors of a word size of 32 bits or more. An exception is the work of Eberle et al. which describes support for ECC over binary extension fields $GF(2^m)$ for the AVR architecture [9]. A custom 8-bit microcontroller for AES has been presented by Chia et al. in [6], with a focus on minimizing code size rather than performance.

3   Overview of AES and AVR

3.1   Short Description of AES

The AES algorithm is a subset of the block cipher Rijndael. The NIST standard fixes the block size to 128 bit and provides three different key sizes: 128, 192, and 256 bit. The 128-bit block is arranged into a logical 4×4-byte matrix, which is commonly denoted as the *AES State*. This State is transformed in a number of identical rounds, each of which consists of the four transformations SubBytes, ShiftRows, MixColumns, and AddRoundKey. An exception is the last round, where the MixColumns transformation is omitted. In each round, a different round key derived from the cipher key is used. SubBytes substitutes individual bytes of the State using a single S-box table, consisting of 256 8-bit entries. ShiftRows rotates the rows of the State, while MixColumns operates on complete State columns, interpreting them as polynomials over $GF(2^8)$. AddRoundKey combines the State and the current round key by means of bitwise exclusive or (XOR). For AES decryption, the inverse transformations InvSubBytes, Inv-ShiftRows, InvMixColumns, and AddRoundKey (which is its own inverse) are applied to the ciphertext block in reverse order. For more details on Rijndael and AES, please refer to [7,16].

On 32-bit processors most of the AES round transformations (SubBytes, ShiftRows, and MixColumns) can be implemented by table lookup, using one or more round lookup (T lookup) tables with 256 32-bit entries. This approach can be scaled down to 8-bit implementations like those of Rinne et al. which we have used for performance comparison [18]

3.2   Description of the AVR Architecture

The Advanced Virtual RISC (AVR) by Atmel is an 8-bit Harvard architecture microcontroller. This means, that the data and program memory are separated. The program memory is implemented as an in-system programmable FLASH memory whose size can vary from 1 kByte to 256 kBytes depending on the model. The available RAM and internal in-system programmable EEPROM also depends on the model. RAM size can vary from 32 bytes to 8 kBytes, whereas the EEPROM size which is a non-volatile memory mainly used to store parameters ranges from 0 kBytes to 4 kBytes.

The AVR instruction set consists of approximately 110 different instructions. Most instructions are encoded with 16 bit and operate on the 32 general-purpose registers of the architecture. Six of these registers can also act as three independent 16-bit pointers for memory access. Most of the instructions require only a single clock cycle to execute. Only a few instructions take two to four clock cycles to finish. The instructions are directly executed from the FLASH memory. Some of the controllers not only support in-system programming but also self-programming is supported. That enables the controllers to reload source code during runtime and supports the flexibility of applications implemented on AVR microcontrollers.

To address the requirements of low-power designs, the supply voltage for the AVR family ranges from 1.8 V to 5.5 V. The controllers are equipped with a sleep controller which supports various modes and the operation frequency can be controlled by software to support power save modes. The AVR family is built to support clock frequencies up to 20 MHz.

The AVR microcontrollers are explicitly designed to be programmed in C. There are various free software development kits available like avr-gcc for compiling C code and AVR Studio including a simulator. The availability of free development tools supports the widespread use of the AVR controllers in various embedded applications like sensor nodes.

4   Our Proposed AES Extensions

All AES extensions proposed so far in the literature try to make full use of the 32-bit datapath of the underlying processor [2,14,21]. Therefore, none of these solutions can be scaled down to an 8-bit architecture in a straight-forward way. As we will show in this section, it is however possible to reuse some of the important concepts of these 32-bit approaches to arrive at a worthwhile solution for small microcontrollers.

4.1   Support for AES Encryption

We propose three instructions to speed up AES encryption, whereby two instructions are intended to speed up the AES round transformations, while the third instruction is conceived for use in the final round and also in the key expansion. The instruction formats fully adhere to the AVR architecture and therefore allow for easy integration. All instructions use similar hardware components and a small and flexible functional unit can be easily designed to reach the maximal speed of current state-of-the-art AVR cores (which ranges at the time of writing at around 20 MHz).

Our basic concept is to use the capability of typical AVR microcontrollers to retrieve two register values per clock cycle [1]. With appropriate selection of the register operands, all four AES round transformations can be executed for two State bytes with only a few instructions. In the best case, a complete round for an AES State column (contained in registers) can be processed and stored back to the original registers in only 15 clock cycles.

The functionality of the two instruction variants AESENC(1) and AESENC(2) is depicted in Figure 1. Note that the symbols $\oplus$ and $\otimes$ denote addition (conforming to bitwise XOR) and multiplication in the Galois field $GF(2^8)$, respectively. These instructions have the same format as the integer multiplication instruction MUL of the basic AVR architecture. First, the values from the two specified registers Rd and Rr are substituted according to the AES (forward) S-box. Depending on the instruction variant, the substituted bytes are multiplied with specific constants from the field $GF(2^8)$ (we use the notation $\{x\}$ to discern such constants from integers). Two of the multiplication results

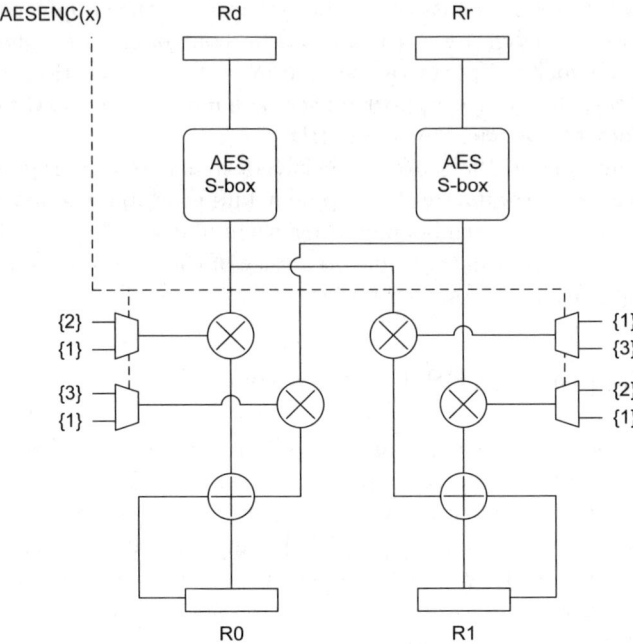

Fig. 1. AES extensions for a "normal" encryption round

are then combined with the values from the registers R0 and R1 by means of an XOR operation. The resulting values are stored to the registers R0 and R1.

The intended use of the AESENC(x) instructions (where $x \in \{1, 2\}$) is to perform all transformations of a single AES round on two bytes of a State column with merely two invocations. The $GF(2^8)$ constants have been chosen carefully from the AES MixColumns matrix. Each invocation of AESENC(x) conforms to the processing of a quadrant of that matrix. Due to the symmetry of the MixColumns matrix, there are only two distinct quadrants. Therefore, the two variants of AESENC(x) instruction are sufficient to transform the complete AES State.

The AESENC(x) instructions can be used to produce two State bytes at the end of a round from the according four State bytes at the start of the round and the corresponding two bytes of the round key. In order to do this, the two bytes of the round key are loaded into R1 and R0 and then AESENC(1) and AESENC(2) are invoked with the according State bytes to produce a half of the resulting State column. The feedback from R1 and R0 into the final XOR stage (cf. Figure 2) has a dual functionality: On the first invocation of AESENC(x), the round key bytes are added to the intermediate result. On the second invocation, this intermediate result is combined with the contribution from the other State bytes in the final XOR stage.

Our approach is similar to the ones of [14] and [2] in that it tries to pack as many operations as possible into a single instruction. It has been shown in [21] that slight modifications can lead to a considerable increase in implementation flexibility. Therefore, we also propose a lightweight variant of the `AESENC(x)` instruction, which can be used in the final round of AES encryption as well as in the key expansion. The functionality of this instruction, which we denote by `AESSBOX` is shown in Figure 2.

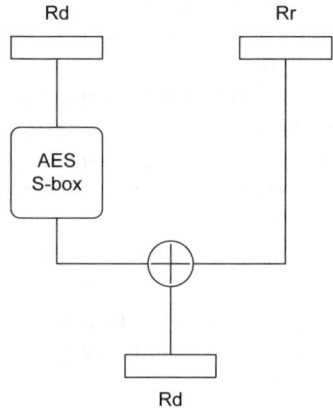

Fig. 2. AES extension for the final round

The `AESSBOX` instruction adheres to the "two-input, one-output" format, which is common to most of the arithmetic and logic instructions of the AVR architecture, e.g., integer addition `ADD` and bitwise exclusive or `EOR`. One of the two input registers (namely `Rd`) is also the target register of the instruction, while the second input register (`Rr`) can be chosen freely. For our proposed `AESSBOX` instruction, the value from register `Rd` is substituted according to the AES S-box and XORed to the value from register `Rr`.

4.2 Support for AES Decryption

Most common modes of operations of block ciphers are defined with the sole use of the according encryption function, e.g., the CTR mode for confidentiality and the CBC-MAC variants for authentication. However, in some situations the decryption function of the block cipher might be of use, e.g., when CBC encryption mode is preferred over CTR mode. For this case we also propose instruction set support for AES decryption, additionally motivated by the following reasons:

- Decryption support can be seamlessly integrated with encryption support with little extra hardware cost.
- With these extensions, decryption speed can be made equal to that of encryption, opening up additional options for more flexible protocol implementations.

Similarly to encryption, decryption support consists of the two instruction variants AESDEC(1) and AESDEC(2), conforming to the two distinct quadrants of the InvMixColumn constant matrix. Another necessary change is the use of the inverse S-box.

Decryption support incurs a slight complication of the implementation in regard to the AddRoundKey transformation. For AESENC(x) instructions, the final XOR stage (cf. Figure 2) performs both AddRoundKey and a combination of intermediate values to yield the State bytes at the end of the round. The AESDEC(x) instructions require AddRoundKey at a different stage (after the inverse S-boxes), due to the slightly changed order of inverse round transformations in AES decryption [16]. One possible solution is to introduce a conditional XOR stage after the inverse S-boxes (for AddRoundKey) and another conditional XOR stage at the end (for combination of intermediate results). The AESDEC(1) instruction can then make use of the first stage and bypass the second stage, whereas AESDEC(2) can do the opposite. By sticking to a fixed order of AESDEC(1) and AESDEC(2) instructions, decryption can be implemented correctly. The functionality of the AESDEC(x) instruction variants is depicted in Figure 3.

For the last round, we propose an instruction AESINVSBOX similar to AESSBOX for encryption. The only difference is the use of the inverse S-box in the case of decryption.

4.3   Performance Enhancement and Implementation Flexibility

Our proposed extensions are designed to improve performance using three main strategies. Firstly, the instructions support AES transformations which are not very well catered for by the microcontroller's native instruction set (especially MixColumns and InvMixColumns). Secondly, two State bytes are transformed simultaneously, which effectively "widens" the 8-bit datapath. And finally, several transformations can be executed by a single instruction invocation.

Compared to typical AES coprocessors, our instruction set extensions allow a more flexible application. The custom instructions support all three key sizes of 128, 192, and 256 bit. All modes of operations can be realized seamlessly, as the AES State can be retained in the register file. In contrast, a coprocessor might require to transfer blocks to and from the processor whenever the chosen mode requires operations which are not supported by the coprocessor. The resulting overhead can be detrimental to the overall performance. Another advantage of our extensions is that they support fast implementations of all variants of Rijndael, which is a superset of AES and which specifies independent block sizes and key sizes between 128 and 256 bit in 32-bit increments. A potential application of Rijndael is as building block for a cryptographic hash function: By setting Rijndael's block and key size equal, it can be applied in a hashing mode of operation to build a hash function with a hash size equal to the block size.

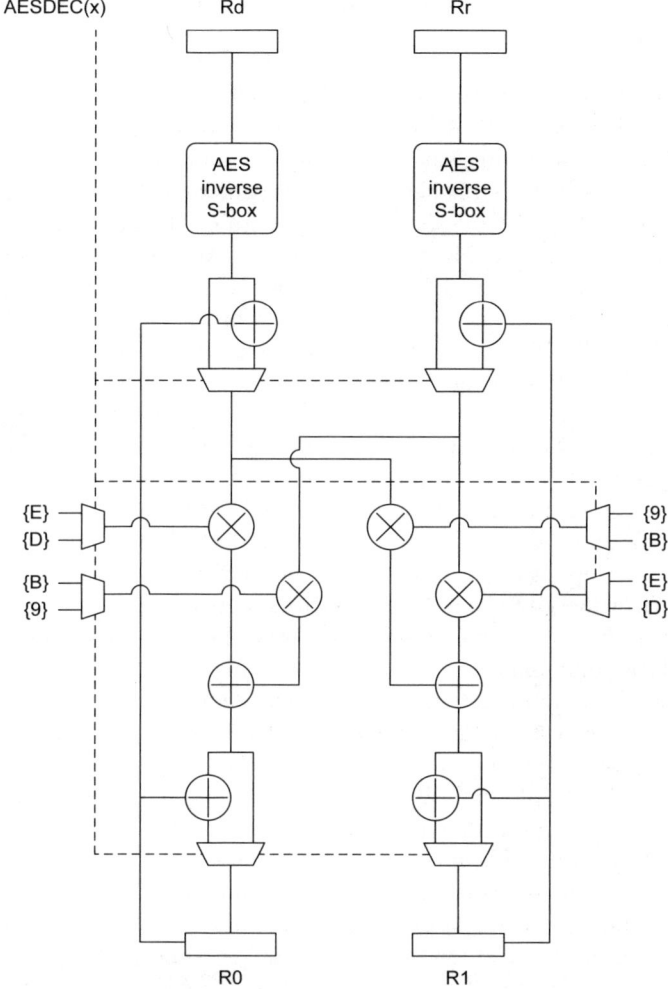

Fig. 3. AES extensions for a "normal" decryption round

5   Implementation Issues

We now give details on possible hardware implementation options for our proposed extensions and different ways to optimize AES software implementations through utilization of those extensions.

5.1   Hardware Implementation of the Proposed Extensions

In this section we outline important implementation issues for the functional units as well as integration issues for the AVR architecture. We will thereby refer to a unified implementation, which is able to provide support for both

AES encryption as well as AES decryption as described in Sections 4.1 and 4.2, respectively.

One important aspect is the support for both the AES S-box and its inverse. In the literature, there have been several proposals for S-box hardware implementations targeting low area, high speed or low power consumption. A comparison of the state-of-the-art regarding their implementation characteristics in standard-cell technology has been published in [20]. An implementation offering a mix of small size and relatively good speed is the design of David Canright [5].

The functional part for MixColumns and InvMixColumns demands multiplication with constants in $GF(2^8)$ under a fixed reduction polynomial [16]. These multiplications are rather easy to implement, as the characteristic two of the finite field allows for addition without carry. This is a very desirable property which makes $GF(2^m)$ multipliers generally much faster than their integer counterparts.

Several implementation options are available to realize the $GF(2^8)$ constant multipliers required by our proposed extensions. The smallest solution would be to integrate fixed multipliers similar to those used by Wolkerstorfer in [24]. Wolkerstorfer's approach reuses the results for MixColumns to perform InvMixColumns, thus keeping the overall size of the multipliers small. In another approach, Elbirt proposed to realize the multipliers in a flexible fashion, so that not only AES, but also other implementations in need of fast $GF(2^m)$ multiplication with a constant could experience an increase in performance [10]. Naturally, this flexibility has to be bought with an increased demand in hardware. Moreover, the multipliers of Elbirt's solution have to be configured for the specific constants and the reduction polynomial at hand, before they can be used.

The highest degree of flexibility is offered by fully-fledged $GF(2^8)$ multipliers which can vary both multiplier and multiplicand at runtime without configuration overhead. Eberle et al. have proposed to integrate an (8×8)-bit multiplier and multiply-accumulate unit for binary polynomials in an AVR microcontroller to accelerate Elliptic Curve Cryptography (ECC) over binary extension fields [9]. Similar synergies for instruction set support for AES and ECC have already been demonstrated in the case of 32-bit architectures [23]. Although this variant would be the most costly option in terms of hardware, the increased flexibility and potential support of both symmetric and asymmetric cryptography could make the integration of such multipliers a worthwhile solution for 8-bit architectures.

5.2   AES Software Implementation Using the Proposed Extensions

In order to check the benefits of the proposed extensions and to have a base for performance estimations, we have implemented AES-128 encryption and decryption in AVR assembly. We have tried to make the best use of the vast amount of 32 general-purpose registers offered by the architecture in order to keep costly memory accesses at an absolute minimum. In our implementation, the 16-byte AES State is kept in 16 registers at all times and an on-the-fly key expansion is used to preserve key agility. Three of the four 32-bit words of the current round key are also kept in 12 additional registers and only a single round key word has to be held in memory. From the remaining four registers, two (namely R0 and R1) are used to receive the result of AESENC(x) or AESDEC(x) instructions and the other two registers are necessary to hold temporary values during round transformation.

A round function is called to perform the four round transformations on the State and to generate the subsequent round key. The transformations are performed in-place on the 16 registers holding the State, i.e. all State columns are written back to the same four registers from which they were originally loaded. The ShiftRows function is not performed explicitly on this "register State", but it is only taken into account by appropriate selection of registers in the round function. As a consequence, a specific State column is contained in a different set of registers after each invocation of the round function. Consequently, we require several different round functions which load the State bytes from the correct registers in conformance to the current layout of the State. Luckily, the layout of the State reverts back to its original form after four invocations of ShiftRows. This property is illustrated in Figure 4, where the four State columns are marked in different colors. Hence, it is sufficient to have four variants of the round function.

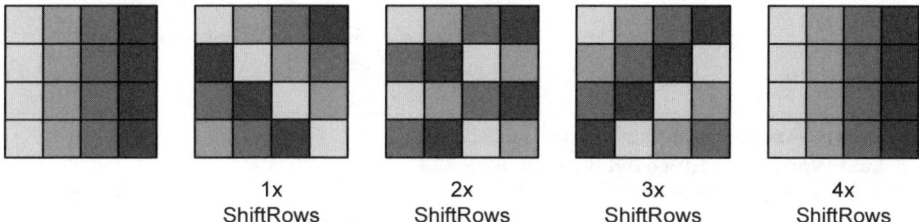

1x
ShiftRows

2x
ShiftRows

3x
ShiftRows

4x
ShiftRows

Fig. 4. Change of AES State layout through ShiftRows for in-place storage

The assembly code performing all four round transformations on a single State column is shown in Figure 5. The update of the first round key word is shown in Figure 6.

The main function is responsible for saving the 32 registers onto the stack at entry. Moreover, the function has to load the AES State and cipher key into the corresponding registers. After nine calls to the appropriate round functions, the final round is performed directly by the main function. At the end, the ciphertext is stored to memory and the registers are restored from stack prior to return.

6   Performance Analysis

This section gives figures on implementation cost of the proposed instruction set extensions, the performance of our optimized AES implementation and its cost in terms on program memory and working memory.

```
; State column in R6, R11, R16, R5
; Round key word in R22-R25
; New State column is written over old column

; Calculate upper half of new column
MOVW R0, R22            ; Move two round key bytes into R0-R1
AESENC(1) R6, R11       ; ShiftRows, SubBytes, MixColumns & AddRoundKey
AESENC(2) R16, R5       ; ShiftRows, SubBytes, MixColumns
MOVW R30, R0            ; Save half column in temporary registers R30-R31

; Calculate lower half of new column
MOVW R0, R24            ; Move the other two round key bytes into R0-R1
AESENC(2) R6, R11       ; ShiftRows, SubBytes, MixColumns & AddRoundKey
AESENC(1) R16, R5       ; ShiftRows, SubBytes, MixColumns

; Store new column over old column
MOV R6, R30
MOV R11, R31
MOV R16, R0
MOV R5, R1
```

Fig. 5. Round transformations for a single State column

```
; First word of old round key in R18-R21
; Last word of old round key in R26-R29
; Rcon located in R30
; New first round key word written over old word

EOR R18, R30           ; Add Rcon
AESSBOX R18, R27       ; RotWord, SubWord, Add to old byte
AESSBOX R19, R28       ; RotWord, SubWord, Add to old byte
AESSBOX R20, R29       ; RotWord, SubWord, Add to old byte
AESSBOX R21, R26       ; RotWord, SubWord, Add to old byte
```

Fig. 6. Update of the first round key word

6.1   Hardware Cost

In order to determine the hardware cost for the proposed extensions, we have
implemented a functional unit capable of supporting all six custom instructions
for AES encryption and decryption. For the AES S-boxes we used the approach
of Canright [5]. We included a pipeline stage in the functional unit to adapt it to
the read-write capabilities of the register file of existing AVR microcontrollers [1].

Our functional unit is depicted in Figure 7. The different sections conforming
to different AES transformations are highlighted. The dashed line represents
configuration information which determines the functionality in dependence on

the actual instruction. The S-boxes are used in forward direction for the instructions for encryption (AESENC(x) and AESSBOX) and in inverse direction for the instructions for decryption (AESDEC(x) and AESINVSBOX). The multiplexors in the AddRoundKey section select the left input for AESENC(x) and AESDEC(2) instructions and the right input for the AESDEC(1) instruction. The multiplexors in the (Inv)MixColumns section also always select the same input. Starting from the top input, the according instructions are AESENC(1), AESENC(2), AESDEC(1), and AESDEC(2). The result for AESENC(x) and AESDEC(x) instructions is delivered into R0 and R1, while the result for AESSBOX and AESINVSBOX instructions appears at the output for Rd.

The GF(2^8) multipliers of the functional units have been hardwired for the constants used in MixColumns and InvMixColumns. Thereby, the two multipliers for a byte have been implemented jointly. A byte b is multiplied with the powers of two, yielding four intermediate results (b, $\{2\}b$, $\{4\}b$, and $\{8\}b$). Depending on the instruction, these intermediate results are added to yield the required multiplication results. Figure 8 shows the implementation for the first byte (i.e. the upper two multipliers transforming the byte from Rd in Figure 7).

We have implemented our functional unit using a 0.35 μm CMOS standard cell library from austriamicrosystems. The synthesized circuit had a size of 1,109 gates with a critical path of 18.3 ns (about 55 MHz). Note that we have optimized the synthesis result towards minimal area, just setting a maximal critical path of 50 ns to match the 20 MHz maximal clock frequency of state-of-the-art AVR microcontrollers. The speed of the circuit could easily be increased by trading off area efficiency.

The smallest AES coprocessor reported in literature so far is by Feldhofer et al. with a size of about 3,400 gates [11]. Our proposed extensions have only a third of this size.

6.2   Performance

Based on our optimized assembly implementation, we have estimated the number of clock cycles for a single AES-128 encryption and decryption (including the complete on-the-fly key expansion). Thanks to the simple and deterministic structure of AVR microcontrollers, this estimation can be done with a high level of accuracy. For all our custom instructions we have assumed a cycle count of 2, which we deem to be realistic for implementation. Executing a single round function (either for encryption or decryption) requires 106 clock cycles. With the overhead from the main function, the cycle count for encryption of a 128-bit block amounts to 1,262 (including the loading of the plaintext from memory and the storing of the ciphertext back to memory). Thanks to the symmetry of the extensions, AES decryption can be equally fast in 1,263 cycles.

We compare our performance to that of an assembly-optimized software implementation of AES for the AVR architecture reported in [18]. It requires 3,766 cycles for encryption and 4,558 cycles for decryption, where the overhead for

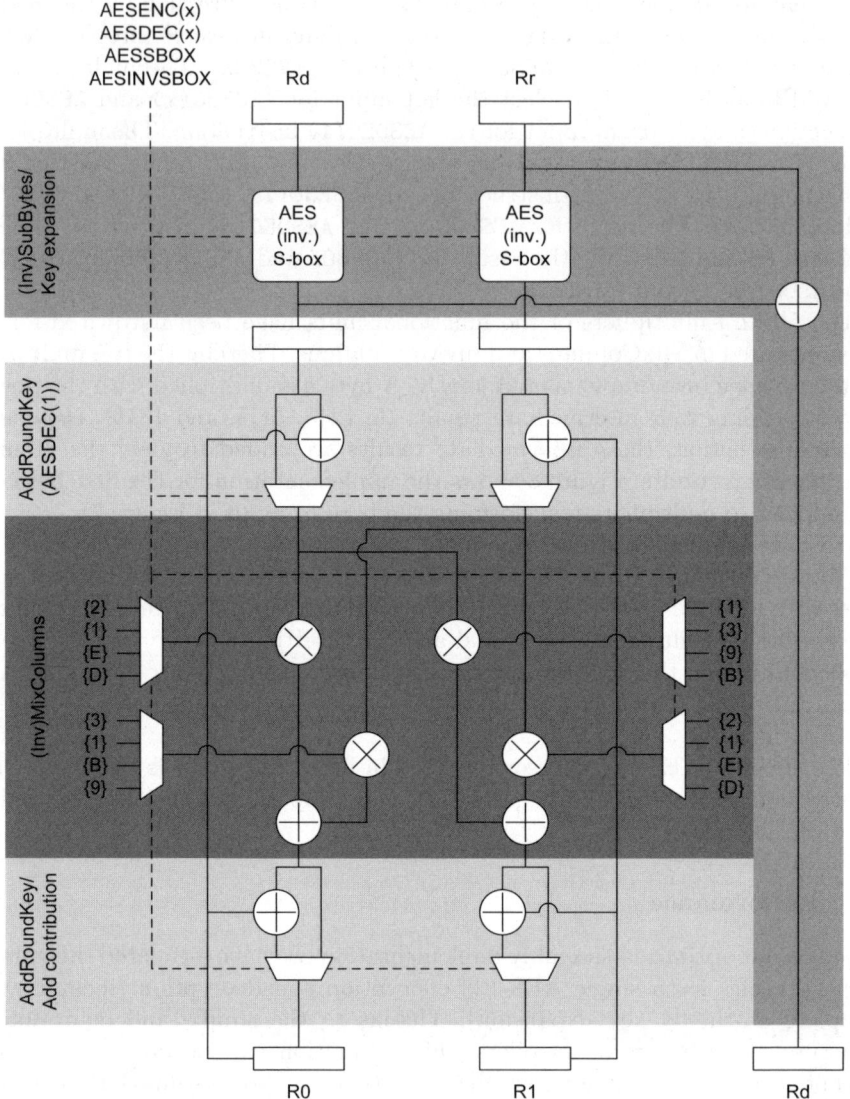

Fig. 7. Implementation of the functional unit for supporting the AES extensions

decryption mainly stems from the more complicated InvMixColumns transformation. The speedup factors for our implementation are therefore about 3 and 3.6, respectively.

The coprocessor of Feldhofer et al. has a performance roughly equivalent to our extensions with a cycle count of 1,032 for encryption and 1,165 for decryption of a single block [11].

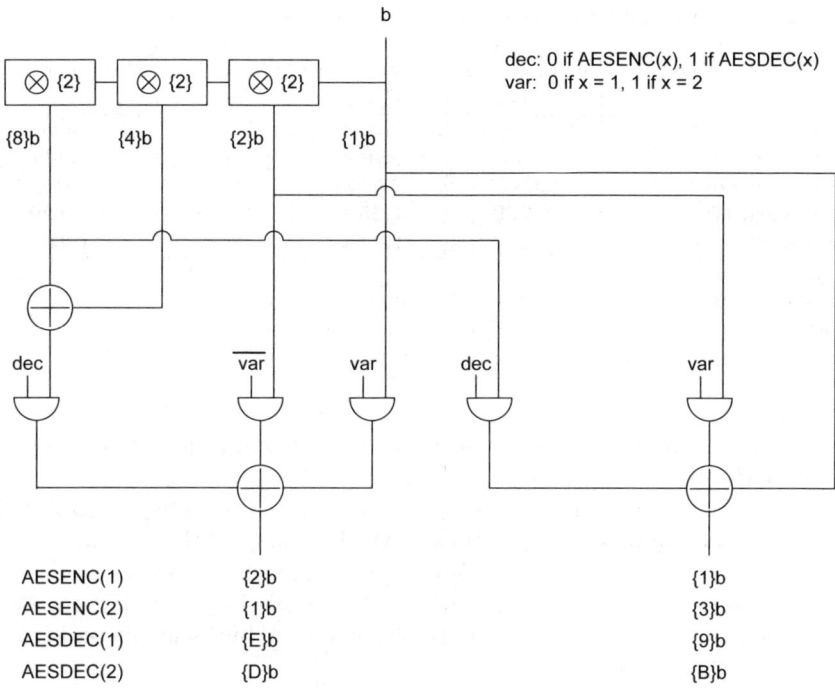

Fig. 8. Implementation of the finite field constant multipliers for the first byte

6.3   Code Size and RAM Requirements

Our assembly implementation of encryption and decryption requires 1,708 bytes of code memory. This size can be further reduced with an explicit ShiftRows at the end of each round function (20 additional MOV instructions requiring 20 cycles). In this case, a single round function for encryption and decryption would suffice, which brings the overall code size down to 840 bytes. However, the number of cycles per encryption and decryption would increase by 180.

In terms of RAM, our implementation requires only four bytes of extra memory in addition to the use of the general-purpose registers. Note that we are not considering the memory from which we load the plaintext at the start of encryption and where we store the ciphertext to at the end.

6.4   Summary of Comparison

Table 1 summarizes our performance figures with those of the optimized software implementation from [18], the custom AES microcontroller from [6] and Feldhofer et al.'s tiny AES coprocessor [11]. We have included both of our implementation variants for maximal speed (fast) and minimal code size (compact), cf. Section 6.3. The cycle count refers to AES-128 encryption or decryption of

Table 1. AES performance characteristics in comparison to related work

Implementation	Encryption Cycles	Decryption Cycles	Code size Bytes	Hardware cost Gate equivalents
AVR software [18]	3,766	4,558	3,410	none
AES coprocessor [11]	1,032	1,165	n/a	3,400
AES microcontroller [6]	2,695 [a]	2,944 [a]	918 [b]	n/a
This work (fast)	**1,259**	**1,259** [c]	**1,708**	**1,109**
This work (compact)	**1,442**	**1,443** [c]	**840**	**1,109**

[a] Excluding cost for precomputed key schedule (2,167 cycles).
[b] Total size for encryption, decryption and key expansion.
[c] Last round key supplied to decryption function.

a single 16-byte block. The code size refers to an implementation which can support both encryption and decryption.

Our proposed solution is considerably faster and requires less code size than the pure-software approach. Nevertheless, the flexibility of the software solution is fully retained. Compared to the coprocessor approach, our solution offers similar performance at much smaller hardware overhead. The AES microcontroller has a similar code size as our compact implementation, but is significantly slower.

7 Conclusions

In this work we have presented a set of small and simple AES instruction set extensions for the 8-bit AVR architecture. We have demonstrated the benefits of these extensions with an optimized AES encryption implementation, which is about three times faster than an optimized assembly implementation using native AVR instructions. Speedup for decryption is even higher, amounting to a factor of about 3.6. As an additional benefit, code size is small and RAM requirements are very low. The hardware cost of our extensions ranges around 1.1 kGates. Compared to the smallest AES coprocessor reported so far, our extensions deliver similar performance at only a third of the hardware cost. All in all, our extensions provide a very good tradeoff between hardware overhead, performance gain and implementation flexibility and position themselves at a favorable section of the design space.

Acknowledgements. The research described in this paper has been supported by the Austrian Science Fund (FWF) under grant number P18321-N15 ("Investigation of Side-Channel Attacks") and by the European Commission under grant number FP6-IST-033563 (Project SMEPP). The information in this document reflects only the authors' views, is provided as is and no guarantee or warranty is given that the information is fit for any particular purpose. The user thereof uses the information at its sole risk and liability.

References

1. Atmel Corporation. 8-bit AVR Microcontroller with 128K Bytes In-System Programmable Flash (August 2007),
   http://www.atmel.com/dyn/resources/prod_documents/doc2467.pdf
2. Bertoni, G., Breveglieri, L., Roberto, F., Regazzoni, F.: Speeding Up AES By Extending a 32-Bit Processor Instruction Set. In: Proceedings of the 17th IEEE International Conference on Application-specific Systems, Architectures and Processors (ASAP 2006), pp. 275–282. IEEE Computer Society Press, Los Alamitos (2006)
3. Buchty, R.: Cryptonite — A Programmable Crypto Processor Architecture for High-Bandwidth Applications. Ph.d. thesis, Technische Universität München, LRR (September 2002),
   http://tumb1.biblio.tu-muenchen.de/publ/diss/in/2002/buchty.pdf
4. Burke, J., McDonald, J., Austin, T.: Architectural Support for Fast Symmetric-Key Cryptography. In: ASPLOS-IX Proceedings of the 9th International Conference on Architectural Support for Programming Languages and Operating Systems, Cambridge, MA, USA, 2000, November 12-15, pp. 178–189. ACM Press, New York (2000)
5. Canright, D.: A Very Compact S-Box for AES. In: Rao, J.R., Sunar, B. (eds.) CHES 2005. LNCS, vol. 3659, pp. 441–455. Springer, Heidelberg (2005)
6. Chia, C.-C., Wang, S.-S.: Efficient Design of an Embedded Microcontroller for Advanced Encryption Standard. In: Proceedings of the 2005 Workshop on Consumer Electronics and Signal Processing (WCEsp 2005) (2005),
   http://www.mee.chu.edu.tw/labweb/WCEsp2005/96.pdf
7. Daemen, J., Rijmen, V.: The Design of Rijndael. In: Information Security and Cryptography, Springer, Heidelberg (2002)
8. Dhem, J.-F.: Design of an efficient public-key cryptographic library for RISC-based smart cards. PhD thesis, Université Catholique de Louvain, Louvain-la-Neuve, Belgium (May 1998)
9. Eberle, H., Wander, A., Gura, N., Chang-Shantz, S., Gupta, V.: Architectural Extensions for Elliptic Curve Cryptography over $GF(2^m)$ on 8-bit Microprocessors. In: Proceedings of the 16th IEEE International Conference on Application-specific Systems, Architectures and Processors (ASAP 2005), July 2005, pp. 343–349. IEEE Computer Society Press, Los Alamitos (2005)
10. Elbirt, A.J.: Fast and Efficient Implementation of AES via Instruction Set Extensions. In: Proceedings of the 21st International Conference on Advanced Information Networking and Applications Workshops (AINAW 2007), May 2007, vol. 1, pp. 396–403. IEEE Computer Society Press, Los Alamitos (2007)
11. Feldhofer, M., Wolkerstorfer, J., Rijmen, V.: AES Implementation on a Grain of Sand. IEE Proceedings on Information Security 152(1), 13–20 (2005)
12. Gonzalez, R.E.: Xtensa: A Configurable and Extensible Processor. IEEE Micro 20(2), 60–70 (2000)
13. McGregor, J.P., Lee, R.B.: Architectural Enhancements for Fast Subword Permutations with Repetitions in Cryptographic Applications. In: Proceedings of the International Conference on Computer Design (ICCD 2001), September 2001, pp. 453–461. IEEE, Los Alamitos (2001)
14. Nadehara, K., Ikekawa, M., Kuroda, I.: Extended Instructions for the AES Cryptography and their Efficient Implementation. In: IEEE Workshop on Signal Processing Systems (SIPS 2004), Austin, Texas, USA, October 2004, pp. 152–157. IEEE Press, Los Alamitos (2004)

15. Nahum, E., O'Malley, S., Orman, H., Schroeppel, R.: Towards High Performance Cryptographic Software. In: Third IEEE Workshop on the Architecture and Implementation of High Performance Communication Subsystems, 1995 (HPCS 1995), August 1995, pp. 69–72. IEEE, Los Alamitos (1995)
16. National Institute of Standards and Technology (NIST). FIPS-197: Advanced Encryption Standard (November 2001), http://www.itl.nist.gov/fipspubs/
17. Ravi, S., Raghunathan, A., Potlapally, N., Sankaradass, M.: System design methodologies for a wireless security processing platform. In: DAC 2002: Proceedings of the 39th Conference on Design Automation, pp. 777–782. ACM Press, New York (2002)
18. Rinne, S., Eisenbarth, T., Paar, C.: Performance Analysis of Contemporary Light-Weight Block Ciphers on 8-bit Microcontrollers (June 2007),
    http://www.crypto.ruhr-uni-bochum.de/imperia/md/content/texte/
    publications/conferences/lw_speed2007.pdf
19. Shi, Z., Lee, R.B.: Bit Permutation Instructions for Accelerating Software Cryptography. In: Proceedings of the 11th IEEE International Conference on Application-specific Systems, Architectures and Processors (ASAP 2000), pp. 138–148. IEEE, Los Alamitos (2000)
20. Tillich, S., Feldhofer, M., Großschädl, J.: Area, Delay, and Power Characteristics of Standard-Cell Implementations of the AES S-Box. In: Vassiliadis, S., Wong, S., Hämäläinen, T.D. (eds.) SAMOS 2006. LNCS, vol. 4017, pp. 457–466. Springer, Heidelberg (2006)
21. Tillich, S., Großschädl, J.: Instruction Set Extensions for Efficient AES Implementation on 32-bit Processors. In: Goubin, L., Matsui, M. (eds.) CHES 2006. LNCS, vol. 4249, pp. 270–284. Springer, Heidelberg (2006)
22. Tillich, S., Großschädl, J.: Power-Analysis Resistant AES Implementation with Instruction Set Extensions. In: Paillier, P., Verbauwhede, I. (eds.) CHES 2007. LNCS, vol. 4727, pp. 303–319. Springer, Heidelberg (2007)
23. Tillich, S., Großschädl, J.: VLSI Implementation of a Functional Unit to Accelerate ECC and AES on 32-bit Processors. In: Carlet, C., Sunar, B. (eds.) WAIFI 2007. LNCS, vol. 4547, pp. 40–54. Springer, Heidelberg (2007)
24. Wolkerstorfer, J.: An ASIC Implementation of the AES-MixColumn operation. In: Rössler, P., Döderlein, A. (eds.) Austrochip 2001, pp. 129–132 (2001); ISBN 3-9501517-0-2
25. Wu, L., Weaver, C., Austin, T.: CryptoManiac: A Fast Flexible Architecture for Secure Communication. In: ISCA 2001: Proceedings of the 28th annual international symposium on Computer architecture, pp. 110–119. ACM Press, New York (2001)

A Fast and Cache-Timing Resistant Implementation of the AES*

Robert Könighofer

Institute for Applied Information Processing and Communications (IAIK),
Graz University of Technology, Inffeldgasse 16a, A-8010 Graz, Austria
`robert.koenighofer@student.tugraz.at`

Abstract. This work presents a fast bitslice implementation of the AES with 128-bit keys on processors with x64-architecture processing 4 blocks of input data in parallel. In contrast to previous work on this topic, our solution is described in detail from the general approach to the actual implementation. As the implementation does not need table-lookups it is immune to cache-timing attacks while being only 5% slower than the widely used optimized reference implementation. Outspeeding other approaches for making an implementation cache-timing resistant, the solution needs 8% less code memory and 93% less data memory than the reference implementation. Further improvements are possible.

1 Introduction

There are two fundamental requirements for the implementation of cryptographic primitives. The implementation should be as fast as possible and it should not contain any weaknesses that could be exploited in an attack. The bitslicing technique can be a good choice for implementing cryptographic algorithms with respect to these requirements.

Bitslicing was first used for software in [1] to implement the DES block cypher. A block of input data is viewed as a collection of slices of one bit. The processor is seen as a SIMD parallel computer, simultaneously operating on a number of registers, each containing one bit. A conventional 64-bit register can be used just like 64 one-bit-registers, with the operations effecting all notional one-bit-registers in the same manner. Different bits of the input data are manipulated in a different way, so the bits are stored in different registers. In order to fully utilize the registers, typically many blocks of input data are processed in parallel. Another characteristic for bitslicing implementations is not to use table-lookups. This leads to a low amount of needed memory, and makes the implementation naturally immune to cache-timing attacks. The work in [2] showed that this attack can be used to even recover the key from measuring the time needed for the encryption of known plaintexts with the AES. Other techniques to prevent cache-timing attacks (e.g. from [3]) often do not provide natural immunity and suffer from performance problems. Our implementation is also immune to cache probing [4] [5] and cache collision attacks [6] [7].

---

* The work described in this paper has been supported in part through the Austrian Science Fund (FWF) under grant number P18321.

T. Malkin (Ed.): CT-RSA 2008, LNCS 4964, pp. 187–202, 2008.

The goal of the work presented in this paper was to implement the AES in a bitsliced manner on a processor with the common x64-architecture. Previous work on bitslice implementations of the AES was done in [8] and [9], but the implementation described in this paper is different. Although details about the actual implementation of [8] and [9] are hard to find out, [8] implements at least the S-box transformation, which is the operation with the most workload, completely differently. [9] is designed to encrypt 64 blocks of input data in parallel. The implementation presented here processes 4 blocks simultaneously.

After introducing the general approach, we discuss the transformation into the bit-slice domain and its implementation. We furthermore derive an efficient bitslice implementation of the S-box transformation from the work done in [10]. Section 5 and 6 address intelligent approaches for the implementation of the ShiftRows and the Mix-Columns step. Finally we present performance results. The implementation is available at http://www.iaik.tugraz.at/research/krypto/AES/index.php#software.

2   The General Approach

The state of the AES-128 is a 128-bit piece of data, which is interpreted as a matrix of dimension 4x4, whereas each element represents one bytes. There are different possibilities for the representation of this state in the bitsliced domain.

All of the 128 bits could be stored in an own variable. To fully utilize the register size, 64 blocks could be processed in parallel on a 64-bit CPU. The disadvantage of this method is, that the bitsliced state can not be kept in registers, because there are only 16 registers defined in the x64-architecture. Keeping the state in memory makes a lot of load and store operations necessary leading to worse performance results.

In the solution presented in this document, only the 8 bits of the matrix elements are stored in different registers. This makes sense, since all matrix elements are manipulated in a very similar manner, so the structure of the algorithm is taken into account at this point. The approach requires 8 variables, each containing one bit of the elements of the state-matrix. These variables can be held in 8 registers during the whole encryption, which reduces data movement and thus increases the performance. The downside is, that MixColumns and ShiftRows require interconnections with bits of elements of other rows or columns. This leads to rotations of the register's content, which do not have to be done in the first approach.

The AES-128 state consists of 16 bytes, so only 16 of the 64 bits of the registers storing the state would be used. To fully utilize the register size, 4 blocks can be worked up simultaneously. There are multiple ways of arranging the bits belonging to different blocks and to different state-matrix elements inside one register. The one leading to the highest efficiency is shown in figure 1.

Access to AES-functionality can be separated into two parts. One function derives the round keys in bitslice representation from the key in normal representation. The round keys can then be used in a second function to encrypt plaintexts. The advantage of this interface is, that the key scheduling needs to be done only once, if large amounts of blocks are encrypted with the same key.

MSB								LSB
row 0				row 1		row 3		
col 0	col 1	col 2	col 3	col 0	col 1		col 2	col 3
block 0 block 1 block 2 block 3	block 0 block 1 block 2 block 3	block 0 block 1 block 2 block 3	block 0 block 1 block 2 block 3	block 0 block 1 block 2 block 3	block 0 block 1 block 2 block 3		block 0 block 1 block 2 block 3	block 0 block 1 block 2 block 3

Fig. 1. The content of a register holding one bit

3 The Transformation Function

Loading the bits of the state-matrix elements into different registers can be done in various ways. Each bit of each element could be shifted into the carry-bit. Then it could be rotated into the right register with a *Rotate with Carry* operation. This would require 8 shifts and 8 rotates per byte, leading to 256 logical operations per transformation. Another and more efficient way is to use a function called *SWAPMOVE* introduced in [11]. It is defined as:

$$T = ((A >> N) \oplus B) \wedge M$$
$$B = B \oplus T$$
$$A = A \oplus (T << N)$$

The bits in B, masked by M, are swapped with the bits in A, masked by $(M << N)$ in 6 logical operations. To get to a bitslice representation as shown in figure 1, 24 *SWAPMOVE* calls, so only 144 logical operations, are necessary. This is shown in listing 1.4 in Appendix A.1. The inverse transformation can be done with exactly the same calls of *SWAPMOVE* but in reverse order.

4 The SubBytes Step

The SubBytes step is defined as replacement of all elements of the matrix with the S-box transformation of the elements. In conventional implementations, this operation is typically done as table-lookup. However in a bitslice implementation, collecting the bits together into one register and spreading the bits again after performing the lookup would cost too much calculation time. The lookup has to be expressed by logical equations instead.

The S-box transformation of one byte is defined as multiplicative inverse in $GF(2^8)$ followed by an affine transformation. There are some solutions for the implementation in hardware. A very common one was presented by Satoh et.al. in [12], which was also taken as starting point for the bitslice implementation of the S-box in [8]. However, there is another version published by Canright in [10] which promises a higher performance. Like in almost all solutions, a change of basis into a domain $\mathcal{X}$, where the inverse is easier to compute, is performed. The transformation back is merged with the affine transformation of the S-box.

Some steps and intermediate results of converting the logical expressions from [10] into an efficient bitslice implementation shall be presented in this section.

4.1   The Transformation into the Domain $\mathcal{X}$

The transformation from the normal domain $\mathcal{A}$ into the domain $\mathcal{X}$ is linear and can be expressed as matrix multiplication with a transformation matrix X:

$$x = \begin{bmatrix} x_7 \\ x_6 \\ x_5 \\ x_4 \\ x_3 \\ x_2 \\ x_1 \\ x_0 \end{bmatrix} = X \cdot a = \begin{bmatrix} 1\,1\,1\,0\,0\,1\,1\,1 \\ 0\,1\,1\,1\,0\,0\,0\,1 \\ 0\,1\,1\,0\,0\,0\,1\,1 \\ 1\,1\,1\,0\,0\,0\,0\,1 \\ 1\,0\,0\,1\,1\,0\,1\,1 \\ 0\,0\,0\,0\,0\,0\,0\,1 \\ 0\,1\,1\,0\,0\,0\,0\,1 \\ 0\,1\,0\,0\,1\,1\,1\,1 \end{bmatrix} \cdot \begin{bmatrix} a_7 \\ a_6 \\ a_5 \\ a_4 \\ a_3 \\ a_2 \\ a_1 \\ a_0 \end{bmatrix} \quad with\ a \in \mathcal{A},\ x \in \mathcal{X}$$

This transformation can be implemented with one MOV and 15 XORs as shown in the lines 22 to 37 in listing 1.5 in Appendix B.1. This implementation of the transformation also does a permutation of the resulting variables. Allowing such permutations reduces the number of MOV-instructions.

4.2   The Transformation from the Domain $\mathcal{X}$ Back to the Domain $\mathcal{A}$

The transformation into the domain $\mathcal{X}$ can be undone by multiplying with X^{-1}. This multiplication can be merged with the affine transformation of the S-box transformation:

$$s = S \cdot a + b = S \cdot X^{-1} \cdot x + b \qquad s, a \in \mathcal{A},\ x \in \mathcal{X}$$

$$with:\ b = \begin{bmatrix} 0 \\ 1 \\ 1 \\ 0 \\ 0 \\ 0 \\ 1 \\ 1 \end{bmatrix} \quad and\ S = \begin{bmatrix} 1\,1\,1\,1\,1\,0\,0\,0 \\ 0\,1\,1\,1\,1\,1\,0\,0 \\ 0\,0\,1\,1\,1\,1\,1\,0 \\ 0\,0\,0\,1\,1\,1\,1\,1 \\ 1\,0\,0\,0\,1\,1\,1\,1 \\ 1\,1\,0\,0\,0\,1\,1\,1 \\ 1\,1\,1\,0\,0\,0\,1\,1 \\ 1\,1\,1\,1\,0\,0\,0\,1 \end{bmatrix} \rightarrow S \cdot X^{-1} = \begin{bmatrix} 0\,0\,1\,0\,1\,0\,0\,0 \\ 1\,0\,0\,0\,1\,0\,0\,0 \\ 0\,1\,0\,0\,0\,0\,0\,1 \\ 1\,0\,1\,0\,1\,0\,0\,0 \\ 1\,1\,1\,1\,1\,0\,0\,0 \\ 0\,1\,1\,0\,1\,1\,0\,1 \\ 0\,0\,1\,1\,0\,0\,1\,0 \\ 0\,1\,0\,1\,0\,0\,1\,0 \end{bmatrix}$$

The XOR with b negates 4 of the 8 bits. This step can be left out, if the round keys are modified accordingly (i.e. the bits 0, 1, 5 and 6 are negated).

The multiplication with $S \cdot X^{-1}$ can be implemented with 2 MOVs and 11 XORs if a permutation of the variables is allowed. As already mentioned, the implementation of the transformations from $\mathcal{A}$ to $\mathcal{X}$ also does a permutation of the variables. Section 4.3 will show that the calculation of the inverse in $GF(2^8)$ does a permutation as well, in order to increase efficiency. At the end of the S-box transformation all the permutations have to be undone for the next elements of the round function to work correctly. The most efficient way of doing that is to modify the transformation back from $\mathcal{X}$ to $\mathcal{A}$ accordingly. The desired permutation can be seen in line 9 and 13 of listing 1.5 in Appendix B.1. The solution is shown in line 42 to 62 of the same listing, needing 10 MOVs and 11 XORs. An explicit permutation at the end would cost 10 additional MOVs instead of the 8 additional MOVs of this solution.

4.3   The Calculation of the Inverse in $GF(2^8)$

The illustration of Canrights equations in C [10] was taken as a starting point for constructing an efficient bitslice implementation. In the first step, the number of MOV-instructions can be reduced by allowing subroutines to return the results with some permutation. A lot of further modifications can be done in order to improve the performance of the subroutines.

Optimizing G16_inv from [10]: If the calls of G4_mul, G4_sq and G4_scl_N as introduced in [10] are inserted, the first part of G16_inv could be written with the following equations:

$$e = (x3 \oplus x2) \wedge (x1 \oplus x0)$$
$$d1 = (x3 \wedge x1) \oplus e \oplus x1 \oplus x3$$
$$d0 = (x2 \wedge x0) \oplus e \oplus x1 \oplus x3 \oplus x0 \oplus x2$$

The variables x1 and x3 are xored to d1 and to d0. The content of e is xored to d1 and to d0 anyway, so it is more efficient to add x1 and x3 to e. Instead of 6 MOVs, 12 logical instructions and 6 temporary variables, these equations could be implemented with 4 MOVs, 11 logical instructions and only 4 temporary variables. However, the interconnections $x3 \oplus x2$ and $x1 \oplus x0$ can be used in the second part of G16_inv, so the values must not be overwritten. The first part of G16_inv, shown in the lines 144 to 147 of listing 1.5 in Appendix B.1, therefore needs 5 MOVs, 11 logical instructions and 5 temporary variables.

The only thing left in the calculation of G16_inv are the two multiplications in $GF(2^4)$. The two calls would need 4 MOVs and 14 logical instructions. The result can be expressed with the following boolean formulas:

$$e1 = (x1 \oplus x0) \wedge (d1 \oplus d0) \quad e2 = (x3 \oplus x2) \wedge (d1 \oplus d0)$$
$$x1 = (x1 \wedge d0) \oplus e1 \qquad\qquad x3 = (x3 \wedge d0) \oplus e2$$
$$x0 = (x0 \wedge d1) \oplus e1 \qquad\qquad x2 = (x2 \wedge d1) \oplus e2$$

With the intermediate values $x3\_x2 = x3 \oplus x2$ and $x1\_x0 = x1 \oplus x0$, calculated for the first part, these equations can be implemented as shown in line 148 to 150 of listing 1.5 in Appendix B.1 with 11 instead of 14 logical operations and without any MOV-instruction.

With the optimizations discussed in this paragraph, an improvement from 10 MOVs, 26 logical instructions and 6 temporary variables to 5 MOVs, 22 logical instructions and 5 temporary variables could be achieved. This means savings of 25% of the instructions for the calculation of the multiplicative inverse in $GF(2^4)$.

Optimizing G16_mul from [10]: The outcome of the first call of G4_mul followed by G4_scl_N can be expressed as:

$$e = (t3 \oplus t2) \wedge (t1 \oplus t0)$$
$$t3 = (t3 \wedge t1) \oplus e \oplus (t2 \wedge t0) \oplus e = (t3 \wedge t1) \oplus (t2 \wedge t0)$$
$$t2 = (t2 \wedge t0) \oplus e$$

The xor with the temporary variable e is irrelevant for the calculation of t3. The above equations can be implemented with 7 logical instructions and only one MOV instruction as shown in line 6 to 7 of listing 1.1.

Listing 1.1. Efficient bitslice implementation of G16_mul

```
 1   // in:   x3 x2 x1 x0, y3 y2 y1 y0      out: x3 x2 x1 x0
 2   void G16_mul(u64* x3, u64* x2, u64* x1, u64* x0, u64* y3, u64* y2, u64* y1,
 3                u64* y0, u64* t4, u64* t3, u64* t2, u64* t1, u64* t0) {
 4     *t3 = *x3;   *t3 ^= *x1;   *t2 = *x2;   *t2 ^= *x0;   *t1 = *y3;   *t1 ^= *y1;
 5     *t0 = *y2;   *t0 ^= *y0;
 6     *t4 = *t3;   *t4 ^= *t2;   *t2 &= *t0;   *t0 ^= *t1;   *t4 &= *t0;   *t3 &= *t1;
 7     *t3 ^= *t2;   *t2 ^= *t4;
 8     G4_mul(x3, x2, y3, y2, t1,  t0);   G4_mul(x1,x0,y1,y0, t1,  t0);
 9     *x3 ^= *t2;   *x2 ^= *t3;   *x1 ^= *t2;   *x0 ^= *t3;
10   }
```

The workload could be reduced from 10 MOVs, 30 logical instructions and 6 temporary variables to 9 MOVs, 29 logical instructions and 5 temporary variables. Especially the savings in the number of temporary variables are of great value, as shown later.

Optimizing G256_inv from [10]: The result of G16_sq_scl is xored with the result of G16_mul. Including the last 4 XOR operations of G16_mul (see listing 1.1), this can be written as:

$$d3 = d3 \oplus a2 \oplus x6 \oplus x2 \oplus x4 \oplus x0 \qquad d2 = d2 \oplus a3 \oplus x7 \oplus x3 \oplus x5 \oplus x1$$
$$d1 = d1 \oplus a2 \oplus x5 \oplus x1 \oplus x4 \oplus x0 \qquad d0 = d0 \oplus a3 \oplus x4 \oplus x0$$

The above equations can be implemented with 2 MOVs and 14 logical instructions as shown in the lines 10 to 12 in listing 1.2. Inserting G16_sq_scl and G16_mul, the number of logical instructions could be reduces from 120 to 119, the number of MOVs could be reduced from 40 to 38.

Listing 1.2. Calculation of the inverse in $GF(2^8)$

```
 1   // in:   x7 x6 x5 x4 x3 x2 x1 x0        out: x3 x2 x1 x0 x7 x6 x5 x4
 2   void G256_inv(u64 *x7, u64 *x6, u64 *x5, u64 *x4, u64 *x3, u64 *x2, u64 *x1,
 3                 u64 *x0, u64 *d3, u64 *d2, u64 *d1, u64 *d0, u64 *a3, u64 *a2,
 4                 u64 *a1, u64 *a0) {
 5     *a3 = *x7;   *a3 ^= *x5;   *a2 = *x6;   *a2 ^= *x4;   *d3 = *x3;   *d3 ^= *x1;
 6     *d2 = *x2;   *d2 ^= *x0;   *d1 = *a3;   *d1 ^= *a2;   *a2 &= *d2;   *d2 ^= *d3;
 7     *d1 &= *d2;   *a3 &= *d3;   *a3 ^= *a2;   *a2 ^= *d1;   *d3 = *x7;   *d2 = *x6;
 8     *d1 = *x5;   *d0 = *x4;
 9     G4_mul(d3, d2, x3, x2, a1, a0);   G4_mul(d1,d0,x1,x0, a1, a0);
10     *a1 = *x0;   *a1 ^= *x4;   *a2 ^= *a1;   *d1 ^= *a2;   *d0 ^= *a3;   *d3 ^= *a2;
11     *d2 ^= *a3;   *d0 ^= *a1;   *a1 = *x1;   *a1 ^= *x5;   *d1 ^= *a1;   *d2 ^= *a1;
12     *d3 ^= *x6;   *d3 ^= *x2;   *d2 ^= *x7;   *d2 ^= *x3;
13     G16_inv(d3,d2,d1,d0, a4, a3, a2, a1, a0);
14     G16_mul(x3,x2,x1,x0, d1,  d0, d3, d2, a4, a3, a2, a1, a0);
15     G16_mul(x7,x6,x5,x4, d1,  d0, d3, d2, a4, a3, a2, a1, a0);
16   }
```

4.4   Final Improvements

Two problems are remaining. Firstly, the number of registers needed is 17. The x64-architecture defines 16 register. 15 of them (all except for the register for the stack pointer) can be used, so the number of temporary registers has to be reduced by 2. Secondly, inefficiencies because of the multiple calculation of some intermediate values in different subroutines remain. Both problems can be solved by using the stack as a container for temporary data. The use of data slots on the stack is illustrated in listing 1.5 in Appendix B.1.

Table 1 compares the solution shown in listing 1.5 with a fully optimized x64-version (see section 7) presented in listing 1.6 and the solution in [8].

Table 1. Comparison of the S-box implementation with [8]

Version	MOVs	Logical operations	Load or Store	x64-instructions	Bytes in memory
S-box (listing 1.5)	37	130	36	193	96
Optimized S-box (listing 1.6)	31	130	31	182	96
S-box in [8]	30	152	21	204	40

5   The ShiftRows Step

In this step, row i of the matrix representing the state in the normal representation is rotated by i positions to the left:

$$ShiftRows \left(\begin{bmatrix} a_{00} & a_{01} & a_{02} & a_{03} \\ a_{10} & a_{11} & a_{12} & a_{13} \\ a_{20} & a_{21} & a_{22} & a_{23} \\ a_{30} & a_{31} & a_{32} & a_{33} \end{bmatrix} \right) = \begin{bmatrix} a_{00} & a_{01} & a_{02} & a_{03} \\ a_{11} & a_{12} & a_{13} & a_{10} \\ a_{22} & a_{23} & a_{20} & a_{21} \\ a_{33} & a_{30} & a_{31} & a_{32} \end{bmatrix}$$

Translated into the bitslice domain, this means, that the lowest 16 bits in all registers storing the state have to be rotated by 12 positions to the left as they correspond to row 3 (see figure 1). The next lowest 16 bits have to be rotated by 8 positions to the left and the bits 32 to 47 are to be rotated by 4 positions to the left. The highest 16 bits belong to the elements in row 0 and stay untouched.

The rotation in the subgroups of 16 bit could be done with quite a lot of bitmasking operations. Luckily, the lowest 16 bits of each of the 64-bit registers are available as independent register. With this feature, the ShiftRows operation could be done with 6 rotations per bit, leading to 48 rotations all together.

The ShiftRows operation can be speeded up by leaving out one rotation, such that the resulting state-registers are rotated by 16 positions to the right compared to the original specification. This needs $5 \cdot 8 = 40$ rotations instead of 48. The additional rotation can be undone in the MixColumns step. For the MixColumns step it does not make any difference in the number of instructions if the result is rotated by a multiple of 16 positions or not.

6   The MixColumns Step

The MixColumns step is defined as multiplication of the state-matrix with a constant matrix:

$$\begin{bmatrix} b_{00} & b_{01} & b_{02} & b_{03} \\ b_{10} & b_{11} & b_{12} & b_{13} \\ b_{20} & b_{21} & b_{22} & b_{23} \\ b_{30} & b_{31} & b_{32} & b_{33} \end{bmatrix} = \begin{bmatrix} 2 & 3 & 1 & 1 \\ 1 & 2 & 3 & 1 \\ 1 & 1 & 2 & 3 \\ 3 & 1 & 1 & 2 \end{bmatrix} \bullet \begin{bmatrix} a_{00} & a_{01} & a_{02} & a_{03} \\ a_{10} & a_{11} & a_{12} & a_{13} \\ a_{20} & a_{21} & a_{22} & a_{23} \\ a_{30} & a_{31} & a_{32} & a_{33} \end{bmatrix}$$

According to the rules for matrix multiplication, the resulting elements can be calculated with $(i + y)_4$ denoting $i+y \bmod 4$ as:

$$b_{ij} = 2 \bullet a_{(i)_4 j} \oplus 3 \bullet a_{(i+1)_4 j} \oplus a_{(i+2)_4 j} \oplus a_{(i+3)_4 j} \tag{1}$$

where $'\bullet'$ stands for a polynomial multiplication modulo 0x11b. On a bit level, the polynomial multiplication with 2 and 3 can be expressed as shown in table 2, where $x|_b$ denotes the b'th bit of x. With the content of table 2, the equation (1) can be decomposed into equations for each bit:

$$b_{ij}|_0 = a_{(i)_4 j}|_7 \qquad \oplus a_{(i+1)_4 j}|_0 \oplus a_{(i+1)_4 j}|_7 \qquad \oplus a_{(i+2)_4 j}|_0 \oplus a_{(i+3)_4 j}|_0$$
$$b_{ij}|_1 = a_{(i)_4 j}|_0 \oplus a_{(i)_4 j}|_7 \quad \oplus a_{(i+1)_4 j}|_0 \oplus a_{(i+1)_4 j}|_1 \oplus a_{(i+1)_4 j}|_7 \quad \oplus a_{(i+2)_4 j}|_1 \oplus a_{(i+3)_4 j}|_1$$
$$b_{ij}|_2 = a_{(i)_4 j}|_1 \qquad \oplus a_{(i+1)_4 j}|_1 \oplus a_{(i+1)_4 j}|_2 \qquad \oplus a_{(i+2)_4 j}|_2 \oplus a_{(i+3)_4 j}|_2$$
$$b_{ij}|_3 = a_{(i)_4 j}|_2 \oplus a_{(i)_4 j}|_7 \quad \oplus a_{(i+1)_4 j}|_2 \oplus a_{(i+1)_4 j}|_3 \oplus a_{(i+1)_4 j}|_7 \quad \oplus a_{(i+2)_4 j}|_3 \oplus a_{(i+3)_4 j}|_3$$
$$b_{ij}|_4 = a_{(i)_4 j}|_3 \oplus a_{(i)_4 j}|_7 \quad \oplus a_{(i+1)_4 j}|_3 \oplus a_{(i+1)_4 j}|_4 \oplus a_{(i+1)_4 j}|_7 \quad \oplus a_{(i+2)_4 j}|_4 \oplus a_{(i+3)_4 j}|_4$$
$$b_{ij}|_5 = a_{(i)_4 j}|_4 \qquad \oplus a_{(i+1)_4 j}|_4 \oplus a_{(i+1)_4 j}|_5 \qquad \oplus a_{(i+2)_4 j}|_5 \oplus a_{(i+3)_4 j}|_5$$
$$b_{ij}|_6 = a_{(i)_4 j}|_5 \qquad \oplus a_{(i+1)_4 j}|_5 \oplus a_{(i+1)_4 j}|_6 \qquad \oplus a_{(i+2)_4 j}|_6 \oplus a_{(i+3)_4 j}|_6$$
$$b_{ij}|_7 = a_{(i)_4 j}|_6 \qquad \oplus a_{(i+1)_4 j}|_6 \oplus a_{(i+1)_4 j}|_7 \qquad \oplus a_{(i+2)_4 j}|_7 \oplus a_{(i+3)_4 j}|_7$$

The bitslice representation illustrated in figure 1 allows a simple implementation of these equations. For each bit, some elements x_{ij} in row i and column j have to be interconnected with other elements $y_{(i+c)_4 j}$ of the same column j, but different row. This can be done by simply interconnecting the value x in bitslice representation with the value y also in bitslice representation, where y is rotated by a multiple of 16 positions. Each rotation of 16 positions to the left makes the bits of y appear in the same column but in the next row. As mentioned in section 5, the ShiftRows step can be done faster if an additional rotation by 16 positions to the right is performed. MixColumns is the step performed after ShiftRows and has to undo this rotation by doing an additional rotation by 16 positions to the left. The according equations are:

$$bit_0 = bit_0 \oplus rl^{16}(bit_7) \qquad \oplus rl^{32}(bit_0 \oplus bit_7) \qquad \oplus rl^{48}(bit_0)$$
$$bit_1 = bit_1 \oplus rl^{16}(bit_0 \oplus bit_7) \oplus rl^{32}(bit_0 \oplus bit_1 \oplus bit_7) \oplus rl^{48}(bit_1)$$
$$bit_2 = bit_2 \oplus rl^{16}(bit_1) \qquad \oplus rl^{32}(bit_1 \oplus bit_2) \qquad \oplus rl^{48}(bit_2)$$
$$bit_3 = bit_3 \oplus rl^{16}(bit_2 \oplus bit_7) \oplus rl^{32}(bit_2 \oplus bit_3 \oplus bit_7) \oplus rl^{48}(bit_3)$$
$$bit_4 = bit_4 \oplus rl^{16}(bit_3 \oplus bit_7) \oplus rl^{32}(bit_3 \oplus bit_4 \oplus bit_7) \oplus rl^{48}(bit_4)$$
$$bit_5 = bit_5 \oplus rl^{16}(bit_4) \qquad \oplus rl^{32}(bit_4 \oplus bit_5) \qquad \oplus rl^{48}(bit_5)$$
$$bit_6 = bit_6 \oplus rl^{16}(bit_5) \qquad \oplus rl^{32}(bit_5 \oplus bit_6) \qquad \oplus rl^{48}(bit_6)$$
$$bit_7 = bit_7 \oplus rl^{16}(bit_6) \qquad \oplus rl^{32}(bit_6 \oplus bit_7) \qquad \oplus rl^{48}(bit_7)$$

Table 2. Polynomial multiplication with 2 and 3 on a bit level

bit	x	$2 \bullet x$	$3 \bullet x$
0	$x\|_0$	$x\|_7$	$x\|_0 \oplus x\|_7$
1	$x\|_1$	$x\|_0 \oplus x\|_7$	$x\|_0 \oplus x\|_0 \oplus x\|_7$
2	$x\|_2$	$x\|_1$	$x\|_0 \oplus x\|_1$
3	$x\|_3$	$x\|_2 \oplus x\|_7$	$x\|_0 \oplus x\|_2 \oplus x\|_7$
4	$x\|_4$	$x\|_3 \oplus x\|_7$	$x\|_0 \oplus x\|_3 \oplus x\|_7$
5	$x\|_5$	$x\|_4$	$x\|_0 \oplus x\|_4$
6	$x\|_6$	$x\|_5$	$x\|_0 \oplus x\|_5$
7	$x\|_7$	$x\|_6$	$x\|_0 \oplus x\|_6$

In a straight forward approach, these equations could be implemented with 10 MOVs, 24 ROLs and 34 XORs if $bit_0 \oplus bit_7$ and $bit_3 \oplus bit_7$ were precalculated. However there is also a better solution, precalculating the following subterms:

$$t0 = bit_0 \oplus rl^{16}(bit_0) \qquad t1 = rl^{16}(bit_1) \oplus rl^{32}(bit_1)$$
$$t2 = bit_2 \oplus rl^{16}(bit_2) \qquad t3 = rl^{16}(bit_3) \oplus rl^{32}(bit_3)$$
$$t4 = bit_4 \oplus rl^{16}(bit_4) \qquad t5 = rl^{16}(bit_5) \oplus rl^{32}(bit_5)$$
$$t6 = bit_6 \oplus rl^{16}(bit_6) \qquad t7 = rl^{16}(bit_7) \oplus rl^{32}(bit_7)$$

The pseudocode in listing 1.3 shows how these values can be used to get an efficient implementation of the MixColumns step in a bitsliced manner.

Listing 1.3. MixColumns in pseudocode

```
 1    bit2  ^= t1       // bit2 = bit2^rl16(bit1)^rl32(bit1)
 2    t1    ^= t0       // t1   = rl16(bit1)^rl32(bit1)^bit0^rl16(bit0)
 3    t1 = rl16(t1)     // t1   = rl32(bit1)^rl48(bit1)^rl16(bit0)^rl32(bit0)
 4    bit1  ^= t1       // bit1 = bit1^rl32(bit1)^rl48(bit1)^rl16(bit0)^rl32(bit0)
 5    t0 = rl32(t0)     // t0   = rl32(bit0)^rl48(bit0)
 6    bit0  ^= t0       // bit0 = bit0^rl32(bit0)^rl48(bit0)
 7    bit4  ^= t3       // bit4 = bit4^rl16(bit3)^rl32(bit3)
 8    t3   ^= t2        // t3   = rl16(bit3)^rl32(bit3)^bit2^rl16(bit2)
 9    t3 = rl16(t3)     // t3   = rl32(bit3)^rl48(bit3)^rl16(bit2)^rl32(bit2)
10    bit3  ^= t3       // bit3 = bit3^rl32(bit3)^rl48(bit3)^rl16(bit2)^rl32(bit2)
11    t2 = rl32(t2)     // t2   = rl32(bit2)^rl48(bit2)
12    bit2  ^= t2       // bit2 finished
13    bit6  ^= t5       // bit6 = bit6^rl16(bit5)^rl32(bit5)
14    t5   ^= t4        // t5   = rl16(bit5)^rl32(bit5)^bit4^rl16(bit4)
15    t5 = rl16(t5)     // t5   = rl32(bit5)^rl48(bit5)^rl16(bit4)^rl32(bit4)
16    bit5  ^= t5       // bit5 finished
17    t4 = rl32(t4)     // t4   = rl32(bit4)^rl48(bit4)
18    bit4  ^= t4       // bit4 = bit4^rl32(bit4)^rl48(bit4)
19    bit0  ^= t7       // bit0 finished
20    bit1  ^= t7       // bit1 finished
21    bit3  ^= t7       // bit3 finished
22    bit4  ^= t7       // bit4 finished
23    t7   ^= t6        // t7   = rl16(bit7)^rl32(bit7)^bit6^rl16(bit6)
24    t7 = rl16(t7)     // t7   = rl32(bit7)^rl48(bit7)^rl16(bit6)^rl32(bit6)
25    bit7  ^= t7       // bit7 finished
26    t6 = rl32(t6)     // bit6 = rl32(bit6)^rl48(bit6)
27    bit6  ^= t6       // bit6 finished
```

This pseudocode can be implemented with 8 MOVs, 20 ROLs and 27 XOR-instruct-ions. Compared to the straight forward approach, 2 MOVs, 4 ROLs and 7 XORs so 13 instructions in total could be saved. These are savings of almost 20% of the instructions of the MixColumns operation.

7 Improving the Performance on Instruction Level

The concepts presented so far where implemented in a standard version of the bitsliced AES. In order to increase the speed of the implementation the following modifications where made for a fast version:

– Whenever possible, the assembler instructions were mixed in a way to minimize data dependencies between neighboring instructions.
– Whenever code contained lots of load and store instructions, they were spread be-tween logical instructions as good as possible.
– For the implementation of the S-box, subroutines were inserted for better opti-mization concerning the previous two points. The usage of temporary registers was changed slightly, to reduce the number of x64-instructions, but the number of logi-cal operations was not affected (see also table 1).
– The MixColumns step was merged with the AddRoundKey step, to avoid lots of load operations after each other in the AddRoundKey step.
– The transformation   was merged with the key addition of round 0 for the same reasons.

8 Performance Results

On a machine with an AMD Opteron^{TM} 146 Processor with 2.00 GHz and 1 GB of RAM, the sources where compiled and executed with Microsoft Visual Studio®2005 Standard Edition under Windows®XP x64 Edition. Timing measurement was done as shown in Appendix A of [8]. The optimised ANSI C code for the Rijndael cipher in ver-sion 3.0[1] from Rijmen, Bosselaers and Barreto was used as reference implementation for benchmarking, as it is widely used in practice (e.g. in openSSL).

The performance results are shown in table 3. The standard version is about 15% slower than the reference implementation but needs 24% less code memory and 93% less data memory. The fast version is only 5% slower than the reference implementation and needs 8% less code memory and 93% less data memory. Table 4 lists the operations necessary for the whole encryption. Supposing that the execution time of every logical instruction is independent from the data it operates on, the execution time of the whole implementation is constant, since no data dependent branch or memory access occurs.

Comparing the results to [8] and [9] is not easy, since both papers only provide absolute performance measures, heavily depending on the processor type and on other components of the test environment. As the AMD Opteron^{TM} 146 can not compete

---

[1] Currently (Jan. 2008) available at http://www.iaik.tugraz.at/research/
krypto/AES/old/ rijmen/rijndael/rijndael-fst-3.0.zip

with modern processors, especially those with multiple cores, comparisons with such results do not give any information on the quality of this implementation either.

[3] investigates other mitigation techniques against cache-based software side channel attacks with a performance loss between a factor of 1.35 and 2.85 (compared to the same reference implementation). The solution presented in this paper clearly outperforms these mitigations.

Table 3. Performance results

Implementation	Cycles/block	Encrypt 100MB	Code size	Data size
Standard bitslice	347	1125 ms	3702 byte	704 byte
Fast bitslice	317	1031 ms	4474 byte	704 byte
Optimized reference	303	922 ms	4887 byte	10280 byte

Table 4. Instruction count

	11x addRoundKey	10x S-box	9x shiftRowsRR16	9x mixColumnsRL16	1x shiftRows	1x mu	1x muInv	overhead	
XOR	88	940		243		72	72		1415
AND		360				24	24		408
ROL			72	180	8				260
ROR			72		16				88
ROL16			216		24				240
SHR						24	24		48
SHL						24	24		48
logical instr.	88	1300	360	423	48	144	144		2507
ADD								10	10
MOV		350		72		24	24		470
Load	88	180				12		20	300
Store		180					12	10	202
CALL								42	42
RET								42	42
x64 instr.	88	1920	360	495	48	180	180	124	3395

9 Further Work

Most modern processors support SSE2-extensions with 128-bit registers. An implementation only using SSE2-instructions could process even more blocks in parallel due to wider registers. Mixing SSE2-instructions with x64-instructions may also be a good way to a solution with higher performance. However, [8] claims, that in most of today's processors, SSE2-instructions are of no real use for bitslice implementations due to poor latency and throughput.

Another possibility for improvements is to completely remove all borders between the different steps of the algorithm. Simplifications may be possible.

The main focus during this work was to achieve a high throughput. No attempts were made to minimize the code size or the needed amount of data. Macros were used often as they are easy to use. Some macros might be replaced by function calls without or with only little decline of the speed of the implementation while decreasing the code size significantly.

10   Conclusion

Bitslicing is an unorthodox technique for the implementation of algorithms in software. The application of this method to the implementation of cryptographic primitives provides some very strong advantages like the low amount of memory needed and the resistance against cache-timing attacks. In this paper we showed, that a bitslice solution can outperform other mitigations to cache-timing attacks.

Further improvements and different tradeoffs in various directions are still possible. The fact that processors with larger word size are more and more upcoming, and the circumstance, that the speed in executing logical operations grows faster than the speed in memory access will make the bitslice approach even more important in the future.

References

1. Biham, E.: A fast new DES implementation in software. In: Biham, E. (ed.) FSE 1997. LNCS, vol. 1267, pp. 260–272. Springer, Heidelberg (1997)
2. Bernstein, D.J.: Cache-timing attacks on AES (April 2005), Revised version of earlier 2004-11 version, http://cr.yp.to/antiforgery/cachetiming-20050414.pdf
3. Brickell, E., Graunke, G., Neve, M., Seifert, J.P.: Software mitigations to hedge AES against cache-based software side channel vulnerabilities. Cryptology ePrint Archive, Report 2006/052 (2006) http://eprint.iacr.org/
4. Neve, M., Seifert, J.-P.: Advances on access-driven cache attacks on AES. In: Biham, E., Youssef, A.M. (eds.) SAC 2006. LNCS, vol. 4356, pp. 147–162. Springer, Heidelberg (2007)
5. Osvik, D.A., Shamir, A., Tromer, E.: Cache attacks and countermeasures: The case of AES. In: Pointcheval, D. (ed.) CT-RSA 2006. LNCS, vol. 3860, pp. 1–20. Springer, Heidelberg (2006)
6. Bonneau, J., Mironov, I.: Cache-collision timing attacks against AES. In: Goubin, L., Matsui, M. (eds.) CHES 2006. LNCS, vol. 4249, pp. 201–215. Springer, Heidelberg (2006)
7. Aciiçmez, O., Schindler, W., Koç, Ç.K.: Cache based remote timing attack on the AES. In: Abe, M. (ed.) CT-RSA 2007. LNCS, vol. 4377, pp. 271–286. Springer, Heidelberg (2006)
8. Matsui, M.: How far can we go on the x64 processors? In: Robshaw, M.J.B. (ed.) FSE 2006. LNCS, vol. 4047, pp. 341–358. Springer, Heidelberg (2006)
9. Rebeiro, C., Selvakumar, A.D., Devi, A.S.L.: Bitslice implementation of AES. In: Pointcheval, D., Mu, Y., Chen, K. (eds.) CANS 2006. LNCS, vol. 4301, pp. 203–212. Springer, Heidelberg (2006)
10. Canright, D.: A very compact Rijndael S-box (revised). Naval Postgraduate School Technical Report, NPS-MA-05-001 (May 2005), http://handle.dtic.mil/100.2/ADA427050

11. May, L., Penna, L., Clark, A.: An implementation of bitsliced DES on the pentium mmx[tm] processor. In: Clark, A., Boyd, C., Dawson, E.P. (eds.) ACISP 2000. LNCS, vol. 1841, pp. 112–122. Springer, Heidelberg (2000)
12. Satoh, A., Morioka, S., Takano, K., Munetoh, S.: A compact Rijndael hardware architecture with S-box optimization. In: Boyd, C. (ed.) ASIACRYPT 2001. LNCS, vol. 2248, pp. 239–254. Springer, Heidelberg (2001)

Appendix A.1

Listing 1.4. The transformation   into the bitslice domain with SWAPMOVE

```
 1  ; R8  =  low  64 bytes  of  block 3      R9  =  low  64 bytes  of  block 2
 2  ; R10 =  low  64 bytes  of  block 1      R11 =  low  64 bytes  of  block 0
 3  ; R12 =  high 64 bytes  of  block 3      R13 =  high 64 bytes  of  block 2
 4  ; R14 =  high 64 bytes  of  block 1      R15 =  high 64 bytes  of  block 0
 5  swapmove R12, R8, 8,00ff00ff00ff00ffh    swapmove R8,R12,16,0000ffff0000ffffh
 6  swapmove R12, R8,32,00000000ffffffffh    swapmove R13, R9, 8,00ff00ff00ff00ffh
 7  swapmove R9,R13,16,0000ffff0000ffffh     swapmove R13, R9,32,00000000ffffffffh
 8  swapmove R14,R10, 8,00ff00ff00ff00ffh    swapmove R10,R14,16,0000ffff0000ffffh
 9  swapmove R14,R10,32,00000000ffffffffh    swapmove R15,R11, 8,00ff00ff00ff00ffh
10  swapmove R11,R15,16,0000ffff0000ffffh    swapmove R15,R11,32,00000000ffffffffh
11  swapmove R14,R15, 1,5555555555555555h    swapmove R12,R13, 1,5555555555555555h
12  swapmove R10,R11, 1,5555555555555555h    swapmove R8, R9, 1,5555555555555555h
13  swapmove R13,R15, 2,3333333333333333h    swapmove R12,R14, 2,3333333333333333h
14  swapmove R9,R11, 2,3333333333333333h     swapmove R8,R10, 2,3333333333333333h
15  swapmove R11,R15, 4,0f0f0f0f0f0f0f0fh    swapmove R10,R14, 4,0f0f0f0f0f0f0f0fh
16  swapmove R9,R13, 4,0f0f0f0f0f0f0f0fh     swapmove R8,R12, 4,0f0f0f0f0f0f0f0fh
```

Appendix B.1

Listing 1.5. Bitsliced S-box transformation in C

```c
 1  void Sbox(u64* x7,u64* x6,u64* x5,u64* x4,u64* x3,u64* x2,u64* x1,u64* x0,
 2           u64* t7,u64* t6,u64* t5,u64* t4,u64* t3,u64* t2,u64* t1,u64* t0) {
 3      // in:  x7 x6 x5 x4 x3 x2 x1 x0
 4      G256_newbasisA2X(x7,x6,x5,x4,x3,x2,x1,x0,t0);
 5      // out: x2 x4 x1 x7 x3 x0 x5 x6
 6      // in:  x2 x4 x1 x7 x3 x0 x5 x6
 7      // in:  x2 x4 x1 x7 x3 x0 x5 x6
 8      G256_inv(x2,x4,x1,x7,x3,x0,x5,x6,t6,t5,t4,t3,t2,t1,t0);
 9      // out: x3 x0 x5 x6 x2 x4 x1 x7
11      // in:  x3 x0 x5 x6 x2 x4 x1 x7
12      G256_newbasisX2S(x3,x0,x5,x6,x2,x4,x1,x7,t3,t2,t1,t0);
13      // out: x7 x6 x5 x4 x3 x2 x1 x0
15      // do the affine transformation (can be skipped if round keys are modified):
16      *x6 = ~ *x6;   *x5 = ~ *x5;   *x1 = ~ *x1;   *x0 = ~ *x0;
17  }
18  // in:  a7 a6 a5 a4 a3 a2 a1 a0      out: a2 a4 a1 a7 a3 a0 a5 a6
19  void G256_newbasisA2X(u64* a7, u64* a6, u64* a5, u64* a4, u64* a3, u64* a2,
20           u64* a1, u64* a0, u64* a1_a7) {
21      // a0 is already OK (=x2)
22      *a1_a7 = *a1;   // a1_a7 = a1
23      *a1_a7 ^= *a7;  // a1_a7 = a1 ^ a7
24      *a5 ^= *a6;     // a5 = a5 ^ a6
25      *a5 ^= *a0;     // a5 = a0 ^ a5 ^ a6 -> OK (=x1)
26      *a6 ^= *a0;     // a6 = a0 ^ a6
27      *a6 ^= *a1;     // a6 = a0 ^ a1 ^ a6
28      *a6 ^= *a2;     // a6 = a0 ^ a1 ^ a2 ^ a6
29      *a6 ^= *a3;     // a6 = a0 ^ a1 ^ a2 ^ a3 ^ a6 -> OK (=x0)
```

```
30    *a7 ^= *a5;          // a7 = a0 ^ a5 ^ a6 ^ a7 -> OK (=x4)
31    *a1 ^= *a5;          // a1 = a0 ^ a1 ^ a5 ^ a6 -> OK (=x5)
32    *a3 ^= *a4;          // a3 = a3 ^ a4
33    *a4 ^= *a5;          // a4 = a0 ^ a4 ^ a5 ^ a6 -> OK (=x6)
34    *a2 ^= *a1_a7;       // a2 = a1 ^ a2 ^ a7
35    *a2 ^= *a5;          // a2 = a0 ^ a1 ^ a2 ^ a5 ^ a6 ^ a7 -> OK (=x7)
36    *a3 ^= *a1_a7;       // a3 = a1 ^ a3 ^ a4 ^ a7
37    *a3 ^= *a0;          // a3 = a0 ^ a1 ^ a3 ^ a4 ^ a7 -> OK (=x3)
38  }
39  // in:   x7 x6 x5 x4 x3 x2 x1 x0        out: x0 x4 x5 x2 x7 x3 x1 x6
40  void G256_newbasisX2S(u64* x7,u64* x6,u64* x5,u64* x4,u64* x3,u64* x2,u64*x1,
41         u64* x0,u64* x6_backup,u64* x2_backup,u64* x1_backup,u64* x0_backup) {
42    *x6_backup = *x6;   // x6_backup = x6
43    *x6 ^= *x4;         // x6 = x4 ^ x6
44    *x1_backup = *x1;   // x1_backup = x1
45    *x1 ^= *x4;         // x1 = x1 ^ x4
46    *x2_backup = *x2;   // x2_backup = x2
47    *x1 ^= *x5;         // x1 = x1 ^ x4 ^ x5 -> OK (=s1)
48    *x4 = *x3;          // x4 = x3
49    *x4 ^= *x7;         // x4 = x3 ^ x7 -> OK (=s6)
50    *x2 = *x4;          // x2 = x3 ^ x7
51    *x2 ^= *x5;         // x2 = x3 ^ x5 ^ x7 -> OK (=s4)
52    *x7 = *x2;          // x7 = x3 ^ x5 ^ x7
53    *x7 ^= *x6;         // x7 = x3 ^ x4 ^ x5 ^ x6 ^ x7 -> OK (=s3)
54    *x0_backup = *x0;   // x0_backup = x0
55    *x6 ^= *x1_backup;  // x6 = x1 ^ x4 ^ x6 -> OK (=s0)
56    *x0 = *x3;          // x0 = x3
57    *x0 ^= *x5;         // x0 = x3 ^ x5 -> OK (=s7)
58    *x5 = *x6_backup;   // x5 = x6
59    *x5 ^= *x0_backup;  // x5 = x0 ^ x6 -> OK (=s5)
60    *x3 = *x0;          // x3 = x3 ^ x5
61    *x3 ^= *x5;         // x3 = x0 ^ x3 ^ x5 ^ x6
62    *x3 ^= *x2_backup;  // x3 = x0 ^ x2 ^ x3 ^ x5 ^ x6 -> OK (=s2)
63  }
64  // in:  a1 a0, b1 b0          out: a1 a0
65  void G4_mul21(u64* a1, u64* a0, u64* b1, u64* b0, u64* e) {
66    *e = load(-64);      // stack[-64] = x7 ^ x6 = r3 ^ r2 = a1 ^ a0 (line 93)
67    *e &= load(-32);     // stack[-32] = d1 ^ d0 = s3 ^ s2 = b1 ^ b0 (line 79)
68    *a1 &= *b1;   *a0 &= *b0;   *a1 ^= *e;   *a0 ^= *e;
69  }
70  // in:  a1 a0, b1 b0          out: a1 a0
71  void G4_mul22(u64* a1, u64* a0, u64* b1, u64* b0, u64* e) {
72    *e = load(-80);      // stack[-80] = x5 ^ x4 = r1 ^ r0 = a1 ^ a0 (line 101)
73    *e &= load(-48);     // stack[-48] = d3 ^ d2 = s1 ^ s0 = b1 ^ b0 (line 86)
74    *a1 &= *b1;   *a0 &= *b0;   *a1 ^= *e;   *a0 ^= *e;
75  }
76  // in:  a1 a0, b1 b0          out: a1 a0
77  void G4_mul11(u64* a1, u64* a0, u64* b1, u64* b0, u64* e) {
78    *e = *b1;   *e ^= *b0;
79    store(*e,-32);       // stack[-32] <- e = b1 ^ b0 = s3 ^ s2 = d1 ^ d0
80    *e &= load(-72);     // stack[-72] = x3 ^ x2 = r3 ^ r2 = a1 ^ a0  (line 95)
81    *a1 &= *b1;   *a0 &= *b0;   *a1 ^= *e;   *a0 ^= *e;
82  }
83  // in:  a1 a0, b1 b0          out: a1 a0
84  void G4_mul12(u64* a1, u64* a0, u64* b1, u64* b0, u64* e) {
85    *e = *b1;   *e ^= *b0;
86    store(*e,-48);       // stack[-48] <- e = b1 ^ b0 = s1 ^ s0 = d3 ^ d2
87    *e &= load(-88);     // stack[-88] = x1 ^ x0 = r1 ^ r0 = a1 ^ a0 (line 103)
88    *a1 &= *b1;   *a0 &= *b0;   *a1 ^= *e;   *a0 ^= *e;
89  }
90  // in:  a1 a0, b1 b0          out: a1 a0
91  void G4_mul_store_x7x6_x3x2(u64* a1,u64* a0,u64* b1,u64* b0,u64* e,u64* e1){
92    *e = *a1;   *e ^= *a0;
93    store(*e,-64);       // stack[-64] <- e = a1 ^ a0 = x7 ^ x6
94    *e1 = *b1;   *e1 ^= *b0;
95    store(*e1,-72);      // stack[-72] <- e1 = b1 ^ b0 = x3 ^ x2
96    *e &= *e1;   *a1 &= *b1;   *a0 &= *b0;   *a1 ^= *e;   *a0 ^= *e;
97  }
```

```
98   // in:   a1 a0, b1 b0         out: a1 a0
99   void G4_mul_store_x5x4_x1x0(u64* a1,u64* a0,u64* b1,u64* b0,u64* e,u64* e1){
100    *e = *a1;   *e ^= *a0;
101    store(*e,-80);       // stack[-80] <- e = a1 ^ a0 = x5 ^ x4
102    *e1 = *b1;   *e1 ^= *b0;
103    store(*e1,-88);      // stack[-88] <- e1 = b1 ^ b0 = x1 ^ x0
104    *e &= *e1;   *a1 &= *b1;   *a0 &= *b0;   *a1 ^= *e;   *a0 ^= *e;
105   }
106   // in:   r3 r2 r1 r0, s3 s2 s1 s0          out: r3 r2 r1 r0
107   void G16_mul1(u64* r3,  u64* r2,  u64* r1,  u64* r0,  u64* s3,  u64* s2,  u64* s1,
108                 u64* s0,  u64* t3,  u64* t2,  u64* t1) {
109    *t3 = load(-24);     // stack[-24] = x3 ^ x1 = r3 ^ r1 (line 160)
110    *t2 = *s3;   *t2 ^= *s1;
111    store(*t2,-24);      // stack[-24] <- t2 = s1 ^ s3 = d1 ^ d3
112    *t3 &= *t2;   *t1 = *s0;   *t1 ^= *s2;
113    store(*t1,-56);      // stack[-56] <- t1 = s0 ^ s2 = d0 ^ d2
114    *t2 ^= *t1;
115    store(*t2,-96);      // stack[-96] <- t2 = s0^s1^s2^s3 = d0 ^ d1 ^ d2 ^ d3
116    *t1 &= load(-32);    // stack[-32] = x2 ^ x0 = r2 ^ r0 (line 162)
117    *t3 ^= *t1;
118    *t2 &= load(-48);    // stack[-48] = x2^x0^x3^x1 = r2 ^ r0 ^ r3 ^ r1 (line 166)
119    *t2 ^= *t1;
120    G4_mul11(r3,  r2,  s3,  s2,  t1);
121    *r3 ^= *t2;   *r2 ^= *t3;
122    G4_mul12(r1,  r0,  s1,  s0,  t1);
123    *r1 ^= *t2;   *r0 ^= *t3;
124   }
125   // in:   r3 r2 r1 r0, s3 s2 s1 s0          out: r3 r2 r1 r0
126   void G16_mul2(u64* r3,  u64* r2,  u64* r1,  u64* r0,  u64* s3,  u64* s2,  u64* s1,
127                 u64* s0,  u64* t3,  u64* t2,  u64* t1) {
128    *t3 = load(-8);      // stack[-8] = x7 ^ x5 = r3 ^ r1 (line 156)
129    *t3 &= load(-24);    // stack[-24] = d1 ^ d3 = s3 ^ s1 (line 111)
130    *t2 = load(-16);     // stack[-26] = x6 ^ x4 = r2 ^ r0 (line 158)
131    *t2 &= load(-56);    // stack[-56] = d0 ^ d2 = s2 ^ s0 (line 113)
132    *t3 ^= *t2;
133    *t1 = load(-40);     // stack[-40] = x7^x4^x5^x4 = r3 ^ r2 ^ r1 ^ r0 (line 164)
134    *t1 &= load(-96);    // stack[-96] = d7^d4^d5^d4 = s3 ^ s2 ^ s1 ^ s0 (line 115)
135    *t2 ^= *t1;
136    G4_mul21(r3,  r2,  s3,  s2,  t1);
137    *r3 ^= *t2;   *r2 ^= *t3;
138    G4_mul22(r1,  r0,  s1,  s0,  t1);
139    *r1 ^= *t2;   *r0 ^= *t3;
140   }
141   // in:   r3 r2 r1 r0         out: r1 r0 r3 r2
142   void G16_inv(u64* r3,  u64* r2,  u64* r1,  u64* r0,  u64* e,  u64* r3_r2,
143                u64* r1_r0,  u64* d1,  u64* d0) {
144    *r3_r2 = *r3;   *r3_r2 ^= *r2;   *r1_r0 = *r1;   *r1_r0 ^= *r0;   *e = *r3_r2;
145    *e &= *r1_r0;   *e ^= *r3;   *e ^= *r1;   *d1 = *r3;   *d1 &= *r1;
146    *d1 ^= *e;   *d0 = *r2;   *d0 &= *r0;   *d0 ^= *e;   *d0 ^= *r2;
147    *d0 ^= *r0;
148    *r3 &= *d0;   *r2 &= *d1;   *r1 &= *d0;   *r0 &= *d1;
149    *d0 ^= *d1;   *r3_r2 &= *d0;   *r3 ^= *r3_r2;   *r2 ^= *r3_r2;   *r1_r0 &= *d0;
150    *r1 ^= *r1_r0;   *r0 ^= *r1_r0;
151   }
152   // in:   x7 x6 x5 x4 x3 x2 x1 x0          out: x3 x2 x1 x0 x7 x6 x5 x4
153   void G256_inv(u64 *x7,u64 *x6,u64 *x5,u64 *x4,u64 *x3,u64 *x2,u64 *x1,u64*x0,
154                 u64 *d3,u64 *d2,u64 *d1,u64 *d0,u64 *a3,u64 *a2,u64 *a1) {
155    *a3 = *x7;   *a3 ^= *x5;
156    store(*a3,-8);       // stack[-8] <- x7 ^ x5
157    *a2 = *x6;   *a2 ^= *x4;
158    store(*a2,-16);      // stack[-16] <- x6 ^ x4
159    *d3 = *x3;   *d3 ^= *x1;
160    store(*d3,-24);      // stack[-24] <- x3 ^ x1
161    *d2 = *x2;   *d2 ^= *x0;
162    store(*d2,-32);      // stack[-32] <- x2 ^ x0
163    *d1 = *a3;   *d1 ^= *a2;
164    store(*d1,-40);      // stack[-40] <- x7 ^ x5 ^ x6 ^ x4
165    *a2 &= *d2;   *d2 ^= *d3;
```

```
166    store(*d2,−48);        //stack[−48] <− x2 ^ x0 ^ x3 ^ x1
167    *d1 &= *d2;   *a3 &= *d3;   *a3 ^= *a2;   *a2 ^= *d1;
168    *d3 = *x7;    *d2 = *x6;    *d1 = *x5;    *d0 = *x4;
169    store(*a2,−56);        //stack[−56] <− a2 (to use a2 as temporary variable)
170    G4_mul_store_x7x6_x3x2(d3, d2, x3, x2, a2, a1);
171    G4_mul_store_x5x4_x1x0(d1, d0, x1, x0, a2, a1);
172    *a2 = load(−56);       //a2 <− stack[−56] (restore a2 again) (line 169)
173    *a1 = *x0;   *a1 ^= *x4;   *a2 ^= *a1;   *d1 ^= *a2;   *d0 ^= *a3;   *d3 ^= *a2;
174    *d2 ^= *a3;   *d0 ^= *a1;   *a1 = *x1;   *a1 ^= *x5;   *d1 ^= *a1;   *d2 ^= *a1;
175    *d3 ^= *x6;   *d3 ^= *x2;   *d2 ^= *x7;   *d2 ^= *x3;
176    store(*x1,−56);        //stack[−48] <− x1 (to use x1 as temporary variable)
177    store(*x0,−96);        //stack[−88] <− x0 (to use x0 as temporary variable)
178    G16_inv(d3, d2, d1, d0, x1, x0, a3, a2, a1);
179    *x1 = load(−56);       //x1 <− stack[−48] (restore x1 again) (line 176)
180    *x0 = load(−96);       //x0 <− stack[−88] (restore x0 again) (line 177)
181    G16_mul1(x3, x2, x1, x0, d1, d0, d3, d2, a3, a2, a1);
182    G16_mul2(x7, x6, x5, x4, d1, d0, d3, d2, a3, a2, a1);
183  }
```

Appendix B.2

Listing 1.6. Fully optimized S-box implementation in x64-MASM Syntax

```
1    SBOX MACRO x7, x6, x5, x4, x3, x2, x1, x0, t6, t5, t4, t3, t2, t1, t0
2        MOV t0,x1   XOR x5,x6   XOR t0,x7   XOR x5,x0   XOR x6,x1   XOR x7,x5   XOR x6,x3
3        XOR x1,x5   XOR x6,x0   XOR x3,x4   XOR x6,x2   XOR x4,x5   XOR x2,t0   XOR x3,x0
4        XOR x2,x5   XOR x3,t0   MOV t1,x4   MOV t2,x2   XOR t1,x7   XOR t2,x1
5        MOV [RSP−16],t1   MOV t4,t1   MOV t5,x0   MOV [RSP−8],t2   MOV t6,x3   XOR t5,x6
6        XOR t6,x5   MOV [RSP−32],t5   XOR t4,t2   AND t1,t5   MOV [RSP−24],t6   AND t2,t6
7        MOV [RSP−40],t4   XOR t5,t6   XOR t2,t1   AND t4,t5   MOV [RSP−48],t5   MOV t6,x2
8        XOR t1,t4   AND t6,x3   MOV [RSP−56],t1   MOV t0,x3   MOV t1,x2   XOR t0,x0
9        XOR t1,x4   MOV t5,x4   MOV [RSP−64],t1   MOV t3,x7   AND t1,t0   AND t3,x6
10       MOV [RSP−72],t0   AND t5,x0   XOR t6,t1   XOR t5,t1   MOV [RSP−96],x6   MOV t1,x1
11       MOV t0,x5   XOR t1,x7   XOR t0,x6   MOV [RSP−80],t1   XOR t5,t2   MOV t4,x1
12       MOV [RSP−104],x5   AND t1,t0   AND t4,x6,x7   MOV [RSP−88],t0
13       XOR t4,t1   XOR t3,t1   XOR x5,x1   MOV t1,[RSP−56]   XOR t3,t2   XOR t5,x5
14       XOR t3,x6   XOR t5,x2   XOR t1,x6   XOR t6,x0   XOR t4,t1   XOR t6,t1   XOR t4,x5
15       XOR t6,x4   MOV t2,t4   XOR t5,x3   XOR t2,t3   MOV x6,t5   MOV t0,t5   XOR x6,t6
16       AND t0,t3   MOV x5,t2   XOR t0,t3   AND x5,x6   XOR t0,t5   XOR x5,t6   MOV t1,t6
17       XOR x5,t4   AND t1,t4   XOR t0,x5   XOR t1,x5   AND t6,t0   AND t4,t0
18       MOV x5,[RSP−104]   AND t5,t1   XOR t0,t1   AND t3,t1   AND x6,t0   AND t2,t0
19       XOR t6,x6   XOR t4,t2   XOR t5,x6   XOR t3,t2   MOV t1,t6   MOV x6,[RSP−96]
20       AND x3,t4   XOR t1,t4   MOV t2,[RSP−24]   AND x2,t4   AND [RSP−8],t1
21       AND x5,t6   XOR t4,t3   AND x1,t6   AND [RSP−64],t4   AND x0,t3   XOR t6,t5
22       AND t4,[RSP−72]   AND x4,t3   AND [RSP−80],t6   AND x7,t5   XOR t3,t5
23       AND t6,[RSP−88]   AND x6,t5   AND t2,t1   AND [RSP−16],t3   XOR x5,t6
24       MOV t5,[RSP−64]   XOR x6,t6   MOV t6,t1   XOR x3,t4   XOR t1,t3   AND t3,[RSP−32]
25       XOR x0,t4   XOR x2,t5   AND [RSP−40],t1   XOR t2,t3   AND t1,[RSP−48]   XOR x0,t2
26       MOV t0,[RSP−8]   XOR t1,t3   XOR x4,t5   MOV t3,[RSP−80]   XOR x6,t2   XOR x3,t1
27       XOR x5,t1   MOV t6,x0   MOV t1,[RSP−16]   XOR x0,x6   XOR x1,t3   XOR t0,t1
28       XOR x7,t3   XOR t1,[RSP−40]   XOR x4,t0   XOR x7,t0   XOR x1,t1   XOR x2,t1
29       MOV t2,x1   MOV t1,x4   XOR x1,x6   MOV x4,x5   MOV x6,x2   XOR x1,x5   XOR x6,x3
30       MOV t0,x7   XOR x4,x6   MOV x3,x0   MOV x7,x5   XOR x3,x4   XOR x7,x2   MOV x5,t6
31       MOV x2,x7   XOR x5,t0   XOR x2,t1   XOR x0,t2   XOR x2,x5
32   ENDM
```

Identity-Based Threshold Key-Insulated Encryption without Random Oracles*

Jian Weng[1], Shengli Liu[1,3], Kefei Chen[1], Dong Zheng[1], and Weidong Qiu[2]

[1] Dept. of Computer Science and Engineering
Shanghai Jiao Tong University, Shanghai 200240, P.R. China
[2] School of Information Security Engineering
Shanghai Jiao Tong University, Shanghai 200240, P.R. China
[3] State Key Laboratory of Information Security
Institute of Software, Chinese Academy of Sciences, Beijing 100080, China
cryptjweng@gmail.com, {slliu,kfchen,dzheng,qiuwd}@sjtu.edu.cn

Abstract. With more and more cryptosystems being deployed on insecure environments such as mobile devices, key exposures appear to be unavoidable. This is perhaps the most devastating attack on a cryptosystem, since it typically means that security is entirely lost. This problem is especially hard to tackle in identity-based encryption (IBE) settings, where the public key is determined as a user's identity and is not desirable to be changed. In this paper, we extend Dodis et al.'s key-insulation idea and present a new paradigm named threshold key-insulation. The new paradigm not only greatly enhances the security of the system, but also provides flexibility and efficiency. To deal with the key-exposure problem in IBE settings, we further propose an identity-based threshold key-insulated encryption (IBTKIE) scheme. The proposed scheme is proved to be semantically secure without random oracles.

Keywords: key-exposure, threshold key-insulation, identity-based encryption, standard model.

1 Introduction

1.1 Background

Identity Based Encryption (IBE), introduced by Shamir [36], provides a public key mechanism where an arbitrary string, such as recipient's identity, can be served as a public key. The ability to use identities as public keys eliminates the need for certificates as used in the traditional pubic key infrastructure (PKI). Although the concept was proposed in 1984, it was only in 2001 that a secure and truly practical IBE scheme was proposed by Boneh and Franklin [6]. Since then, a series of papers have been devoted to IBE systems, e.g. [4,5,37,24,26,16,25,2,7].

With more and more cryptographic primitives being applied to insecure environments such as mobile devices, key-exposure seems to be inevitable. This

---

* Supported by the National Science Foundation of China under Grant Nos. 90704004, 60673077, 60573030, and 60707030.

T. Malkin (Ed.): CT-RSA 2008, LNCS 4964, pp. 203–220, 2008.

problem is perhaps the most devastating attack on a cryptosystem, since it typically means that security is lost. This problem is especially hard to tackle in identity-based scenarios, since the public key is determined as a user's identity and is not desirable to be changed. One exemplification as shown in [28] is the application of IBE systems in a mobile phone scenario, where the phone number represents a user's identity, and it will be simple and convenient for the mobile phone users to identify and communicate with each other by their phone numbers only.

In Eurocrypt'02, Dodis et al. [21] initially introduced a paradigm named key-insulation to deal with the key-exposure problem for public key encryption systems. Concretely, in a key-insulated system, the life time of the system is broken into discrete time periods, while the private key is split into two parts: a *temporary private key*, hold by the user on a powerful but insecure device (e.g., a mobile device), and a *helper key*, stored in a physically-secure but computationally-limited device named "helper". The public key remains unchanged throughout the lifetime of the system, while the temporary private key is refreshed at every period via the interaction between the user and the helper. Decryption operations in a given period only involve the corresponding temporary private key without further access to the helper. Exposure of the temporary private keys at some periods will not compromise the security of the remaining periods. Even if the helper key is compromised, the security is still ensured as long as none of the temporary private keys is exposed. As a result, the damage caused by key-exposure is minimized.

However, for the key-insulation mechanism with a single helper, there exists some situations hard to deal with. For instance, to increase the system tolerance against key-exposure, the temporary private key has to be updated at short intervals. Unfortunately, this in turn increases the frequency of helper's connection to insecure environments, and the risk of helper key-exposure is accordingly increased. It should be noted that, if the helper is exposed, the security of the system is entirely lost as long as one of the temporary private key is compromised. Next, let's consider another example: Suppose a person works in several branches of a company (for an easy explanation, consider the case of two branches), e.g., he works in branch B1 during the odd days, while during the even days he works in branch B2. He decides to update the temporary private key once per day. For instance, suppose today is an odd day, he updates the temporary private key in branch B1, and tomorrow he will update it in branch B2. Now, some problems happen: He must remind himself to bring the helper to-and-from branches B1 and B2; Also, bringing the helper back and forth means a frequent connection to insecure environments, and puts the helper key in a higher risk of exposure.

To deal with the above problems, in PKC'06, Hanaoka et al. [27] presented a very clever method named parallel key-insulation, where two distinct helpers are alternately used to update the temporary private key. Since the two helpers are independent of each other, even if the frequency of temporary private key-updates is twice as that in original key-insulated systems, the risk of helper key-exposure is still the same as the latter. As a result, the security of the system is enhanced. Furthermore, as to the aforementioned person who works in branches B1 and B2, he can put the first helper key at B1 and the second one at B2. Then during the odd days when he works in branch B1, he can update his temporary private keys using the first helper key, and when he stays in the branch B2 during the even days, he can use the second one to do so. Now, there is no need for him to bring the helpers to-and-from branches B1 and B2, and hence the risk of helper key-exposure will not be increased.

Parallel key-insulation is a good solution to the key-exposure problem. Nevertheless, there still exist some situations hard for it to tackle. Again, take the aforementioned person working in branches B1 and B2 as an example. However, suppose due to certain reason, now he works in the two branches without regularity, e.g., he may works in branch B1 either in even or odd days, and the similar situation holds for his working in branch B2. It is worth noting that, according the model of key-insulation described in [27], this person can only use *the first helper key* (not the second one) to update his temporary private keys for *odd days*, and vice versa. Now, a troublesome problem occurs: suppose in an *odd day* he is working in branch B2, then he *can not* use *the second helper key* at hand to update his temporary private keys! Another situation hard to deal with is that, according to the model of parallel key-insulation in [27], it only supports *sequential key updates*, i.e., the temporary private key for period t can only be updated from that for period $t-1$. Therefore, the user is unable to decrypt old ciphertexts. In other words, during period t, it is impossible for him to decrypt ciphertexts encrypted for any period t' with $t' < t$. Hence, a natural question is to introduce a variant of the key-insulation paradigm, which not only can enhance the security of the system but also does not suffer from the above problems.

1.2   Our Results

In this paper, we introduce a new paradigm named threshold key-insulation. Concretely, for a (k, n) threshold key-insulated system, at least k out of n helpers are needed to update the user's temporary private keys. The security intentions for a (k, n) threshold key-insulated system are:

1. If none of the helper is compromised, similar to the original key-insulated systems, exposure of any of temporary private keys does not compromise the security of the non-exposed periods.
2. Even if up to $k-1$ helpers are compromised in addition to the exposure of any of temporary private keys, it still can not compromise the security of the non-exposed periods.
3. Even if all the n helpers are compromised, security of all periods is still ensured as long as none of temporary private keys is exposed.

Obviously, compared with original key-insulated systems, the security of threshold key-insulated systems is greatly enhanced. Moreover, compared with parallel key-insulated systems, threshold key-insulated systems provide more flexibilities. Recall the aforementioned person who works in branches B1 and B2 without regularity. In our threshold key-insulated paradigm, he can put k helper keys in branch B1, and another k helper keys in branch B2. Interestingly, even if he works in branch B1 (B2, resp.) during even (odd, resp.) days, he can still use the k helper keys at hand to update his temporary private keys. Therefore, there is no need for him to bring the helpers to-and-from branches B1 and B2, and hence it prevents the helper key from a higher exposure risk. Besides, as shown in Section 3.1, threshold key-insulation can support *randomness key-updates* i.e., a temporary private key for period t can be updated from a temporary private key for any period t' in a single step. Consequently, it allows the user to decrypt old ciphertexts.

We further apply the threshold key-insulated paradigm to identity-based scenarios, and propose an identity-based threshold key-insulated encryption (IBTKIE) scheme. Our IBTKIE scheme is provably secure in the standard model, due to the inheritance of Waters' IBE. It is worth noting that the ciphertext length, as well as the computational cost of encryption and decryption, is constant and independent of the threshold parameter k and the total number n of helpers. In contrast, as to Libert et al.'s parallel key-insulated public key encryption (PKIPKE) scheme [34] and Weng et al.'s identity-based parallel key-insulated encryption (IBPKIE) scheme [38] without random oracles, their ciphertext lengths, together with the computational cost of encryption and decryption, grow linearly with the number of helper keys. Of course, the key-update operation in threshold key-insulated systems is slightly more time-consuming than that of the parallel key-insulated system. However, keep in mind that, the encryption/decryption operation is run more frequently than the key-update operation, and hence the overall computational cost of threshold key-insulated systems is obviously less than that of parallel key-insulated systems.

Identity-based threshold decryption (IBTD), introduced by Baek and Zheng [12], distributes the private key across multiple servers so that at least k out of n decryption shares can recover plaintexts. Usually, it is difficult for an adversary to corrupt up to k servers, and hence it can be viewed as another try to deal with the key-exposure problem. Compared with IBTD systems, IBTKIE systems have the following attractive features:

- In IBTD systems, each decryption operation requires the cooperation of at least k servers. In contrast, in IBTKIE systems, the decryption can be done by the user himself, and only the key-update operation needs the cooperation of k helpers. Again, since the decryption operation is run more frequently than the key-update operation, IBTKIE systems save much more computational cost than IBTD systems.
- For an IBTD system, if up to k severs are corrupted, its security is completely lost. In contrast, even if all of a user's helper keys are compromised, security is still ensured as long as none of his temporary private keys is compromised. Hence the security is enhanced.

1.3   Related Works

A long line of research has focused on handling the threat of key-exposure. One approach is to distribute the private key across multiple servers in order to make key-exposure more difficult. This paradigm includes threshold cryptography [17] and proactive cryptography [35]. However, such solutions are some what costly, and most importantly, once sufficient private key shares are (simultaneously) exposed, the security of the system is entirely lost.

Another approach is to evolve the private key with time. Forward-security cryptosystems [3,9,13] is the first solution in this vein. In a forward-secure system, time is divided into discrete periods and the private key is updated at the beginning of every period. Exposure of the current key does not render usages of previous keys insecure, but security of the future periods is lost.

Key-insulated cryptosystems, introduced by Dodis et al. [21] in Eurocrypt'02, combine key evolution with key splitting ideas, and provide stronger security level than forward-secure systems. Up to now, several key-insulated encryption/signature schemes have been proposed, e.g., [11,15,23,22,20,39,28]. In PKC'06, Hanaoka et al. [27] presented the idea of parallel key-insulation and proposed a parallel key-insulated public key encryption (PKIPKE) scheme secure in the random oracle model. In PKC'07, Libert et al. [34] proposed an elegant PKIPKE scheme without random oracles. Anh et al. [1] generalized the notion of PKIPKE and proposed a new paradigm called key-insulated public key encryption with auxiliary helper.

The key-insulated model has been further extended and strengthened by Itkis and Reyzin [30] to yield the notion of intrusion-resilience. The main strength of intrusion-resilient schemes, as opposed to prior notions, is that they remain secure even after arbitrarily many compromises of both helper key and temporary private keys, as long as the compromises are not simultaneous. Moreover, even if both keys are exposed simultaneously, it is still impossible to compromise the security of previous periods. Up to now, several intrusion-resilient systems have been proposed, e.g., [29,18,19,33]. Indeed, the intrusion-resilient model appears to provide the maximum possible security in the face of key-exposure, whereas it becomes less convenient, since it only supports *sequential key updates* and is unable to decrypt old ciphertexts.

To deal with the key-exposure problem in identity-based scenarios, Hanaoka et al. [28] introduced the hierarchial key-insulation paradigm where the helpers are hierarchically structured. This is another efficient method to enhance the security of the key-insulated systems. However, as to the aforementioned person who works in the two branches without regularity, he still has to bring the first-level helper from-and-to the two branches. Weng et al. [38] extended the parallel insulation paradigm to identity-based settings, and proposed an identity-based parallel key-insulated encryption scheme without random oracles.

2   Preliminaries

2.1   Notations

Throughout this paper, let $\mathbb{Z}_p$ denote the set $\{0, 1, 2, \cdots, p - 1\}$, and $\mathbb{Z}_p^*$ denote $\mathbb{Z}_p \backslash \{0\}$. For a finite set S, $x \xleftarrow{\$} S$ means choosing an element x from S with a uniform distribution.

2.2 Bilinear Pairings

Let $\mathbb{G}$ and $\mathbb{G}_T$ be two cyclic multiplicative groups with the same prime order p. A bilinear pairing is a map $e : \mathbb{G} \times \mathbb{G} \to \mathbb{G}_T$ with the following properties:

- Bilinearity: $\forall g_1, g_2 \in \mathbb{G}, \forall a, b \in \mathbb{Z}_p^*$, we have $e(g_1^a, g_2^b) = e(g_1, g_2)^{ab}$;
- Non-degeneracy: There exist $g_1, g_2 \in \mathbb{G}$ such that $e(g_1, g_2) \neq 1$;
- Computability: There exists an efficient algorithm to compute $e(g_1, g_2)$ for $\forall g_1, g_2 \in \mathbb{G}$.

2.3 Complexity Assumptions

Let $\mathbb{G}$ and $\mathbb{G}_T$ be two cyclic multiplicative groups with the same prime order p, and $e : \mathbb{G} \times \mathbb{G} \to \mathbb{G}_T$ be a bilinear paring. The decisional bilinear Diffie-Hellman (DBDH) problem in groups $(\mathbb{G}, \mathbb{G}_T)$ is, given a tuple $(g, g^a, g^b, g^c, Z) \in \mathbb{G}^4 \times \mathbb{G}_T$, to decide whether $Z = e(g, g)^{abc}$ holds.

Definition 1. *We say that a polynomial-time adversary $\mathcal{B}$ has advantage ϵ in solving the DBDH problem in groups $(\mathbb{G}, \mathbb{G}_T)$ if*

$$\left| Pr\left[\mathcal{B}(g, g^a, g^b, g^c, e(g, g)^{abc}) = 1\right] - Pr\left[\mathcal{B}(g, g^a, g^b, g^c, e(g, g)^z) = 1\right] \right| \geq 2\epsilon,$$

where the probability is taken over the randomly chosen a, b, c, z and the random bits consumed by $\mathcal{B}$.

Definition 2. *We say that (t, ϵ)-DBDH assumption holds in $(\mathbb{G}, \mathbb{G}_T)$ if no t-time adversary has advantage at least ϵ in solving the DBDH problem in $(\mathbb{G}, \mathbb{G}_T)$.*

3 Model of IBTKIE

3.1 Definition

Before formalizing the definition for IBTKIE systems, we first give an overview for IBTKIE. As original key-insulated systems, the lifetime of IBTKIE systems is divided into discrete time periods. A user's identity acts as his public key and is fixed throughout the lifetime, while his private key is updated in every period. Each user ID may have a number n_{ID} of helper keys, with respect to a threshold parameter k_{ID}, and each helper key is stored in different helper devices. At the beginning of a period t, each helper key owned by user ID is used to generate a key-update information share. Combining at least k_{ID} key-update information shares (holding the common identity ID and common period t) with one temporary private key corresponding to another period t', user ID can derive the temporary private key for the current period t, while less than k_{ID} key-update information shares are unable to do so.

Recall the aforementioned person who works in branches B1 and B2 without regularity. Now, he can put k_{ID} helper keys in branch B1, and another k_{ID} helper keys in B2. Then, even if he works in branch B1 (B2, resp.) during even (odd, resp.) days, he can still use the k_{ID} helper keys at hand to update his temporary private keys. Therefore, there is no need for him to bring the helpers from-and-to B1 and B2, and hence it prevents the helper keys from a higher exposure risk. Note that it can be easily extended to support arbitrary number of branches. That is, for a number m of branches, this can be done by simply letting $n_{ID} = m \cdot k_{ID}$.

Next, we give a formal definition for IBTKIE. For an easy explanation, in the subsequent depiction, we assume that all the users share the same number n of helpers and the same threshold parameter k. Note that it can easily be adapted to allow different n_{ID} (or k_{ID}) for different user IDs. Formally, an IBTKIE scheme consists of the following six algorithms.

Setup (κ): The setup algorithm takes as input a security parameter κ. It generates the public parameters *param* and the master secret key *msk*.

Extract (msk, ID): The key extraction algorithm takes as input the master secret key *msk* and an identity ID. It outputs an initial private key $TK_{ID,0}$ and a number n of helper keys $\{HK_{ID,i}\}_{1 \leq i \leq n}$.

Here the threshold parameter corresponding to n is k. Each helper key $HK_{ID,i}$ is kept by the i-th helper, while the initial private key $TK_{ID,0}$ is kept by the user ID.

HelperUpt $(t, ID, HK_{ID,i})$: The helper key-update algorithm takes as input a period index t, an identity ID and his i-th $(1 \leq i \leq n)$ helper key $HK_{ID,i}$. It outputs the i-th key-update information share $KU_{ID,t,i}$ with respect to identity ID and period t.

UserUpt $(ID, TK_{ID,t'}, \{KU_{ID,t,i}\}_{i \in S})$: The user key-update algorithm takes as input an identity ID, his temporary private key $TK_{ID,t'}$ for period t', and a set $\{KU_{ID,t,i}\}_{i \in S}$ of key-update information shares, where $S \subseteq \{1, \cdots, n\}$ and $|S| \geq k$. It returns this user's temporary private key $TK_{ID,t}$ for period t, and deletes $TK_{ID,t'}$ and $\{KU_{ID,t,i}\}_{i \in S}$.

Encrypt $(param, ID, M, t)$: The encryption algorithm takes as input the public parameters *param*, an identity ID, a plaintext M and a period t. It outputs a ciphertext C encrypted under ID and t.

Decrypt $(C, TK_{ID,t})$: The decryption algorithm takes as input a ciphertext C under identity ID and period t, and the temporary private key $TK_{ID,t}$. It outputs the corresponding plaintext M.

Roughly speaking, for correctness, we require that any set of at least k key-update information shares (holding the shares of a common identity ID and a common period t), together with a temporary private key $TK_{ID,t'}$ for period t', should be able to generate a valid temporary private key $TK_{ID,t}$. Moreover, the *decryption consistency* requires that, for any message M, any identity ID and any period t, $\mathsf{Decrypt}(C, TK_{ID,t}) = M$ always holds, where $C = \mathsf{Encrypt}(param, ID, M, t)$.

We remark that, the above definition corresponds to schemes supporting *randomness key-updates* [21]; that is, one can update $TK_{ID,t'}$ to $TK_{ID,t}$ in one "step" for any period indices t' and t. A weaker definition allows $t = t' + 1$ only.

3.2 Security Notions for IBTKIE

Key-insulated security. The key-insulated security notion in the original key-insulated encryption [21] captures the intuition that, if an adversary does not compromise the helper key, exposure of temporary private key does not enable him to derive the remaining temporary private keys. To model this security notion in IBTKIE scenarios, we consider a more powerful adversary: first, he can issue extraction queries as in IBE settings; second, he is allowed access to any of the temporary private keys (of course, not the target key); further, he is even allowed to compromise up to $k-1$ helper keys.

Formally, for an IBTKIE scheme $\mathcal{E}$, its *semantic security against an adaptive chosen ciphertext attack under an adaptive chosen identity and adaptive chosen key-exposure attack* (IND-ID-KE-CCA2) can be defined via the following game between an adversary $\mathcal{A}$ and a challenger $\mathcal{C}$:

Setup. The challenger $\mathcal{C}$ runs algorithm $\mathsf{Setup}(\kappa)$, and gives $\mathcal{A}$ the resulting public parameters *param*, keeping the master key *msk* to itself.

Query Phase 1. $\mathcal{A}$ adaptively issues queries $q_1, \cdots, q_m$ where query q_i is one of the following:

- Extraction query $\langle \mathsf{ID} \rangle$: $\mathcal{C}$ first runs algorithm $\mathsf{Extract}$ to obtain the initial private key $\mathrm{TK}_{\mathsf{ID},0}$ and n helper keys $\{\mathrm{HK}_{\mathsf{ID},i}\}_{1 \le i \le n}$. It then sends these results to $\mathcal{A}$.
- Helper key query $\langle \mathsf{ID}, i \rangle$: $\mathcal{C}$ responds by running algorithm $\mathsf{Extract}$ to generate $\mathrm{HK}_{\mathsf{ID},i}$ and sends it to $\mathcal{A}$.
- Temporary private key query $\langle \mathsf{ID}, t \rangle$: $\mathcal{C}$ responds by running algorithms $\mathsf{HelperUpt}$ and $\mathsf{UserUpt}$ to generate $\mathrm{TK}_{\mathsf{ID},t}$. It then returns it to $\mathcal{A}$.
- Decryption query $\langle (t, C), \mathsf{ID} \rangle$: $\mathcal{C}$ responds by running algorithms $\mathsf{HelperUpt}$ and $\mathsf{UserUpt}$ to generate $\mathrm{TK}_{\mathsf{ID},t}$. It then runs algorithm $\mathsf{Decrypt}$ using $\mathrm{TK}_{\mathsf{ID},t}$ and sends the resulting plaintext to $\mathcal{A}$.

Challenge. Once $\mathcal{A}$ decides that Phase 1 is over, it outputs a target identity ID^*, a period index t^*, and two equal-length plaintexts $M_0, M_1 \in \mathcal{M}$ on which it wishes to be challenged. $\mathcal{C}$ flips a random coin $\gamma \in \{0,1\}$, and sets the challenge ciphertext to be $C^* = \mathsf{Encrypt}(param, \mathsf{ID}^*, M_\gamma, t^*)$, which is sent to $\mathcal{A}$.

Phase 2. $\mathcal{A}$ continues to issue additional queries as in Phase 1, and $\mathcal{B}$ responds these queries as in Phase 1.

Guess. Finally, $\mathcal{A}$ outputs a guess $\gamma' \in \{0,1\}$.

We refer to the above game as an IND-ID-KE-CCA2 game. In the above game, it is also mandated that the following conditions are simultaneously satisfied:

1. $\langle \mathsf{ID}^* \rangle$ does not appear in extraction queries;
2. $\langle \mathsf{ID}^*, t^* \rangle$ does not appear in temporary private key queries;
3. $\mathcal{A}$ can only corrupt up to $k-1$ helper keys with respect to identity ID^*;
4. $\langle (t^*, C^*), \mathsf{ID}^* \rangle$ does not appear in decryption queries.

We refer to adversary $\mathcal{A}$ as an IND-ID-KE-CCA2 adversary. We define his advantage in attacking scheme $\mathcal{E}$ as $\mathrm{Adv}_{\mathcal{E},\mathcal{A}} = \left| \Pr[\gamma' = \gamma] - \frac{1}{2} \right|$, where the probability is taken over the random coins consumed by the challenger and the adversary. As usual, we can define the chosen plaintext security similarly to the above game except that the adversary is not allowed to issue any decryption queries. We call this adversary IND-ID-KE-CPA *adversary*.

Definition 3. *We say that an IBTKIE scheme $\mathcal{E}$ is $(t, q_e, q_h, q_t, q_d, \epsilon)$-IND-ID-KE-CCA2 secure if for any t-time IND-ID-KE-CCA2 adversary $\mathcal{A}$ that makes at most q_e extraction queries, at most q_h helper key queries, at most q_t temporary private key queries and at most q_d decryption queries, we have that $\mathrm{Adv}_{\mathcal{E},\mathcal{A}} < \epsilon$. Also, we say that $\mathcal{E}$ is $(t, q_e, q_h, q_t, \epsilon)$-IND-ID-KE-CPA secure if it is $(t, q_e, q_h, q_t, 0, \epsilon)$-IND-ID-KE-CCA2 secure.*

Strong key-insulated security[1]. The strong key-insulated security considers attacks which compromise the physically-secure helpers. Informally, a strong key-insulated secure cryptosystem should ensure that even if all the n helpers corresponding to an identity ID are compromised, it is still impossible for the adversary to derive any of ID's temporary private keys. To model this security notion, we define another game named *strong* IND-ID-KE-CCA2, which is identical to IND-ID-KE-CCA2 with the following exceptions: (i). temporary private key queries are no longer provided for $\mathcal{A}$, (ii). $\mathcal{A}$ is said to win the *strong* IND-ID-KE-CCA2 game if $\gamma = \gamma'$ and the following conditions are simultaneously satisfied:

1. $\langle \mathsf{ID}^* \rangle$ does not appear in extraction queries;
2. $\langle (t^*, C^*), \mathsf{ID}^* \rangle$ does not appear in decryption queries.

We refer to the above adversary $\mathcal{A}$ as a *strong* IND-ID-KE-CCA2 adversary. Similarly, we can define the *strong* IND-ID-KE-CPA adversary, and here as well, the adversary is not allowed to issue any decryption queries.

Definition 4. *We say that an IBTKIE scheme $\mathcal{E}$ is $(t, q_e, q_h, q_d, \epsilon)$ strongly IND-ID-KE-CCA2 secure if for any strong IND-ID-KE-CCA2 adversary $\mathcal{A}$ who makes at most q_e extraction queries, at most q_h helper key queries and at most q_d decryption queries, we have that $\mathrm{Adv}_{\mathcal{E},\mathcal{A}} < \epsilon$. Also, we say that $\mathcal{E}$ is (t, q_e, q_h, ϵ) strongly IND-ID-KE-CPA secure if it is $(t, q_e, q_h, 0, \epsilon)$ strongly IND-ID-KE-CCA2 secure.*

4   Proposed IBTKIE Scheme

In this section, based on Waters' IBE scheme [37], we present an IBTKIE scheme. We also prove its security under the DBDH assumption.

---

[1] The term "strong key-insulated" is borrowed from [21]. Note that we can not say that the strong key-insulated security is "stronger" than the key-insulated security. In fact, they are orthogonal in some sense.

4.1 Construction

Let $\mathbb{G}$ and $\mathbb{G}_T$ be two groups with prime order p of size κ, and let e be a bilinear map such that $e : \mathbb{G} \times \mathbb{G} \rightarrow \mathbb{G}_T$. Identities and period indices will be represented as bitstrings of length n_u and n_w respectively(We can also let identities and period indices be bitstrings of arbitrary length and n_u, n_w be the output lengths of collision-resistant hash functions, $H_1 : \{0,1\}^* \rightarrow \{0,1\}^{n_u}$, $H_2 : \{0,1\}^* \rightarrow \{0,1\}^{n_w}$). Hereafter, for an identity ID, we use $\mathcal{V}_{\mathsf{ID}} \subseteq \{1, \cdots, n_u\}$ to denote the set of indices for which the bitstring ID is set to 1. Similarly, for a period index t, we use $\mathcal{W}_t \subseteq \{1, \cdots, n_w\}$ to denote the set of indices for which the bitstring t is set to 1. We also define the Lagrange coefficient $\Delta_{i,S}(x) = \prod_{\substack{v \in S \\ v \neq i}} \frac{x-v}{i-v}$ for $i \in \mathbb{Z}_p$ and a set S of elements in $\mathbb{Z}_p$. The proposed IBTKIE system consists of the following algorithms:

Setup(κ): Given a security parameter κ, the PKG works as below:

1. Generate $\mathbb{G}, \mathbb{G}_T$ and e;

2. Pick $\alpha \xleftarrow{\$} \mathbb{Z}_p^*$, $g, g_2 \xleftarrow{\$} \mathbb{G}$, and set $g_1 = g^\alpha$;

3. Pick $u' \xleftarrow{\$} \mathbb{G}$ and a random n_u-length vector $\boldsymbol{U} = (u_i)$ whose elements are randomly chosen from $\mathbb{G}$;

4. Pick $w' \xleftarrow{\$} \mathbb{G}$ and a random n_w-length vector $\boldsymbol{W} = (w_i)$ whose elements are randomly chosen from $\mathbb{G}$;

5. Output the public parameters and the master secret key as

$$param = (g, g_1, g_2, u', w', \boldsymbol{U}, \boldsymbol{W}), \qquad msk = g_2^\alpha.$$

For convenience, we define two Waters' hash function as below[2]:

$$H_u(\mathsf{ID}) = u' \prod_{j \in \mathcal{V}_{\mathsf{ID}}} u_j, \qquad H_w(t) = w' \prod_{j \in \mathcal{W}_t} w_j.$$

Extract (ID, msk): Given an identity ID, the PKG uses the master key msk to generate the initial private key and a number n of helper keys for identity ID as follows:

1. Pick $\beta \xleftarrow{\$} \mathbb{Z}_p^*$, compute $R = g_2^\beta$ and set the initial private key for identity ID to be

$$\mathrm{TK}_{\mathsf{ID},0} = (R, -, -, -). \tag{1}$$

2. For each index $i \in \{1, \cdots, k-1\}$, pick $c_i, r_i \xleftarrow{\$} \mathbb{Z}_p^*$ and set the i-th helper key to be

$$\mathrm{HK}_{\mathsf{ID},i} = (g_2^{c_i} H_u(\mathsf{ID})^{r_i}, g^{r_i}). \tag{2}$$

---

[2] We thank the anonymous review's pointing out that, the technique from [16] can be applied to recycle the w_i's in the definition of H_w and thus the public parameters can be shortened.

3. Let $S' = \{0, 1, \cdots, k-1\}$. Pick $r \xleftarrow{\$} \mathbb{Z}_p^*$. For each remaining index $i \in \{k, \cdots, n\}$, set the i-th helper key $\mathrm{HK}_{\mathsf{ID},i}$ to be

$$\left(\left(g_2^{\alpha-\beta} H_u(\mathsf{ID})^r \right)^{\Delta_{i,S'}(0)} \prod_{j=1}^{k-1} \left(\mathrm{HK}_{\mathsf{ID},j}^{\langle 1 \rangle} \right)^{\Delta_{i,S'}(j)}, (g^r)^{\Delta_{i,S'}(0)} \prod_{j=1}^{k-1} \left(\mathrm{HK}_{\mathsf{ID},j}^{\langle 2 \rangle} \right)^{\Delta_{i,S'}(j)} \right) \tag{3}$$

4. Give $\mathrm{HK}_{\mathsf{ID},i}$ to user ID's i-th helper for each $i \in \{1, \cdots, n\}$. While the initial private key $\mathrm{TK}_{\mathsf{ID},0}$ is kept by user ID himself.

Here we claim that the helper keys in Eq. (3) have the same form as those in Eq. (2). To see this, let $f_1(x)$ denote the $(k-1)$-degree polynomial such that $f_1(0) = \alpha - \beta$ and $f_1(i) = c_i$ for each $i \in \{1, \cdots, k-1\}$. Also let $f_2(x)$ denote the $(k-1)$-degree polynomial such that $f_2(0) = r$ and $f_2(i) = c_i$ for each $i \in \{1, \cdots, k-1\}$. Besides, for each $i \in \{k, \cdots, n\}$, we let $f_1(i)$ and $f_2(i)$ denote c_i and r_i respectively. Then, for each $i \in \{k, \cdots, n\}$, we have

$$\mathrm{HK}_{\mathsf{ID},i}^{\langle 1 \rangle} = \left(g_2^{\alpha-\beta} H_u(\mathsf{ID})^r \right)^{\Delta_{i,S'}(0)} \prod_{j=1}^{k-1} \left(\mathrm{HK}_{\mathsf{ID},j}^{\langle 1 \rangle} \right)^{\Delta_{i,S'}(j)}$$

$$= \left(g_2^{\alpha-\beta} H_u(\mathsf{ID})^r \right)^{\Delta_{i,S'}(0)} \prod_{j=1}^{k-1} \left(g_2^{c_j} H_u(\mathsf{ID})^{r_j} \right)^{\Delta_{i,S'}(j)}$$

$$= g_2^{(\alpha-\beta)\cdot\Delta_{i,S'}(0) + \sum_{j=1}^{k-1} c_j \cdot \Delta_{i,S'}(j)} H_u(\mathsf{ID})^{r\cdot\Delta_{i,S'}(0) + \sum_{j=1}^{k-1} r_j \cdot \Delta_{i,S'}(j)}$$

$$= g_2^{f_1(0)\cdot\Delta_{i,S'}(0) + \sum_{j=1}^{k-1} f_1(j)\cdot\Delta_{i,S'}(j)} H_u(\mathsf{ID})^{f_2(0)\cdot\Delta_{i,S'}(0) + \sum_{j=1}^{k-1} f_2(j)\cdot\Delta_{i,S'}(j)}$$

$$= g_2^{\sum_{j\in S'} f_1(j)\cdot\Delta_{i,S'}(j)} H_u(\mathsf{ID})^{\sum_{j\in S'} f_2(j)\cdot\Delta_{i,S'}(j)}$$

$$= g_2^{f_1(i)} H_u(\mathsf{ID})^{f_2(i)}$$

$$= g_2^{c_i} H_u(\mathsf{ID})^{r_i}.$$

Similarly, for each $i \in \{k, \cdots, n\}$, we have $\mathrm{HK}_{\mathsf{ID},i}^{\langle 2 \rangle} = g^{r_i}$.

HelperUpt $(t, \mathsf{ID}, \mathrm{HK}_{\mathsf{ID},i})$: Given a period index t, an identity ID and his i-th helper key $\mathrm{HK}_{\mathsf{ID},i}$, this algorithm works as below:

1. Parse $\mathrm{HK}_{\mathsf{ID},i}$ as $(\mathrm{HK}_{\mathsf{ID},i}^{\langle 1 \rangle}, \mathrm{HK}_{\mathsf{ID},i}^{\langle 2 \rangle})$;

2. Pick $s_i \xleftarrow{\$} \mathbb{Z}_p^*$ and output user ID's i-th key-update information share $\mathrm{KU}_{\mathsf{ID},t,i}$ for period t as

$$\left(\mathrm{HK}_{\mathsf{ID}}^{\langle 1 \rangle} \cdot H_w(t)^{s_i}, \mathrm{HK}_{\mathsf{ID}}^{\langle 2 \rangle}, g^{s_i} \right) = \left(g_2^{c_i} H_u(\mathsf{ID})^{r_i} H_w(t)^{s_i}, g^{r_i}, g^{s_i} \right). \tag{4}$$

UserUpt $(\mathsf{ID}, \mathrm{TK}_{\mathsf{ID},t'}, \{\mathrm{KU}_{\mathsf{ID},t,i}\}_{i\in S})$: Given an identity ID, a temporary private key $\mathrm{TK}_{\mathsf{ID},t'}$ for period t', and a set $\{\mathrm{KU}_{\mathsf{ID},t,i}\}_{i\in S}$ of key-update information shares for period t, where $S \subseteq \{1, \cdots, n\}$ and $|S| \geq k$ (for convenience, we assume $|S| = k$), this algorithm works as below:

1. Parse $\mathrm{TK}_{\mathsf{ID},t'}$ as $\left(R, \mathrm{TK}_{\mathsf{ID},t'}^{\langle 2\rangle}, \mathrm{TK}_{\mathsf{ID},t'}^{\langle 3\rangle}, \mathrm{TK}_{\mathsf{ID},t'}^{\langle 4\rangle}\right)$;
2. Parse $\mathrm{KU}_{\mathsf{ID},t,i}$ as $\left(\mathrm{KU}_{\mathsf{ID},t,i}^{\langle 1\rangle}, \mathrm{KU}_{\mathsf{ID},t,i}^{\langle 2\rangle}, \mathrm{KU}_{\mathsf{ID},t,i}^{\langle 3\rangle}\right)$;
3. Set user ID's temporary private key $\mathrm{TK}_{\mathsf{ID},t}$ for period t to be

$$
\left(R, \prod_{i\in S}\left(\mathrm{KU}_{\mathsf{ID},t,i}^{\langle 1\rangle}\right)^{\Delta_{0,S}(i)}, \prod_{i\in S}\left(\mathrm{KU}_{\mathsf{ID},t,i}^{\langle 2\rangle}\right)^{\Delta_{0,S}(i)}, \prod_{i\in S}\left(\mathrm{KU}_{\mathsf{ID},t,i}^{\langle 3\rangle}\right)^{\Delta_{0,S}(i)}\right).
$$

4. Delete $\mathrm{TK}_{\mathsf{ID},t'}$ and $\{\mathrm{KU}_{\mathsf{ID},t,i}\}_{i\in S}$. Return $\mathrm{TK}_{\mathsf{ID},t}$.

Note that if let $r = \sum_{i\in S}\Delta_{0,S}(i)\cdot r_i$ and $s = \sum_{i\in S}\Delta_{0,S}(i)\cdot s_i$, then the temporary private key $\mathrm{TK}_{\mathsf{ID},t}$ is always set to be

$$
\left(R, \mathrm{TK}_{\mathsf{ID},t}^{\langle 2\rangle}, \mathrm{TK}_{\mathsf{ID},t}^{\langle 3\rangle}, \mathrm{TK}_{\mathsf{ID},t}^{\langle 4\rangle}\right) = \left(g_2^{\beta}, g_2^{(\alpha-\beta)}H_u(\mathsf{ID})^r H_w(t)^s, g^r, g^s\right). \tag{5}
$$

This can be seen from the following:

$$
\begin{aligned}
\mathrm{TK}_{\mathsf{ID},t}^{\langle 2\rangle} &= \prod_{i\in S}\left(\mathrm{KU}_{\mathsf{ID},t,i}^{\langle 1\rangle}\right)^{\Delta_{0,S}(i)} = \prod_{i\in S}\left(g_2^{c_i}H_u(\mathsf{ID})^{r_i}H_w(t)^{s_i}\right)^{\Delta_{0,S}(i)}\\
&= \prod_{i\in S}(g_2^{c_i})^{\Delta_{0,S}(i)}\prod_{i\in S}(H_u(\mathsf{ID})^{r_i})^{\Delta_{0,S}(i)}\prod_{i\in S}(H_w(t)^{s_i})^{\Delta_{0,S}(i)}\\
&= g_2^{\sum_{i\in S}\Delta_{0,S}(i)\cdot c_i}H_u(\mathsf{ID})^{\sum_{i\in S}\Delta_{0,S}(i)\cdot r_i}H_w(t)^{\sum_{i\in S}\Delta_{0,S}(i)\cdot s_i}\\
&= g_2^{\sum_{i\in S}\Delta_{0,S}(i)\cdot f_1(i)}H_u(\mathsf{ID})^r H_w(t)^s\\
&= g_2^{f_1(0)}H_u(\mathsf{ID})^r H_w(t)^s\\
&= g_2^{\alpha-\beta}H_u(\mathsf{ID})^r H_w(t)^s,\\
\mathrm{TK}_{\mathsf{ID},t}^{\langle 3\rangle} &= \prod_{i\in S}\left(\mathrm{KU}_{\mathsf{ID},t,i}^{\langle 2\rangle}\right)^{\Delta_{0,S}(i)} = \prod_{i\in S}(g^{r_i})^{\Delta_{0,S}(i)} = g^{\sum_{i\in S}\Delta_{0,S}(i)\cdot r_i} = g^r,\\
\mathrm{TK}_{\mathsf{ID},t}^{\langle 4\rangle} &= \prod_{i\in S}\left(\mathrm{KU}_{\mathsf{ID},t,i}^{\langle 3\rangle}\right)^{\Delta_{0,S}(i)} = \prod_{i\in S}(g^{s_i})^{\Delta_{0,S}(i)} = g^{\sum_{i\in S}\Delta_{0,S}(i)\cdot s_i} = g^s.
\end{aligned}
$$

Encrypt $(param, \mathsf{ID}, m, t)$: In period t, to encrypt a message $M \in \mathbb{G}_T$ under an identity ID, the sender picks $z \xleftarrow{\$} \mathbb{Z}_p^*$ and outputs

$$
C = (C_1, C_2, C_3, C_4) = (g^z, M\cdot e(g_1, g_2)^z, H_u(\mathsf{ID})^z, H_w(t)^z). \tag{6}
$$

Decrypt $(C, \mathrm{TK}_{\mathsf{ID},t})$: To decrypt ciphertext $C = (C_1, C_2, C_3, C_4)$ under identity ID for period t, the recipient with private key $\mathrm{TK}_{\mathsf{ID},t} = \left(R, \mathrm{TK}_{\mathsf{ID},t}^{\langle 2\rangle}, \mathrm{TK}_{\mathsf{ID},t}^{\langle 3\rangle}, \mathrm{TK}_{\mathsf{ID},t}^{\langle 4\rangle}\right)$

outputs $M = \dfrac{C_2\cdot e\left(C_3, \mathrm{TK}_{\mathsf{ID},t}^{\langle 3\rangle}\right)\cdot e\left(C_4, \mathrm{TK}_{\mathsf{ID},t}^{\langle 4\rangle}\right)}{e\left(C_1, R\cdot \mathrm{TK}_{\mathsf{ID},t}^{\langle 2\rangle}\right)}.$

4.2   Correctness

In the above description of the scheme, it has been shown that, any set of at least k key-update information shares (holding the shares of a common identity ID and a common period t), together with a temporary private key $\text{TK}_{\text{ID},t'}$ for period t', can correctly generate a valid temporary private key $\text{TK}_{\text{ID},t}$. Furthermore, assuming the ciphertext is well-formed for ID and t, we have

$$
\frac{C_2 \cdot e\left(C_3, \text{TK}_{\text{ID},t}^{\langle 3 \rangle}\right) \cdot e\left(C_4, \text{TK}_{\text{ID},t}^{\langle 4 \rangle}\right)}{e\left(C_1, R \cdot \text{TK}_{\text{ID},t}^{\langle 2 \rangle}\right)}
$$

$$
= \frac{M \cdot e(g_1, g_2)^z \cdot e\left(H_u(\text{ID})^z, g^r\right) \cdot e\left(H_w(t)^z, g^s\right)}{e\left(g^z, g_2^\beta \cdot g_2^{(\alpha - \beta)} H_u(\text{ID})^r H_w(t)^s\right)}
$$

$$
= \frac{M \cdot e(g_1, g_2)^z \cdot e\left(g^z, H_u(\text{ID})^r\right) \cdot e\left(g^z, H_w(t)^s\right)}{e\left(g^z, g_2^\alpha H_u(\text{ID})^r H_w(t)^s\right)}
$$

$$
= \frac{M \cdot e(g_1, g_2)^z \cdot e\left(g^z, H_u(\text{ID})^r\right) \cdot e\left(g^z, H_w(t)^s\right)}{e\left(g^z, g_2^\alpha\right) \cdot e\left(g^z, H_u(\text{ID})^r\right) \cdot e\left(g^z, H_w(t)^s\right)}
$$

$$
= M.
$$

4.3   Security

The above IBTKIE scheme is IND-ID-KE-CPA secure in the standard model.

Theorem 1. *Our IBTKIE scheme is IND-ID-KE-CPA secure in the standard model, assuming the DBDH assumption holds in groups $(\mathbb{G}, \mathbb{G}_T)$. Concretely, if there exists a $(t, q_e, q_h, q_t, \epsilon)$-IND-ID-KE-CPA adversary $\mathcal{A}$ against our scheme, then there exists an efficient algorithm $\mathcal{B}$ which can solve the (t', ϵ')-DBDH assumption in groups $(\mathbb{G}, \mathbb{G}_T)$, where*

$$
t' = t + \mathcal{O}(\epsilon^{-2} \ln(\epsilon^{-1})\lambda^{-1} \ln(\lambda^{-1})), \qquad \epsilon' \geq \frac{\epsilon}{2\lambda^{-1}}.
$$

Here $\lambda = \dfrac{1}{16(q_e + q_t)q_t(n_u + 1)(n_w + 1)}.$

The proof technique is mainly borrowed from [37,31], and the detailed proof can be founded in the full version of this paper.

Theorem 2. *Our IBTKIE scheme is strongly IND-ID-KE-CPA secure in the standard model, assuming the DBDH assumption holds in groups $(\mathbb{G}, \mathbb{G}_T)$. Concretely, if there exists a (t, q_e, q_h, ϵ) strong IND-ID-KE-CPA adversary $\mathcal{A}$ against our scheme, then there exists an efficient algorithm $\mathcal{B}$ which can solve the (t', ϵ')-DBDH assumption in groups $(\mathbb{G}, \mathbb{G}_T)$, where*

$$
t' = t + \mathcal{O}(\epsilon^{-2} \ln(\epsilon^{-1})\lambda^{-1} \ln(\lambda^{-1})), \qquad \epsilon' \geq \frac{\epsilon}{2\lambda^{-1}}.
$$

Here $\lambda = \dfrac{1}{4q_e(n_u + 1)}.$

Detailed proof can be founded in the full version of this paper. We here briefly explain why the proposed IBTKIE scheme inherently provides the strong key-insulated security. From a high level, the intuition comes from the fact that, in algorithm Extract, the secret $g_2^\alpha H_u(\mathsf{ID})^r$ is broken into two parts, the first part g_2^β hold by the user, and the second part $g_2^{\alpha-\beta} H_u(\mathsf{ID})^r$ which is masked by some random elements and is shared by the helpers. Concretely, consider the following target temporary private key

$$\mathrm{TK}_{\mathsf{ID}^*,t} = \left(R, \mathrm{TK}^{\langle 2 \rangle}_{\mathsf{ID}^*,t^*}, \mathrm{TK}^{\langle 3 \rangle}_{\mathsf{ID}^*,t^*}, \mathrm{TK}^{\langle 4 \rangle}_{\mathsf{ID}^*,t^*} \right)$$
$$= \left(g_2^\beta, g_2^{(\alpha-\beta)} H_u(\mathsf{ID}^*)^r H_w(t^*)^s, g^r, g^s \right).$$

Given enough helper keys, indeed, an adversary $\mathcal{A}$ can derive the last three components, say $\mathrm{TK}^{\langle 2 \rangle}_{\mathsf{ID}^*,t^*}, \mathrm{TK}^{\langle 3 \rangle}_{\mathsf{ID}^*,t^*}$ and $\mathrm{TK}^{\langle 4 \rangle}_{\mathsf{ID}^*,t^*}$. However, as an important fact should be noted, since adversary $\mathcal{A}$ is disallowed to issue extraction query $\langle \mathsf{ID}^* \rangle$ and any temporary private key queries on behalf of ID^*, the first component $R = g_2^\beta$ is unknown to adversary $\mathcal{A}$ even if he compromises all of ID^*'s helper keys. Therefore, adversary $\mathcal{A}$ can not derive the whole target key $\mathrm{TK}_{\mathsf{ID}^*,t^*}$, and hence the strong IND-ID-KE-CPA security is ensured.

4.4   Chosen-Ciphertext Security

Generic results from Canetti, Halevi and Katz [14], further improved upon by Boneh and Katz [8], can be applied to our IBTKIE scheme and achieve the chosen-ciphertext security in the standard model. The non-generic but more efficient technique of Boyen, Mei and Waters [10], can be also used to achieve the chosen-ciphertext security. However, these methods involve some overhead to the ciphertext. Interestingly, recent result from Kiltz [32] can be similarly applied to our IBTKIE scheme to achieve the chosen-ciphertext security without ciphertext overhead. Due to the space limit, we here do not provide the detailed construction.

5   Prevent Cheating from Compromised Helpers

There may be situations that, after a helper has been compromised, the user can not immediately be aware of this fact, and he may still use the compromised helper key to update his temporary private key. Now, an adversary, who does not corrupt enough helper keys to compromise the security of the system, can still cheat the user as follows: he may control the compromised helper to output some incorrectly generated key-update information shares. Then it will render the user to derive an invalid temporary private key, which can not be used for decryption. Fortunately, we can slightly modify our proposed IBTKIE scheme to prevent such a cheating. Concretely, only algorithms Extract and UserUpt need to be changed as below:

Extract: Steps 2 to 4 are replaced by the following:

2. For each index $i \in \{1, \cdots, k-1\}$, pick $c_i, r_i \xleftarrow{\$} \mathbb{Z}_p^*$, set the i-th helper key to be $\mathrm{HK}_{\mathsf{ID},i} = (g_2^{c_i} H_u(\mathsf{ID})^{r_i}, g^{r_i})$, and the i-th verification key $\mathrm{VK}_{\mathsf{ID},i}$ to be $\mathrm{VK}_{\mathsf{ID},i} = g^{c_i}$.

3. Let $S' = \{0, 1, \cdots, k-1\}$. Pick $r \xleftarrow{\$} \mathbb{Z}_p^*$. For each remaining index $i \in \{k, \cdots, n\}$, set the i-th helper key $\mathrm{HK}_{\mathsf{ID},i}$ to be

$$\left(\left(g_2^{\alpha-\beta} H_u(\mathsf{ID})^r \right)^{\Delta_{i,S'}(0)} \prod_{j=1}^{k-1} \left(\mathrm{HK}_{\mathsf{ID},j}^{\langle 1 \rangle} \right)^{\Delta_{i,S'}(j)}, (g^r)^{\Delta_{i,S'}(0)} \prod_{j=1}^{k-1} \left(\mathrm{HK}_{\mathsf{ID},j}^{\langle 2 \rangle} \right)^{\Delta_{i,S'}(j)} \right),$$

and set the i-th verification key to be $\mathrm{VK}_{\mathsf{ID},i} = (g^{\alpha-\beta})^{\Delta_{i,S'}(0)} \prod_{j=1}^{k-1} \mathrm{VK}_{\mathsf{ID},j}^{\Delta_{i,S'}(j)}$.

4. Give $\mathrm{HK}_{\mathsf{ID},i}$ to user ID's i-th helper for each $i \in \{1, \cdots, n\}$. While the initial private key $\mathrm{TK}_{\mathsf{ID},0}$ and all the verification keys $\{\mathrm{VK}_{\mathsf{ID},i}\}_{i \in \{1, \cdots, n\}}$ are kept by user ID himself.

UserUpt: The following step is added between Step 2 and Step 3:

2-3. Ensure that all the key-update information shares $\{\mathrm{KU}_{\mathsf{ID},t,i}\}_{i \in S}$ are valid. To do this, for each $i \in S$, check whether the following equation holds:

$$e\left(g, \mathrm{KU}_{\mathsf{ID},t,i}^{\langle 1 \rangle}\right) = e(\mathrm{VK}_{\mathsf{ID},i}, g_2) e\left(\mathrm{KU}_{\mathsf{ID},t,i}^{\langle 2 \rangle}, H_u(\mathsf{ID})\right) e\left(\mathrm{KU}_{\mathsf{ID},t,i}^{\langle 3 \rangle}, H_w(t)\right).$$

If one of the verifications fails, output **"Invalid Shares"** and abort.

Indeed, for each correctly generated key-update information share $\mathrm{KU}_{\mathsf{ID},t,i}$, we have

$$
\begin{aligned}
e\left(g, \mathrm{KU}_{\mathsf{ID},t,i}^{\langle 1 \rangle}\right) &= e\left(g, g_2^{c_i} H_u(\mathsf{ID})^{r_i} H_w(t)^{s_i}\right) \\
&= e\left(g, g_2^{c_i}\right) e\left(g, H_u(\mathsf{ID})^{r_i}\right) e\left(g, H_w(t)^{s_i}\right) \\
&= e\left(g^{c_i}, g_2\right) e\left(g^{r_i}, H_u(\mathsf{ID})\right) e\left(g^{s_i}, H_w(t)\right) \\
&= e(\mathrm{VK}_{\mathsf{ID},i}, g_2) e\left(\mathrm{KU}_{\mathsf{ID},t,i}^{\langle 2 \rangle}, H_u(\mathsf{ID})\right) e\left(\mathrm{KU}_{\mathsf{ID},t,i}^{\langle 3 \rangle}, H_w(t)\right).
\end{aligned}
$$

6   Conclusions

We introduced a new paradigm named threshold key-insulation, where at least k out of n helpers are needed to update a user's temporary private key. The new paradigm not only enhances the security of the system, but also provides flexibility and efficiency. To deal with the key-exposure problem in identity-based settings, we further proposed an identity-based threshold key-insulated encryption scheme, whose security does not rely on the random oracle methodology.

Acknowledgements

The authors would like to thank the anonymous referees and Junzuo Lai for their helpful comments and suggestions.

References

1. Anh, P.T.L., Hanaoka, Y., Hanaoka, G., Matsuura, K., Imai, H.: Reducing the Spread of Damage of Key Exposures in Key-Insulated Encryption. In: Nguyên, P.Q. (ed.) VIETCRYPT 2006. LNCS, vol. 4341, pp. 366–384. Springer, Heidelberg (2006)
2. Abdalla, M., Kiltz, E., Neven, G.: Generalized Key Delegation for Hierarchical Identity-Based Encryption. In: Biskup, J., López, J. (eds.) ESORICS 2007. LNCS, vol. 4734, pp. 139–154. Springer, Heidelberg (2007)
3. Anderson, R.: Two Remarks on Public-Key Cryptology. Invited lecture. In: Proc. of CCCS 1997 (1997), http://www.cl.cam.ac.uk/users/rja14/
4. Boneh, D., Boyen, X.: Efficient selective-ID secure identity-based encryption without random oracles. In: Cachin, C., Camenisch, J.L. (eds.) EUROCRYPT 2004. LNCS, vol. 3027, pp. 223–238. Springer, Heidelberg (2004)
5. Boneh, D., Boyen, X.: Secure identity based encryption without random oracles. In: Franklin, M. (ed.) CRYPTO 2004. LNCS, vol. 3152, pp. 443–459. Springer, Heidelberg (2004)
6. Boneh, D., Franklin, M.: Identity based encryption from the Weil pairing. In: Kilian, J. (ed.) CRYPTO 2001. LNCS, vol. 2139, pp. 213–229. Springer, Heidelberg (2001)
7. Boneh, D., Gentry, C., Hamburg, M.: Space-Efficient Identity Based Encryption Without Parings. In: Proc. of FOCS 2007, pp. 647–657 (2007)
8. Boneh, D., Katz, J.: Improved efficiency for CCA-secure cryptosystems built using identity-based encryption. In: Menezes, A. (ed.) CT-RSA 2005. LNCS, vol. 3376, pp. 87–103. Springer, Heidelberg (2005)
9. Bellare, M., Miner, S.: A Forward-Secure Digital Signature Scheme. In: Wiener, M.J. (ed.) CRYPTO 1999. LNCS, vol. 1666, pp. 431–448. Springer, Heidelberg (1999)
10. Boyen, X., Mei, Q., Waters, B.: Simple and efficient CCA2 security from IBE techniques. In: Proc. of ACM CCS 2005, pp. 320–329. ACM Press, New-York (2005)
11. Bellare, M., Palacio, A.: Protecting against key-exposure: strongly key-insulated encryption with optimal threshold. In: Proc. of AAECC 2006, pp. 379–396 (2006)
12. Baek, J., Zheng, Y.: Identity-based threshold decryption. In: Bao, F., Deng, R., Zhou, J. (eds.) PKC 2004. LNCS, vol. 2947, pp. 248–261. Springer, Heidelberg (2004)
13. Canetti, R., Halevi, S., Katz, J.: A Forward-Secure Public-Key Encryption Scheme. In: Biham, E. (ed.) EUROCRYPT 2003. LNCS, vol. 2656, pp. 255–271. Springer, Heidelberg (2003)
14. Canetti, R., Halevi, S., Katz, J.: Chosen-ciphertext security from identity-based encryption. In: Cachin, C., Camenisch, J.L. (eds.) EUROCRYPT 2004. LNCS, vol. 3027, pp. 207–222. Springer, Heidelberg (2004)
15. Cheon, J.H., Hopper, N., Kim, Y., Osipkov, I.: Timed-Release and Key-Insulated Public Key Encryption. In: Di Crescenzo, G., Rubin, A. (eds.) FC 2006. LNCS, vol. 4107, pp. 191–205. Springer, Heidelberg (2006)
16. Chatterjee, S., Sarkar, P.: HIBE with Short Public Parameters Secure in the Full Model Without Random Oracles. In: Lai, X., Chen, K. (eds.) ASIACRYPT 2006. LNCS, vol. 4284, pp. 145–160. Springer, Heidelberg (2006)
17. Desmedt, Y., Frankel, Y.: Threshold cryptosystems. In: Brassard, G. (ed.) CRYPTO 1989. LNCS, vol. 435, pp. 307–315. Springer, Heidelberg (1990)

18. Dodis, Y., Franklin, M., Katz, J., Miyaji, A., Yung, M.: Intrusion-Resilient Public-Key Encryption. In: Joye, M. (ed.) CT-RSA 2003. LNCS, vol. 2612, pp. 19–32. Springer, Heidelberg (2003)
19. Dodis, Y., Franklin, M., Katz, J., Miyaji, A., Yung, M.: A Generic Construction for Intrusion-Resilient Public-Key Encryption. In: Okamoto, T. (ed.) CT-RSA 2004. LNCS, vol. 2964, pp. 81–98. Springer, Heidelberg (2004)
20. Dodis, Y., Katz, J., Xu, S., Yung, M.: Strong key-insulated signature schemes. In: Desmedt, Y.G. (ed.) PKC 2003. LNCS, vol. 2567, pp. 130–144. Springer, Heidelberg (2002)
21. Dodis, Y., Katz, J., Xu, S., Yung, M.: Key-insulated public-key cryptosystems. In: Knudsen, L.R. (ed.) EUROCRYPT 2002. LNCS, vol. 2332, pp. 65–82. Springer, Heidelberg (2002)
22. González-Deleito, N., Markowitch, O., Dall'lio, E.: A new key-insulated signature scheme. In: López, J., Qing, S., Okamoto, E. (eds.) ICICS 2004. LNCS, vol. 3269, pp. 465–479. Springer, Heidelberg (2004)
23. Dodis, Y., Yung, M.: Exposure-resilience for free: the hierarchical ID-based encryption case. In: Proc. of IEEE SISW 2002, pp. 45–52 (2002)
24. Gentry, C.: Practical identity-based encryption without random oracles. In: Cachin, C., Camenisch, J.L. (eds.) EUROCRYPT 2004. LNCS, vol. 3027, pp. 445–464. Springer, Heidelberg (2004)
25. Green, M., Hohenberger, S.: Blind Identity-Based Encryption and Simulatable Oblivious Transfer. In: Kurosawa, K. (ed.) ASIACRYPT 2007. LNCS, vol. 4833, Springer, Heidelberg (2007)
26. Galindo, D., Kiltz, E.: Chosen-Ciphertext Secure Threshold Identity-Based Key Encapsulation Without Random Oracles. In: De Prisco, R., Yung, M. (eds.) SCN 2006. LNCS, vol. 4116, pp. 173–185. Springer, Heidelberg (2006)
27. Hanaoka, G., Hanaoka, Y., Imai, H.: Parallel key-insulated public key encryption. In: Yung, M., Dodis, Y., Kiayias, A., Malkin, T.G. (eds.) PKC 2006. LNCS, vol. 3958, pp. 105–122. Springer, Heidelberg (2006)
28. Hanaoka, Y., Hanaoka, G., Shikata, J., Imai, H.: Unconditionally secure key insulated cryptosystems: models, bounds and constructions. In: Deng, R.H., Qing, S., Bao, F., Zhou, J. (eds.) ICICS 2002. LNCS, vol. 2513, pp. 85–96. Springer, Heidelberg (2002)
29. Itkis, G.: Intrusion-Resilient Signatures: Generic Constructions, or Defeating a Strong Adversary with Minimal Assumptions. In: Cimato, S., Galdi, C., Persiano, G. (eds.) SCN 2002. LNCS, vol. 2576, pp. 102–118. Springer, Heidelberg (2003)
30. Itkis, G., Reyzin, L.: SiBIR: Signer-base intrusion-resilient signatures. In: Yung, M. (ed.) CRYPTO 2002. LNCS, vol. 2442, pp. 499–514. Springer, Heidelberg (2002)
31. Kiltz, E., Galindo, D.: Direct Chosen-Ciphertext Secure Identity-Based Key Encapsulation without Random Oracles. In: Batten, L.M., Safavi-Naini, R. (eds.) ACISP 2006. LNCS, vol. 4058, pp. 336–347. Springer, Heidelberg (2006), http://eprint.iacr.org/2006/034
32. Eike Kiltz. Chosen-ciphertext secure identity-based encryption in the standard model with short ciphertexts. Cryptology ePrint Archive, Report, 2006/122 (2006), http://eprint.iacr.org/
33. Libert, B., Quisquater, J., Yung, M.: Efficient Intrusion-Resilient Signatures Without Random Oracles. In: Lipmaa, H., Yung, M., Lin, D. (eds.) Inscrypt 2006. LNCS, vol. 4318, pp. 27–41. Springer, Heidelberg (2006)
34. Libert, B., Quisquater, J.J., Yung, M.: Parallel Key-Insulated Public Key Encryption Without Random Oracles. In: Okamoto, T., Wang, X. (eds.) PKC 2007. LNCS, vol. 4450, pp. 298–314. Springer, Heidelberg (2007)

35. Ostrovsky, R., Yung, M.: How to withstand mobile virus attacks. In: Proc. of PODC 1991, pp. 51–59. ACM (1991)
36. Shamir, A.: Identity-based cryptosystems and signature schemes. In: Blakely, G.R., Chaum, D. (eds.) CRYPTO 1984. LNCS, vol. 196, pp. 47–53. Springer, Heidelberg (1984)
37. Waters, B.: Efficient identity-based encryption without random oracles. In: Cramer, R.J.F. (ed.) EUROCRYPT 2005. LNCS, vol. 3494, pp. 114–127. Springer, Heidelberg (2005)
38. Weng, J., Liu, S., Chen, K., Ma, C.: Identity-Based Parallel Key-Insulated Encryption Without Random Oracles: Security Notions and Construction. In: Barua, R., Lange, T. (eds.) INDOCRYPT 2006. LNCS, vol. 4329, pp. 409–423. Springer, Heidelberg (2006)
39. Yum, D.H., Lee, P.J.: Efficient key updating signature schemes based on IBS. In: Paterson, K.G. (ed.) Cryptography and Coding 2003. LNCS, vol. 2898, pp. 16–18. Springer, Heidelberg (2003)

CCA2 Secure IBE: Standard Model Efficiency through Authenticated Symmetric Encryption

Eike Kiltz[1,*] and Yevgeniy Vahlis[2]

[1] Cryptology and Information Security Theme
CWI Amsterdam
The Netherlands
kiltz@cwi.nl
[2] University of Toronto
Canada
evahlis@cs.toronto.edu

Abstract. We propose two constructions of chosen-ciphertext secure identity-based encryption (IBE) schemes. Our schemes have a security proof in the standard model, yet they offer performance competitive with all known random-oracle based schemes. The efficiency improvement is obtained by combining modifications of the IBE schemes by Waters [38] and Gentry [21] with authenticated symmetric encryption.

1 Introduction

An Identity-Based Encryption (IBE) scheme is a public-key encryption scheme where any string is a valid public key. In particular, email addresses and dates can be public keys. The ability to use identities as public keys avoids the need to distribute public key certificates — which is one of the main technical difficulties when setting up a public-key infrastructure. An efficient construction of an IBE was not given until almost two decades after Shamir posed the initial open question in [35] regarding the existence of such cryptographic primitives. The first efficient IBEs appeared in 2001, given separately by Boneh and Franklin [10, 11], and Sakai et al. [33]. In particular, Boneh and Franklin [10, 11] proposed formal security notions for IBE systems and designed a fully functional secure IBE scheme using bilinear maps. This scheme and the tools developed in its design have been successfully applied in numerous cryptographic settings, transcending by far the identity based cryptography framework.

Despite its only recent invention, IBE is already used extensively in practice. Two companies — Voltage security and Identum — are specialized in identity-based security solutions. This is one of the reasons why IBE is currently in the process of getting standardized — the new IEEE P1363.3 standard for "Identity-Based Cryptographic Techniques using Pairings" is currently in preparation [25].

---

* Supported by the research program Sentinels (http://www.sentinels.nl). Sentinels is being financed by Technology Foundation STW, the Netherlands Organization for Scientific Research (NWO), and the Dutch Ministry of Economic Affairs.

T. Malkin (Ed.): CT-RSA 2008, LNCS 4964, pp. 221–238, 2008.

The schemes that are currently in consideration are the one by Boneh and Franklin [11]; the one by Boneh and Boyen [7, 12]; and the one by Kasahara and Sakai [33, 16].

All the above IBE schemes provide security against *chosen-ciphertext attacks*. In a chosen ciphertext attack [32, 11], the adversary is given access to a decryption oracle that allows to obtain the decryptions of ciphertexts of his choosing. Intuitively, security in this setting means that an adversary obtains (effectively) no information about encrypted messages, provided the corresponding ciphertexts are never submitted to the decryption oracle. Since the dramatic attack by Bleichenbacher [5], the notion of chosen-ciphertext security is commonly agreed as the "right" notion of security for encryption schemes [37]. We stress that, in general, chosen-ciphertext security is a much stronger security requirement than semantic security, where in the latter an attacker is not given access to the decryption oracle.

RANDOM ORACLES. The drawback of all the IBE schemes [11, 7, 33, 16] that are currently under submission to the new IEEE P1363.3 standard is that their security can only be guaranteed in the *random oracle* model [3], i.e. in an idealized world where all parties get black-box access to a truly random function. Unfortunately a proof in the random oracle model can only serve as a heuristic argument and, admittedly using quite contrived constructions, has been shown to possibly lead to insecure schemes when the random oracles are implemented in the standard model (see, e.g., [14]). More importantly, there exist results [20] indicating that even certain standardized cryptographic schemes (such as full-domain hash signatures) will always remain in the grey area of schemes having a proof in the random oracle yet are "provably unprovable" in the standard model.

IBE WITHOUT RANDOM ORACLES. Waters [38] presents the first practical IBE that is chosen-plaintext secure without random oracles. It fits the category of "commutative-blinding" IBE schemes from Boneh and Boyen [7] and its chosen-plaintext security can be reduced to the Bilinear Decisional Diffie-Hellman (BDDH) assumption. Based on Waters scheme several chosen-ciphertext secure IBE schemes were proposed starting with generic constructions [9] whose specific instantiations were later improved [13, 28]. Today's most efficient variant is due to Kiltz and Galindo who successfully applied "direct chosen-ciphertext" techniques from [13, 27] to Waters' IBE scheme. More recently, Gentry [21] proposed yet another practical chosen-ciphertext secure IBE scheme based on the class of "inversion-based" IBE schemes from [7], offering interesting efficiency trade-offs compared to the commutative-blinding schemes [28].

RANDOM ORACLES: THEORY VS. PRACTICE. The above mentioned drawbacks of the random oracle model readily leads to the question why random-oracle based schemes are sometimes chosen over schemes with a rigorous proof in the standard model. The answer is straight-forward: it is common knowledge that schemes in the random-oracle model are usually much more efficient than schemes in the standard model. As long as the "theoretical problems" from [14, 20] do not lead to an actual break of a non-artificial scheme, using random-oracle schemes seems justifiable in practice. On the other hand it is in the belief of the authors that this general perception about random oracles will change when alternative random-oracle free schemes become available that offer competitive performance.

1.1 Our Contributions

In this paper we demonstrate that there exist identity-based encryption schemes that are provably secure in the standard model, yet their performance is competitive with the best schemes in the random oracle model. We propose two constructions of chosen-ciphertext secure IBE schemes which outperform all such existing standard-model schemes, and have performance comparable to the random-oracle based schemes that were described above.

SCHEME I. Our first IBE scheme is based on Waters' semantically secure IBE. Our approach to protecting a ciphertext against chosen ciphertext attacks bears some resemblance to the one used by Cramer and Shoup [18, 19] to obtain chosen ciphertext secure public key encryption. More precisely, we use the more efficient "encrypt-then-mac" or "authenticated symmetric encryption" variant proposed by Kurosawa and Desmedt [30]. More precisely, in our construction decryption of ill-formed ciphertexts (i.e. ciphertexts that could not have been generated by the encryption algorithm) uses randomness which is built into the user private key (and is independent of the master public key). Such ill-Formed ciphertexts can be detected using extra-information that is algebraically encoded into the "identity-carrying" part of the ciphertext (similar to the HIBE construction from [8]). Overall this allows us to obtain a CCA secure IBE scheme by only adding *one exponentiation* to the encryption/decryption algorithm of Waters' scheme, which is secure only against chosen plaintext attacks. We give a standard-model security proof reducing the intractability of the *modified Bilinear Decisional Diffie-Hellman* (mBDDH) problem (a problem closely related to BDDH) to breaking the CCA security of our scheme.

SCHEME II. Our second construction is a variant of Gentry's chosen-ciphertext secure IBE scheme. Here our new contribution is to use authenticated symmetric encryption [30, 23] to reduce ciphertext expansion and encryption/decryption cost compared to Gentry's original schemes. We prove chosen-ciphertext security of our scheme with respect to the decisional augmented bilinear Diffie-Hellman exponent (q-ABDHE) assumption [21] in the standard model. We remark that the proof technique is different from the one used for the first scheme.

1.2 Comparison

We carefully review all known chosen-ciphertext secure IBE constructions and make an extensive comparison with our schemes. Our studies also incorporate all relevant practical issues when making a comparison, including the tightness of the security reduction with respect to different assumptions and instantiating the schemes in asymmetric pairing groups. To obtain concrete comparison

	Size (bits)		Cost (relative)	
Scheme	Ciphertext	Public Key	Encrypt	Decrypt
Standard model				
Ours: IBE_1 (§4)	422	2376	39	216
Ours: IBE_2 (§5)	1277	2223	110	222
KG [28]	513	2565	40	360
Gentry [21]	2223	3249	146	408
Random Oracle model				
BF [10]	331	171	187	151
BB_1 [7]	502	1386	39	217
KS [16]	331	171	38	152

Fig. 1. Efficiency comparison for CCA-secure IBE schemes in the standard/random oracle model for MNT/80-bit security level. Timings are relative to one exponentiation in group $\mathbb{G}$.

values we estimate ciphertext expansion and encryption/decryption cost when implemented in different pairing groups using recent (independent) timing data from [12]. This includes pairing groups based on super-singular curves and MNT curves.

The numerical results of our comparison for 80 bits MNT curves are given in Fig. 1 (For 80 bits super-singular curves the results are similar. We refer the reader to Fig. 5 in Section 6.) The figure shows that our schemes outperform all known IBE schemes in the standard model. Most notably, compared to the standard-model scheme KG from [28] decryption cost and ciphertext expansion is reduced by approximately one third, whereas encryption cost is the same. More importantly, in comparison with the random-oracle based schemes BF from [11], BB_1 from [7, 12], and KS from [33, 16] our schemes offer competitive performance in all parameters, yet are provably secure in the standard model.

1.3   Related Work

A special class of authenticated symmetric encryption schemes which is obtained using the "encrypt-then-mac" primitive was recently successfully applied to public-key encryption schemes by Kurosawa and Desmedt [30, 2] who greatly improved efficiency of the original Cramer-Shoup encryption scheme [19]. Their result was generalized to cover arbitrary authenticated encryption schemes [23]. In fact, our second IBE scheme can be seen as the "Kurosawa-Desmedt variant" of the original CCA secure scheme by Gentry. A variant of it was also sketched in independent work by Boneh, Gentry and Hamburg [6] using their general framework of "hash proof systems". In connection with IBE, authenticated encryption was first used in [34]. This paper is an extended version of an unpublished manuscript [26] by the first author.

2   Preliminaries

2.1   Notation

If x is a string, then $|x|$ denotes its length, while if S is a set then $|S|$ denotes its size. If $k \in \mathbb{N}$ then 1^k denotes the string of k ones. If S is a set then $s \leftarrow_R S$ denotes the operation of picking an element s of S uniformly at random. Unless otherwise indicated, algorithms are randomized and polynomial time. By $z \leftarrow_R \mathsf{A}^{\mathcal{O}_1, \mathcal{O}_2, \cdots}(x, y, \ldots)$ we denote the operation of running algorithm A with inputs $x, y, \ldots$ and access to oracles $\mathcal{O}_1, \mathcal{O}_2, \ldots$, and letting z be the output. An adversary is an algorithm or a tuple of algorithms.

2.2   Identity Based Encryption

An IBE scheme consists of four algorithms: Setup, KeyGen, Enc, and Dec. Setup generates the global public and private keys; KeyGen uses the global private key to generate an individual private key PRI_{id} for a given identity; Enc uses the global public key to encrypt a message to a given identity; and Dec uses the individual private key to decrypt ciphertexts.

The strongest and commonly accepted notion of security for an identity-based key encryption is that of indistinguishability against an adaptive chosen cipher-text attack [11]. This notion, denoted IND-ID-CCA (or simply CCA), is captured by defining the following advantage function for an adversary $\mathsf{A} = (\mathsf{A}_1, \mathsf{A}_2)$, and for an IBE scheme IBE:

$$\mathsf{Adv}^{\mathsf{CCA}}_{\mathsf{IBE},\mathsf{A}}(k) \;=\; \left| \Pr[\mathsf{Exp}^{\mathsf{CCA}}_{\mathsf{IBE},\mathsf{A}}(k) = 1] - 1/2 \right|$$

where $\mathsf{Exp}^{\mathsf{CCA}}_{\mathsf{IBE},\mathsf{A}}(k)$ is defined by the following experiment.

> Experiment $\mathsf{Exp}^{\mathsf{CCA}}_{\mathsf{IBE},\mathsf{A}}(k)$
> $(\mathsf{PUB}, \mathsf{PRI}) \leftarrow_R \mathsf{Setup}(1^k)$
> $(id^*, m_0, m_1, St) \leftarrow_R \mathsf{A}_1^{\mathsf{KeyGen}(\cdot), \mathsf{Dec}(\cdot, \cdot)}(\mathsf{PUB})$
> $b \leftarrow_R \{0, 1\}; \quad C^* \leftarrow_R \mathsf{Enc}(\mathsf{PUB}, id^*, m_b)$
> $b' \leftarrow_R \mathsf{A}_2^{\mathsf{KeyGen}(\cdot), \mathsf{Dec}(\cdot, \cdot)}(C^*, St)$
> If $b = b'$ Return 1 else return 0

The oracle KeyGen($\cdot$) on input id generates a new private key for the identity id and returns it. The oracle Dec($\cdot, \cdot$) on input id and C first generates a new private key for id and then uses it to decrypt C. When A_1 outputs id^* it must not be any of the identities that the adversary queried to the KeyGen($\cdot$) oracle. Furthermore, A_2 is not allowed to query the KeyGen($\cdot$) oracle on id^*, and is not allowed to query the Dec($\cdot, \cdot$) oracle on (id^*, C^*). The variable St represents some internal state information of adversary A and can be any (polynomially bounded) string.

Definition 1. *An IBE scheme IBE is secure against chosen-ciphertext attacks (CCA secure) if for all adversaries A the advantage function $\mathsf{Adv}^{\mathsf{CCA}}_{\mathsf{IBE},\mathsf{A}}(\cdot)$ is negligible.*

For a more precise analysis of the tightness of reduction we will sometimes use the following more detailed notation. For integers k, t, q_x, q_d, $\mathsf{Adv}^{\mathsf{CCA}}_{\mathsf{IBE},t,q_x,q_d}(k) = \max_{\mathsf{A}} \mathsf{Adv}^{\mathsf{CCA}}_{\mathsf{IBE},\mathsf{A}}(k)$, where the maximum is over all adversaries A that make at most t computational steps, q_x key-derivation, and q_d decryption queries. Here we make the convention to count all decryption queries for $id \neq id^*$ as a key-derivation query.

2.3   Symmetric Encryption

A symmetric encryption scheme $\mathsf{SE} = (\mathsf{E}, \mathsf{D})$ is specified by its encryption algorithm E (encrypting $m \in MsgSp(k)$ with keys $K \in \mathcal{K}(k)$) and decryption algorithm D (returning $m \in MsgSp(k)$ or $\perp$). Here we restrict ourselves to deterministic algorithms E and D.

The most common notion of security for symmetric encryption is that of ciphertext indistinguishability, which requires that all efficient adversaries fail to distinguish between the encryptions of two messages of their choice. Another common security requirement is *ciphertext authenticity*. Ciphertext authenticity requires that no efficient adversary can produce a new valid ciphertext under some key when given one encryption of a message of his choice under the same key. A symmetric encryption scheme which satisfies *both* requirements simultaneously is called secure in the sense of authenticated encryption (AE-OT secure). Note that AE-OT security is a stronger notion than chosen-ciphertext security. Formal definitions and constructions are provided in the full version [29].

3   Intractability Assumptions

3.1   Bilinear Groups

Our schemes will be parameterized by a *pairing parameter generator*. This is an algorithm $\mathcal{G}$ that on input 1^k returns the description of an multiplicative cyclic group $\mathbb{G}$ of prime order p, where $2^k < p < 2^{k+1}$, the description of a multiplicative cyclic group $\mathbb{G}_T$ of the same order, and a non-degenerate bilinear pairing $\hat{e} : \mathbb{G} \times \mathbb{G} \to \mathbb{G}_T$. See [11] for a description of the properties of such pairings. We use $\mathbb{G}^*$ to denote $\mathbb{G} \setminus \{1\}$, i.e. the set of all group elements except the neutral element. Throughout the paper we use $\mathbb{PG} = (\mathbb{G}, \mathbb{G}_T, p, \hat{e}, g, g_T)$ as shorthand for the description of bilinear groups, where g is a generator of $\mathbb{G}$ and $g_T = \hat{e}(g, g) \in \mathbb{G}_T$.

3.2   The Modified BDDH Assumption

Let $\mathbb{PG}$ be the description of pairing groups. The Bilinear Decisional Diffie-Hellman (BDDH) assumption [11] states that the two distributions $(g^x, g^y, g^z, \hat{e}(g,g)^{xyz})$ and $(g^x, g^y, g^z, \hat{e}(g,g)^r)$, for $x, y, z, r \leftarrow_{\mathrm{R}} \mathbb{Z}_p$ are indistinguishable for any adversary. For the modified BDDH assumption we furthermore provide the adversary with the element $g^{(y^2)}$. More formally we define the advantage function $\mathsf{Adv}_{\mathcal{G},\mathsf{B}}^{\mathrm{mbddh}}(k)$ of an adversary B as

$$
\left| \begin{array}{l}
\Pr[\mathsf{B}(\mathbb{PG}, g^x, g^y, g^{y^2}, g^z, \hat{e}(g,g)^{xyz}) = 1] \\
- \Pr[\mathsf{B}(\mathbb{PG}, g^x, g^y, g^{y^2}, g^z, \hat{e}(g,g)^r) = 1]
\end{array} \right| ,
$$

where $x, y, z, r \leftarrow_{\mathrm{R}} \mathbb{Z}_p$ and $\mathbb{PG} \leftarrow_{\mathrm{R}} \mathcal{G}(1^k)$. We say that the *modified Bilinear Decision Diffie-Hellman (mBDDH) assumption relative to generator* $\mathcal{G}$ holds if $\mathsf{Adv}_{\mathcal{G},\mathsf{B}}^{\mathrm{mbddh}}(\cdot)$ is negligible for all adversaries B.

3.3   The Truncated q-ABDHE Assumption

Let $q = q(k)$ be a polynomial. The q-BDDHI assumption [7] states that the two distributions $(g^x, \ldots, g^{x^q}, \hat{e}(g,g)^{1/x})$ and $(g^x, \ldots, g^{x^q}, \hat{e}(g,g)^r)$, for $x, r \leftarrow_{\mathrm{R}} \mathbb{Z}_p$ are indistinguishable for any adversary. In [21] Gentry proposed the related truncated decisional augmented bilinear Diffie-Hellman exponent (truncated q-ABDHE) assumption which augments the q-BDDHI assumption with additional information to the adversary. We define the advantage function $\mathsf{Adv}_{\mathcal{G},\mathsf{B}}^{q\text{-}\mathrm{abdhe}}(k)$ of an adversary B as

$$
\left| \begin{array}{l}
\Pr[\mathsf{B}(\mathbb{PG}, g^x, \ldots, g^{x^q}, g^z, g^{zx^{q+2}}, \hat{e}(g,g)^{zx^{q+1}}) = 1] \\
- \Pr[\mathsf{B}(\mathbb{PG}, g^x, \ldots, g^{x^q}, g^z, g^{zx^{q+2}}, \hat{e}(g,g)^r) = 1]
\end{array} \right| ,
$$

where $x, z, r \leftarrow_{\mathrm{R}} \mathbb{Z}_p$ and $\mathbb{PG} \leftarrow_{\mathrm{R}} \mathcal{G}(1^k)$. We say that the *truncated q-ABDHE assumption relative to generator* $\mathcal{G}$ holds if $\mathsf{Adv}_{\mathcal{G},\mathsf{B}}^{q\text{-}\mathrm{abdhe}}(\cdot)$ is negligible for all B.

3.4   Relations

The next lemma classifies the strength of the modified BDDH assumption we introduced between the well known *standard pairing-based assumptions* BDDH and 2-BDDHI. Here "A $\leq$ B" means that assumption B implies assumption A (in a generic sense), i.e. assumption B is a stronger assumption than A.

Lemma 1. *BDDH $\leq$ mBDDH $\leq$ 2-BDDHI $\leq \ldots \leq$ q-BDDHI $\leq$ truncated q-ABDHE*

The simple proof will be given in the full version [29]. We remark that the complexity of q-BDDHI (as well as truncated q-ABDHE) in the in the generic-group model [36] is roughly $\Omega(\sqrt{p/q})$ [7, 21] which matches the recent attack due to Cheon [17].

4   IBE Scheme I

In this section we present our first CCA secure IBE scheme. It is based on the Boneh-Boyen "commutative-blinding" IBE scheme [7] in its full-identity secure variant of Waters [38] which is chosen-plaintext secure. We construct a CCA secure IBE by adding a redundant group element to the ciphertext, and authenticating the two group elements both explicitly, using target collision resistant hash function, and implicitly by using the same randomness to generate both elements.

A similar technique was already used by Cramer and Shoup to obtain chosen-ciphertext secure public-key encryption and later also successfully applied in [13, 27, 28]. All the above works make a distinction between ciphertexts that can be generated by the encryption algorithm (well-formed ciphertexts), and strings that the encryption algorithm would never output (ill-formed ciphertexts) in their security analysis. The first CCA secure IBE that applies this methodology is [28]. The IBE of [28] handles ill-formed ciphertexts by decrypting them to a fresh random value chosen by the decryption algorithm ("implicit rejection"). This approach is sufficient for obtaining CCA security, but is prohibitively expensive as it requires the decryption algorithm to be randomized, and to compute several exponentiations of group elements to handle ill-formed ciphertexts.

We avoid this additional computation by exploiting the fact that in our IBE the decryption of an ill-formed ciphertext depends on the randomness of the private key that was used for the decryption. In other words, we decrypt ill-formed ciphertexts in the same way as we would decrypt well-formed ciphertext, but for a well formed ciphertext the outcome of the decryption is independent of the randomness in the private key. As a result our decryption algorithm is deterministic and significantly faster than [28]. Furthermore, our scheme also has one group element less in the ciphertext than [28]. This is achieved by algebraically integrating the implicit ciphertext consistency check into the part of the ciphertext that carries the information about the recipient's identity.

4.1   The IBE Construction

We assume that $\mathbb{PG} = (\mathbb{G}, \mathbb{G}_T, p, \hat{e}, g, g_T)$ are public system parameters obtained by running the group parameter algorithm $\mathcal{G}(1^k)$ (that may be shared among multiple systems).

We review the hash function $H : \{0,1\}^n \to \mathbb{G}$ used in Waters' identity based encryption schemes [38]. On input of $\mathbb{G}$ and an integer n, the randomized hash key generator $\mathsf{HGen}(\mathbb{G}; n)$ chooses $n + 1$ random group elements $h_0, \ldots, h_n \in \mathbb{G}$ and returns $h = (h_0, h_1, \ldots, h_n) \in \mathbb{G}^{n+1}$ as the public description of the hash function. The algebraic hash function $H : \{0,1\}^n \to \mathbb{G}$ is evaluated on a string $id = (id_1, \ldots, id_n) \in \{0,1\}^n$ as the product

$$H(id) = h_0 \prod_{i=1}^{n} h_i^{id_i} \in \mathbb{G}.$$

Let $TCR : \mathbb{G} \to$ be a target collision-resistant hash function and $\mathsf{SE} = (\mathsf{E}, \mathsf{D})$ be a symmetric encryption scheme with key-space $\mathcal{K} = \mathbb{G}_T$. Our IBE scheme IBE_1 with identity space $IDSp = \{0,1\}^n$ is described in Fig. 2. Here it is understood that decryption rejects if the ciphertext C does not parse to (c_1, c_2, c_3) with $c_1 \in \mathbb{G}$ and $c_2 \in \mathbb{G}^*$. An IBE scheme with arbitrary identity space $IDSp = \{0,1\}^*$

can be obtained by applying a collision-resistant hash function to the identities. (The choice of $n = 2k$ is due to the birthday paradox.)

Setup(1^k)	KeyGen(PRI, id)
$\alpha, u \leftarrow_R \mathbb{G}$; $z \leftarrow \hat{e}(g, \alpha)$	$s \leftarrow_R \mathbb{Z}_p$
$H \leftarrow_R$ HGen($\mathbb{G}; n$)	$\text{PRI}_{id} \leftarrow (\alpha \cdot H(id)^s, g^{-s}, u^s) \in \mathbb{G}^3$
PUB $\leftarrow (H, u, z)$; PRI $\leftarrow \alpha$	Return PRI_{id}
Return (PUB, PRI)	
Enc(PUB, id, m)	Dec(PUB, id, PRI_{id}, C)
$r \leftarrow_R \mathbb{Z}_p$; $c_1 \leftarrow g^r$	Parse C as $(c_1, c_2, c_3) \in \mathbb{G} \times \mathbb{G}^* \times \{0,1\}^*$
$t \leftarrow TCR(c_1)$; $c_2 \leftarrow (H(id) \cdot u^t)^r$	Parse PRI_{id} as $(d_1, d_2, d_3) \in \mathbb{G}^3$
$K \leftarrow z^r \in \mathbb{G}_T$; $c_3 \leftarrow \mathsf{E}_K(m)$	$t \leftarrow TCR(c_1)$; $K \leftarrow \hat{e}(c_1, d_1 \cdot d_3^t) \cdot \hat{e}(c_2, d_2)$
Return ciphertext $C = (c_1, c_2, c_3)$	Return $m \leftarrow \mathsf{D}_K(c_3)$

Fig. 2. Our first CCA-secure IBE scheme IBE_1

We now show correctness of the scheme, i.e. that the symmetric key K computed in the encryption algorithm matches the key K computed in the decryption algorithm.[1] A correctly generated secret key for identity id has the form $\text{PRI}_{id} = (d_1, d_2, d_3) = (\alpha \cdot H(id)^s, g^{-s}, u^s)$ for some $s \in \mathbb{Z}_p$. Therefore the decryption algorithm computes the symmetric key K as

$$\begin{aligned}
K &= \hat{e}(c_1, d_1 \cdot d_3^t) \cdot \hat{e}(c_2, d_2) \\
&= \hat{e}(g^r, \alpha \cdot H(id)^s \cdot (u^s)^t) \cdot \hat{e}((H(id) \cdot u^t)^r, g^{-s}) \\
&= \hat{e}(g^r, \alpha) \cdot \hat{e}(g^r, H(id)^s \cdot (u^s)^t) \cdot \hat{e}((H(id) \cdot u^t)^r, g^{-s}) \\
&= z^r \cdot \hat{e}(g^r, (H(id) \cdot u^t)^s) \cdot \hat{e}((H(id) \cdot u^t)^{-s}, g^r) \\
&= z^r,
\end{aligned}$$

which is the same as the key computed in the encryption algorithm. Now correctness of the scheme is implied by correctness of SE.

4.2   Security

Theorem 1. *Assume TCR is a target collision resistant hash function and* (E, D) *is a AE-OT-secure symmetric scheme. Under the modified Bilinear Decisional Diffie-Hellman (mBDDH) assumption relative to generator $\mathcal{G}$, the IBE scheme IBE_1 is CCA secure. In particular, for $\varepsilon(k) = \text{Adv}^{CCA}_{\mathsf{IBE}_1, t, q_x, q_d}(k)$ and $\tilde{\varepsilon}(k) = \text{Adv}^{mbddh}_{\mathcal{G}, \tilde{t}}(k)$ we have*

$$\varepsilon(k) \leq (\text{Adv}^{IND}_{SE, \tilde{t}}(k) + \tilde{\varepsilon}(k)) \cdot 10nq + \text{Adv}^{TCR}_{TCR, t}(k) + q_d \cdot \text{Adv}^{CT\text{-}INT}_{SE, t}(k) + 2q_d^2/p;$$

$$t \geq \tilde{t} - \mathcal{O}(\tilde{\varepsilon}^{-2}(k) \cdot \ln(\tilde{\varepsilon}^{-1}(k)) + q_d + q_x) .$$

---

[1] Decryption rejects all ciphertexts with $c_2 = 1 \in \mathbb{G}$. We can assume that encryption does not generate ciphertexts with $c_2 = 1$. In case it does encryption can pick fresh randomness r.

The full proof is given in the full version [29]. We give a brief overview here. Our proof for this system has many similarities with [28] (which in turn is based on [38]). The key difference between the two proofs is the treatment of ill-formed ciphertexts. [28] use the fact that anyone that has the global public key can check whether a ciphertext is well-formed. Then, if the ciphertext is ill-formed the decryption algorithm chooses a random value for K, and uses it to attempt and decrypt the symmetric ciphertext. Thus, the adversary himself could have decrypted any ill-formed ciphertext, and does not gain any information from querying the decryption oracle on such ciphertexts.

Our approach to dealing with ill-formed ciphertexts is different. We do not rely on the ability of anyone who has the global public key to check whether a ciphertext is well-formed. Instead, we make the observation that an ill-formed ciphertext, i.e. a ciphertext of the form $C = (g^r, (H(id) \cdot u^t)^{r'}, c_3)$, where $r \neq r'$, decrypts in the following way:

1. The intermediate key K is computed: $K = z^r \cdot \hat{e}(g, H(id) \cdot u^t)^{(r-r')s}$, where s is the random value that was used to generate the private key.
2. K is used to attempt and decrypt the AE ciphertext.

Now, the adversary makes a polynomial number of decryption queries with ill-formed ciphertexts. We show that the first such query is likely to decrypt as "reject", and each query after the first is likely to decrypt as "reject" given that all previous ill-formed queries decrypted as reject, which completes the proof. The idea is that the value s remains random in the view of the adversary as he makes decryption queries with valid ciphertexts, or ciphertexts that decrypt as "reject". Since s is random, K is also a random element of $\mathbb{G}_T$. Thus, by the authenticity property of the AE encryption, c_3 will be decrypted to "reject" when the random element K is used as the key.

4.3   Extensions

TRADING PUBLIC KEY SIZE AND SECURITY REDUCTION. As independently discovered in [15, 31], there exists an interesting trade-off between key-size of Waters' hash H and the security reduction of the IBE schemes. The construction modifies Waters hash H as follows: Let the integer $l = l(k)$ be a new parameter of the scheme. In particular, we represent an identity $id \in \{0,1\}^n$ as an n/l-dimensional vector $id = (id_1, \ldots, id_{n/l})$, where each id_i is an l bit string. Waters hash is then redefined to $H : \{0,1\}^n \to \mathbb{G}$, with $H(id) = h_0 \prod_{i=1}^{n/l} h_i^{id_i}$ for random public elements $h_0, h_1, \ldots, h_{n/l} \in \mathbb{G}$. Waters' original hash function is obtained as the special case $l = 1$. It is easy to see that using this modification in our IBE scheme (i) reduces the size of the public key from $n+2$ to $n/l+2$ elements in $\mathbb{G}$, whereas (ii) it adds another multiplicative factor of 2^l to the security reduction of the IBE scheme (Theorem 1).

For concreteness we propose the following value for l (our choice will become clear in Section 6). For a scheme implemented in groups offering 80 bits of security we have $n = 2 \cdot 80 = 160$ bits and use 128. This shrinks the public-key size to reasonable $n/l + 2 \approx 10$ elements in $\mathbb{G}$ (plus one element in $\mathbb{G}_T$).

We further remark that in the random-oracle model we can replace Waters' hash $H : \{0,1\}^* \rightarrow \mathbb{G}$ with $H(id) = h_0 \cdot h_1^{RO(id)}$, where $RO : \{0,1\}^* \rightarrow \mathbb{Z}_p$ is a cryptographic hash function which is modeled as a random oracle [3] in the security analysis.

HIERARCHICAL IDENTITIES. Hierarchical identity-based encryption (HIBE) is a generalization of IBE to identities supporting hierarchical structures [24]. In a HIBE, identities are hierarchical and take the form $id = [id_1.id_2.id_3]$. This particular hierarchical identity has depth 3, and is subordinate to $[id_1]$, $[id_1.id_2]$, but not to $[id_1.id_2.id_3']$. Each user in the hierarchy may act as a local key-generation authority for all subordinate hierarchical identities.

By the relation to Waters IBE scheme it is easy to see that our technique can also be used to obtain a chosen-ciphertext secure HIBE. Using a technique from [8] it is furthermore possible to reduce the HIBE ciphertext size to three elements, i.e. it is independent of the hierarchy's depth. To be more precise, the IBE from Section 4.1 is modified to a HIBE supporting maximal d hierarchies as follows. The setup algorithm chooses d different and independent hash functions $H_i \leftarrow_{\mathrm{R}} \mathsf{HGen}(\mathbb{G}; n)$, for $1 \leq i \leq d$. The user secret key for the hierarchical identity $id = [id_1.\cdots.id_\mu]$ of depth $\mu \leq d$ is defined as $\mathsf{PRI}_{id} = (d_1, d_2, d_3, (d_{ij})_{\mu+i \leq j \leq d, 0 \leq i \leq n}) \in \mathbb{G}^{3+(n+1)\cdot(d-\mu-1)}$, where $d_1 = \alpha \cdot (\prod_{j=1}^{\mu} H_i(id^{(j)}))^r$, $d_2 = g^{-r}$, $d_3 = u^r$, and $d_{ij} = ((h_i^{(j)})^r)$. We remark that the latter $(n+1) \cdot (d - \mu - 1)$ elements d_{ij} are only needed for hierarchical key delegation (and may be not included in PRI_{id} if such a feature is not wanted). Encryption of m with respect to id computes the two ciphertext elements $c_1 = g^r$ and $c_2 = (u^t \prod_{j=1}^{\mu} H_i(id^{(j)}))^r$ and uses the key $K = z^r$ to compute the symmetric ciphertext (using an AE-OT-secure scheme). Decryption uses $K = \hat{e}(d_1 \cdot d_3^t, c_1) \cdot \hat{e}(d_2, c_2)$ to reconstruct the plaintext from the symmetric ciphertext. Note that this only needs two pairing operations, independent of the depth of the hierarchy d. (In contrast the HIBE from [28] needs $d + 1$ pairings.)

Security can be proved with respect to the *d-modified BDDH assumption*, where compared to the mBDDH assumption the adversary gets the values g^y, $g^{y^2}, \ldots, g^{y^{d+1}}$ (instead of just g^y, g^{y^2}). As in [22, 38] the security reduction is exponential in the depth d of the hierarchy, i.e. it introduces, roughly, a multiplicative factor of $(nq)^d$. Hence the scheme can only be considered practical for small hierarchies, say of depth $d = 4$.

TRADING CIPHERTEXT SIZE FOR EFFICIENCY. A variant of our IBE scheme can be combined with CCA-secure symmetric encryption. CCA-secure symmetric encryption is less demanding than authenticated encryption and, in particular, strong pseudorandom permutations imply CCA-secure symmetric encryption without any redundancy. This has the advantage of more compact ciphertexts while decryption has to perform some algebraic consistency checks and is therefore less efficient.

5   IBE Scheme II

In this section we present our second chosen-ciphertext secure IBE scheme from the q-ABDHE assumption. It is based on the Boneh-Boyen "exponent inversion" IBE scheme [7] in its full-identity secure variant of Gentry [21]. Gentry also presents a chosen-ciphertext secure variant of his basic chosen-plaintext secure scheme. Our main improvement is to combine it with a strongly secure symmetric encryption scheme to considerably reduce ciphertext size and encryption/decryption cost.

5.1   The IBE Construction

Let $\mathbb{PG} = (\mathbb{G}, \mathbb{G}_T, p, \hat{e}, g, g_T = \hat{e}(g, g))$ be a pairing group. Let $TCR : \mathbb{G} \times \mathbb{G} \to \mathbb{Z}_p$ be a target collision-resistant hash function. Let (E, D) be a symmetric cipher. Our IBE scheme $\mathsf{IBE}_2 = (\mathsf{Setup}, \mathsf{KeyGen}, \mathsf{Enc}, \mathsf{Dec})$ with identity space $IDSp = \mathbb{Z}_p$ is depicted in Fig. 3.

$\mathsf{Setup}(1^k)$	$\mathsf{KeyGen}(\mathsf{PRI}, id)$
$x, y_1, y_2 \leftarrow_{\mathrm{R}} \mathbb{Z}_p$	$s_1, s_2 \leftarrow_{\mathrm{R}} \mathbb{Z}_p$
$u \leftarrow g^x;\ v_1 \leftarrow g_T^{y_1};\ v_2 \leftarrow g_T^{y_2}$	$d_1 \leftarrow g^{\frac{y_1 - s_1}{x - id}};\ d_2 \leftarrow g^{\frac{y_2 - s_2}{x - id}}$
$\mathsf{PUB} \leftarrow (u, v_1, v_2);\ \mathsf{PRI} \leftarrow (x, y_1, y_2)$	$\mathsf{PRI}_{id} \leftarrow (d_1, s_1, d_2, s_2)$
Return $(\mathsf{PUB}, \mathsf{PRI})$	Return user secret-key PRI_{id}
$\mathsf{Enc}(\mathsf{PUB}, id, m)$	$\mathsf{Decaps}(\mathsf{PUB}, id, \mathsf{PRI}_{id}, C)$
$r \leftarrow_{\mathrm{R}} \mathbb{Z}_p\ c_1 \leftarrow (ug^{-id})^r;\ c_2 \leftarrow g_T^r$	Parse C as $(c_1, c_2, c_3) \in \mathbb{G} \times \mathbb{G}_T \times \{0,1\}^*$
$t \leftarrow TCR(c_1, c_2);\ K \leftarrow (v_1^t v_2)^r$	Parse PRI_{id} as (d_1, s_1, d_2, s_2)
$c_3 \leftarrow \mathsf{E}_K(m)$	$t \leftarrow TCR(c_1, c_2)\ ;\ K \leftarrow \hat{e}(c_1, d_1^t d_2) \cdot c_2^{s_1 t + s_2}$
Return ciphertext $C = (c_1, c_2, c_3)$	Return $m \leftarrow \mathsf{D}_K(c_3)$

Fig. 3. Our CCA-secure IBE scheme IBE_2

To show correctness consider a ciphertext (c_1, c_2, c_3) generated for identity id that gets decrypted with a valid user secret key $\mathsf{PRI}_{id} = (d_1, d_2, s_1, s_2)$ by computing the symmetric key K as follows

$$
\begin{aligned}
K &= \hat{e}(c_1, d_1^t d_2) \cdot c_2^{s_1 t + s_2} \\
&= \hat{e}(g^{(x - id)r}, g^{\frac{(y_1 - s_1)t + (y_2 - s_2)}{x - id}}) \cdot \hat{e}(g, g)^{(s_1 t + s_2)r} \\
&= \hat{e}(g^r, g^{y_1 t + y_2}) \\
&= (v_1^t v_2)^r,
\end{aligned}
$$

as in the encryption algorithm.

Theorem 2. *Assume TCR is a target collision resistant hash function and* (E, D) *is a AE-OT-secure symmetric scheme. Let* $q = q_x + 1$, *where* q_x *is the* *number of key-derivation queries. Under the truncated q-ABDHE assumption relative to generator $\mathcal{G}$, the IBE scheme* IBE_2 *is IND-CCA secure. In particular, we have*

$$\mathsf{Adv}^{\mathsf{CCA}}_{\mathsf{IBE}_2,t,q_x,q_d}(k)$$

$$\leq \mathsf{Adv}^{\mathsf{q\text{-}abdhe}}_{\mathcal{G},t}(k) + \mathsf{Adv}^{\mathsf{TCR}}_{TCR,t}(k) + 2q_d \cdot \mathsf{Adv}^{\mathsf{CT\text{-}INT}}_{\mathsf{SE},t}(k) + \mathsf{Adv}^{\mathsf{IND}}_{\mathsf{SE},t}(k) + \frac{q_d}{p} \ .$$

The proof of Theorem 2 will be given in the full version [29]. We give some intuition why the scheme is IND-CCA secure. First, the proof of Gentry [21] can be used to show that user secret-key queries, as well as *consistent* decryption queries for the challenge identity id^* are basically useless for an adversary attacking the scheme (unless it can efficiently solve the q-ABDHE problem). However, inconsistent decryption queries with respect to the challenge identity id^* may leak information about the hidden bit b. Here we use a Cramer-Shoup argument. The idea is that the user secret-key $\mathsf{PRI}_{id^*} = (d_1^*, s_1^*, d_2^*, s_2^*)$ used to answer such decryption queries contains some internal randomness $(s_1, s_2) \in \mathbb{Z}_p^2$ that is initially hidden from the adversary's view. During the simulation of the IND-CCA environment the challenge ciphertext will leak (in an information-theoretic sense) one linear equation on the hidden randomness (s_1^*, s_2^*). Decryption queries of inconsistent ciphertexts will use a key K for symmetric decryption that is computed as a linear equation in s_1^*, s_2^* which is linearly independent from the equation the adversary knows. Hence, one single key K is uniformly distributed over $\mathbb{G}_T$. By the ciphertext authenticity property of SE the adversary will not be able to come up with an inconsistent ciphertext (c_1, c_2, c_3) such that $\mathsf{D}_K(c_3)$ does not reject. Consequently, all inconsistent ciphertext will get rejected by the scheme.

5.2   Extensions

Using techniques from [1] it is further possible to prove IBE_2 anonymous in the sense that the ciphertext does not leak any information about the sender's identity. This property has recently proved useful in the area of public-key encryption with keyword search [1].

We remark that in contrast to the IBE construction from Section 4 it is not possible to trade algebraic consistency checks for a weaker symmetric encryption scheme. In general, the class of inversion-based IBE schemes are less versatile than the commutative-blinding IBE schemes; for example, adding extensions like hierarchical key delegation to inversion-based IBE schemes seems a difficult task.

6   Comparison

6.1   Considered Schemes

For our comparison we consider the following standard-model IBE schemes.

IBE_1: Our scheme from Section 4 with the shorter public-parameters. See Section 4.3 for details.

IBE$_2$: Our scheme from Section 5.
KG: The scheme from Kiltz and Galindo [28].
Gentry: The scheme from Gentry [21] (IND-CCA variant).

We furthermore consider the following three IBE schemes that only have a proof in the random-oracle model. All of them are currently in submission for the IEEE1363.3 standardization project [25].

BF: The (FullIdent) scheme from Boneh and Franklin [11].
BB$_1$: The scheme from Boneh and Boyen [7] in its "hashed identities" variant [12].
KS: The scheme from Kasahara and Sakai [33] as described in [16].

We remark that when assuming the interactive *gap Bilinear Diffie-Hellman* (gap-BDH) assumption efficiency of BF and BB$_1$ can be further improved [12]. Due to the strong assumption we will not consider those schemes.

6.2   Security Reductions

For determining the parameters of the compared schemes, we make the following assumptions, most of the are conservative towards the efficiency of our new schemes. For $k = 80$ bit security we estimate (following Bellare and Rogaway [4]) the number of (random oracle) hash queries as $q_H = 2^{50}$. This seems reasonable since a hash function is in the hand of an adversary and can be attacked offline. Similar to signatures schemes we think that a reasonable estimate for the number of key-derivation queries is $q_x \approx 2^{25}$. This is much smaller than the number of hash queries since key-derivation queries can only be made online, in interaction with the system. In practice it is easy to limit the number of key-derivation queries.

The IBE schemes IBE$_1$ and KG have two additional integer parameters: n, l. Parameter $n = 2k$ resembles the bit size $n = 160 \approx 2^7$ of the identity space and $l(k)$ defines the tradeoff between public parameters and security-reduction (cf. Section 4.3). We choose $l = 18$ to obtain a security loss of $2^{18+7+25} = 2^{50} = q_H$. This explains our choice of $l(k)$: it is chosen such that the security loss of the above schemes matches the one of all random-oracle schemes.

The concrete security reductions are given in Fig. 4. For a fair comparison the security reductions of the random-oracle based schemes are given relative to the respective decisional assumption (e.g., BDDH instead of BCDH for BB$_1$). We note that the two schemes IBE$_2$ and Gentry have a tight security reduction to a much stronger security assumption. Due to the recent attacks by Cheon [17] it seems reasonable that the q-xxxx assumption are $\sqrt{q}$ times "less secure" than the BDDH assumption. This in particular implies (by Lemma 1) that we can treat the mBDDH assumption as "as secure" as the BDDH assumption. To simplify the comparison we make the conservative assumption that all the above schemes with the given parameters have the same security loss with respect to the BDDH assumption.

Scheme	Standard Assumption Model?		Security reduction	
			Bounds	concrete ($k = 80$)
IBE$_1$	$\checkmark$	mBDDH	$2^l n q_x$	2^{50}
IBE$_2$	$\checkmark$	q-ABDHE	1	1
KG	$\checkmark$	BDDH	$2^l n q_x$	2^{50}
Gentry	$\checkmark$	q-ABDHE	1	1
BF	—	BDDH	$> q_H$	2^{50}
BB$_1$	—	BDDH	q_H	2^{50}
KS	—	q-BDDHI	q_H^3	$\gg 2^{50}$

Fig. 4. Security assumptions and (concrete) reduction factors for IBE schemes

6.3   Results

A comparison with concrete timing values from Boyen [12] is carried out in Fig. 1 (Section 1) and Fig. 5. Ciphertext overhead represents the difference (in bits) between the ciphertext length and the message length. All timings are given in multiplicative factors relative to one exponentiation in $\mathbb{G}$. As usual, all symmetric operations (cryptographic hash function, symmetric encryption, etc) are ignored. All schemes come with a security proof based on different security assumption, furthermore introducing a different loss of security in the reduction, depending on several system parameters. A high loss in the security reduction reduces the real-world efficiency of the scheme by making it necessary to increase the size of the groups for any given security level. In order not to compare apples with pears, attempted to pick the parameters (in particular the parameter l for IBE$_1$ and KG) such that we obtain the *same concrete security reduction* for all schemes. We refer to the full version [29] for more details of the comparison.

We conclude that our schemes are the most efficient chosen-ciphertext secure IBE schemes in the standard model. Furthermore its performance and

Scheme	Size (bits)		Cost (relative)	
	Ciphertext	Public Key	Encrypt	Decrypt
Standard model				
Ours: IBE$_1$ (§4)	1104	6144	8	25
Ours: IBE$_2$ (§5)	1616	2560	14	25
KG [28]	1536	5632	9	29
Gentry [21]	2560	3584	18	49
Random Oracle model				
BF [10]	672	512	22	21
BB$_1$ [7]	1184	2048	7	29
KS [16]	672	512	6	22

Fig. 5. Efficiency comparison for CCA-secure IBE schemes in the standard/random oracle model for SS/80-bit security level. Timings are relative to one exponentiation in group $\mathbb{G}$.

ciphertext expansion seems comparable to the known random-oracle based schemes, in particular to the one by Boneh and Franklin which is intensively used in practice (see, e.g., http://www.voltage.com).

Acknowledgement

We thank Charles Rackoff, Ian Blake, and the anonymous CT-RSA reviewers for useful comments.

References

[1] Abdalla, M., Bellare, M., Catalano, D., Kiltz, E., Kohno, T., Lange, T., Malone-Lee, J., Neven, G., Paillier, P., Shi, H.: Searchable encryption revisited: Consistency properties, relation to anonymous IBE, and extensions. In: Shoup, V. (ed.) CRYPTO 2005. LNCS, vol. 3621, pp. 205–222. Springer, Heidelberg (2005)

[2] Abe, M., Gennaro, R., Kurosawa, K., Shoup, V.: Tag-KEM/DEM: A new framework for hybrid encryption and a new analysis of Kurosawa-Desmedt KEM. In: Cramer, R.J.F. (ed.) EUROCRYPT 2005. LNCS, vol. 3494, pp. 128–146. Springer, Heidelberg (2005)

[3] Bellare, M., Rogaway, P.: Random oracles are practical: A paradigm for designing efficient protocols. In: Ashby, V. (ed.) ACM CCS 1993, pp. 62–73. ACM Press, New York (1993)

[4] Bellare, M., Rogaway, P.: The exact security of digital signatures: How to sign with RSA and Rabin. In: Maurer, U.M. (ed.) EUROCRYPT 1996. LNCS, vol. 1070, pp. 399–416. Springer, Heidelberg (1996)

[5] Bleichenbacher, D.: Chosen ciphertext attacks against protocols based on the RSA encryption standard PKCS #1. In: Krawczyk, H. (ed.) CRYPTO 1998. LNCS, vol. 1462, pp. 1–12. Springer, Heidelberg (1998)

[6] Boneh, D., Gentry, C., Hamburg, M.: Space-efficient identity based encryption without pairings. In: Proceedings of FOCS 2007, pp. 647–657. IEEE, Los Alamitos (2007)

[7] Boneh, D., Boyen, X.: Efficient selective-ID secure identity based encryption without random oracles. In: Cachin, C., Camenisch, J.L. (eds.) EUROCRYPT 2004. LNCS, vol. 3027, pp. 223–238. Springer, Heidelberg (2004)

[8] Boneh, D., Boyen, X., Goh, E.-J.: Hierarchical identity based encryption with constant size ciphertext. In: Cramer, R.J.F. (ed.) EUROCRYPT 2005. LNCS, vol. 3494, pp. 440–456. Springer, Heidelberg (2005)

[9] Boneh, D., Canetti, R., Halevi, S., Katz, J.: Chosen-ciphertext security from identity-based encryption. SIAM Journal on Computing 5(36), 1301–1328 (2006)

[10] Boneh, D., Franklin, M.K.: Identity-based encryption from the Weil pairing. In: Kilian, J. (ed.) CRYPTO 2001. LNCS, vol. 2139, pp. 213–229. Springer, Heidelberg (2001)

[11] Boneh, D., Franklin, M.K.: Identity based encryption from the Weil pairing. SIAM Journal on Computing 32(3), 586–615 (2003)

[12] Boyen, X.: The BB1 identity-based cryptosystem: A standard for encryption and key encapsulation. Submitted to IEEE 1363.3, (August 2006), http://grouper.ieee.org/groups/1363/

[13] Boyen, X., Mei, Q., Waters, B.: Direct chosen ciphertext security from identity-based techniques. In: Atluri, V., Meadows, C., Juels, A. (eds.) ACM CCS 2005, pp. 320–329. ACM Press, New York (2005)

[14] Canetti, R., Goldreich, O., Halevi, S.: The random oracle methodology, revisited. In: 30th ACM STOC, pp. 209–218. ACM Press, New York (1998)

[15] Chatterjee, S., Sarkar, P.: Trading time for space: Towards an efficient IBE scheme with short(er) public parameters in the standard model. In: Won, D.H., Kim, S. (eds.) ICISC 2005. LNCS, vol. 3935, pp. 424–440. Springer, Heidelberg (2006)

[16] Chen, L., Cheng, Z., Malone-Lee, J., Smart, N.P.: An efficient ID-KEM based on the Sakai-Kasahara key construction. IEE Proceedings Information Security 152, 19–26 (2006)

[17] Cheon, J.H.: Security analysis of the strong Diffie-Hellman problem. In: Vaudenay, S. (ed.) EUROCRYPT 2006. LNCS, vol. 4004, pp. 1–11. Springer, Heidelberg (2006)

[18] Cramer, R., Shoup, V.: A practical public key cryptosystem provably secure against adaptive chosen ciphertext attack. In: Krawczyk, H. (ed.) CRYPTO 1998. LNCS, vol. 1462, pp. 13–25. Springer, Heidelberg (1998)

[19] Cramer, R., Shoup, V.: Design and analysis of practical public-key encryption schemes secure against adaptive chosen ciphertext attack. SIAM Journal on Computing 33(1), 167–226 (2003)

[20] Dodis, Y., Oliveira, R., Pietrzak, K.: On the generic insecurity of the full domain hash. In: Shoup, V. (ed.) CRYPTO 2005. LNCS, vol. 3621, pp. 449–466. Springer, Heidelberg (2005)

[21] Gentry, C.: Practical identity-based encryption without random oracles. In: Vaudenay, S. (ed.) EUROCRYPT 2006. LNCS, vol. 4004, pp. 445–464. Springer, Heidelberg (2006)

[22] Gentry, C., Silverberg, A.: Hierarchical ID-based cryptography. In: Zheng, Y. (ed.) ASIACRYPT 2002. LNCS, vol. 2501, pp. 548–566. Springer, Heidelberg (2002)

[23] Hofheinz, D., Kiltz, E.: Secure hybrid encryption from weakened key encapsulation. In: Menezes, A. (ed.) CRYPTO 2007. LNCS, vol. 4622, pp. 553–571. Springer, Heidelberg (2007)

[24] Horwitz, J., Lynn, B.: Toward hierarchical identity-based encryption. In: Knudsen, L.R. (ed.) EUROCRYPT 2002. LNCS, vol. 2332, pp. 466–481. Springer, Heidelberg (2002)

[25] IEEE P1363.3 Committee. IEEE 1363.3 — standard for identity-based cryptographic techniques using pairings (April 2007), http://grouper.ieee.org/groups/1363/

[26] Kiltz, E.: Chosen-ciphertext secure identity-based encryption in the standard model with short ciphertexts. Cryptology ePrint Archive, Report 2006/122 (2006), http://eprint.iacr.org/

[27] Kiltz, E.: Chosen-ciphertext security from tag-based encryption. In: Halevi, S., Rabin, T. (eds.) TCC 2006. LNCS, vol. 3876, pp. 581–600. Springer, Heidelberg (2006)

[28] Kiltz, E., Galindo, D.: Direct chosen-ciphertext secure identity-based key encapsulation without random oracles. In: Batten, L.M., Safavi-Naini, R. (eds.) ACISP 2006. LNCS, vol. 4058. Springer, Heidelberg (2006)

[29] Kiltz, E., Vahlis, Y.: CCA2 Secure IBE: standard model efficiency through authenticated symmetric encryption. Cryptology ePrint Archive, Report 2008 (2008), http://eprint.iacr.org/

[30] Kurosawa, K., Desmedt, Y.: A new paradigm of hybrid encryption scheme. In: Franklin, M. (ed.) CRYPTO 2004. LNCS, vol. 3152, pp. 426–442. Springer, Heidelberg (2004)

[31] Naccache, D.: Secure and *practical* identity-based encryption. Cryptology ePrint Archive, Report 2005/369 (2005), http://eprint.iacr.org/

[32] Rackoff, C., Simon, D.R.: Non-interactive zero-knowledge proof of knowledge and chosen ciphertext attack. In: Feigenbaum, J. (ed.) CRYPTO 1991. LNCS, vol. 576, pp. 433–444. Springer, Heidelberg (1992)

[33] Sakai, R., Ohgishi, K., Kasahara, M.: Cryptosystems based on pairing over elliptic curve (in japanese). In: Proceedings of the Symposium on Cryptography and Information Security — SCIS 2001 (Janurary 2001)

[34] Sarkar, P., Chatterjee, S.: Transforming a CPA-secure HIBE protocol into a CCA-secure hibe protocol without loss of security. Cryptology ePrint Archive, Report 2006/362 (2006), http://eprint.iacr.org/

[35] Shamir, A.: Identity-based cryptosystems and signature schemes. In: Blakely, G.R., Chaum, D. (eds.) CRYPTO 1984. LNCS, vol. 196, pp. 47–53. Springer, Heidelberg (1985)

[36] Shoup, V.: Lower bounds for discrete logarithms and related problems. In: Fumy, W. (ed.) EUROCRYPT 1997. LNCS, vol. 1233, pp. 256–266. Springer, Heidelberg (1997)

[37] Shoup, V.: Why chosen ciphertext security matters. IBM Research Report RZ 3076 (November 1998)

[38] Waters, B.R.: Efficient identity-based encryption without random oracles. In: Cramer, R.J.F. (ed.) EUROCRYPT 2005. LNCS, vol. 3494, pp. 114–127. Springer, Heidelberg (2005)

Public-Key Encryption with Non-interactive Opening

Ivan Damgård[1], Dennis Hofheinz[2], Eike Kiltz[2,*], and Rune Thorbek[1]

[1] BRICS, Aarhus
[2] CWI, Amsterdam

Abstract. We formally define the primitive of public-key encryption with non-interactive opening (PKENO), where the receiver of a ciphertext C can, convincingly and without interaction, reveal what the result was of decrypting C, without compromising the scheme's security. This has numerous applications in cryptographic protocol design, e.g., when the receiver wants to demonstrate that some information he was sent privately was not correctly formed. We give a definition based on the UC framework as well as an equivalent game-based definition. The PKENO concept was informally introduced by Damgård and Thorbek who suggested that it could be implemented based on Identity-Based Encryption. In this paper, we give direct and optimized implementations, that work without having to keep state information, unlike what one obtains from directly using IBE.

1 Introduction

MOTIVATION. Consider the following extremely common scenario from cryptographic protocol design: Player A sends a secret message to player B who (perhaps at some later time) checks what he receives against some public information. For instance, it may be that the message is supposed to be information for opening a commitment that A established earlier. If the check is OK, B will be able to proceed, but otherwise some "exception handling" must be done. The standard solution to this is to have B broadcast a complaint, and A must then broadcast what he claims to have sent privately, allowing all players to check the information. This is secure, since the conflict can only occur if at least one of A,B is corrupt, so the adversary already knows what is broadcast. But it has the important drawback that interaction is required, in particular A must be present to help resolve the conflict. In many cases, one cannot rely on this assumption. For instance, suppose A is one of many clients who want to provide some input to a set of servers, who will then do a secure computation on the inputs. It is highly desirable that this can be done without interaction, in particular that the servers can decide efficiently among themselves which clients provided well-formed input.

---

* Supported by the research program Sentinels (http://www.sentinels.nl). Sentinels is being financed by Technology Foundation STW, the Netherlands Organization for Scientific Research (NWO), and the Dutch Ministry of Economic Affairs.

T. Malkin (Ed.): CT-RSA 2008, LNCS 4964, pp. 239–255, 2008.

PUBLIC-KEY ENCRYPTION WITH NON-INTERACTIVE OPENING. An alternative solution was suggested by Damgård and Thorbek in [12], namely *public-key encryption with non-interactive opening* (PKENO). This is based on the observation that in practice, the private communication from A to B would typically be implemented using public key encryption, i.e., A sends a ciphertext C encrypted under B's public key pk_B. PKENO now means that if B chooses to reveal the result m of decrypting C (typically, if he is unhappy about m), he can do so, convincingly and without interaction[1]. That is, he can broadcast m, π where π is a proof that can be checked against pk_B and C and demonstrates that indeed decrypting C using the secret key matching pk_B results in m. Of course, this must be done such that other ciphertexts remain secure, and this excludes the trivial solution of revealing B's secret key. Clearly, if PKENO can be implemented efficiently, we have a nice general tool for removing interaction from cryptographic protocols.

DIFFICULTY OF PKENO. Note that having the receiver open a ciphertext is less trivial than having the sender do so: the sender could always be asked to simply reveal the plaintext and the random coins used to construct the ciphertext. This does not work when the receiver does the opening: one has to consider the fact that the sender might be corrupt and hence C is adversarially constructed. It may not even be a valid ciphertext, in which case "the coins used to construct C" is not a well defined concept.

INEFFICIENT CONSTRUCTIONS. A few straightforward solutions for implementing PKENO do exist which, however, have various drawbacks: In principle, one can implement PKENO if a common reference string can be reliably set up. Then the receiver B can commit to his secret key initially and π would be a non-interactive zero-knowledge proof that the secret key committed to matches pk_B and produces m when used to decrypt C. Unfortunately, with the known techniques for non-interactive zero-knowledge, this solution is very inefficient and essentially useless in practice. Efficient solutions are easy to find in the random oracle model, since one can take known efficient and interactive zero-knowledge proofs and make them non-interactive using the Fiat-Shamir heuristic. However, it is unclear what security guarantees in the random oracle model mean for the real world, so in this paper, we will concentrate on efficient solutions that do not use random oracles.

KNOWN CONSTRUCTIONS AND THEIR LIMITATIONS. In [12], the PKENO notion was informally introduced, and it was suggested that it could be implemented based on identity-based encryption (IBE). The idea here is that pk_B would be the public master key of an IBE system, and the secret key sk_B would be the secret master key. To encrypt m, one chooses an identity id (see below for details on how id is chosen), and encrypts m to this identity. Thus, the ciphertext C is the pair $C = (id, \mathsf{IBEenc}(id, m))$. The receiver B uses sk_B to generate the IBE user secret key $usk[id]$ corresponding to id and can then decrypt. To open C, B

---

[1] Note that m may not be a meaningful message, it may be a special reject symbol if C was rejected as invalid by the decryption algorithm.

simply reveals the decryption result m and $usk[id]$, this allows anyone to do the decryption and check that the result is m. Note that efficient IBE schemes exist (under specific assumptions) that do not require random oracles [20].

It follows directly from the properties of IBE that revealing $usk[id]$ does not compromise security of ciphertexts directed to other identities, not even if id is adversarially chosen. This solution is therefore secure if we can guarantee that identities cannot be reused — but only then. This would be the case if it is used in a protocol that assigns unique labels to all ciphertexts to be sent. Then these labels can be used as identities. But note that these labels must be different in different instances of the same protocol. Alternatively, all players could keep state information allowing to test if a label has been used before. This puts some rather heavy demands on the implementation and hence, using IBE in this straightforward way is not satisfactory in general.

An alternative construction of PKENO can be obtained by using *public-key encryption with witness-recovering decryption* (PKEWR) [19]. Here the receiver (i. e., the holder of the secret key) can efficiently reconstruct the "randomness" that was used for encryption. This randomness then serves as the proof. Verification performs (deterministic) re-encrypting using the randomness and the messages. The proof is valid if the result equals the ciphertext. There exists construction of PKEWR from the Decisional Diffie-Hellman assumption and from an assumption related to lattices. However, both constructions are relatively inefficient since the ciphertext size is linear in the message length.

OUR CONTRIBUTIONS. In this paper, we make two contributions: first, we give a formal definition of PKENO, in fact we give two equivalent definitions, one based on the UC framework, and a game-based definition. This allows to show that an implementation is secure using the game-based definition (which is usually easier than with UC), while at same time being guaranteed the composition properties that follows from the UC theorem. We assume — for simplicity — a trusted key set-up, i.e., all key pairs are correctly generated. We emphasize, however, that this assumption is not inherent to the PKENO concept. The definitions can be modified to do without it and some implementations do not need it.

Second, we show some concrete implementations of PKENO. One of our techniques gives a simple and general solution to the problem with unique identities in the IBE implementation, allowing a stateless solution. To this end we use a technique by Naor and Yung [17] that was also used more recently by Canetti, Halevi, and Katz [9] in a transformation of any chosen-plaintext secure IBE scheme into a chosen-ciphertext secure PKE scheme. We adopt the latter transformation to construct PKENO from IBE. The idea is to use, for each PKENO encryption, a fresh random verification key of a one-time signature scheme as the "identity" id for IBE encryption. In order to tie the IBE ciphertext to this verification key it is signed using the corresponding signing key. This ensures the uniqueness of the identity and hence allows a stateless solution of PKENO.

Another technique gives a more direct implementation that is not based on IBE and hence is more efficient. We use a modification of the pairing-based chosen-ciphertext secure PKE scheme which was proposed by Boyen, Mei,

Waters [5] and Kiltz [15]. We show that it is possible to update their scheme with a non-interactive opening functionality without compromising its security. Security of this scheme can be reduced to the Bilinear Decisional Diffie-Hellman (BDDH) assumption.

2   Preliminaries

2.1   Notational Conventions

If x is a string, then $|x|$ denotes its length, while if S is a set then $|S|$ denotes its size. If $k \in \mathbb{N}$ then 1^k denotes the string of k ones. If S is a set then $s \leftarrow_R S$ denotes the operation of picking an element s of S uniformly at random. Unless otherwise indicated, algorithms are randomized and polynomial time. An adversary is an algorithm or a tuple of algorithms. A function $f : \mathbb{N} \to \mathbb{R}$ is *negligible* iff there exists $c < 0$ such that $|f(k)| < k^c$ for all sufficiently large k. We write $f \approx g$ if $f - g$ is negligible.

2.2   The UC Model

Canetti's Universal Composability (UC) framework [6, 7] for multi-party computation allows to formulate security and composition of multi-party protocols in a very general way. The idea of the UC model is to compare a protocol to an idealization of the respective protocol task. Security means that the protocol "looks like" the idealization even in face of arbitrary attacks and in arbitrary protocol environments. This notion of security is very strict [8, 2, 13], but implies useful compositional properties [6]. In fact, in a certain sense, this notion is even necessary for secure composition of protocols [16].

THE REAL MODEL. We shortly outline the framework for multi-party protocols defined in [6, 7]. First of all, parties (denoted by P_1 through P_n) are modeled as interactive Turing machines (ITMs) (cf. [7]) and are supposed to run some fixed protocol (i.e., program) Π. There also is an adversary, denoted $\mathcal{A}$ and modeled as an ITM as well, that carries out attacks on protocol Π. Therefore, $\mathcal{A}$ may corrupt parties (in which case it learns the party's state and controls its future actions), and intercept or inject messages sent between parties. If $\mathcal{A}$ corrupts parties only before the actual protocol run of Π takes place, $\mathcal{A}$ is called non-adaptive, otherwise $\mathcal{A}$ is said to be adaptive. In this work, we only consider non-adaptive corruptions. The respective local inputs for all parties of protocol Π are supplied by an environment machine (modeled as an ITM and denoted $\mathcal{Z}$), which may also read all protocol outputs locally made by the parties and communicate with the adversary.

THE IDEAL MODEL. The model we have just described is called the real model of computation. In contrast to this, the ideal model of computation is defined just like the real model, with the following exceptions: all party ITMs are replaced with one single ideal functionality $\mathcal{F}$. The ideal functionality may not be corrupted by the adversary, yet may send messages to and receive messages from it. Finally, the adversary in the ideal model is called "simulator" and denoted $\mathcal{S}$. The only means of attack the simulator has in the ideal model are corruptions (in which case $\mathcal{S}$ may supply inputs to and read outputs from $\mathcal{F}$ in the name of the corrupted party), delaying or suppressing outputs of $\mathcal{F}$, and all actions that are explicitly specified in $\mathcal{F}$. However, $\mathcal{S}$ has no access to the inputs $\mathcal{F}$ gets and to the outputs $\mathcal{F}$ generates, nor are there any protocol messages to intercept. Intuitively, the ideal model of computation (or, more precisely, the ideal functionality $\mathcal{F}$ itself) should represent what one ideally expects the protocol to do. In fact, for a number of standard tasks, there are formulations as such ideal functionalities (see, e.g., [6]).

SECURITY DEFINITION. To decide whether or not a given protocol Π fulfills the requirements of our ideal specification $\mathcal{F}$, the framework of [6] uses a simulatability-based approach: at a time of its choice, $\mathcal{Z}$ may halt and generate output. The random variable describing the first bit of $\mathcal{Z}$'s output will be denoted by $\text{REAL}_{\Pi,\mathcal{A},\mathcal{Z}}(k,z)$ when $\mathcal{Z}$ is run with security parameter $k \in \mathbb{N}$ and initial input $z \in \{0,1\}^*$ in the real model of computation, and $\text{IDEAL}_{\mathcal{F},\mathcal{S},\mathcal{Z}}(k,z)$ when $\mathcal{Z}$ is run in the ideal model. Now Π is said to *securely realize* $\mathcal{F}$ iff for any real adversary $\mathcal{A}$, there exists a simulator $\mathcal{S}$ such that for any environment $\mathcal{Z}$ and any (possibly non-uniform) family of initial inputs $z = (z_k)_k$, we have

$$\Pr\left[\text{REAL}_{\Pi,\mathcal{A},\mathcal{Z}}(k,z_k) = 1\right] \approx \Pr\left[\text{IDEAL}_{\mathcal{F},\mathcal{S},\mathcal{Z}}(k,z_k) = 1\right]. \tag{1}$$

This slightly differs from the original formulations in [6, 7], but is equivalent and eases our presentation. Intuitively, Equation 1 means that any attack against the protocol can be simulated in the ideal model. Hence, any weakness of the real protocol is already contained in the ideal specification (that does not contain an "actual" weakness by definition). Interestingly, the "worst" real attack possible is the one carried out by the dummy adversary $\tilde{\mathcal{A}}$ that simply follows $\mathcal{Z}$'s instructions. That means that for security, it actually suffices to demand existence of a simulator that simulates attacks carried out by $\tilde{\mathcal{A}}$.

COMPOSITION OF PROTOCOLS. To formalize the composition of protocols, there also exists a model "in between" the real and ideal model of computation. Namely, the hybrid model of computation is identical to the real model, except that parties have access to (multiple instances of) an ideal functionality that aids in running the protocol. This is written as $\varphi^{\mathcal{F}}$ for the actual protocol φ and the ideal functionality $\mathcal{F}$. Instances of $\mathcal{F}$ are distinguished via session identifiers (short: session ids, or *sid*s). Note that syntactically, instances of $\mathcal{F}$ can be implemented by a protocol Π geared towards realizing $\mathcal{F}$. And indeed, the universal composition theorem [6, 7] guarantees that if one protocol instance of Π is secure, then many protocol instances are, even when used in arbitrary protocols φ. More concretely, if Π securely realizes $\mathcal{F}$, then φ^{Π} securely realizes $\varphi^{\mathcal{F}}$ for any protocol φ. Here, $\varphi^{\mathcal{F}}$ denotes that φ uses (up to polynomially many) instances of $\mathcal{F}$ as a subprimitive, and φ^{Π} denotes that φ uses instances of Π instead.

CONDITIONAL SECURITY AND COMPOSABILITY. Universal composability is a very strict notion. So sometimes (e.g., in the case of bit commitments), it is not possible to achieve full UC security. Hence, several weakenings of the notion have been proposed. One method that will be useful in our case is to consider only protocol environments that conform to certain rules (see [18, 1]). Concretely, secure realization with respect to a certain class $\mathfrak{Z}$ of environments means that in Equation 1, we quantify only over environments in $\mathfrak{Z}$. This relaxed security notion still gives precisely those compositional guarantees one would expect: secure composition with larger protocols that can be seen as restricted environments from $\mathfrak{Z}$ (see [18, 1] for details).

3 Public-Key Encryption with Non-interactive Opening

3.1 A UC-Based Definition

Figure 1 depicts our ideal functionality for public-key encryption with non-interactive openings. $\mathcal{F}_{\text{PKENO}}$ is an extension of the $\mathcal{F}_{\text{PKE}}$ functionality [6, 10, 14] that captures IND-CCA secure public-key encryption. The most notable difference to $\mathcal{F}_{\text{PKE}}$ are the additional Prove and Verify queries, which allow the receiver to open a ciphertext and every party to verify openings. Also, we dropped public keys, since we assume a trusted PKI (i.e., keypair setup) for a realization.

DISCUSSION OF $\mathcal{F}_{\text{PKENO}}$. First, note that the session id *sid* already determines the distinguished receiving party P_{recv}. Any party may ask for encryptions, but only P_{recv} may ask for decryptions. As for the encryption of a message m, the adversary may determine a unique tag C via the algorithm Enc. However, note that C depends only on the length $|m|$ of m, but not on m itself (except if the receiver is corrupted, in which case we obviously cannot guarantee secrecy). This reflects that ideally, encryptions reveal only the length of the message. Decryption takes care that correctness is ensured, i.e., ciphertexts are mapped back to the encrypted messages. (For this, $\mathcal{F}_{\text{PKENO}}$ stores a list of ciphertexts and associated messages.)

Opening and verifying openings is a bit trickier. For any ciphertext, the receiver P_{recv} can obtain a proof π that should ideally prove what message was encrypted. Formally, π is determined by the adversary (in form of a pre-stored algorithm Prove) to ensure that during the simulation, at least the shape of π matches the one of a possible real implementation. However, $\mathcal{F}_{\text{PKENO}}$ ensures that verification (via Verify queries) satisfies some natural and crucial requirements. Namely, an honestly (i.e., via $\mathcal{F}_{\text{PKENO}}$) generated encryption C of m cannot be proven to contain a different message $m' \neq m$. Also, honestly (i.e., via $\mathcal{F}_{\text{PKENO}}$) generated proofs are always accepted. In all cases left open by this (and in particular, if a wrong public key is used with Verify), the adversary is free to determine the verification outcome in order to simulate a real implementation.

Note that from the functionality's perspective, ciphertexts and proofs are merely tags and do not carry any semantics. The adversary is free to determine these tags, but the functionality takes care that decryptions and proofs are handled as ideally expected. (E.g., the ciphertext tags do not depend on the messages, or honestly generated proofs verify correctly.)

WHY KEY MANAGEMENT IS OUTSOURCED. Also note that there are no public or secret keys in the functionality. This is unlike, e.g., in the $\mathcal{F}_{\text{PKE}}$ modelings from [6, 10, 14], which do contain a public key. This simplification is possible, since we will consider keys to be already set up, which corresponds to running a public-key encryption scheme protocol in the $\mathcal{F}_{\text{PKI}}$-hybrid model (see below).

The reason *why* we opted to outsource key management into $\mathcal{F}_{\text{PKI}}$ is the following: if the receiving party P_{recv} was allowed to take care of key generation on its own, then a corrupted P_{recv} could generate keys in a dishonest way. (E.g., if the public key contains a common reference string for a non-interactive zero-knowledge proof, then P_{recv} could generate this CRS along with a *trapdoor* that allows P_{recv} to cheat in the proofs. That would not have been possible with an honest generation of keys.) While our concrete scheme from Section 6 is secure even if a dishonest P_{recv} chooses its keys arbitrarily, our game-based formulation (Definition 1) guarantees nothing in that setting. Of course, an adaptation of both Definition 1 and $\mathcal{F}_{\text{PKENO}}$ is possible, such that a dishonest choice of keys is reflected; we chose *not* to do so because be believe that an honest generation of keys is more realistic.

INTERPRETING A PUBLIC-KEY ENCRYPTION SCHEME AS A PROTOCOL. If we assume that the public/secret keys have been set up already, then, syntactically, any public-key encryption scheme $\mathsf{PKENO} = (\mathsf{Gen}, \mathsf{Enc}, \mathsf{Dec}, \mathsf{Prove}, \mathsf{Ver})$ with non-interactive opening can be interpreted as a protocol aimed at realizing $\mathcal{F}_{\text{PKENO}}$. Namely, every party executes $\mathsf{Enc}_{pk}(m)$ upon $(\mathtt{Encrypt}, sid, m)$ inputs, and similarly executes $\mathsf{Ver}_{pk}(C, m, \pi)$ upon $(\mathtt{Verify}, sid, C, m, \pi)$ inputs. In addition, the receiving party P_{recv} (which is uniquely determined by the session id $sid = (recv, sid')$) honors $\mathtt{Decrypt}$ and $\mathtt{Prove}$ inputs by using the Dec and Prove algorithms with P_{recv}'s private sk. Note that although $\mathcal{Z}$ is free to choose sid, a machine can never be invoked with two different sids (even across invocations), so there are not going to be two different secret keys that would need to be managed by one receiving party.

It remains to concretize how we imagine a trusted key setup. We do so by considering a helper functionality $\mathcal{F}_{\text{PKI}}$, as depicted in Figure 2. Note that $\mathcal{F}_{\text{PKE}}$ is parametrized over a key-generation algorithm Gen. That means if we consider a scheme PKENO as a protocol, we actually mean the protocol described above, run in the $\mathcal{F}_{\text{PKI}}^{\mathsf{Gen}}$-hybrid model for the key-generation algorithm Gen of PKENO.

3.2   A Game-Based Definition

A public-key encryption scheme with non-interactive opening is a tuple $\mathsf{PKENO} = (\mathsf{Gen}, \mathsf{Enc}, \mathsf{Dec}, \mathsf{Prove}, \mathsf{Ver})$ of algorithms such that:

- The key generation algorithm Gen takes as input a security parameter 1^k and outputs a public key pk and a secret key sk. We write $(pk, sk) \leftarrow_{\text{R}} \mathsf{Gen}(1^k)$. The public key pk specifies the message space $\mathcal{M}_{pk} \leftarrow \mathsf{MSpc}(pk)$ by a mapping MSpc.
- The encryption algorithm Enc takes as input a public key pk and a message $m \in \mathcal{M}_{pk}$ and outputs a ciphertext C. We write $C \leftarrow_{\text{R}} \mathsf{Enc}_{pk}(m)$.

Functionality $\mathcal{F}_{\text{PKENO}}$

$\mathcal{F}_{\text{PKENO}}$ proceeds as follows, running with parties $P_1, \ldots, P_n$ and an adversary $\mathcal{S}$. All session-ids sid used in the following are expected to be of the form $sid = (recv, sid')$, such that sid uniquely determines a receiving party P_{recv}.

1. Upon the first activation (no matter with which input), first:
   (a) Hand $(\texttt{KeyGen}, sid)$ to the adversary.
   (b) Receive descriptions of the plaintext domain $\mathcal{M}$, randomized algorithms $\overline{\text{Encrypt}}$, $\overline{\text{Prove}}$, and deterministic algorithms $\overline{\text{Decrypt}}$, $\overline{\text{Verify}}$ from the adversary.
   Then proceed to handle the actual query as described below.
2. Upon receiving $(\texttt{Encrypt}, sid, m)$ from some party P_j:
   (a) If $m \notin \mathcal{M}$ then output an error message to P_j.
   (b) If P_{recv} is not corrupted, set $C \leftarrow_{\text{R}} \overline{\text{Encrypt}}(\text{length}, |m|)$. If P_{recv} is corrupted, $C \leftarrow_{\text{R}} \overline{\text{Encrypt}}(\text{message}, m)$.
   (c) Hand C to P_j and store the tuple $(\overline{\text{Encrypt}}, C, m)$. If there already is a stored tuple $(\overline{\text{Encrypt}}, C, m')$ for some different message $m \neq m'$, then halt.
3. Upon receiving $(\texttt{Decrypt}, sid, C)$ from P_{recv} (and P_{recv} only):
   (a) If there is a tuple $(\overline{\text{Encrypt}}, C, m')$ (for some m') stored then set $m := m'$. Otherwise, set $m \leftarrow \overline{\text{Decrypt}}(C)$.
   (b) Hand m to P_{recv}.
4. Upon receiving a value $(\texttt{Prove}, sid, C)$ from P_{recv} (and P_{recv} only):
   (a) If there is a tuple $(\overline{\text{Encrypt}}, C, m')$ (for some m') stored then set $m := m'$. Otherwise, set $m \leftarrow \overline{\text{Decrypt}}(C)$.
   (b) Set $\pi \leftarrow_{\text{R}} \overline{\text{Prove}}(C, m)$ and hand π to P_{recv}. Also, store the tuple $(\overline{\text{Prove}}, C, m, \pi)$; if the tag π already appears in a previously stored $\overline{\text{Prove}}$ tuple then halt.
5. Upon receiving a value $(\texttt{Verify}, sid, C, m, \pi)$ from some party P_j, determine res as follows:
   (a) If there is a stored tuple $(\overline{\text{Prove}}, C, m, \pi)$, then set $res := \texttt{accept}$.
   (b) Else, if there is a tuple $(\overline{\text{Encrypt}}, C, m')$ for some $m' \neq m$, then set $res := \texttt{reject}$.
   (c) In all other cases, set $res \leftarrow \overline{\text{Verify}}(C, m, \pi)$.
   Finally, hand res to P_j.

Fig. 1. Functionality $\mathcal{F}_{\text{PKENO}}$ for public-key encryption with non-interactive openings

- The deterministic decryption algorithm Dec takes as input a ciphertext C and a secret key sk. It returns a message $m \in \mathcal{M}_{pk}$ or the distinguished symbol $\bot \notin \mathcal{M}_{pk}$. We write $m \leftarrow \text{Dec}_{sk}(C)$.
- The proving algorithm Prove takes as input a ciphertext C and a secret key sk. It returns a proof π. We write $\pi \leftarrow_{\text{R}} \text{Prove}_{sk}(C)$.
- The deterministic verification algorithm Ver takes as input a tuple (C, m, π, pk), consisting of a ciphertext C, a plaintext m, a proof π, and a public key pk. It returns a result $res \in \{\texttt{accept}, \texttt{reject}\}$. We write $res \leftarrow \text{Ver}_{pk}(C, m, \pi)$.

We require that with probability overwhelming in the security parameter k, an honestly generated keypair $(pk, sk) \leftarrow_{\text{R}} \text{Gen}(1^k)$ satisfies the following:

- **Correctness.** For all $m \in \mathcal{M}_{pk}$, we have $\Pr[\text{Dec}_{sk}(\text{Enc}_{pk}(m)) = m] = 1$.
- **Completeness.** For all ciphertexts C and all possible $\pi \leftarrow \text{Prove}_{sk}(C)$, we have that for $m \leftarrow \text{Dec}_{sk}(C)$, algorithm $\text{Ver}_{pk}(C, m, \pi)$ accepts.[2]

---

[2] Note that m may be $\bot$.

Functionality $\mathcal{F}_{\text{PKI}}^{\text{Gen}}$

$\mathcal{F}_{\text{PKI}}^{\text{Gen}}$ proceeds as follows, running with parties $P_1, \ldots, P_n$ and an adversary $\mathcal{S}$. All session-ids sid used in the following are expected to be of the form $sid = (recv, sid')$, such that sid uniquely determines a receiving party P_{recv}. Furthermore, $\mathcal{F}_{\text{PKI}}$ is parametrized over a key-generation algorithm Gen.

1. Upon the first activation (no matter with which input), first run $(pk, sk) \leftarrow_{\text{R}}$ Gen(1^k) to generate a public key pk along with a secret key sk.
2. Upon any input from some party P_j or the adversary, send pk to P_j. In addition, if $j = recv$, send also sk to P_j.

Fig. 2. Functionality $\mathcal{F}_{\text{PKI}}$ that captures a trusted key setup

Definition 1 (PKENO security). *A scheme* PKENO *is* PKENO-*secure if it is* IND-CCPA *secure and satisfies computational proof soundness. We define those two below:*

IND-CCPA SECURITY. *For an adversary* A, *consider the following game:*

1. Gen(1^k) *outputs* (pk, sk). *Adversary* A *is given* 1^k *and* pk.
2. *The adversary may make polynomially many queries to a decryption oracle* Dec$_{sk}(\cdot)$ *and a proof oracle* Prove$_{sk}(\cdot)$.
3. *At some point,* A *outputs two equal-length messages* m_0, m_1. *A bit* b *is randomly chosen and the adversary is given the challenge ciphertext* $C^* \leftarrow$ Enc$_{pk}(m_b)$.
4. A *may continue to query its decryption and its proof oracle, except that it may not query either with* C^*.
5. *Finally,* A *outputs a guess* b'.

Denote A*'s advantage in guessing* b' *by*

$$\text{Adv}_{\text{PKENO,A}}^{\text{ind-ccpa}}(k) := |\Pr[b = b'] - 1/2|.$$

Scheme PKENO *is called* indistinguishable against chosen-ciphertext and prove attacks (IND-CCPA secure) *if for every adversary* A, $\text{Adv}_{\text{PKENO,A}}^{\text{ind-ccpa}}(\cdot)$ *is negligible.*

PROOF SOUNDNESS. *For an adversary* A, *consider the following game:*

1. Gen(1^k) *outputs* (pk, sk). *Adversary* A *is given* 1^k *and* (pk, sk).
2. *The adversary chooses a message* $m \in \{0, 1\}^*$ *and gives it to an encryption oracle which returns* $C \leftarrow_{\text{R}} Enc_{pk}(m)$.
3. *The adversary returns* (m', π').

Denote A*'s probability to forge a proof by*

$$\text{Adv}_{\text{PKENO,A}}^{\text{snd}}(k) := \Pr[\text{accept} \leftarrow \text{Ver}_{pk}(C, m', \pi') \wedge m' \neq m].$$

Scheme PKENO *is said to satisfy* computational proof soundness *if for every adversary* A, $\text{Adv}_{\text{PKENO,A}}^{\text{snd}}(\cdot)$ *is negligible.*

4 Equivalence

We will show that PKENO security is equivalent to universal composability in the sense of realizing $\mathcal{F}_{\text{PKENO}}$. The idea is simple: the guarantees that $\mathcal{F}_{\text{PKENO}}$ gives are precisely the properties that Definition 1 requires. However, there is one catch: our simulation breaks down once proofs are asked in a situation in which both sender and receiver are honest. Technically, this stems from a commitment problem the simulation runs into: if sender and receiver are honest, $\mathcal{F}_{\text{PKENO}}$ demands as secrecy guarantee that a ciphertext C in the system does not depend on the associated message m. However, if later on a *proof* is requested that C really decrypts to m, we would need to break —ironically— exactly proof soundness for a good simulation. There seems no easy way to change $\mathcal{F}_{\text{PKENO}}$ itself to prevent this: if $\mathcal{F}_{\text{PKENO}}$ behaves differently depending on whether, e.g., the receiver is corrupted or not, the sender can deduce whether the receiver is indeed corrupted or not. This however would lead to an unachievable functionality (since the receiver might be passively corrupted).

OPTIMISTIC ENVIRONMENTS. To establish equivalence of the definitions, we hence restrict to UC-environments that do not ask for proofs if both sender and receiver are uncorrupted. We call such environments *optimistic*. It is natural to assume that any larger protocol context that uses a PKENO scheme is optimistic: proofs are only requested upon conflicts, which should not happen if both parties are honest.

Theorem 1. *Say that* PKENO *is a public-key encryption scheme with non-interactive opening. Then* PKENO *is* PKENO-*secure (in the sense of Definition 1) if and only if* PKENO *(interpreted as a protocol as described in Section 3.1) securely realizes* $\mathcal{F}_{\text{PKENO}}$ *in the* $\mathcal{F}_{\text{PKI}}^{\text{Gen}}$-*hybrid model, with respect to non-adaptive adversaries and optimistic environments.*

A formal proof will be given in the full version. Here, we give some intuition.

To show that universal composability implies PKENO security, attacks on PKENO's IND-CCPA and proof soundness properties must be translated into attacks on PKENO's indistinguishability from $\mathcal{F}_{\text{PKENO}}$. Suppose A successfully attacks PKENO's IND-CCPA property. We build an environment $\mathcal{Z}$ that internally simulates A and the whole IND-CCPA experiment. In this, $\mathcal{Z}$ obtains decryptions and proofs via its own protocol interface (i.e., via PKENO, resp. $\mathcal{F}_{\text{PKENO}}$), and the challenge message m_b is encrypted with an Encrypt query. In the real model, this yields a true encryption of m_b, and in the ideal model results in something independent of b by definition of $\mathcal{F}_{\text{PKENO}}$. Hence the output distribution of the internally simulated A is correlated with b in the real model, and independent of b in the ideal model, which allows to distinguish. The translation of attacks on PKENO's proof soundness property works similarly.

To show that PKENO security implies universal composability, we describe a simulator $\mathcal{S}$ that, in the ideal setting with $\mathcal{F}_{\text{PKENO}}$, simulates attacks performed with the dummy adversary $\tilde{\mathcal{A}}$ on PKENO. Essentially, $\mathcal{S}$ only provide algorithms for $\mathcal{F}_{\text{PKENO}}$'s Encrypt, Decrypt, Prove, and Verify answers. (Of course,

$\mathcal{F}_{\mathrm{PKENO}}$ enforces several rules with its answers, like proof soundness guarantees, but apart from that, $\mathcal{S}$'s algorithms determine these answers.) Algorithms for decryption, proofs, and verifications are chosen just as in the real model. (Note that $\mathcal{S}$ is free to make up a $\mathcal{F}_{\mathrm{PKI}}^{\mathsf{Gen}}$ instance on its own, so $\mathcal{S}$ knows and in fact chooses the secret keys.) The encryption algorithm for the case the sender is uncorrupted is simply yields encryptions of $1^{|m|}$ (i.e., all-one encryptions of the right length), whereas encryptions in case the sender is corrupted can be performed faithfully as in the real model (in this case, the encryption may depend on the full message, since so secrecy is guaranteed then). The proof that this simulation is sound proceeds by transforming real into ideal model, while showing that this transformation preserves $\mathcal{Z}$'s view:

1. The substitution of m-encryptions with $1^{|m|}$-encryptions can be justified with PKENO's IND-CCPA property.
2. $\mathcal{F}_{\mathrm{PKENO}}$'s list-based decryption of known ciphertexts is simply an enforced correctness, which can be justified with PKENO's correctness.
3. $\mathcal{F}_{\mathrm{PKENO}}$'s verification rules can be justified with PKENO's proof soundness.

This sketches why the simulation that $\mathcal{S}$ provides is correct, and hence the theorem is proven.

ACHIEVING FULL UC SECURITY. It is natural to ask whether $\mathcal{F}_{\mathrm{PKENO}}$ can be realized unconditionally, i.e., without restricting $\mathcal{Z}$. (This corresponds to composability in arbitrary protocol contexts.) As sketched above, to put up a successful simulation here, it must be possible to produce special ciphertexts (sent between an honest sender and an honest verifier) that can be opened to an *arbitrary*, a-priori unknown message. Intuitively, this seems to break proof soundness; however, this *is* possible in principle, since in the ideal model, the simulator has control over the generation of the used keypair (pk, sk). (Note that PKENO security only gives guarantees if this keypair is *honestly* generated.)

To be more concrete, consider the (inefficient) non-interactive zero-knowledge based scheme from the introduction. By, e.g., producing a CRS in pk with knowledge of a trapdoor, $\mathcal{S}$ is able to give fake proofs that some ciphertext really encrypts a message m. We stress that this can *not* be used to break the intuitive guarantees that $\mathcal{F}_{\mathrm{PKENO}}$ provides: $\mathcal{F}_{\mathrm{PKENO}}$ still checks that the verification of this proof succeeds only for the "right" message that is associated with a ciphertext.

5   Implementation of PKENO Using IBE

5.1   Identity-Based Encryption

We first define syntax and required security properties of an identity-based encryption (IBE) scheme.

SYNTAX. An IBE scheme is a tuple $\mathsf{IBE} = (\mathsf{IBEgen}, \mathsf{KeyGen}, \mathsf{IBEenc}, \mathsf{IBEdec})$ of algorithms along with a family $\mathcal{M} = (\mathcal{M}_k)_k$ of message spaces such that:

- The key generation algorithm IBEgen takes as input a security parameter 1^k and outputs a public key pk and a secret key sk. We write $(pk, sk) \leftarrow_R$ IBEgen(1^k).
- The encryption algorithm IBEenc takes as input a public key pk, an identity $id \in \{0,1\}^*$ and a message $m \in \mathcal{M}_k$ and outputs a ciphertext c. We write $c \leftarrow_R$ IBEenc$_{pk}(id, m)$.
- The deterministic decryption algorithm IBEdec takes as input a cipher-text c, an identity $id \in \{0,1\}^*$ and a user secret key $usk[id]$. It returns a message $m \in \mathcal{M}_k$ or the distinguished symbol $\perp \notin \mathcal{M}_k$. We write $m \leftarrow$ IBEdec$_{usk[id]}(c)$.
- The deterministic user secret key algorithm KeyGen takes as input an identity $id \in \{0,1\}^*$ and a secret key sk. It returns a user secret key $usk[id]$. We write $usk[id] \leftarrow$ KeyGen$_{sk}(id)$.[3]

CONSISTENCY. We require that for every honestly generated keypair $(pk, sk) \leftarrow_R$ IBEgen(1^k), for all identities $id \in \{0,1\}^*$ and messages $m \in \mathcal{M}_k$ we have IBEdec$_{\text{KeyGen}(sk,id)}$(IBEenc$_{pk}(id, m)) = m$ with probability one.

Here we also require a non-standard soundness property that it is efficiently verifiable if a given user secret key $usk[id]$ was properly generated for identity id.[4] We write $\{\texttt{accept}, \texttt{reject}\} \leftarrow$ IBEver$_{pk}(id, usk[id])$. We require for all hon-estly generated keypair $(pk, sk) \leftarrow_R$ IBEgen(1^k) satisfies the following: For all identities $id \in \{0,1\}^*$ and strings $s \in \{0,1\}^*$ we have IBEver$_{pk}(id, s) = \texttt{accept}$ iff $s = usk[id]$, where $usk[id] \leftarrow$ KeyGen$_{sk}(id)$.

SECURITY. We only require a relatively weak security property, namely indis-tinguishability against selective-ID chosen-plaintext attacks (IND-sID-CPA) [3]. Formally, for an adversary A, consider the following game:

1. Adversary A is given 1^k and outputs a target identity id^*
2. IBEgen(1^k) outputs (pk, sk). Adversary A is given 1^k and pk.
3. The adversary may make polynomially many queries to a user secret-key oracle KeyGen$_{sk}(\cdot)$, except that it may not query for id^*
4. At some point, A outputs two equal-length messages m_0, m_1. A bit b is randomly chosen and the adversary is given the challenge ciphertext $C^* \leftarrow_R$ IBEenc$_{pk}(id^*, m_b)$.
5. A may continue to query its user secret-key oracle, except that it may not query for id^*.
6. Finally, A outputs a guess b'.

Denote A's advantage in guessing b' by

$$\text{Adv}_{\text{IBE,A}}^{\text{sid-cpa}}(k) := |\Pr[b = b'] - 1/2|.$$

---

[3] We can always assume the user secret key algorithm KeyGen to be deterministic. If it is not, the owner of the secret key ensures using the same randomness for each identity either by maintaining a state or by deriving the randomness using a PRF applied to the identity.

[4] It is *not* sufficient to check whether, e.g., some random encryptions decrypt correctly. A given alleged user secret key might misbehave on precisely one ciphertext.

Scheme IBE is called IND-sID-CPA secure if $\mathrm{Adv}_{\mathsf{IBE,A}}^{\mathrm{sid\text{-}cpa}}(\cdot)$ is negligible for every PPT adversary A. We remark that there exist efficient IND-sID-CPA secure IBE schemes without random oracle [3].

5.2   From IBE to PKENO

We use an adaptation of the IBE-to-PKE transformation by Canetti, Halevi and Katz [9]. Let $\mathsf{IBE} = (\mathsf{IBEgen}, \mathsf{KeyGen}, \mathsf{IBEenc}, \mathsf{IBEdec})$ be an IBE scheme and $\mathsf{OTS} = (\mathsf{SGen}, \mathsf{SSign}, \mathsf{SVer})$ be a one-time signature scheme which we require to be strongly unforgeable against one-time attacks. (Syntax and security properties of OTS can be looked up in [9].) We construct a PKENO scheme $\mathsf{PKENO} = (\mathsf{Gen}, \mathsf{Enc}, \mathsf{Dec}, \mathsf{Prove}, \mathsf{Ver})$ as follows.

$\mathsf{Gen}(1^k)$. The key generation algorithm runs the IBE key generation algorithm $(pk, sk) \leftarrow_R \mathsf{IBEgen}(1^k)$ and returns the key-pair (pk, sk).

$\mathsf{Enc}_{pk}(m)$. The encryption algorithm first generates a key-pair of the one-time signature scheme by running $(vk, sigk) \leftarrow_R \mathsf{SGen}(1^k)$. Next, it IBE encrypts m with "identity" vk to obtain $c \leftarrow_R \mathsf{IBEenc}_{pk}(vk, m)$. Finally, it signs the IBE ciphertext $\sigma \leftarrow \mathsf{SSign}_{sigk}(c)$. and returns the PKENO ciphertext $C = (vk, c, \sigma)$.

$\mathsf{Dec}_{sk}(C)$. The decryption algorithm parses C as the tuple (vk, c, σ). Next, it verifies if σ is a correct signature on c by running $\mathsf{SVer}_{vk}(c)$. If not, it returns $\bot$. Otherwise, it computes $usk[vk] \leftarrow \mathsf{KeyGen}_{sk}(vk)$ and IBE decrypts c by running $m \leftarrow \mathsf{IBEdec}_{usk[vk]}(c)$. Finally, it returns $m \in \mathcal{M}_k \cup \{\bot\}$.

$\mathsf{Prove}_{sk}(C)$. The prove algorithm parses C as the tuple (vk, c, σ). Next, it verifies if σ is a correct signature on c by running $\mathsf{SVer}_{vk}(c)$. If not, it returns $\bot$. Otherwise, it computes $usk[vk] \leftarrow \mathsf{KeyGen}_{sk}(vk)$ and returns $\pi \leftarrow usk[vk]$ as the proof.

$\mathsf{Ver}_{pk}(C, m, \pi)$. The verification algorithm parses C as the tuple (vk, c, σ). Next it verifies if σ is a correct signature on c with respect to verification key vk by running $\mathsf{SVer}_{vk}(c)$. If not, it returns $\mathtt{reject}$. Otherwise, it checks if π is a properly generated user secret-key for "identity" vk by running $\mathsf{IBEver}_{pk}(vk, \pi)$. If not, it returns $\mathtt{reject}$. Otherwise, it IBE decrypts c by running $\hat{m} \leftarrow \mathsf{IBEdec}_{\pi}(vk, c)$, where $\hat{m} \in \mathcal{M}_k \cup \{\bot\}$. If $\hat{m} \neq m$, it returns $\mathtt{reject}$. Otherwise it returns $\mathtt{accept}$.

It is easy to check that the above scheme satisfies correctness and completeness.

Theorem 2. *Assume* IBE *is IND-sID-CPA secure and* OTS *is SUF-OT secure. Then* PKENO *constructed above is* PKENO *secure.*

First note that IBE soundness directly implies perfect proof soundness of PKENO. This is since the proof algorithm makes sure that the proof $\pi = usk[vk]$ is a properly generated user secret key for the the "identity" vk from the ciphertext by running the verification algorithm. Hence by consistency of the IBE scheme the decrypted message $\hat{m}$ will always equal the real message m of the ciphertext and hence verification accepts.

Let us now give some intuition why PKENO is IND-CCPA secure. A formal proof (following [9]) will be given in the full version. Let (c^*, vk^*, σ^*) be the challenge ciphertext in the IND-CCPA security experiment. It is clear that, without any oracle queries, the value of the bit b remains hidden to the adversary. This is so because c^* is output by IBEenc which is IND-sID-CPA secure, vk^* is independent of the message, and σ^* is the result of applying the one-time signing algorithm to c^*.

We claim that decryption and proof oracle queries cannot further help the adversary in guessing the value of b. First note that a proof for some ciphertext enables the adversary to decrypt the same ciphertext without making the decryption query. It remains to consider an arbitrary proof query $(c, vk, \sigma) \neq (c^*, vk^*, \sigma^*)$ made by the adversary during the experiment. If $vk = vk^*$ then $(c, \sigma) \neq (c^*, \sigma^*)$ and the proof oracle will answer $\bot$ since the adversary is unable to forge a new valid signature σ with respect to vk^*. If $vk \neq vk^*$ then the proof query will not help the adversary since the the proof $\pi = usk[vk]$ is an IBE user secret key for the "identity" vk distinct from vk^*.

6   Direct Implementation of PKENO in Bilinear Group

6.1   Bilinear Groups and Assumptions

Our schemes will be parametrized by a *pairing parameter generator*. This is an algorithm $\mathcal{G}$ that on input 1^k returns the description of an multiplicative cyclic group $\mathbb{G}$ of prime order p, where $2^k < p < 2^{k+1}$, the description of a multiplicative cyclic group $\mathbb{G}_T$ of the same order, and a non-degenerate bilinear pairing $\hat{e} : \mathbb{G} \times \mathbb{G} \to \mathbb{G}_T$. We use $\mathbb{G}^*$ to denote $\mathbb{G} \setminus \{1\}$, i.e. the set of all group elements except the neutral element. The pairing has to be satisfy the following two conditions.

Non-degenerate: for all $g \in \mathbb{G}^*$, $\hat{e}(g, g) \neq 1 \in \mathbb{G}_T$.
Bilinear: for all $g \in \mathbb{G}^*$, $x, y \in \mathbb{Z}_p$, $\hat{e}(g^x, g^y) = \hat{e}(g, g)^{xy}$.

We use $\mathbb{PG} = (\mathbb{G}, \mathbb{G}_T, p, \hat{e}, g, g_T)$ as shorthand for the description of bilinear groups, where g is a generator of $\mathbb{G}$ and $g_T = \hat{e}(g, g) \in \mathbb{G}_T$. The Bilinear Decisional Diffie-Hellman (BDDH) assumption [4] states that the two distributions $(g^x, g^y, g^z, \hat{e}(g, g)^{xyz})$ and $(g^x, g^y, g^z, \hat{e}(g, g)^r)$, for $x, y, z, r \leftarrow_R \mathbb{Z}_p$ are indistinguishable for any adversary. More formally we define the advantage function $\text{Adv}_{\mathcal{G},A}^{\text{bddh}}(k)$ of an adversary A as

$$| \Pr[A(\mathbb{PG}, g^x, g^y, g^z, \hat{e}(g, g)^{xyz}) = 1] - \Pr[A(\mathbb{PG}, g^x, g^y, g^z, \hat{e}(g, g)^r) = 1]|$$

where $\mathbb{PG} \leftarrow_R \mathcal{G}(1^k)$ and $x, y, z, r \leftarrow_R \mathbb{Z}_p$. We say that the Bilinear Decision Diffie-Hellman (BDDH) assumption holds relative to $\mathcal{G}$ if for every adversary A, $\text{Adv}_{\mathcal{G},A}^{\text{bddh}}(\cdot)$ is negligible.

6.2   The PKENO Scheme

Our scheme uses the "direct chosen ciphertext technique" which results in an adaptation of the IND-CCA secure PKE scheme from [5, 15]. Let $\mathsf{TCR} : \mathbb{G} \to \mathbb{Z}_p$ be a hash function that we assume to be target collision resistant [11]. Let $\mathbb{P}\mathbb{G} \leftarrow_R \mathcal{G}(k)$ be a pairing group that is contained in the system parameters. Let (E, D) be a symmetric encryption scheme that we assume to be chosen-ciphertext secure.[5] We assume that uses elements of the target group $\mathbb{G}_T$ as secret keys. We construct a PKENO scheme $\mathsf{PKENO} = (\mathsf{Gen}, \mathsf{Enc}, \mathsf{Dec}, \mathsf{Prove}, \mathsf{Ver})$ as follows.

$\mathsf{Gen}(1^k)$. The key generation algorithm picks random exponents $x_1, x_2, y \in Z_p$. The secret key is $sk = (x_1, x_2, y) \in \mathbb{Z}_p^3$ and the public key is $pk = (X_1, X_2, Y) \in \mathbb{G}^2 \times \mathbb{G}_T$, where

$$X_1 = g^{x_1} \in \mathbb{G}, \quad X_2 = g^{x_2} \in \mathbb{G}, \quad Y = \hat{e}(g,g)^y \in \mathbb{G}_T.$$

$\mathsf{Enc}_{pk}(m)$. The encryption algorithm first picks a random $r \in \mathbb{Z}_p$. The ciphertext is the tuple (c_1, c_2, c_3), where

$$c_1 = g^r, \quad t = \mathsf{TCR}(c_1), \quad c_2 = (X_1^t X_2)^r, \quad K \leftarrow Y^r, \quad c_3 \leftarrow \mathsf{E}_K(m)$$

$\mathsf{Dec}_{sk}(C)$. The decryption algorithm parses C as the tuple (c_1, c_2, c_3). Next, it computes $t = \mathsf{TCR}(c_1)$ and checks if $c_1^{x_1 t + x_2} \overset{?}{=} c_2$. If not, it returns $\perp$ meaning the ciphertext is inconsistent. Otherwise, it computes

$$K \leftarrow \hat{e}(c_1, g^y)$$

and returns $m \leftarrow \mathsf{D}_K(c_3) \in \mathcal{M} \cup \{\perp\}$.

$\mathsf{Prove}_{sk}(C)$. The prove algorithm parses C as the tuple (c_1, c_2, c_3). Next, it computes $t = \mathsf{TCR}(c_1)$ and checks if $c_1^{x_1 t + x_2} = c_2$. If not, it returns $\perp$. Otherwise, it picks $s \leftarrow_R \mathbb{Z}_p$. The proof consists of $\pi = (d_1, d_2) \in \mathbb{G}^2$, where

$$d_1 = g^s, \quad d_2 = g^y \cdot (X_1^t X_2)^s . \tag{2}$$

$\mathsf{Ver}_{pk}(C, m, \pi)$. The verification algorithm parses C as the tuple (c_1, c_2, c_3) and π as the tuple (d_1, d_2). Next, it computes $t = \mathsf{TCR}(c_1)$ and checks if

$$\hat{e}(c_2, g) \overset{?}{=} \hat{e}(c_1, X_1^t X_2) \quad \text{and} \quad \hat{e}(g, d_2) \overset{?}{=} Y \cdot \hat{e}(X_1^t X_2, d_1) . \tag{3}$$

If one of the checks fails, it returns **reject**. Otherwise, it computes

$$\hat{K} \leftarrow \hat{e}(c_1, d_2)/\hat{e}(c_2, d_1),$$

and $\hat{m} \leftarrow \mathsf{D}_{\hat{K}}(c_3) \in \mathcal{M}_k \cup \{\perp\}$. It returns **accept** if $\hat{m} = m$ and **reject**, otherwise.

It is easy to check that the above scheme satisfies correctness and completeness.

---

[5] A symmetric encryption scheme is chosen-ciphertext secure if the encryptions of two adversarially-chosen messages under a random hidden key K remain indistinguishable even relative to a decryption oracle. We refer to [11] for a formal definition.

254   I. Damgård et al.

6.3   Security

Theorem 3. *Assume the BDDH assumption holds relative to* $\mathcal{G}$, TCR *is a target collision-resistant hash function, and* (E, D) *is a chosen-ciphertext secure symmetric encryption scheme. Then* PKENO *constructed above is* PKENO *secure.*

The proof of IND-CCPA security is similar to the one from [5, 15] and omitted here.

We verify proof soundness. Fix a key-pair and let $C = (c_1 = g^r, c_2 = (X_1^t X_2)^r, c_3 = E_K(m))$ be a proper encryption of a message m, where $K = Y^r$ is the symmetric key used for encrypting m. Now consider the verification algorithm run with C, a message $m' \neq m$ and an arbitrary proof $\pi' = (d'_1, d'_2)$. The right check of (3) implies that $\pi' = (d'_1, d'_2)$ is a properly generated proof of the form (2), for some $s \in \mathbb{Z}_p$ and for $t = \mathsf{TCR}(c_1)$. Hence, for the symmetric key $\hat{K}$ we have

$$\hat{K} = \hat{e}(c_1, d'_2)/\hat{e}(c_2, d'_1) = \hat{e}(g^r, g^y \cdot (X_1^t X_2)^s)/\hat{e}((X_1^t X_2)^r), g^s) = Y^r = K$$

By consistency of the symmetric scheme the recovered message $\hat{m} = D_K(c_3)$ equals $m \neq m'$, hence verification always outputs `reject`.

References

[1] Backes, M., Dürmuth, M., Hofheinz, D., Küsters, R.: Conditional reactive simulatability. In: Gollmann, D., Meier, J., Sabelfeld, A. (eds.) ESORICS 2006. LNCS, vol. 4189, pp. 424–443. Springer, Heidelberg (2006), http://eprint.iacr.org/2006/132.ps

[2] Backes, M., Pfitzmann, B.: Limits of the cryptographic realization of Dolev-Yao-style XOR. In: de Capitani di Vimercati, S., Syverson, P.F., Gollmann, D. (eds.) ESORICS 2005. LNCS, vol. 3679, pp. 178–196. Springer, Heidelberg (2005), http://eprint.iacr.org/2005/220.ps

[3] Boneh, D., Boyen, X.: Efficient selective-ID secure identity based encryption without random oracles. In: Cachin, C., Camenisch, J.L. (eds.) EUROCRYPT 2004. LNCS, vol. 3027, pp. 223–238. Springer, Heidelberg (2004)

[4] Boneh, D., Franklin, M.K.: Identity based encryption from the Weil pairing. SIAM Journal on Computing 32(3), 586–615 (2003)

[5] Boyen, X., Mei, Q., Waters, B.: Direct chosen ciphertext security from identity-based techniques. In: Atluri, V., Meadows, C., Juels, A. (eds.) ACM CCS 2005, pp. 320–329. ACM Press, New York (2005)

[6] Canetti, R.: Universally composable security: A new paradigm for cryptographic protocols. In: 42th Annual Symposium on Foundations of Computer Science, Proceedings of FOCS 2001, pp. 136–145. IEEE Computer Society, Los Alamitos (2001), http://www.eccc.uni-trier.de/eccc-reports/2001/TR01-016/revisn01.ps

[7] Canetti, R.: Universally composable security: A new paradigm for cryptographic protocols. IACR ePrint Archive, Online (January 2005), http://eprint.iacr.org/2000/067.ps

[8] Canetti, R., Fischlin, M.: Universally composable commitments. In: Kilian, J. (ed.) CRYPTO 2001. LNCS, vol. 2139, pp. 19–40. Springer, Heidelberg (2001), http://eprint.iacr.org/2001/055.ps

[9] Canetti, R., Halevi, S., Katz, J.: Chosen-ciphertext security from identity-based encryption. In: Cachin, C., Camenisch, J.L. (eds.) EUROCRYPT 2004. LNCS, vol. 3027, pp. 207–222. Springer, Heidelberg (2004)

[10] Canetti, R., Krawczyk, H., Nielsen, J.B.: Relaxing chosen-ciphertext security. In: Boneh, D. (ed.) CRYPTO 2003. LNCS, vol. 2729, pp. 565–582. Springer, Heidelberg (2003), http://eprint.iacr.org/2003/174.ps

[11] Cramer, R., Shoup, V.: Design and analysis of practical public-key encryption schemes secure against adaptive chosen ciphertext attack. SIAM Journal on Computing 33(1), 167–226 (2003)

[12] Damgård, I., Thorbek, R.: Non-interactive proofs for integer multiplication. In: Naor, M. (ed.) EUROCRYPT 2007. LNCS, vol. 4515, pp. 412–429. Springer, Heidelberg (2007), http://eprint.iacr.org/2007/086

[13] Datta, A., Derek, A., Mitchell, J.C., Ramanathan, A., Scredrov, A.: Games and the impossibility of realizable ideal functionality. In: Halevi, S., Rabin, T. (eds.) TCC 2006. LNCS, vol. 3876, pp. 360–379. Springer, Heidelberg (2006), http://eprint.iacr.org/2005/211.pdf

[14] Hofheinz, D., Müller-Quade, J., Steinwandt, R.: On modeling IND-CCA security in cryptographic protocols. 14 pages. Tatra Mountains Mathematical Publications (to be published, 2005)

[15] Kiltz, E.: Chosen-ciphertext security from tag-based encryption. In: Halevi, S., Rabin, T. (eds.) TCC 2006. LNCS, vol. 3876, pp. 581–600. Springer, Heidelberg (2006)

[16] Lindell, Y.: General composition and universal composability in secure multi-party computation. In: 44th Annual Symposium on Foundations of Computer Science, Proceedings of FOCS 2003, pp. 394–403. IEEE Computer Society, Los Alamitos (2003), http://eprint.iacr.org/2003/141.ps

[17] Naor, M., Yung, M.: Public-key cryptosystems provably secure against chosen ciphertext attacks. In: 22nd ACM STOC, May 1990, ACM Press, New York (1990)

[18] Nielsen, J.B.: On Protocol Security in the Cryptographic Model. PhD thesis, University of Aarhus (2003), http://www.brics.dk/~buus/jbnthesis.ps.gz

[19] Peikert, C., Waters, B.: Lossy trapdoor functions and their applications. Cryptology ePrint Archive, Report 2007/279 (2007), http://eprint.iacr.org/

[20] Waters, B.R.: Efficient identity-based encryption without random oracles. In: Cramer, R.J.F. (ed.) EUROCRYPT 2005. LNCS, vol. 3494, pp. 114–127. Springer, Heidelberg (2005)

A Vulnerability in RSA Implementations Due to Instruction Cache Analysis and Its Demonstration on OpenSSL

Onur Acıiçmez[1] and Werner Schindler[2]

[1] Samsung Information Systems America, Samsung Electronics
95 West Plumeria Drive, San Jose, CA 95134, USA
onur.aciicmez@gmail.com
[2] Bundesamt für Sicherheit in der Informationstechnik (BSI)
Godesberger Allee 185–189, 53175 Bonn, Germany
Werner.Schindler@bsi.bund.de

Abstract. MicroArchitectural Analysis (MA) techniques, more specifically Simple Branch Prediction Analysis (SBPA) and Instruction Cache Analysis, have the potential of disclosing the entire execution flow of a software-implemented cryptosystem ([5,2]). In this paper we will show that one can completely break RSA in the original *unpatched* OpenSSL version (v.0.9.8e) even if the most secure configuration is in place, including all countermeasures against side-channel and MicroArchitectural analysis (in particular, base blinding). We also discuss (known) countermeasures that prevent this attack.

In a first step we apply an instruction cache attack to reveal which Montgomery operations require extra reductions. To exploit this information we model the timing behavior of the modular exponentiation algorithm by a stochastic process. Its analysis provides the optimal guessing strategy, which reveals the secret key $(\bmod\, p_1)$ and finally the factorization of the RSA modulus $n = p_1 p_2$. For the instruction cache attack we applied a spy process that was embedded in the target process (OpenSSL), which clearly facilitates the experimental part. This simplification yet does not nullify our results since in cache attacks empirical results from embedded spy processes and (suitably implemented) standalone spy processes are very close to each other [16] and, moreover, our guessing strategy is fault-tolerant. Interestingly, the second step of our attack is related to that of a particular combined power and timing attack on smart cards [23] (see also [27,22]).

Before we published our result [1] we informed the OpenSSL development team who included a patch into the stable branch of v.0.9.7e ([31,32]) and CERT which informed software vendors ([33,34,35]). In particular, this countermeasure is included in the current version 0.9.8f. We have only analyzed OpenSSL, thus we currently do not know the strength of other cryptographic libraries.

Keywords: RSA, Montgomery Multiplication, Instruction-Cache Attack, MicroArchitectural Analysis, Side Channel Analysis, Stochastic Process.

T. Malkin (Ed.): CT-RSA 2008, LNCS 4964, pp. 256–273, 2008.

1   Introduction

MicroArchitectural Attacks (MA), which exploit the microarchitectural behavior of modern computer systems, form a new group of side-channel analysis. Data Cache (e.g.[9,18,20,7]), Instruction Cache ([2]), Branch Prediction ([5,6]), and Shared Functional Unit Attacks ([3]) are different types of MA in the literature. Branch Prediction Analysis (BPA) and its very powerful variant Simple Branch Prediction Analysis (SBPA) have been introduced by Acıiçmez et. al. [5,6]. They showed that a carefully written spy-process running simultaneously with an RSA-process, is able to collect during one *single* RSA signing execution almost all of the secret key bits. They call such an attack, analyzing the CPU's Branch Predictor states through spying on a single quasi-parallel computation process, a *Simple Branch Prediction Analysis (SBPA)* attack — sharply differentiating it from those one relying on statistical methods and requiring many computation measurements under the same key. Following this interesting research vector, Acıiçmez has developed another MA attack based on the functionality of instruction cache (I-cache), which is another major processor component [2]. This new attack, called I-cache Analysis, is also aiming to reveal the instruction flow of cryptosystems just like SBPA.

The major cryptographic libraries, especially OpenSSL which is widely utilized in most of the security software today (according to an estimation from NTT made in November 2006, more than 60% of the web servers worldwide has OpenSSL toolkit installed [30]), have gone under several revisions to mitigate different MA attacks immediately after the announcements of their feasibilities. However, despite of all these efforts spent to provide better protections against MA attacks, the original (unpatched) version 0.9.8e of OpenSSL still had a vulnerability, and also the originally (before we shared our attack with the OpenSSL development team) planned SPBA countermeasures in v. 0.9.8f would not have prevented our attack. In fact, even if all these countermeasures were turned on, one could completely break this RSA implementation[1].

The most secure configuration of the unpatched OpenSSL version v.0.9.8e employed fixed-window exponentiation, which was implemented as a mitigation technique against cache analysis presented in [20]. OpenSSL also handles the RSA structures in a special way to avoid cache based threats. These two techniques come in a bundle, i.e. used together to protect enhanced security against cache analysis. Besides various other countermeasures for MA vulnerabilities OpenSSL v.0.9.8e used base blinding to prevent pure timing attacks. Base-blinding had first been implemented as an optional protection and later on became a default protection mechanism after the publication of Brumley and Boneh Attack [10], see also [8].

---

[1] The side-channel related countermeasures in OpenSSL are optional. Some of them are turned on by default, while others are not. The systems built upon OpenSSL can choose which countermeasures to be active by setting the corresponding OpenSSL flags. It is possible for a system to activate all of these countermeasures, which naturally come with some performance loss, or totally ignore/deactivate all of them.

The source of our attack was the implementation of Montgomery Multiplication (MM), which still had the extra reduction step in the unpatched v.0.9.8e, although it had been shown many times that this particular step cause serious security vulnerabilities, cf. [8,10,11,26,25,27,23,22]. It also had been known for a long time that MM can be easily implemented without extra reduction step, cf. [28,29,12]. In this paper we will show that even if we utilize all of these mitigation techniques in the original OpenSSL v.0.9.8e, we can still completely break the RSA implementation, due to the extra reduction step in MM. Base blinding does not prevent our attack.

It is important to note that OpenSSL implementation was modified to eliminate extra reduction step in MM after we showed the feasibility of completely breaking even the most secure configuration of the library ([31]). This change was also applied to the current OpenSSL version 0.9.8f. We also contacted CERT who informed software vendors. The US CERT assigned the vulnerability explained in this paper CVE name CVE-2007-3108 and CERT vulnerability number VU#724968, and they issued a vulnerability note ([33,34,35]).

Although SBPA and I-cache Analysis have the potential to reveal the entire execution flow of an RSA cipher, our focus in this paper is only on the extra reduction step of the Montgomery multiplication algorithm. We show that MA can be used to determine which Montgomery multiplication operations perform an extra reduction during the entire execution of RSA. The importance of this information was demonstrated in [27,23,22]. Walter et. al. developed a non-optimal attack on RSA (only for fixed window size of 2 bits) which allow to extract the secret exponent if the occurrences of extra reduction steps for a sample of RSA decryption/signing under the same RSA key are known to the attacker [27]. Later, Schindler generalized and optimized this attack for arbitrary window sizes ([23]), and then Schindler and Walter extended this attack to a variant of Montgomery's multiplication algorithm ([22]). References [27,23,22] considered fixed-window algorithms that are not equal but related to the implementation in OpenSSL. We need to mention that these attacks may be considered as "theoretical" because neither Walter et. al. nor Schindler implemented actual attacks on real systems but instead they assumed that side-channel information obtained via power and timing analysis would reveal such occurrences of extra reduction step. The second part of these attacks, the exploitation of the gathered information, was practically verified by simulation studies.

In this paper, we combine the practicality of MicroArchitectural Analysis (Instruction Cache Analysis in particular) and the theory of [23] to show that the use of extra reduction step in Montgomery multiplication leads to a total break of current RSA software implementations. We also suggest several countermeasures that must be employed in software implementations of RSA.

2   Actual Practical Implementations of the Attack on Extra Reduction Via MA

Motivated from the attacks given in [27,23,22], we decided to realize these attacks via MicroArchitectural Analysis. It was already proven in [5,2] that SBPA and

I-cache Analysis can reveal the execution flow of RSA, which is exactly what we need in order to bring these theoretical extra reduction attacks into practice. We preferred to use the original OpenSSL v.0.9.8.e as our target software RSA implementation and I-cache analysis as our tool to detect the occurrences of extra reduction steps. The results of our experiments are given in the following section. In this section, we outline the basics of I-cache analysis and describe our approach of using I-cache analysis on OpenSSL to gather er-vector values.

2.1   Overview of I-Cache Analysis

I-cache analysis relies on the fact that instruction cache misses increase the execution time of code section. Instruction cache (a.k.a. I-cache) is a small buffer between the main memory and the processor core which provides the processor fast and easy access to the most frequently executed instructions. When the processor needs to read some instructions from the main memory, it first checks to see if they are already in I-cache. If they are already in I-cache (a.k.a. cache hit), the processor immediately uses these "cached" instructions instead of accessing the main memory, which has a significantly longer latency compared to an I-cache. Otherwise (a.k.a. cache miss), the instructions are read from the memory and a copy of them is stored in I-cache. Each I-cache miss mandates an access to a higher level memory, i.e., a higher level cache or main memory, and thus results in additional execution time delays.

In I-cache analysis, an adversary needs to execute a spy code, which keeps track of the changes in the state of I-cache, i.e., metadata, during the execution of a cipher process. A spy code / process can run simultaneously or quasi-parallel with the cipher process and determine which instructions are executed by the cipher. To give a concrete example, [2] takes advantage of the fact that sliding windows exponentiation generates a key dependent sequence of modular operations[2]. Furthermore, OpenSSL uses different functions to compute modular multiplications and square operations. [2] shows that if an adversary can run a spy routine and evict either one of these functions, he can easily determine the operation sequence (squaring / multiplication) of RSA.

In the attack scenario of [2] a "protected" crypto process executes RSA signing/decryption operations and an adversary executes a spy process simultaneously or quasi-parallel with this cipher. The spy routine

1. continuously executes a number of dummy instructions, and
2. measures the overall execution time of all of these instructions

---

[2] OpenSSL implements both sliding window and fixed window exponentiations. Sliding window exponentiation is the default algorithm in OpenSSL. Fixed window exponentiation (which is slower than the sliding window) were implemented as an optional protection to cache attacks (and now it also provides protection against branch prediction attacks). The choice whether to turn the cache attack and/or branch prediction attack countermeasures on (including the fixed window exponentiation) is given to the user.

in such a way that these dummy instructions precisely maps to the same I-cache location with the instructions of multiplication function. In other words, the adversary creates a conflict between the instructions of the multiplication function and the spy routine. Because of this I-cache conflict, either the spy or multiplication instructions can be stored in I-cache at any time. Therefore, whenever the cipher process executes the multiplication function, the instructions of the spy routine have to be evicted from I-cache. This eviction can easily be detected by the spy routine because when it reexecutes its instructions the overall execution time will suffer from I-cache misses. Thus, the spy can determine when the multiplication function is executed. This information directly reveals the operation sequence (multiplcation / squaring) of RSA. For the square & multiply exponentiation algorithm this discovers the whole secret key while for sliding windows exponentiation the attacker learns more than half of the exponent bits anyway. For further details of I-cache analysis and this particular attack, we refer the reader to [2].

2.2   Our Approach

Our approach in this paper is similar to the approach presented in [2]. The positions of squarings and multiplications are known in fixed-windows exponentiation (cf. Subsect. 3.2) but we create I-cache conflicts between the spy routine and the instructions executed during extra reduction step to reveal the er-vectors. An er-vector is an ordered tuple of 1 and 0 values, e.g., $(0, 0, 1, 0, 1, ..., 0)$, which shows the occurrences of extra reduction steps. We denote the occurrence of an extra reduction step with a 1 in er-vectors and the value 0 indicates that extra reduction is not performed for that Montgomery operation. See further [23] and Section 3.2. We point out that the overall attack is much more complicated than in [2] since a single er-vector is not sufficient to recover the secret RSA exponent. OpenSSL library employs Montgomery multiplication algorithm as stated above. During a Montgomery operation, OpenSSL first calls either multiplication or square functions from BIGNUM library [3] and then reduces the result to the modulus via Montgomery reduction function. At the end of the reduction after each Montgomery operation, whether it is a multiplication or square, the intermediate result is compared with the modulus to decide if an extra reduction is needed.

The extra reduction step consists of a multiprecision subtraction and OpenSSL realizes it as a call to BIGNUM library's unsigned multiprecision subtraction function, i.e., BN_usub() function. This function is used/called only for the extra reduction step during the course of the RSA signing/decryption operation. In other words, there is not any other location in OpenSSL's code for RSA decryption that calls this function. Therefore, whenever a cipher process executes this function, it must be performing extra reduction step. Hence, it is sufficient

---

[3] BIGNUM is a software library integrated into OpenSSL and it is responsible for multiprecision operations.

to create a spy routine that has I-cache conflicts with this function in order to detect the occurrences of extra reduction steps[4].

Following this basic approach, we implemented such a spy function as described in [2] and also considering the exact details of RSA implementation in OpenSSL. Our spy function executes some dummy instructions and measures their overall running time. These dummy instructions have a conflict with the extra reduction routine BN_usub() of OpenSSL, which allows the spy to detect the occurrences of extra reduction steps, i.e., er-vector values. After enough er-vector vectors are gathered, it becomes feasible to break the cipher. One has to transfer the attack from [23], which has already been proven to be correct there, to the concrete situation and apply it on the gathered er-vectors.

However, the spy measurements are not perfectly clean, i.e. there is a noise factor to consider. It is already shown in [5] that a carefully-written spy process can get very clean results. But [5] also states that such clean results cannot always be collected and an adversary needs to make some trials to get clean results. Therefore, an adversary will have to deal with this noise factor. As pointed out below our attack tolerates some errors. Thus the problem for an adversary becomes how clearly he can detect the extra reduction steps, i.e., gathering er-vectors with a low enough error rate. The error rate affects the necessary sample size of the attack, and high error rates may make it infeasible to compromise the cipher. However, we will show in the next section that the measurements collected by a spy function is clean enough to practically apply these attacks with a relatively small sample size.

3 Experimental Details, Mathematical Background and Empirical Results

We performed two different phases of experiments in this project. The objective of the first phase was to gather er-vectors via MA, i.e., I-cache Analysis in this case. The second phase consisted of determining the success rate of our attack for different sample sizes, and different error rates, applying the optimal decision strategy developed in Subsect. 3.2.

3.1 Experimental Details

We compiled the RSA decryption function of OpenSSL (unpatched version 0.9.8e) with all the available countermeasures enabled. We disassembled the executable

---

[4] In fact, an adversary also needs to be sure that instructions of the spy routine do not get completely evicted by other functions such as BIGNUM's multiplication and square functions. This possible situation was not a problem in our case, because the executable code did not have such conflicts between these functions when we compiled OpenSSL using gcc compiler. A compiler is expected to remove possible internal cache conflicts to increase the performance of the code. Internal cache conflicts, in this context, indicate the conflicts occur between the instructions that belong to the same process.

file to determine the logical addresses of BIGNUM multiprecision unsigned subtraction function instructions. GNU Project debugger (i.e. gdb) has two functions, "info line" and "disas", that we used for this task. Then we implemented our spy routine according to these logical addresses and also considered the parameters of the I-cache architecture on our platform.

Then we carried out the first experimental phase by letting the spy function run during the execution of RSA signing operation. To simplify our setup and reduce the necessary experimental efforts, we used a simple trick in this phase. Instead of using a stand-alone spy process and relying on Operating System functionalities as done in [17] or exploiting SMT-capability of processor architectures as done in [20,18,5], we called the spy routine inside the RSA process with a certain frequency, i.e., after each exponentiation step. The other options would require special manual handling of each single measurement and would drastically increase the necessary efforts spent in this experiment, because we needed to analyze a large number of measurements to get an accurate estimation of the error rates. In more realistic attack scenarios that use stand-alone spy processes, an adversary may (and most likely will) encounter an error rate higher than our experimental results. However, according to an analysis on cache attacks from Neve which is given in [16], the theoretical and actual results taken from such a spy process are indeed very close to each other. Therefore, we expect the actual error rates in a spy process's measurements to be only slightly higher than our estimated error rates.

We must also mention that these error rates depend on several parameters including the actual platform, operating system and other software components of the system, the implementation of the spy process as well as the cipher. Higher error rates do not necessarily nullify the validity of these attacks since the optimal guessing strategy for correctly observed er-vectors (cf. Theorem 1) seems to tolerate error rates of 4 - 5%, and even guessing strategies exist that take classification errors explicitly into account (cf. Theorem 2). Thus, *we do not claim that our results perfectly reflect the performance of these attacks in every possible scenario. We only prove in this paper that er-vector attacks coupled with MA techniques create a valid threat to software systems.*

We used the possibility of performing the same exact measurement many times and taking the average to decrease the measurement noise, i.e., error rate. OpenSSL uses the same blinding factor 32 times by default before updating it. Therefore, an adversary can force the system to perform RSA decryption for the same base and identical blinding values $t \leq 32$ many times and measure each of these t operations. For example, he can exploit SSL handshake protocol as done in [10,8]. Using the average of more than 1 measurement decreases the noise and thus reduces the error rate as we will show in the next subsection. The second phase of our experiments was to determine the success rate of er-vector attack for several sample sizes and to estimate the effect of different error rates.

In [1] we considered both of the exponentiation algorithms implemented in current OpenSSL version: fixed windows and sliding windows exponentiation ([15], 14.82 and 14.85). The size of the windows used in both exponentiation algorithms depends on the key size in OpenSSL. For common key sizes (1024 and 2048 bits), the windows are 5-bit long, and thus we will focus on this particular case in our paper. Moreover, the Chinese Remainder Theorem (CRT), Montgomery's multiplication algorithm ([15], 14.36) and base blinding are applied. Due to the lack of space, we focus on fixed-windows exponentiation in the following.

3.2 Mathematical Background and Experimental Results

In this subsection we formulate the optimal guessing strategy, explain the mathematical background and present experimental results.

Fixed Window Exponentiation Algorithm. As usual, $n = p_1 p_2$ and R denotes the Montgomery constant while $\mathrm{MM}(a, b; n) := abR^{-1}(\mathrm{mod}\, n)$ stands for the Montgomery multiplication of a and b. The computation of $x^d(\mathrm{mod}\, n)$ is carried out in several phases:

1. Base blinding: $x_b := xA(\mathrm{mod}\, n)$ where A and $B := A^{-d}(\mathrm{mod}\, n)$ are the current blinding values.
2. Compute $x_b^d(\mathrm{mod}\, p_1)$
   a) group the binary representation of $d_{(1)} := d(\mathrm{mod}\,(p_1 - 1)) = (d'_{w-1}, \ldots, d'_0)_2$ into non-overlapping blocks of length $wsize$, starting from the least significant bit d'_0. This gives wsize-bit integers $D_{v-1}, \ldots, D_0$ with $v := \lceil w/wsize \rceil$.
   b) $x_{b,1} :\equiv x_b(\mathrm{mod}\, p_1)$
   c) Exponentiation algorithm **1**: Fixed windows
   $$u_0 := \mathrm{MM}(1,\ R^2(\mathrm{mod}\ p_1); p_1) \qquad (= R\,(\mathrm{mod}\, p_1))$$
   $$u_1 := \mathrm{MM}(x_{b;1},\ R^2(\mathrm{mod}\ p_1); p_1) \qquad (= x_{b;1}R\,(\mathrm{mod}\, p_1))$$
   ```
   for j := 2 to 2^{wsize} − 1 do    u_j := MM(u_{j−1}, u_1; p_1)
   temp := u_0
   for i := v − 1 downto 0 do {
       for j := 1 to wsize do    temp := MM(temp, temp; p_1)
       temp := MM(temp, u_{D_i}; p_1)}
   return MM(temp, 1)            (= x_{b;1}^d(mod p_1) = x_b^d(mod p_1))
   ```
   Note: $u_0, \ldots, u_{2^{wsize}-1}$ denote the table values.
3. Compute $x_b^d(\mathrm{mod}\, p_2)$ analogously to Step 2
4. Compute $x^d(\mathrm{mod}\, n)$
   a) CRT step: Compute $x_b^d(\mathrm{mod}\, n)$ from $x_b^d(\mathrm{mod}\, p_1)$ and $x_b^d(\mathrm{mod}\, p_2)$.
   b) "Remove" blinding: $y := x_b^d B(\mathrm{mod}\, n)$
5. (eventually) update A and B

The adversary's goal is clearly to determine the secret exponent d. We first note that it is sufficient to determine $d_{(1)}$ since

$$y - x^{d_{(1)}} \equiv 0(\mathrm{mod}\, p_1), \quad \text{hence } gcd(y - x^{d_{(1)}}(\mathrm{mod}\, n), n) = p_1 \qquad (1)$$

for any known plaintext / ciphertext pair $(x, y = x^d(\mathrm{mod}\, n))$. The types of operation $T(1), T(2), \ldots, T(M)$ of the Montgomery operations during the exponentiation phase are equivalent to knowing $d_{(1)}$. To increase the readability of the paper we concentrate on $wsize = 5$, the case we are interested in. Of course, all assertions can immediately be transferred to arbitrary window size. Essentially, it remains to substitute 31 by $2^{wsize} - 1$.

Clearly, $T(i) \in \Theta = \{\text{'}S\text{ '}, \text{'}M_0\text{'}, \text{'}M_1\text{'}, \dots, \text{'}M_{31}\text{'}\}$ where 'S' says that the ith Montgomery operation in the exponentiation phase is a squaring while 'M_k' stands for the multiplication with table entry u_k. The adversary grounds his decisions on the exponentiation of bases $x_1, \dots, x_N$. For now, assume that the adversary knows which operations in the table initialization phase and in the exponentiation phase require extra reductions (ERs). More formally, $w'_{i(k)} = 1$, resp. $w'_{i(k)} = 0$ mean that the ith operation in the table initialization phase (= computation of u_i for base x_k) requires an extra reduction, resp. requires no extra reduction. Similarly, $w_{i(k)} = 1$, resp. $w_{i(k)} = 0$ mean that the ith operation in the exponentiation phase requires an extra reduction, resp. requires no extra reduction. In the following we outline the general procedure, explain the main steps.

We interpret the observations $w'_{i(k)}$ and $w_{i(k)}$ as realizations of suitably defined random variables $W'_{i(k)}$ and $W_{i(k)}$. As already shown in [26] and [24] we have

$$\text{Prob}(W_{i(k)} = 1) = \begin{cases} \frac{1}{3}\frac{p_1}{R} & \text{if } T(i) = \text{'}S\text{ '} \\ \frac{u_j}{2p_1}\frac{p_1}{R} & \text{if } T(i) = \text{'}M_j\text{'}. \end{cases} \tag{2}$$

We note that both probabilities depend on the ratio p_1/R which is yet unknown to the attacker. Recall that the positions of the $\#sq = 5\lceil \log_2(d_{(1)})/5 \rceil$ many squarings are well-known in this fixed windows variant since they do not depend on $d_{(1)}$, and hence the first line of (2) can be used to estimate p_1/R. (For the bases $x_1, \dots, x_N$ count the ERs within all squarings in the sample and multiply this number by $3(N\#sq)^{-1}$.) Using the second line of (2) the game was easy if the adversary knew the ratios $u_{j(1)}/p_1, \dots, u_{j(N)}/p_1$ for each $0 \le j \le 31$. Due to base blinding (and the use of CRT) the adversary yet does not know these values.

The key is a formal treatment which analyzes the distribution of the random variables $W'_{1(k)}, \dots, W'_{31(k)}$ and $W_{1(k)}, W_{2(k)}, \dots$. Further, $s'_{0(k)} := x_{b,1}/p_1$, $s'_{i(k)} := u_{i(k)}/p_1$ for $i \in \{1, \dots, 31\}$ and $s_{i(k)} := temp_{i(k)}/p_1$ assume values in the unit interval $[0, 1)$ where $temp_{i(k)}$ stands for the ith temp value in the exponentiation phase for the base x_k. In particular, $s_{0(k)} = u_{0(k)}/p_1 = R(\bmod p_1)/p_1$. We assume that the values $s'_{0(k)}, \dots, s'_{31(k)}$ and $s_{1(k)}, s_{2(k)}, \dots$ are taken on by $[0, 1)$-valued random variables $S'_{0(k)}, \dots, S'_{31(k)}$ and $S_{1(k)}, \dots, S_{M(k)}$. Let M denote the number of Montgomery operations in the exponentiation phase. Lemma 1(iii) in [23] implies that the random variables

$$S'_{0(k)}, \dots, S'_{31(k)}, S_{7(k)}, S_{8(k)}, \dots, S_{M(k)} \tag{3}$$
are independent and uniformly distributed on $[0, 1)$

which matches with the intuition that the intermediate temp values 'spread' wildly over Z_{p_1}. (Note that $S_{0(k)} = \dots = S_{5(k)} = u_{0(k)}/p_1$ since $\text{MM}(u_{0(k)}, u_{0(k)}; p_1) = u_{0(k)}$, and $S_{6(k)} = u_{D_v-1(k)}/p_1$; cf. Remark 1.) It is even more interesting that the random variables $W'_{i(k)}$ and $W_{i(k)}$ can be expressed in terms of $S'_{i-1(k)}$ and $S'_{i(k)}$, resp in terms of $S_{i-1(k)}$ and $S_{i(k)}$. More precisely, Lemma 1(iii) in [23] implies

$$W'_{i(k)} := \begin{cases} 1_{S'_{1(k)} < S'_0 (R^2 \ (\mathrm{mod}\ p_1)/p_1)(p_1/R)} & \text{for } i = 1 \\ 1_{S'_{i(k)} < S'_{i-1(k)} S'_{1(k)} p_1/R} & \text{for } 2 \le i \le 31 \quad and \end{cases} \quad (4)$$

$$W_{i(k)} := \begin{cases} 1_{S_{i(k)} < S^2_{i-1(k)} p_1/R} & \text{if } T(i) = \text{`}S\text{'} \\ 1_{S_{i(k)} < S_{i-1(k)} S'_{j(k)} p_1/R} & \text{if } T(i) = \text{`}M_j\text{'}. \end{cases} \quad (5)$$

Here $1_A(x)$ denotes the indicator function, i.e. $1_A(x) = 1$ iff $x \in A$ and $= 0$ otherwise. Our task is to analyse the stochastic process $W_{8(k)}, W_{9(k)}, \dots, W_{M(k)}$, which is non-stationary and dependent. Nevertheless, exploiting (3), (4) and (5) yet allows the exact solution of our problem. Since the positions of the squarings are known our next goal is to compute the joint probabilities

$$p_\theta(w'_{1(k)}, \dots, w'_{31(k)}, w_{i(k)}) \quad (6)$$
$$:= \mathrm{Prob}_\theta(W'_{1(k)} = w'_{1(k)}, \dots, W'_{31(k)} = w'_{31(k)}, W_{i(k)} = w_{i(k)})$$

for all possible types of operation $\theta \in \Theta_0 := \{\text{`}M_0\text{'}, \text{`}M_1\text{'}, \dots, \text{`}M_{31}\text{'}\}$. Due to (4) and (5) the values $w'_{1(k)}, \dots, w'_{31(k)}$ and $w_{i(k)}$ can be characterized by conditions on $s'_{0(k)}, \dots, s'_{31(k)}$ and $s_{i-1(k)}, s_{i(k)}$. For instance, $w'_{j(k)} = 1$ iff $s'_{j(k)} < s'_{j-1(k)} s'_{1(k)} p_1/R$, while the impact of $w_{i(k)}$ on $s_{i-1(k)}$ and $s_{i(k)}$ clearly depends on θ (cf. (5)). Altogether, if $T(i) = \theta$, observing $(w'_{1(k)}, \dots, w'_{31(k)}, w_{i(k)})$ is equivalent to $(s'_{0(k)}, s'_{1(k)}, \dots, s'_{31(k)}, s_{i-1(k)}, s_{i(k)}) \in A_\theta(w'_{1(k)}, \dots, w'_{31(k)}, w_{i(k)})$ for a well-defined subset $A_\theta(w'_{1(k)}, \dots, w'_{31(k)}, w_{i(k)}) \subset [0, 1)^{34}$ that only depends on $(w'_{1(k)}, \dots, w'_{31(k)}, w_{i(k)})$ and the hypothesis θ. The table entry $u_{0(k)}$ plays an exceptional role as it does not depend on the basis x_k but is constant for all bases (see (9) below). Due to (3) the joint probability in (6) is given by the volume of the set $A_\theta(w'_{1(k)}, \dots, w'_{31(k)}, w_{i(k)})$.

$$p_\theta(w'_{1(k)}, \dots, w'_{31(k)}, w_{i(k)}) = \int_0^1 \int_{a'_1}^{b'_1} \cdots \int_{a'_{31}}^{b'_{31}} \int_0^1 \int_{a_{\theta,i}}^{b_{\theta,i}} 1 \, ds_i ds_{i-1} ds'_{31} \cdots ds'_0 \quad (7)$$

with integration boundaries

$$(a'_1, b'_1) = \begin{cases} (0, s'_0(R^2(\mathrm{mod}\ p_1))/R) & \text{if } w'_{1(k)} = 1 \\ (s'_0(R^2(\mathrm{mod}\ p_1))/R, 1) & \text{if } w'_{1(k)} = 0. \end{cases} \quad (8)$$

$$i \in \{2, \dots, 31\} : \ (a'_i, b'_i) = \begin{cases} (0, s'_{i-1} s'_1 p_1/R) & \text{if } w'_{i(k)} = 1 \\ (s'_{i-1} s'_1 p_1/R, 1) & \text{if } w'_{i(k)} = 0. \end{cases}$$

$$(a_{\theta,i}, b_{\theta,i}) = \begin{cases} (0, s^2_{i-1} p_1/R) & \text{if } \theta = \text{`}S\text{'} \text{ and } w_{i(k)} = 1 \\ (s^2_{i-1} p_1/R, 1) & \text{if } \theta = \text{`}S\text{'} \text{ and } w_{i(k)} = 0 \\ (0, s_{i-1} s'_j p_1/R) & \text{if } \theta = \text{`}M_j\text{'}, j > 0 \text{ and } w_{i(k)} = 1 \\ (s_{i-1} s'_j p_1/R, 1) & \text{if } \theta = \text{`}M_j\text{'}, j > 0, \text{ and } w_{i(k)} = 0. \end{cases}$$

For $\theta =' M_0'$ the er-value $w_{i(k)}$ does not depend on $w'_{1(k)}, \ldots, w'_{31(k)}$, and the term $p'_{M_0'}(w'_{1(k)}, \ldots, w'_{31(k)}, w_{i(k)})$ simplifies to

$$p'_{M_0'}(w'_{1(k)}, \ldots, w'_{31(k)}, w_{i(k)})$$
$$= \mathrm{Prob}(W'_{j(k)} = w'_{j(k)} \text{ for } 1 \leq j \leq 31) \cdot \mathrm{Prob}'_{M_0'}(W_{i(k)} = w_{i(k)}) \quad \text{with} \quad (9)$$

$$\mathrm{Prob}(W'_{j(k)} = w'_{j(k)} \text{ for } 1 \leq j \leq 31) = \int_0^1 \int_{a_1'}^{b_1'} \cdots \int_{a_{31}'}^{b_{31}'} 1 \, ds'_{31} \cdots ds'_0 , (10)$$

and from (2) we obtain $\mathrm{Prob}'_{M_0'}(W_{i(k)} = 1) = R(\mathrm{mod}\, p_1)/2R = 1(\mathrm{mod}\, (p_1/R))/2$. We have already explained how to estimate the ratio p_1/R.

Since all table entries are equally likely and since all guessing errors are equally harmful it follows from Theorem 1 in [21] that the optimal decision strategy is given by the maximum likelihood estimator. Recall that the adversary knows the positions where the squarings are located.

Theorem 1. [wsize=5] Let $\boldsymbol{w}_{i(k)} := (w'_{1(k)}, \ldots, w'_{31(k)}, w_{i(k)})$. Assume that all observed vectors $\boldsymbol{w}_{i(1)}, \ldots, \boldsymbol{w}_{i(N)}$ are correct and that in Montgomery operation $i \geq 8$ the temp value is multiplied by an unknown table value u_j, i.e. index i is a multiple of 6. The optimal strategy to guess $T(i)$ is to decide for that hypothesis $\theta' \in \Theta_0 = \Theta \setminus \{{}^\prime S{}^\prime\}$ that maximizes

$$\prod_{k=1}^N p_\theta(\boldsymbol{w}_{i(k)}). \tag{11}$$

Remark 1
(i) Theorem 1 does not cover the guessing of $T(6) =' M_{D_{v-1}}'$ since (3) does not apply to $S_{5(k)}, S_{6(k)}$ (recall that $S_{5(k)} = u_{0(k)}/p_1$ and $S_{6(k)} = u_{D_{v-1(k)}}/p_1$). Type $T(6)$ can be determined by exhaustive search after $T(12), T(18), \ldots$ have been guessed.
(ii) For Exponentiation algorithms 2 and 3 in [1] (3) even holds with $S_{1(k)}, S_{2(k)}, \ldots$ in place of $S_{7(k)}, S_{8(k)}, \ldots$.

Recall that the er-vectors $ER_{(k)} = (w'_{1(k)}, \ldots, w'_{31(k)}, w_{1(k)}, \ldots, w_{M(k)})$ are the adversary's only information ($k = 1, \ldots, N$). Unlike in [21] (and [23,22]) in our attack we are faced with erroneous observations, i.e. with flipped components of the er-vector. The terms $\mu(0 \mid 1)$ and $\mu(1 \mid 0)$ denote the probabilities to observe $w_{i(k)} = 0$ although an extra reduction is performed, resp. the probability to observe $w_{i(k)} = 1$ although no extra reduction is necessary. Practical experiments showed that the misclassification rate does not depend on the position of the respective Montgomery operation. The default setting in the original OpenSSL library v.0.9.8e kept the blinding values A and B constant for 32 consecutive exponentiations. Consequently, for any base x_k the adversary may repeat measurements $t \leq 32$ times under identical blinding values, i.e. under identical conditions. The results shown in Table 1 are average error rates calculated based on the measurements collected during 1000 decryptions with random ciphertext

Table 1. The average error rates for different values of t

t	$\mu(0 \mid 1)$	$\mu(1 \mid 0)$
1	0.1052	0.0021
2	0.0872	0.0010
4	0.0536	0.0582
8	0.0294	0.0239
16	0.0080	0.0063
32	0.0026	0.0062

under each of 10 different random 1024-bit RSA keys Increasing the value of t reduces the error rates, which become very close to 0 when t reaches 32.

With $t = 16$ we generated $N = 2000$ er-vectors (i.e., we performed $tN = 16 \cdot 2000 = 32000$ measurements), belonging to different bases $x_1, \ldots, x_{2000}$. We randomly selected subsets of size $N_1 < N$ and performed our attack on basis of the er-vectors that were contained in these subsets. Table 2 shows our results, underlining the practical feasibility of our attack even with only 10000 measurements ($N_1 = 600$, $t = 16$). For a 512-bit prime p_1 the attacker has to guess $T(6)$ (cf. Remark 1) and 102 (in rare cases 101) types $T(12), T(18), \ldots$, applying Theorem 1. Of course, 1 error (also 2, or maybe even 3 errors) within these 102 guesses can be corrected by exhaustive search. Whether all guesses are correct can be checked using (1). Note that if a false candidate θ^* maximizes (11) the corresponding correct hypothesis θ is usually also highly ranked.

Table 2. Practical experiments (Guessing strategy according to Theorem 1)

N_1	guessing errors (average)	# attacks with ≤ 1 guessing errors
500	1.90	44/100
600	0.83	85/100
700	0.25	99/100
800	0.12	99/100
900	0.01	100/100
1000	0.02	100/100

For window size 5 the adversary has to compute $|\Theta \setminus \{'S'\}| \cdot N \cdot 2 = 32 \cdot N \cdot 2$ joint probabilities $p_\theta(w'_{1(k)}, \ldots, w'_{31(k)}, j)$ (with $j \in \{0, 1\}$). Since $Prob_\theta(W_{i(k)} = 0 \mid \cdot) = 1 - Prob_\theta(W_{i(k)} = 1 \mid \cdot)$ this essentially requires the computation of $32 \cdot N$ conditional probabilities $p_\theta(0 \mid w'_{1(k)}, \ldots, w'_{31(k)})$, i.e. of $31N$ many 34-dimensional integrals of type (7) and of N many 32-dimensional integrals of type (10). For window size 5 these computations constitute the essential part of the workload. Principally, these computations are not difficult since (7) and (10) split into 34, resp. into 32, consecutive one-dimensional integrations of polynomials. Since each $w'_j = 0$ principally doubles the number of monomials for window size 5 the computations are yet memory- and time-consuming. In particular, identical monomials should be summarized regularly. Fortunately, the computed probabilities can be used for all relevant positions, i.e. for $i = 12, 18, \ldots \leq M$.

Simulation studies indicate that the optimal decision strategy from Theorem 1 seems to tolerate misclassification rates of 4 - 5% (cf. also Table 3 and Table 4 in [1], which yet refer to a related exponentiation algorithm; in the present case the situation is more favourable for an adersary since he only has to perform a sixth of the guesses). Interestingly, if the adversary (roughly) knows the misclassification rates he can take them explicitly into account.

Theorem 2. *[wsize=5] Assume that in Montgomery operation $i \geq 8$ the temp value is multiplied by an unknown table value u_j, i.e. index i is a multiple of 6. Further, $\mu_{01} := \mu(0 \mid 1), \mu_{10} := \mu(1 \mid 0) \geq 0$. For $\boldsymbol{a} = (a_1, \ldots, a_{32}) \in \{0,1\}^{32}$ define complementary subsets $C_0(\boldsymbol{a}) := \{j \leq 32 \mid a_j = 0\}, C_1(\boldsymbol{a}) := \{j \leq 32 \mid a_j = 1\} \subseteq \{1, \ldots, 32\}$ The optimal strategy to guess $T(i)$ is to decide for that hypothesis $\theta' \in \Theta \setminus \{'S'\}$ that maximizes*

$$\prod_{k=1}^{N} \sum_{\boldsymbol{w}^* \in \{0,1\}^{32}} p_\theta(\boldsymbol{w}^*) \times \tag{12}$$

$$\times \; \mu_{10}^{|C_0(\boldsymbol{w}_{i(k)}) \setminus C_0(\boldsymbol{w}^*)|} (1 - \mu_{10})^{|C_0(\boldsymbol{w}_{i(k)}) \cap C_0(\boldsymbol{w}^*)|} \mu_{01}^{|C_1(\boldsymbol{w}_{i(k)}) \setminus C_1(\boldsymbol{w}^*)|} (1 - \mu_{01})^{|C_1(\boldsymbol{w}_{i(k)}) \cap C_1(\boldsymbol{w}^*)|}.$$

Proof. Due to the properties of the stochastic process $S'_{1(k)}, \ldots, S_{M(k)}$, $k = 1, \ldots, N$, we may assume that er-vectors which belong to different bases or to different blinding values are independent. The optimal decision strategy maximizes

$$\prod_{k=1}^{N} Prob_\theta(\boldsymbol{w}_{i(k)} \text{ observed}) \qquad \text{with}$$

$$Prob_\theta(\boldsymbol{w}_{i(k)} \text{observed}) = \sum_{\boldsymbol{w}^* \in \{0,1\}^{32}} Prob_\theta(\boldsymbol{w}^* \text{correct}) Prob_\theta(\boldsymbol{w}_{i(k)} \text{ observed} \mid \boldsymbol{w}^* \text{ correct}).$$

Clearly, $Prob_\theta(\boldsymbol{w}^* \text{ correct}) = p_\theta(\boldsymbol{w}^*)$ while the second term does not depend on θ but only on $\boldsymbol{w}_{i(k)}$, $\boldsymbol{w}^*$, μ_{01} and μ_{10}. Elementary considerations complete the proof of (12).

We first note that for $\mu_{01} = \mu_{10} = 0$ formula (12) coincides with (11). The practical drawback of the (12) is that it requires the computation of $32 \cdot 2^{32}$ probabilities $p_\theta(\boldsymbol{w}^*)$ which is gigantic. Recall that we concentrated on $wsize = 5$ to increase the readability of the document. Clearly, Theorem 1 and Theorem 2 can immediately be adjusted to arbitrary window size (31 corresponds to $2^{wsize} - 1$ in the general case). We point out that for window size 4, for instance, the situation is much better since it requires only $16 \cdot 2^{16}$ probabilities, and each probability can be calculated much faster. For window size 2 only $4 \cdot 2^4$ such probabilities are necessary. On the other hand, if the misclassification rates μ_{01}, μ_{10} are moderate those $\boldsymbol{w}^*$ with large Hamming distance to $\boldsymbol{w}_{i(k)}$ give only little contribution since $Prob(\boldsymbol{w}_{i(k)} \text{ observed} \mid \boldsymbol{w}^* \text{ correct })$ is very small. Consequently, the adversary may only consider those $\boldsymbol{w}^*$ in (12) that have Hamming distance 1 or 2 to the observed er-vector, giving an *approximately optimal decision strategy*.

Of course, in 'real-life' attacks the adversary does not know the values μ_{01} and μ_{10}. However, this does not constitute a serious problem. As already pointed out the computational bottleneck is the computation of the probabilities $p_\theta(\boldsymbol{w}^*)$. Once these probabilities have been computed the adversary can experiment with different values $\widetilde{\mu}_{01}$ and $\widetilde{\mu}_{10}$ since only the conditional probabilities $Prob(\boldsymbol{w}_{i(k)}$ is observed $|\ \boldsymbol{w}^*$ correct) (i.e., the right-hand part of (12)) have to re-calculated which is an easy task. (This concerns both (12) and the approximate decision strategy proposed in the previous paragraph.) The efficiency of the attack may serve as a simple quality measure for the suitability of the guessed values $\widetilde{\mu}_{01}$ and $\widetilde{\mu}_{10}$, i.e. whether these values are sufficiently close to μ_{01} and μ_{10}.

Attacks on other modular exponentiation algorithms. The fixed windows exponentiation algorithm specified above can be speeded up somewhat since $\mathrm{MM}(temp, u_{0(k)}; p_1) = temp$ has no computational effect, and also the beginning of the exponentiation phase can be implemented more efficiently. Such a fixed-windows algorithm is referenced as Exponentiation algorithm 2 and investigated in [1]. Also the attacks in [23,21,22] refer to that related fixed windows exponentiation algorithm. Due to the lack of space we do not go into detail but refer the interested reader to [1], p. 14–16. We just mention that the adversary does not even know the positions of the squarings in this case, which makes the attack somewhat more complicated. Interestingly, the maximum-likelihood-estimator is yet not optimal for this exponentiation algorithm. The efficiency of the attack can be increased by the calculus of statistical decision theory (cf. [21], Sect. 2). Moreover, in [1], p. 16–18 we treat the sliding windows exponentiation algorithm, which is implemented in OpenSSL v.0.9.8e and v.0.9.8f, too.

4   Possible Countermeasure Suggestions

There are several ways to achieve a "protected" RSA implementation against our MicroArchitectural Analysis presented in this paper. Instruction Cache Analysis and also Branch Prediction Analysis techniques exploit input-dependent execution flows of cryptosystems. Therefore, the implementations with fixed execution flows are intrinsically secure against these potential threats; hence they constitute a "risk-free" way of protection for these attacks.

[4] proposes a fixed execution flow implementation of RSA, which removes the extra reduction step completely. This method does not only provide an enhanced security, but also improves the performance of the RSA software compared to the current OpenSSL implementation. The idea of avoiding extra reduction steps is not novel to [4] and it was first given in [28,29].

Walter proved in [28] that if the Montgomery parameters were chosen appropriately, we could avoid every extra reduction during an RSA exponentiation. He proved this property by analyzing the upper bounds on the values of intermediate Montgomery Multiplication results throughout the exponentiation. The parameters suggested by Walter necessitate to use more machine-words to hold the operands than necessary. In other words, his method increases the number of words in the operands and sets the Montgomery constant, which is usually denoted by R, accordingly and performs the multiplications in larger sizes.

Later, Hachez & Quisquater improved the results of Walter [13]. They proved the same concept with tighter bounds and thus showed that the increase in the word size could be relatively smaller. Their method improves the speed of the Montgomery Multiplication and therefore the overall exponentiation compared to Walter's method.

Although we think it is a better practice to remove extra reduction step from the RSA computations, which had been implemented in OpenSSL as a response to our results in this paper, we need to mention that these attacks can also be avoided using alternative methods. For example, exponent blinding, which is a well-known countermeasure against side-channel analysis, is among these alternatives. This method adds a random multiple of the group-order, i.e., a random multiple of $\phi(n)$, where n is the public RSA modulus, to the secret exponent. This random multiple needs to be updated frequently, preferably for each RSA operation. The disadvantage of this method is the increased number of Montgomery operations due to the increase in the exponent size. We also need to mention that this method is covered by a patent, c.f. [14].

5   Conclusions

We have identified a potential serious weakness in RSA implementations. We have developed an attack by leveraging MicroArchitectural analysis techniques and adapted early studies on extra-reduction based attacks (in particular, [23]). We have demonstrated that the RSA implementation in the original OpenSSL version 0.9.8e can completely be broken, even if all of the available counter-measures were turned on, and we proposed effective countermeasures. Before we published our attack [1] we shared our results with the OpenSSL development team who included a patch in a stable branch of v.0.9.8.e which also affects the current version v.0.9.8f.

We believe that the gravity of the vulnerability we identified here mandates revisions of the affected RSA implementations, which had already been done in OpenSSL. We have only focused on OpenSSL library in this paper due to its wide acceptance, however, other libraries and RSA implementations may also be vulnerable to our attack. Due to resource and time limitations, we could not conduct a comprehensive investigation on current security systems. We leave this task to other researchers.

Our attack is another example of bringing an initially smart-card oriented side-channel attack into the arena of general purpose computers. [10] demon-strated on real software-based systems the practicality of a pure timing attack introduced in [26]. As a response, base blinding technique was implemented as a default countermeasure in OpenSSL. As in [23] our attack exploits the knowl-edge of er-vectors although the techniques to gather these information are very different (power analysis vs. instruction cache analysis). Due to our attack the extra reduction step in Montgomery Multiplication was removed. Day by day, we see more instances of "thought to be theoretical" or "smart-card only" side-channel attacks coming into practice. One may expect that it is only a matter of time or a matter of finding new ways to gather required side-channel information to make many theoretical side-channel attack a bare reality. To be on the safe side it seems to be beneficial for software developers also to observe new developments in side-channel analysis on smart cards, just to understand the sources of potential attacks. This might help to anticipate threats and to prevent them by effective countermeasures, even if concrete attacks are not known at this time.

References

1. Acıiçmez, O., Schindler, W.: A Major Vulnerability in RSA Implementations due to MicroArchitectural Analysis Threat. Cryptology ePrint Archive, Report 2007/336 (August 2007)
2. Acıiçmez, O.: Yet Another MicroArchitectural Attack: Exploiting I-cache. In: ACM Workshop on Computer Security Architecture, pp. 11–18. ACM Press, New York (2007)
3. Acıiçmez, O., Seifert, J.-P.: Cheap Hardware Parallelism Implies Cheap Security. In: 4th Workshop on Fault Diagnosis and Tolerance in Cryptography — FDTC 2007, pp. 80–91. IEEE Computer Society, Los Alamitos (2007)
4. Acıiçmez, O., Gueron, S., Seifert, J.-P.: New Branch Prediction Vulnerabilities in OpenSSL and Necessary Software Countermeasures. In: Galbraith, S.D. (ed.) Cryptography and Coding 2007. LNCS, vol. 4887, pp. 185–203. Springer, Heidelberg (2007), Cryptology ePrint Archive, Report 2007/039, (February 2007)
5. Acıiçmez, O., Koç, Ç.K., Seifert, J.-P.: On The Power of Simple Branch Prediction Analysis. In: Deng, R., Samarati, P. (eds.) ACM Symposium on InformAtion, Computer and Communications Security (ASIACCS 2007), pp. 312–320 (2006); Cryptology ePrint Archive, Report 2006/351 (October 2006)
6. Acıiçmez, O., Koç, Ç.K., Seifert, J.-P.: Predicting Secret Keys via Branch Prediction. In: Abe, M. (ed.) CT-RSA 2007. LNCS, vol. 4377, pp. 225–242. Springer, Heidelberg (2006), Cryptology ePrint Archive, Report 2006/288, (August 2006)
7. Acıiçmez, O., Schindler, W., Koç, Ç.K.: Cache Based Remote Timing Attack on the AES. In: Abe, M. (ed.) CT-RSA 2007. LNCS, vol. 4377, pp. 271–286. Springer, Heidelberg (2006)
8. Acıiçmez, O., Schindler, W., Koç, Ç.K.: Improving Brumley and Boneh Timing Attack on Unprotected SSL Implementations. In: Meadows, C., Syverson, P. (eds.) Proceedings of the 12th ACM Conference on Computer and Communications Security, pp. 139–146. ACM Press, New York (2005)
9. Bernstein, D. J.: Cache-timing attacks on AES. Technical Report, 37 pages, (April 2005), http://cr.yp.to/antiforgery/cachetiming-20050414.pdf
10. Brumley, D., Boneh, D.: Remote Timing Attacks are Practical. In: Proceedings of the 12th Usenix Security Symposium, pp. 1–14 (2003)
11. Dhem, J.-F., Koeune, F., Leroux, P.-A., Mestré, P.-A., Quisquater, J.-J., Willems, J.-L.: A Practical Implementation of the Timing Attack. In: Schneier, B., Quisquater, J.-J. (eds.) CARDIS 1998. LNCS, vol. 1820, pp. 175–191. Springer, Heidelberg (2000)

12. Gueron, S.: Enhanced Montgomery Multiplication. In: Kaliski Jr., B.S., Koç, Ç.K., Paar, C. (eds.) CHES 2002. LNCS, vol. 2523, pp. 46–56. Springer, Heidelberg (2003)

13. Hachez, G., Quisquater, J.-J.: Montgomery Exponentiation with no Final Subtractions: Improved Results. In: Paar, C., Koç, Ç.K. (eds.) CHES 2000. LNCS, vol. 1965, pp. 91–100. Springer, Heidelberg (2000)

14. Kocher, P.C., Jaffe, J.M.: Secure Modular Exponentiation with Leak Minimization for Smartcards and other Cryptosystems. United States Patent, Patent No.: US 6,298,442 B1 (October 2001)

15. Menezes, A.J., van Oorschot, P.C., Vanstone, S.C.: Handbook of Applied Cryptography. CRC Press, New York (1997)

16. Neve, M.: Cache-based Vulnerabilities and SPAM Analysis. Ph.D. Thesis, Applied Science, UCL (July 2006)

17. Neve, M., Seifert, J.-P.: Advances on Access-driven Cache Attacks on AES. In: Biham, E., Youssef, A.M. (eds.) SAC 2006. LNCS, vol. 4356, pp. 147–162. Springer, Heidelberg (2007)

18. Osvik, D.A., Shamir, A., Tromer, E.: Cache Attacks and Countermeasures: The Case of AES. In: Pointcheval, D. (ed.) CT-RSA 2006. LNCS, vol. 3860, pp. 1–20. Springer, Heidelberg (2006)

19. Page, D.: Theoretical Use of Cache Memory as a Cryptanalytic Side-Channel. Technical Report, Department of Computer Science, University of Bristol (June 2002)

20. Percival, C.: Cache missing for fun and profit. BSDCan 2005, Ottawa (2005), http://www.daemonology.net/hyperthreading-considered-harmful/

21. Schindler, W.: On the Optimization of Side-Channel Attacks by Advanced Stochastic Methods. In: Vaudenay, S. (ed.) PKC 2005. LNCS, vol. 3386, pp. 85–103. Springer, Heidelberg (2005)

22. Schindler, W., Walter, C.D.: More Detail for a Combined Timing and Power Attack against Implementations of RSA. In: Paterson, K.G. (ed.) Cryptography and Coding 2003. LNCS, vol. 2898, pp. 245–263. Springer, Heidelberg (2003)

23. Schindler, W.: A Combined Timing and Power Attack. In: Naccache, D., Paillier, P. (eds.) PKC 2002. LNCS, vol. 2274, pp. 263–279. Springer, Heidelberg (2002)

24. Schindler, W.: Optimized Timing Attacks against Public Key Cryptosystems. Statistics and Decisions 20, 191–210 (2002)

25. Schindler, W., Koeune, F., Quisquater, J.-J.: Improving Divide and Conquer Attacks Against Cryptosystems by Better Error Detection / Correction Strategies. In: Honary, B. (ed.) Cryptography and Coding 2001. LNCS, vol. 2260, pp. 245–267. Springer, Heidelberg (2001)

26. Schindler, W.: A Timing Attack against RSA with the Chinese Remainder Theorem. In: Paar, C., Koç, Ç.K. (eds.) CHES 2000. LNCS, vol. 1965, pp. 110–125. Springer, Heidelberg (2000)

27. Walter, C.D., Thompson, S.: Distinguishing Exponent Digits by Observing Modular Subtractions. In: Naccache, D. (ed.) CT-RSA 2001. LNCS, vol. 2020, pp. 192–207. Springer, Heidelberg (2001)

28. Walter, C.D.: Montgomery exponentiation needs no final subtractions. IEE Electronics Letters 35(21), 1831–1832 (1999)

29. Walter, C.D.: Montgomery's Multiplication Technique: How to Make It Smaller and Faster. In: Koç, Ç.K., Paar, C. (eds.) CHES 1999. LNCS, vol. 1717, pp. 80–93. Springer, Heidelberg (1999)

30. http://www.ntt.co.jp/news/news06e/0611/061108a.html
31. http://cvs.openssl.org/chngview?cn=16275
32. ftp://ftp.openssl.org/snapshot/
33. http://cve.mitre.org/cgi-bin/cvename.cgi?name=CVE-2007-3108
34. http://www.cert.org/
35. US CERT vulnerability note, http://www.kb.cert.org/vuls/id/724968

Fault Analysis Study of IDEA

Christophe Clavier[1], Benedikt Gierlichs[2], and Ingrid Verbauwhede[2]

[1] Gemalto, Security Labs
Avenue du Jujubier, ZI Athélia IV, 13705 La Ciotat Cedex, France
christophe.clavier@gemalto.com
[2] K.U. Leuven, ESAT/SCD-COSIC
Kasteelpark Arenberg 10, B-3001 Leuven-Heverlee, Belgium
firstname.lastname@esat.kuleuven.be

Abstract. We present a study of several fault attacks against the block cipher IDEA. Such a study is particularly interesting because of the target cipher's specific property to employ operations on three different algebraic groups while not using substitution tables. We observe that the attacks perform very different in terms of efficiency. Although requiring a restrictive fault model, the first attack can not reveal a sufficient amount of key material to pose a real threat, while the second attack requires a large number of faults in the same model to achieve this goal. In the general random fault model, *i.e.* we assume that the fault has a random and *a priori* unknown effect on the target value, the third attack, which is the first Differential Fault Analysis of IDEA to the best of our knowledge, recovers 93 out of 128 key bits exploiting about only 10 faults. For this particular attack, we can also relax the assumption of cycle accurate fault injection to a certain extend.

Keywords: Collision Fault Analysis, Ineffective Fault Analysis, Differential Fault Analysis, IDEA, Random Fault Model.

1   Introduction

The International Data Encryption Algorithm (IDEA) is a block cipher with an iterated round function, which encrypts 64-bit blocks of plaintext to 64-bit blocks of ciphertext using a 128-bit key. It was introduced by Xuejia Lai and James Massey in 1991 [9]. Today, this cipher is commonly available in cryptographic software packages such as PGP, SSH, and OpenSSL and is used in (embedded) cryptographic systems deployed by some operators of GSM networks and PayTV systems [13].

IDEA was designed to resist differential cryptanalysis [3,9] in particular, but seems to be an algorithm difficult to cryptanalyze in general, even in reduced rounds versions. A part of IDEA's security is based on the fact that it applies operations on three different algebraic groups. Rather than comprising substitution tables, IDEA uses a multiplication operation to generate confusion. As a result, the progress in finding weaknesses in the cipher is relatively slow. The best known attack against IDEA at this time has been presented in 2007 by Biham, Dunkelman, and Keller [2]. It breaks 6 out of the 8.5 rounds of IDEA with

T. Malkin (Ed.): CT-RSA 2008, LNCS 4964, pp. 274–287, 2008.

$2^{64} - 2^{52}$ pairs of plaintext/ciphertext and a time complexity equivalent to $2^{126.8}$ encryptions.

Since the algorithm is robust against classical cryptanalysis and widely deployed at the same time we think it is an interesting task to study its security under the threat of physical attacks. The fact that there exists only very little literature on this topic, e.g. [10], underlines the need for such a study.

In this work we do not explore all possible paths for physical attacks. While [10] presents some results for differential side channel analysis, we focus on fault analysis. We observe that Collision Fault Analysis (CFA) [7], which requires a restrictive fault model, does not allow to reveal a sufficient amount of key material to pose a real threat. Further, we show that Ineffective Fault Analysis (IFA) [5,6], requiring the same fault model, allows to recover the entire key at the cost of a large number of fault injections. In this light, we propose a Differential Fault Analysis (DFA) [4] which seems to be of particular interest to an attacker, as it allows to retrieve a substantial part of the key from a small number of fault injections in the most general fault model. To the best of our knowledge, this is the first described Differential Fault Analysis of IDEA. The fact that we can also relax the assumption of cycle accurate fault injection to a certain extend renders this attack even more practical.

We describe IDEA in Section 2 and briefly recapitulate some fault analysis techniques in Section 3. The two following sections are respectively dedicated to the application of collision and Ineffective Fault Analysis to IDEA. Thereafter, we propose an efficient Differential Fault Analysis of this algorithm in Section 6 and provide results from extensive simulations of this attack in Section 7. We conclude our work in Section 8.

2 IDEA

IDEA consists of 8 identical rounds and an output transformation (reduced round). Each round is composed of XOR operations, additions modulo 2^{16} denoted by $\boxplus$, and multiplications modulo $(2^{16} + 1)$ with 0 associated to 2^{16} denoted by $\odot$. Round n ($n = 1, \ldots, 8$) involves six 16-bit subkeys Z_i^n ($i = 1, \ldots, 6$). The output

Round	Z_1^n	Z_2^n	Z_3^n	Z_4^n	Z_5^n	Z_6^n
$n = 1$	0-15	16-31	32-47	48-63	64-79	80-95
$n = 2$	96-111	112-127	25-40	41-56	57-72	73-88
$n = 3$	89-104	105-120	121-8	9-24	50-65	66-81
$n = 4$	82-97	98-113	114-1	2-17	18-33	34-49
$n = 5$	75-90	91-106	107-122	123-10	11-26	27-42
$n = 6$	43-58	59-74	100-115	116-3	4-19	20-35
$n = 7$	36-51	52-67	68-83	84-99	125-12	13-28
$n = 8$	29-44	45-60	61-76	77-92	93-108	109-124
$n = 9$	22-37	38-53	54-69	70-85		

Fig. 1. The IDEA key schedule

transformation employs four additional subkeys Z_i^9 $(i = 1 \ldots, 4)$. Each subkey is composed of selected key bits according to IDEA's key schedule depicted in Figure 1.

The input of each round n $(n = 1, \ldots, 9)$ is split into four 16-bit words X_i^n $(i = 1, \ldots, 4)$ which are processed by the IDEA round function according to Figure 2. The output transformation produces the ciphertext, denoted by C_i $(i = 1, \ldots, 4)$.

The decryption and encryption functions are identical except for the subkeys involved. The decryption subkeys are derived from the encryption ones by considering them in reverse order and computing their respective inverses regarding the operations $\boxplus$ and $\odot$.

3   Fault Analysis Techniques and Fault Models

We now briefly recapitulate the three fault analysis techniques our attacks are based on and their respective fault models. The specific application of these techniques to the IDEA algorithm will be described in detail in the next sections. For all our attacks, we consider an unprotected software implementation of IDEA on a 16-bit platform, but the strategies can be easily adapted to 8-bit and 32-bit implementations. We assume that an adversary is able to induce at most one fault per encryption in a cycle accurate manner.

The first two types of fault analysis we consider both infer information about the key from the fact that two observed ciphertexts (resulting from one normal and one faulty execution) are identical. For these two analysis techniques we assume a fault model where an adversary is able to set the output of an arithmetic operation, $\boxplus$ or $\odot$ in the sequel, to a fixed and known (or guessable) value a. The particular value of this constant does not influence the attack and is usually assumed to be equal to zero.

Applying the first method, named Collision Fault Analysis (CFA) [1,7], the adversary first obtains a faulty ciphertext corresponding to some arbitrary plaintext. Then she exhaustively searches for another input whose encryption eventually collides with this output. CFA is so a chosen message fault analysis which is particularly efficient since a single fault injection is usually sufficient to reveal some piece of information about the key.

Nevertheless, it is not always possible to obtain a collision with a faulted ciphertext by searching a message that provokes it. This is particularly the case if the message's influence on the ciphertext extends the single faulted operation. To bypass this problem, it is necessary that the encryptions with and without fault injection operate on pairs of equal messages. The two ciphertexts of the pair will be different except for the rare case where the normal value (without fault) of the faulted instruction's output is precisely equal to the value a imposed by the fault. The obtained collision, or more precisely the identity of the ciphertexts, makes evident that the induced fault had *no effect*. That is, the considered data *a priori* possessed the value which the fault was going to impose. The observation and exploitation of such coincidences is so called Ineffective Fault Analysis

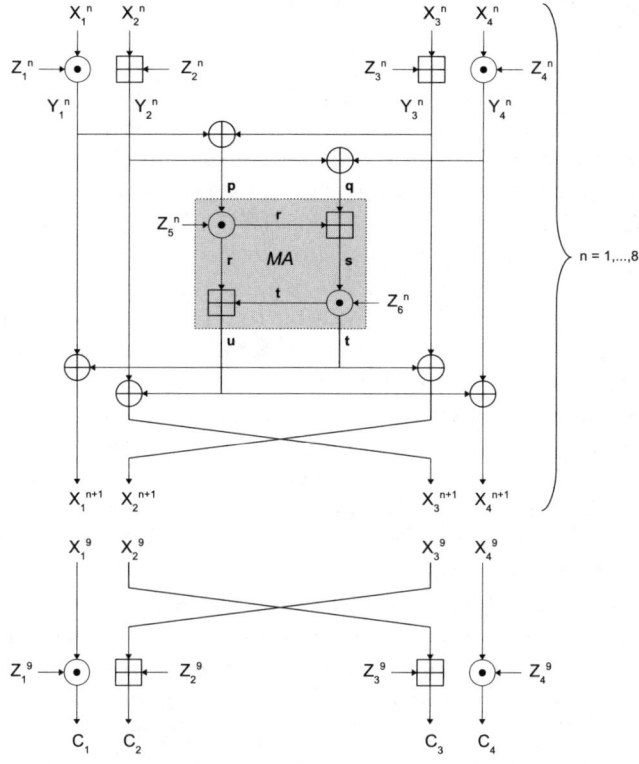

Fig. 2. IDEA round function and output transformation

(IFA) [5,6]. This chosen message technique needs much more fault injections than CFA but allows to target operations deeper in the algorithm. Note that in some sense IFA may also be seen as a particular case of Safe Error Analysis (SEA) techniques [8,11,12].

Our third attack is based on the classical Differential Fault Analysis (DFA) first introduced by Biham and Shamir [4] who applied it to the DES block cipher. DFA is a general known plaintext technique which exploits the difference between a normal and a faulty ciphertext stemming from encryptions of the same message. As for the attack described in [4], we analyze a fault occurring near the end of the algorithm and assume the general random fault model where the fault modifies the processed data in a random way. The adversary does not need to know *a priori* the random value the fault imposed on the data, and, as will be shown later, the assumption of cycle accurate fault induction can be relaxed to a certain extend. Furthermore, our attack needs only very few fault injections, which makes it both practical and very efficient.

4   Collision Fault Analysis of IDEA

Assuming a fault model where an adversary is able to set the output of the $\boxplus$ operation to a fixed and known value a, CFA allows to recover the subkeys Z_2^1 and Z_3^1 at the cost of only two fault injections and on average 2×2^{15} encryptions. Given a faulty ciphertext where the fault occurred on the $X_2^1 \boxplus Z_2^1$ operation, the adversary just has to exhaustively search for that value of X_2^1 which leads to the same output. Then she knows that $X_2^1 \boxplus Z_2^1 = Y_2^1 = a$ and infers the subkey value $Z_2^1 = a \boxminus X_2^1$ (where $\boxminus$ denotes subtraction modulo 2^{16}). Faulting on the $X_3^1 \boxplus Z_3^1$ operation will reveal Z_3^1 in the same manner.

Following the same principle, it is possible to retrieve Z_1^1 and Z_4^1 assuming that the same fault model also holds for the $\odot$ operation. We note that for a general purpose microprocessor it is doubtable whether it is possible to force the output of the multiplication to a fixed and known value in the same manner as for the addition operation. While the addition is probably implemented using a single instruction (at least on 16 or 32 bit systems) which can be faulted relatively easy, the multiplication is necessarily composed of several instructions. However, one could assume that the fault model allows to simply set the output of the last instruction in the sequence to a.

In the best case, an adversary might thus recover 64 key bits using this strategy. We do not see any opportunity to recover more key bits applying CFA[1]. Suppose for a moment that the adversary was able to retrieve Z_i^1 ($i = 1, \ldots, 4$). In this case, it is possible to determine the inputs p and q of the first round's multiplication-addition (MA) layer. However, a fault injection on any operation of the MA layer does not allow to find another message that provokes a collision in the ciphertext. For that, it would be necessary that the part of the message which the adversary varies in order to find the collision (for example X_1^1 or X_3^1 if the fault is induced on $p \odot Z_5^1$) only influences the result of the faulted operation (here $p \odot Z_5^1$). Clearly, this condition is not met here because the message entering the first round (and thus the MA layer) is involved again at the end of the same round when it is XOR-ed with the output of the MA layer. Therefore, it is impossible to find a collision that allows to derive Z_5^1 and Z_6^1 by means of CFA.

In the next section, we discuss how this attack can be extended in order to extract supplementary key material using a more elaborate method.

5   Ineffective Fault Analysis of IDEA

We consider the case where it is possible to set the output of the addition and multiplication operations to a known and constant value a. The 64 bits of the subkeys $Z_1^1, \ldots, Z_4^1$ may have been previously determined by CFA such that the adversary is capable to calculate and control $Y_1^1, \ldots, Y_4^1$ as well as p and q

---

[1] Except, however, for an adversary able to also decrypt chosen ciphertexts. Such an adversary can retrieve both Z_1^1 to Z_4^1 and Z_1^9 to Z_4^9 which amounts to 86 out of the 128 key bits.

entering the first round's *MA* layer. Successively encrypting pairs of messages (normal/faulted) where the fault is induced on the multiplication $(p \odot Z_5^1)$, it is possible to determine the value of $(p \odot Z_5^1)$ as soon as an identity[2] of ciphertexts is observed. Such a *winning pair* allows to reveal the value of Z_5^1 and is obtained after encrypting 2^{15} pairs on average since the adversary is able to control the multiplication's input p.

Knowing Z_5^1, it is now possible to control the input s of the operation $(s \odot Z_6^1)$, and to determine Z_6^1 in the same manner. 96 key bits are hence known and the remaining 32 bits can be found by means of exhaustive search. The attack requires four faults for the CFA phase and on average 2×2^{15} additional faults for the IFA phase.

6   Differential Fault Analysis of IDEA

Our attack applies to the last 1.5 rounds of IDEA and recovers 93 out of 128 key bits. The remaining 35 key bits may be recovered by exhaustive search. Most symbols used in the description refer to the notation introduced in Section 2. However, since the attack involves the output transformation, we depict the construction considered in Figure 3. Note that the figure takes into account the double permutation of the branches two and three at the end of round eight and the beginning of round nine, leading to a permutation of X_3^9 and X_2^9. The attack is split into three successive phases.

6.1   Phase 1: Finding the Subkeys of the Output Transformation

The first phase of the attack aims at recovering the subkeys $Z_1^9, \ldots, Z_4^9$ which are used in the output transformation (cf. Figure 3). Taking into account that our attack does not allow to recover the most significant bits of Z_2^9 and Z_3^9, we thus reveal 62 key bits.

The fault injection targets at the last round's *MA* layer, highlighted by a gray box in Figure 3. A fault may be induced on either p, q, r, or s whereas the approach is identical in either case. Note that any modification of one of these values provokes *a priori* a XOR-difference Δ_u on u and Δ_t on t. These alter the reference ciphertext (C_1, C_2, C_3, C_4) into the faulty ciphertext $(C_1^*, C_2^*, C_3^*, C_4^*)$. One can observe that

$$X_1^9 \oplus X_1^{9*} = X_2^9 \oplus X_2^{9*} = \Delta_t \tag{1}$$

holds, since $Y_1^8 = Y_1^{8*}$ and $Y_3^8 = Y_3^{8*}$, *i.e.* Y_1^8 and Y_3^8 are not affected by the fault. Similarly, also

$$X_3^9 \oplus X_3^{9*} = X_4^9 \oplus X_4^{9*} = \Delta_u \tag{2}$$

holds, because Y_2^8 and Y_4^8 are not affected by the fault.

The relation in Equation 1 in conjunction with a pair of reference and faulty ciphertexts allows to infer a relation between Z_1^9 and Z_3^9 which can be exploited

---

[2] We consciously avoid the term *collision* since the algorithm's inputs are the same.

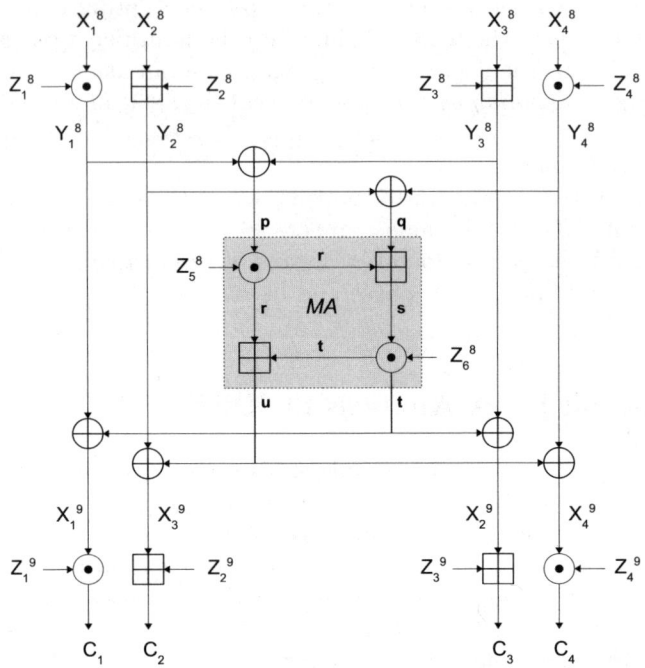

Fig. 3. Last 1.5 rounds of IDEA

to restrict a list of possible candidates for the value of the couple (Z_1^9, Z_3^9). Given the pair (C_1, C_1^*) an adversary is able to derive

$$\Delta_t = (C_1 \odot (Z_1^9)^{-1}) \oplus (C_1^* \odot (Z_1^9)^{-1})$$

for any arbitrary guess on the value of Z_1^9. Using the pair (C_3, C_3^*) she can generate a list of candidate values for Z_3^9 which are compatible with this Δ_t. More precisely, the candidate values for Z_3^9 need to satisfy

$$(C_3 \boxminus Z_3^9) \oplus (C_3^* \boxminus Z_3^9) = \Delta_t . \tag{3}$$

This procedure leads to a list of tuples (Z_1^9, Z_3^9) which are compatible under the observed fault injection.

Applying the above procedure to multiple pairs of reference and faulty ciphertexts obtained from independently performed fault injections, we can progressively further restrict this list of possible candidates for the couple more and more since the correct candidate necessarily appears in the intersection of the lists deduced from each pair of ciphertexts. With only a few faults (cf. Section 7) it is hence possible to definitely identify the value of Z_1^9 and to determine Z_3^9 except for its most significant bit. This bit can not be identified because if Z_3^9 verifies Equation 3 then this also holds for $Z_3^9 \oplus 0x8000$. This simply comes from the fact that adding (or subtracting) $0x8000$ modulo 2^{16} or XOR-ing with $0x8000$ are just equivalent operations.

The same approach can be mounted to derive the values of Z_2^9 and Z_4^9 using the relation in Equation 2. Given the pair (C_4, C_4^*) the adversary derives

$$\Delta_u = (C_4 \odot (Z_4^9)^{-1}) \oplus (C_4^* \odot (Z_4^9)^{-1})$$

for any arbitrary guess on Z_4^9. Then she uses

$$(C_2 \boxminus Z_2^9) \oplus (C_2^* \boxminus Z_2^9) = \Delta_u \tag{4}$$

to build the list of candidates for the tuple (Z_2^9, Z_4^9) compatible with the observed fault. For the same reason as mentioned above, building the intersection of several lists deduced from multiple pairs of ciphertexts allows to definitely identify Z_4^9 and to determine Z_2^9 except for its most significant bit.

Note that each fault injection observed during this phase can be used independently to generate a list of candidates for the tuple (Z_1^9, Z_3^9) as well as for the tuple (Z_2^9, Z_4^9). This fact makes phase one very efficient in terms of recovered bits per fault injection. At the end of this phase we have derived four candidates for the value of $(Z_1^9, Z_2^9, Z_3^9, Z_4^9)$ which corresponds to knowing the key bits 22 to 85 except for bits 38 and 54.

6.2   Phase 2: Finding the Subkey Z_6^8

The second phase of the attack aims at recovering the 16 key bits $109 - 124$ which are used as subkey Z_6^8 in the penultimate round (cf. Table 1).

A fault may be induced on Y_2^8, or any other preceding value as long as Y_1^8, Y_3^8, and Y_4^8 remain unmodified, leading to the XOR-difference Δ_q on q (as well as the XOR-differences Δ_t and Δ_u). Note that the attack also works equivalently when faulting on Y_4^8 or related preceding values as stated above for Y_2^8. Given a pair of reference and faulty ciphertexts and the subkey candidates revealed in the previous phase, an adversary can compute Δ_t and Δ_u as defined in Equations (3,4). Note that the existence of two candidates for Z_2^9 does not matter here, since both candidates lead to the same Δ_u and Δ_t. Knowing these values allows to calculate Δ_q being the XOR-difference $(Y_2^8 \oplus Y_2^{8*})$ introduced by the fault. The potential to calculate Δ_q is interesting in two ways. It will not only enable the success of this phase but can also be useful in the more general setting of device characterization as it allows to precisely determine the effect of the fault.

The adversary generates the list TR of all tuples (t, r) that are compatible with the observed Δ_t and Δ_u. A compatible tuple (t, r) verifies

$$(r \boxplus t) \oplus (r \boxplus (t \oplus \Delta_t)) = \Delta_u .$$

For any arbitrary guess on Z_6^8 the adversary checks whether there exists at least one tuple (t, r), such that

$$(s \boxminus r) \oplus (s^* \boxminus r) = \Delta_q$$

with $s = t \odot (Z_6^8)^{-1}$ and $s^* = (t \oplus \Delta_t) \odot (Z_6^8)^{-1}$. If no such (t, r) exists, the guessed Z_6^8 is discarded from the list of subkey candidates.

Applying the above procedure to multiple pairs of reference and faulty ciphertexts obtained from independently performed fault injections, the list of remaining candidates for Z_6^8 reduces rapidly when building the intersection. Finally, the adversary reveals the correct subkey value.

6.3 Phase 3: Finding the Subkey Z_5^8

The third phase of the attack aims at recovering the 16 key bits $93 - 108$ which are used as subkey Z_5^8 in the penultimate round (cf. Table 1).

A fault may be induced on Y_1^8 leading to the XOR-difference Δ_p on the input p to the MA layer (as well as the XOR-differences Δ_t and Δ_u on its outputs). As for phase 2, any timing location of the fault which results in a modification of either Y_1^8 or Y_3^8 would be equivalently exploitable. Given a pair of reference and faulty ciphertexts and the previously derived subkeys $Z_1^9, \ldots, Z_4^9$, and Z_6^9 an adversary can compute Δ_t, Δ_u, and Δ_p as described for the second phase. We exhaustively consider all possible values for the tuple (t, r). For each of them, we compute $t^* = t \oplus \Delta_t$, $r^* = ((r \boxplus t) \oplus \Delta_u) \boxminus t^*$, $s = t \odot (Z_6^8)^{-1}$, and $s^* = t^* \odot (Z_6^8)^{-1}$. We discard a candidate for (t, r) if it does not verify $\Delta_q = 0$, *i.e.* a valid candidate for the tuple must satisfy

$$s \boxminus r = s^* \boxminus r^*.$$

For all values Z_5^8 we check whether at least one of the remaining candidates for (t, r) satisfies

$$(r \odot (Z_5^8)^{-1}) \oplus (r^* \odot (Z_5^8)^{-1}) = \Delta_p.$$

If this is not the case, we discard this candidate for Z_5^8.

As before, applying the above procedure to multiple pairs of reference and faulty ciphertexts obtained from independently performed fault injections progressively eliminates wrong candidate values.

While simulating this phase of the attack, we observed that a very small number of faults (approximately five) always suffices to reduce the list of possible values for Z_5^8 to only two candidates. Contrarily to the phenomenon regarding the most significant bits of Z_2^9 and Z_3^9 pointed out in the first phase, the ambiguity between these two remaining values is not absolute. In our experiments we observed that it might be removed, but it seems to be difficult and often happens only after several tens of faults[3].

To simplify our analysis, we hence consider the third phase a success a soon as at most two candidates for Z_5^8 remain. Consequently, we stipulate that this phase reveals only 15 additional key bits.

---

[3] Without having found a satisfactory explication for this phenomenon, we have however observed that for each value of Z_5^8 there exists a companion $\widetilde{Z_5^8}$ having the property that the values $\Delta_r = (p \odot Z_5^8) \oplus (p^* \odot Z_5^8)$ and $\widetilde{\Delta_r} = (p \odot \widetilde{Z_5^8}) \oplus (p^* \odot \widetilde{Z_5^8})$, if p and p^* are chosen randomly, are astonishingly close in terms of Hamming Distance and often equal. This observation is worthy of a deeper analysis and it would be interesting to study if it could open new paths for the cryptanalysis of IDEA.

6.4   Summary

Put together, the three phases of our attack allow to recover 93 key bits. The following section presents simulation results which validate our approach and give a precise estimation about how much fault injections each phase requires.

Remark 1. A trick allows to reduce the required number of faults. Indeed, one may notice that faults injected on Y_2^8 which are used for the second phase may also be partly useful for the first phase. This is because any change of Y_2^8 leads to a differential Δ_t which may be exploitable in phase 1 as neither Y_1^8 nor Y_3^8 is modified. In the same way, a fault modifying Y_1^8 in phase 3 is usable in phase 1 by exploiting the differential Δ_u. As a result, injecting n faults for each one of phase 2 and phase 3 allows to save n faults for the first phase.

7   Simulation Results

In this section we evaluate the efficiency of all three phases of our attack by means of simulation. The results we present here are based on about $10\,000$ simulations for each of the two first phases and on about $3\,000$ simulations for the third phase, which requires a longer calculation time.

  We use two metrics to judge the efficiency of each phase of the attack. The first is the *mean residual entropy* of the subkey(s) considered after having exploited m faults. It is defined as

$$h_m(Z_i^n) = \langle \log_2(\#S_m) \rangle$$

where Z_i^n denotes the subkey(s) concerned, $\langle \cdot \rangle$ is the expectation operator which we evaluate by means of the empirical mean, and $\#S_m$ denotes the size of the set of remaining candidates for Z_i^n after m faults. An attack aims at reducing this uncertainty.

  The second is a kind of *success probability*, defined as the probability

$$p(\#S_m \leqslant d)$$

that the number of remaining candidates is less or equal than d after having exploited m faults. An attack aims at increasing this probability.

  Note that both metrics are sound since the filtering processes can not exclude the correct candidate(s) from the list of possible values. Hence, both $h_m(Z_i^n) = 0$ and $\#S_m \leqslant 1$ are equivalent to exactly identifying the correct subkey value.

7.1   Phase 1

Here the fault may occur on either p, q ,r, or s, *i.e.* the fault location is flexible to a certain extend. We present detailed results only for the Z_1^9 and Z_3^9 branch since the results for Z_2^9 and Z_4^9 are virtually identical. Further, we also present results for the entire set Z_{1-4}^9 as this is of most interest. Figure 4 shows the mean residual entropies $h_m(Z_1^9)$, $h_m(Z_3^9)$, $h_m((Z_1^9, Z_3^9))$, and $h_m(Z_{1-4}^9)$ after faulting

$m = 1, 2, \ldots, 10$ times, where each time a random message has been used. One can observe that, as previously argued, revealing the value of Z_3^9 is harder than finding the value of Z_1^9. Therefore, the entropy of the tuple (Z_1^9, Z_3^9) is dominated by the entropy of Z_3^9. Further, $h(Z_{1-4}^9)$ is exactly the double of $h(Z_1^9, Z_3^9)$ which is obviously due to the mutual independence of the two branches. Note that a small number of faults suffices to rapidly approach the residual entropy's intrinsic limit of two bits (the two most significant bits of Z_2^9 and Z_3^9). After only five faults, the initial entropy of 64 bits is reduced to 2.38.

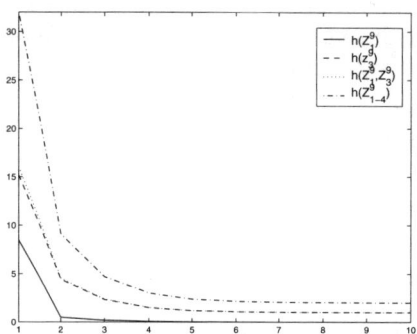

Fig. 4. Mean residual entropy against number of fault injections

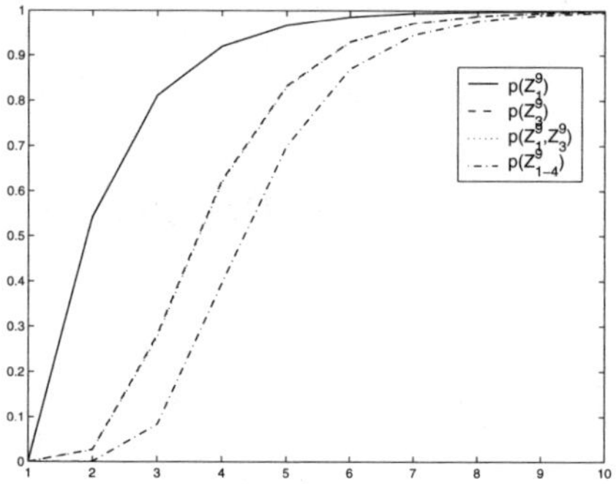

Fig. 5. Probability of $\#S_m \leqslant d$ for Z_i^9 against number of fault injections

Figure 5 depicts the success probabilities $p(\#S_m \leqslant 1)$ for Z_1^9, $p(\#S_m \leqslant 2)$ for Z_3^9, $p(\#S_m \leqslant 2)$ for (Z_1^9, Z_3^9), and $p(\#S_m \leqslant 4)$ for (Z_{1-4}^9) after faulting $m = 1, 2, \ldots, 10$ times. After only five faults these probabilities are 97%, 83%, 83%, and 70% respectively. Still exploiting only five faults, the probability $p(\#S_m \leqslant 8)$ for Z_{1-4}^9 is 93%.

Another interesting question is whether one of the fault locations, *i.e.* p, q, r, and s, is particularly efficient and hence preferable for an attack. Our simulations show that this is not the case. The results for fixed fault locations differ only at a negligible order of magnitude and can clearly be considered noise that will vanish for a huge number of simulations.

We also tested whether using a fixed message for the consecutive fault injections during one simulation could be advantageous. We observed that this is not the case. For 2 000 simulations with a fixed message and 2 000 simulations with a randomly chosen message we obtain, for the set of subkeys Z_{1-4}^9, an entropy of 6.44 bits in the case of a fixed message and only 4.72 bits in the case of a randomly chosen message after exploiting three faults.

7.2 Phase 2

For the second phase, just like for the first one, a small number of fault injections suffices to rapidly reduce the number of remaining candidates for subkey Z_6^8.

Figure 6 shows the mean residual entropy $h_m(Z_6^8)$ after faulting $m = 1, 2, \ldots, 10$ times, where each time a random message has been used. One can observe that after only five faults there remain less than four candidates in the average case. After eight faults, the mean residual entropy is no more than 0.41 bits and the correct subkey is unambiguously identified in 62% of the cases.

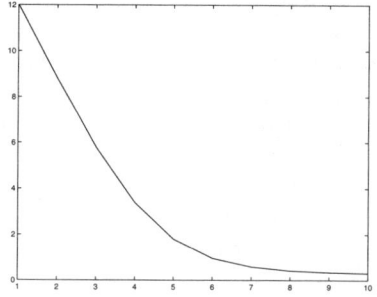 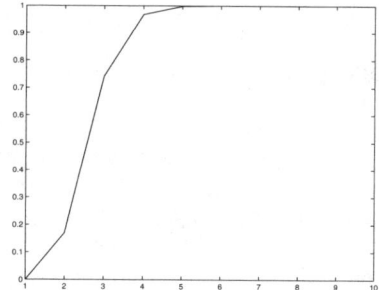

Fig. 6. Mean residual entropy of Z_6^8 against number of fault injections

Fig. 7. Probability of $\#S_m \leqslant 1$ for Z_6^8 against number of fault injections

7.3   Phase 3

The third phase of our attack is particularly efficient. As mentioned in the description of the attack, we consider this phase a success as soon as the number of remaining candidates for subkey Z_5^8 is at most two. As can be seen in Figure 9 only four faults suffice to obtain an even better result in the average case, *i.e.* the mean residual entropy is less than one bit. Exploiting only three faults, the probability that the number of remaining candidates for Z_5^8 is at most two is greater than 74%.

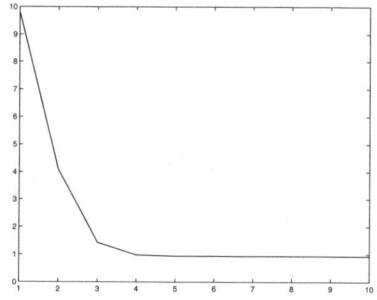 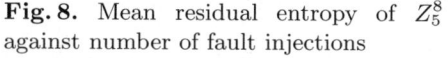

Fig. 8. Mean residual entropy of Z_5^8 against number of fault injections

Fig. 9. Probability of $\#S_m <= 2$ for Z_5^8 against number of fault injections

8   Conclusion

In this work we have studied the applicability and efficiency of several fault attacks against the IDEA block cipher. We conclude that CFA can only infer 64 key bits which renders the necessary exhaustive key search for the remaining 64 key bits practically infeasible. On the other hand, IFA allows to retrieve the entire key in the same restrictive fault model but requires a large number of fault injections. For these practical reasons, the third attack presented is of particular interest. Our DFA which is to the best of our knowledge the first Differential Fault Analysis of IDEA requires no hypothesis on the effect of the fault and efficiently extracts 93 out of 128 key bits exploiting about only 10 faults. Further, we can also relax the assumption of cycle accurate fault injection to a certain extend which makes the attack even more practical.

IDEA's design makes it by default more resistant to CFA/IFA than other ciphers (e.g. based on SP networks like AES). However, as demonstrated in this paper, implementations of IDEA clearly need to be protected against fault attacks.

Acknowledgments

The work described in this document has been partly financially supported by the European Commission through the IST Program under Contract IST-2002-507932 ECRYPT.

It was also supported in part by the IAP Programme P6/26 BCRYPT of the Belgian State (Belgian Science Policy), by FWO projects EMA G.0475.05 and BBC G.0300.07, by the European Commission FP6 MIRG project SESOC, number MIRG-CT-2004-516568, and by the K.U.Leuven-BOF (OT/06/40).

The information in this document reflects only the author's views, is provided as is and no guarantee or warranty is given that the information is fit for any particular purpose. The user thereof uses the information at its sole risk and liability.

References

1. Amiel, F., Clavier, C., Tunstall, M.: Fault Analysis of DPA-Resistant Algorithms. In: Breveglieri, L., Koren, I., Naccache, D., Seifert, J.-P. (eds.) FDTC 2006. LNCS, vol. 4236, pp. 223–236. Springer, Heidelberg (2006)
2. Biham, E., Dunkelman, O., Keller, N.: A New Attack on 6-Round IDEA. In: Biryukov, A. (ed.) FSE 2007. LNCS, vol. 4593, pp. 211–224. Springer, Heidelberg (2007)
3. Biham, E., Shamir, A.: Differential Cryptanalysis of DES-Like Cryptosystems. In: Menezes, A., Vanstone, S.A. (eds.) CRYPTO 1990. LNCS, vol. 537, pp. 2–21. Springer, Heidelberg (1991)
4. Biham, E., Shamir, A.: Differential Fault Analysis of Secret Key Cryptosystems. In: Kaliski Jr., B.S. (ed.) CRYPTO 1997. LNCS, vol. 1294, pp. 513–525. Springer, Heidelberg (1997)

5. Blömer, J., Seifert, J.-P.: Fault Based Cryptanalysis of the Advanced Encryption Standard (AES). In: Wright, R.N. (ed.) FC 2003. LNCS, vol. 2742, pp. 162–181. Springer, Heidelberg (2003)
6. Clavier, C.: Secret External Encodings Do not Prevent Transient Fault Analysis. In: Paillier, P., Verbauwhede, I. (eds.) CHES 2007. LNCS, vol. 4727, pp. 181–194. Springer, Heidelberg (2007)
7. Hemme, L.: A Differential Fault Attack Against Early Rounds of (Triple-)DES. In: Joye, M., Quisquater, J.-J. (eds.) CHES 2004. LNCS, vol. 3156, pp. 254–267. Springer, Heidelberg (2004)
8. Joye, M., Quisquater, J.-J., Yen, S.-M., Yung, M.: Observability Analysis – Detecting When Improved Cryptosystems Fail. In: Preneel, B. (ed.) CT-RSA 2002. LNCS, vol. 2271, pp. 17–29. Springer, Heidelberg (2002)
9. Lai, X., Massey, J.L.: Markov Ciphers and Differentail Cryptanalysis. In: Davies, D.W. (ed.) EUROCRYPT 1991. LNCS, vol. 547, pp. 17–38. Springer, Heidelberg (1991)
10. Lemke, K., Schramm, K., Paar, C.: DPA on n-Bit Sized Boolean and Arithmetic Operations and Its Application to IDEA, RC6, and the HMAC-Construction. In: Joye, M., Quisquater, J.-J. (eds.) CHES 2004. LNCS, vol. 3156, pp. 205–219. Springer, Heidelberg (2004)
11. Yen, S.-M., Joye, M.: Checking Before Output Not Be Enough Against Fault-Based Cryptanalysis. IEEE Transactions on Computers 49(9), 967–970 (2000)
12. Yen, S.-M., Kim, S., Lim, S., Moon, S.-J.: A Countermeasure against One Physical Cryptanalysis May Benefit Another Attack. In: Kim, K.-c. (ed.) ICISC 2001. LNCS, vol. 2288, pp. 414–427. Springer, Heidelberg (2002)
13. (September 17, 2007), http://www.mediacrypt.com/

Susceptibility of UHF RFID Tags to Electromagnetic Analysis*

Thomas Plos

Institute for Applied Information Processing and Communications (IAIK)
Graz University of Technology, Inffeldgasse 16a, 8010 Graz, Austria
Thomas.Plos@iaik.tugraz.at

Abstract. The number of applications that use radio-frequency identification (RFID) technology has grown continually in the last few years. Current RFID tags are mainly used for identification purposes and do not include crypto functionality. Therefore, classical RFID tags are not designed as secure devices and do not contain countermeasures against side-channel analysis (SCA). The lack of such countermeasures makes RFID tags vulnerable to attacks relying on electromagnetic (EM) analysis. When attaching crypto functionality to future RFID tags which is considered for many use cases like forgery protection of goods, SCA becomes a concern. In this work we show the susceptibility of UHF RFID tags to EM analysis by using differential-EM analysis attacks. We have examined commercially-available passive UHF RFID tags with a microchip. The results show that a simple dipole antenna and a digital-storage oscilloscope connected to a computer are enough to determine data-dependent emanation of the microchip of passive UHF RFID tags at distances up to 1 m. Enhancement of RFID tags with crypto functionality therefore requires re-design of the whole tag architecture with respect to SCA.

Keywords: Side-channel analysis (SCA), radio-frequency identification (RFID), EPC Generation 2 standard, ultra-high frequency (UHF), differential electromagnetic analysis (DEMA).

1 Introduction

During the last few years the application of radio-frequency identification (RFID) technology has become more and more important. Ticketing, electronic passports, immobilizers, and supply-chain management are only an outline of a long list of applications that already use RFID systems. The integration of RFID technology can make applications more convenient, more effective, and secure.

The main components of a basic RFID system are an RFID reader that is connected to a back-end database and at least one RFID tag. RFID reader and

---

* This work has been supported by the European Commission under the Sixth Framework Programme (Project BRIDGE, Contract Number IST-FP6-033546) and by the Austrian Science Fund (FWF Project Number P18321).

T. Malkin (Ed.): CT-RSA 2008, LNCS 4964, pp. 288–300, 2008.

RFID tag communicate wirelessly by using a radio frequency (RF) field. The RF field is generated by the RFID reader via an antenna and modulated according to the data that should be transmitted to the RFID tag. The RFID tag itself is equipped with an antenna which it uses to extract data and energy from the RF field. Three types of RFID tags can be distinguished: passive RFID tags, semi-passive RFID tags, and active RFID tags. Passive RFID tags are the most prevalent obtaining their power supply directly from the RF field, semi-passive tags and active tags are supplied by a battery. A typical RFID tag consists of an antenna and a microchip. The microchip contains an analog part and a digital part, whereas the analog part of an RFID tag is responsible for demodulating the RF field and modulating the response of the RFID tag. In addition, the analog part of passive RFID tags is also responsible for extracting the power supply from the RF field. The digital part is more or less complex depending on the application. For security-enhanced applications like contactless smart cards, the digital part of an RFID tag contains a microcontroller with non-volatile memory, less sophisticated applications may only use a state machine with read-only memory.

RFID systems can be classified by the frequency of the RF field and the coupling method. The frequencies used by RFID systems range from about 125 kHz in the low-frequency range up to 5.8 GHz in the microwave range [1]. Deployed coupling methods are: electric coupling, magnetic coupling, and electromagnetic coupling. This work focuses on electromagnetic-coupled systems in the UHF range operating at a frequency of 868 MHz. In contrast to electric coupling and magnetic coupling which operate in the near field, electromagnetic coupling operates in the far field by using electromagnetic waves.

Responsible for the existence of electromagnetic waves is the limited propagation speed of the electromagnetic field. At a certain distance from the antenna the electromagnetic field can no longer follow the voltage changes at the antenna. The electromagnetic field separates from the antenna and propagates as an electromagnetic wave. The region where the electromagnetic field is separated from the antenna is named far field [1]. For UHF RFID tags operating at a frequency of 868 MHz, the far field starts at a distance of about 5.5 cm from the antenna. The simplest antenna shape that is used for generating electromagnetic waves is the dipole antenna which consists of two wires. Since the attenuation of the RF field in the far field is less than in the near field, electromagnetic-coupled systems achieve longer read ranges. Typically, read ranges of 2 to 3 m and more can be achieved, depending on the power of the RFID reader.

An important protocol for electromagnetic-coupled RFID systems in the UHF range is the Electronic Product Code (EPC) Generation 2 standard [2]. The EPC Generation 2 standard is planned to be the future replacement for conventional bar codes. The vision of the inventors of the EPC Generation 2 standard is to attach an RFID tag to each individual product. For now, RFID tags are still too expensive to place them on each individual product, rather they are placed on groups of products like pallets. Equipping pallets with RFID tags allows to increase the efficiency and to reduce costs in supply-chain management.

The driving force behind the introduction of the EPC Generation 2 standard is EPCglobal which is a not-for-profit organization that has been founded by GS1 in 2003. GS1 has emerged from Uniform Code Council (UCC) and European Article Number (EAN) International which are the two organizations that are responsible for managing the bar code systems. Large distributors such as Wal-Mart, Tesco, and Metro have already integrated RFID technology that uses the EPC Generation 2 standard into their supply-chain management [3].

Usually, RFID tags have to be fairly cheap and therefore can only integrate limited functionality which strongly affects the utilized protocol. Thus, protocols like the EPC Geneneration 2 standard neglect to include cryptographic security. The lack of cryptographic security makes the EPC Geneneration 2 standard vulnerable to various attacks such as cloning or revealing secrets like the kill password. However, when using EPC Generation 2 tags to prevent valuable goods from forgery, a higher tag price is acceptable. Pharmacy for example is a use case where valuable goods are involved. Another important aspect is the technological progress that allows to integrate more and more functionality to future RFID tags. There exist various proposals that deal with enhancing the security of RFID protocols which furthermore enforce to include crypto functionality to RFID tags (see [4, 5, 6, 7]). As soon as RFID tags contain crypto functionality, vulnerability against side-channel analysis becomes a concern.

This work is organized as follows. Section 2 provides an overview of the related work with respect to power analysis and electromagnetic analysis. In Section 3, the UHF RFID tags that have been examined in this work are described, followed by a description of the measurement setup in Section 4. Section 5 presents the results of the side-channel analysis that have been conducted. The conclusion of this work is given in Section 6.

2   Related Work

With the introduction of power analysis by Kocher *et al.* in 1998, a wide field for new and effective side-channel attacks has been opened [8]. Power analysis makes use of the fact that the power consumption of CMOS devices is dependent on the data and instructions that are executed. Measuring the power consumption of a CMOS device and deploying statistical methods allows to reveal secrets from cryptographic devices like smart cards. Some years later, the EM radiation of CMOS devices has also been found useful for side-channel attacks. In [9], Gandolfi *et al.* describe the practical implementation of EM attacks and furthermore compare them with conventional power analysis. Thereby, the authors come to the conclusion that EM measurements, although they are noisier, lead to better differentials than power measurements. As explained by Mangard [10], EM attacks are not limited to the near field, they can also be successful in the far field.

Hutter *et al.* [11] describe how to use EM measurements to attack passive RFID devices which are running at 13.56 MHz. Two RFID prototype devices with a cryptographic primitive implemented on them, one in software and one in hardware, are attacked by applying power measurements and EM measurements. In both cases the attacks were successful. A focus on UHF RFID devices and EM measurements is given by the work of Oren and Shamir [12]. There, the authors describe a new attack called *parasitic backscatter attack*. This attack is possible since the amount of power that is reflected by UHF RFID tags is related to the power consumption of its internal circuit. Furthermore, the authors explain how the *parasitic backscatter attack* can be used to extract the secret kill password from EPC Generation 1 tags. Relying on the results in [12], our work goes a step further and focuses on determining the susceptibility of EPC Generation 2 tags to differential electromagnetic analysis (DEMA).

3   Examined UHF RFID Tags

For analyzing the side-channel leakage, two different types of UHF RFID tags have been used. Firstly a self-made prototype of a UHF RFID tag that operates semi passively, and secondly commercially-available UHF RFID tags that operate passively. The self-made prototype which has initially been built to evaluate current UHF RFID protocols has also been found useful for providing the trigger signal when performing measurements on passive UHF RFID tags.

3.1   Description of the UHF Tag Prototype

The first EM measurements presented in this work have been done by using a self-made UHF tag prototype. When evaluating and enhancing the security of current UHF RFID protocols, it is helpful to have a programmable UHF RFID tag. A programmable UHF RFID tag can be used to easily integrate additional functionality such as new security mechanisms and new commands. Furthermore, the additional functionality can be verified and tested, showing a proof of concept. Standard UHF RFID tags do not provide the possibility to integrate additional functionality because their functionality is implemented in silicon.

Unlike most UHF RFID tags, the UHF tag prototype operates semi passively. Like a passive RFID tag, a semi-passive RFID tag only uses the RF field of the reader for communication, but uses an extra battery for power supply like an active RFID tag. Our UHF tag prototype is a printed circuit board (PCB) with discrete components. As shown in Figure 1, the UHF tag prototype can be divided into four parts: an antenna, an analog front end, a digital part, and a protocol implementation.

Antenna. For the UHF tag prototype a simple dipole antenna has been selected which consists of two wires [1]. The dipole antenna of the UHF tag prototype is directly printed on the layout of the PCB. With the help of the antenna, energy is extracted from the RF field. The voltage induced in the antenna is furthermore fed to the input of the analog front end.

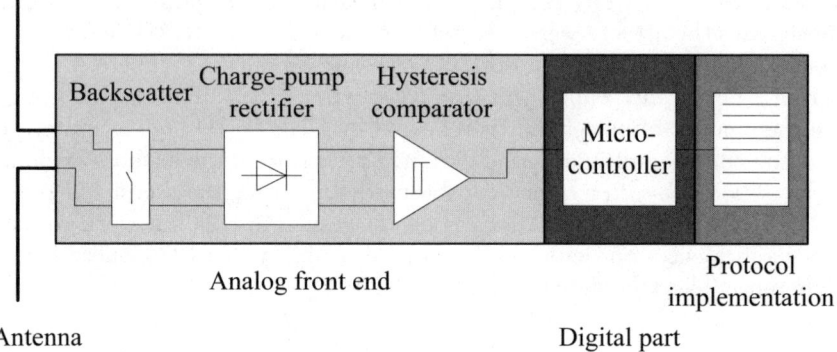

Fig. 1. Architecture of the UHF tag prototype

Analog Front End. The analog front end of the UHF tag prototype contains a charge-pump rectifier, a hysteresis comparator, and a backscatter. Thereby, the charge-pump rectifier demodulates and multiplies the voltage coming from the antenna. Afterwards, the hysteresis comparator turns the analog signal from the charge-pump rectifier into a "clean" digital signal for entering the digital part. The backscatter consists of a resistor and a capacitor forming an impedance that is switched in parallel to the antenna via a fast switching transistor allowing backscatter modulation. UHF RFID tags transmit their reply via backscatter modulation [13].

Digital Part. In order to have the UHF tag prototype programmable, its digital part is realized as a microcontroller. The deployed microcontroller is an Atmel ATMega128 which is an 8-bit microcontroller operating at 16 MHz. Programming and in-system debugging of the microcontroller is done via a JTAG interface.

Protocol Implementation. The EPC Generation 2 standard has been selected as protocol for the UHF tag prototype. The protocol is implemented in software which is written to the program memory of the UHF tag prototype's microcontroller. Besides the implementation of the mandatory functionality of the EPC Generation 2 standard, the UHF tag prototype has also integrated secure tag authentication that uses a 128-bit AES encryption scheme according to [6].

3.2 Description of Passive UHF RFID Tags

In addition to the UHF tag prototype we have used passive UHF RFID tags that are commercially available. In contrast to the UHF tag prototype, passive UHF RFID tags are completely powered by the RF field of the RFID reader requiring no extra battery. A passive UHF RFID tag consists of an antenna and a microchip that comprises an analog part and a digital part. Typically, the protocol handling is implemented via a state machine in dedicated hardware [1].

A comparison with the previous section shows that the overall structure of a passive UHF RFID tag is not that different from the structure of the UHF tag prototype.

In order to obtain read ranges of several meters, passive UHF RFID tags should consume very little power. The power consumption of passive UHF RFID tags is in the range of some microwatts [14]. For detecting data-dependent emanation we have used passive UHF RFID tags from various tag vendors. All examined passive UHF RFID tags have shown data-dependent emanation.

4   Measurement Setup for UHF RFID Tags

This section describes the measurement setup that has been used to reveal data-dependent emanation of UHF RFID tags. Measurements have been done some centimeters away from the UHF RFID tags in the near field and up to 1 m away from the UHF RFID tags in the far field. The RF field of the RFID reader has been switched on during all the measurements. Initial measurements have detected data-dependent emanation of the UHF tag prototype. Measurements on passive UHF RFID tags could also reveal the side-channel leakage.

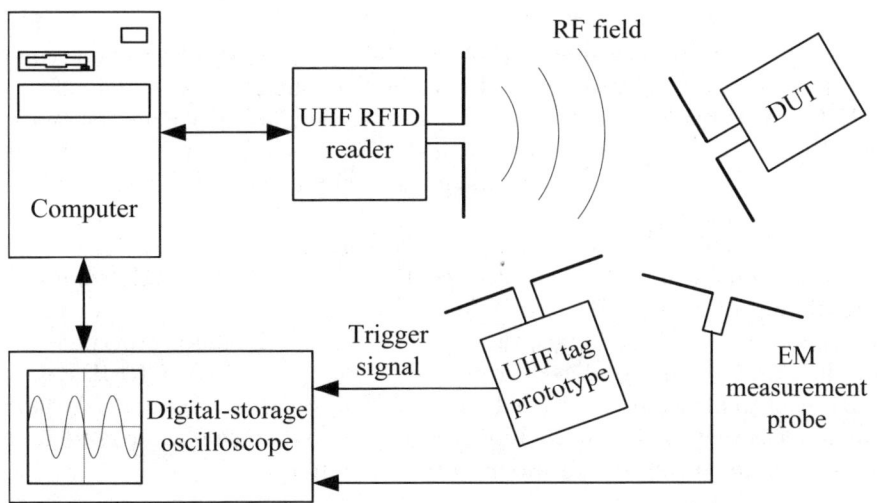

Fig. 2. Measurement setup for examining the emanation of a passive UHF RFID tag (DUT) in the far field

The automation of the measurement setup is important for performing side-channel analysis. Only an automated measurement setup allows to gather thousands of individual measurements within an acceptable time. Besides the examined UHF RFID tag which we call device under test (DUT), the main components of the measurement setup are a digital-storage oscilloscope, a UHF RFID reader that is compliant to the EPC Generation 2 standard, and an EM

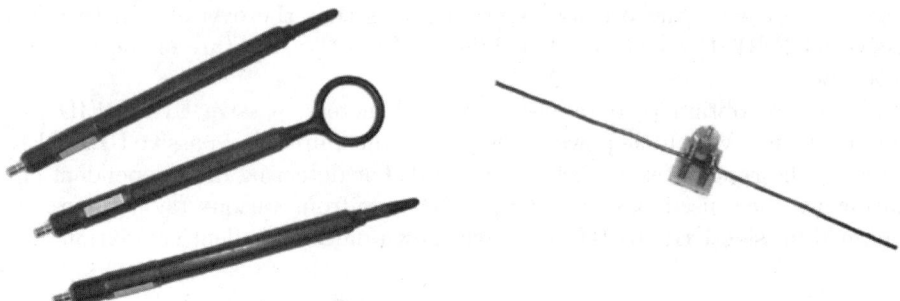

Fig. 3. Near-field probes that have been used for the measurements

Fig. 4. Self-made dipole antenna that has been used for the measurements

measurement probe. The digital-storage oscilloscope and the UHF RFID reader are connected to a computer. A program on the computer controls the whole measurement flow and performs the subsequent analysis of the recorded data. Depending on the measurement, different EM measurement probes are used to obtain the EM signal that is radiated from the DUT. Figure 2 shows the measurement setup for examining the emanation of a passive UHF RFID tag in the far field.

Acquiring a single measurement follows always the same scheme and requires several steps. After initializing the DUT, the computer sends the command to the UHF RFID reader that is used for detecting the data-dependent emanation. The UHF RFID reader in turn communicates with the DUT via the air interface. While the DUT processes this command, its radiated EM field is recorded by the digital-storage oscilloscope with the help of an EM measurement probe. The data acquisition of the digital-storage oscilloscope is started by a trigger signal. When examining the UHF tag prototype, the trigger signal directly comes from the UHF tag prototype. Passive UHF RFID tags are not suitable for directly providing a trigger signal. Thus, the software of the UHF tag prototype is modified such that it can be placed in parallel to the passive UHF RFID tag into the RF field to provide the external trigger signal (compare Figure 2). Acquiring a single measurement is finalized by transferring and storing the recorded data from the digital-storage oscilloscope to the computer.

4.1 Near-Field Measurements

For measuring the emanation of UHF RFID tags in the near field, we have used special near-field probes. Near-field probes are available in various sizes and shapes depending on the frequency range and the application they are dedicated for. During our measurements we have used the three near-field probes in Figure 3 which are designated for detecting magnetic fields. One near-field probe works for frequencies from 100 kHz to 50 MHz, the other two near-field probes work for frequencies from 30 MHz to 3 GHz.

Since the signal amplitudes that can be obtained with near-field probes are rather small, we have deployed an additional preamplifier. The preamplifier has a voltage gain of 30 dB and is connected between the output of the near-field probe and the input of the digital-storage oscilloscope. When doing measurements in the range of some tenths of megahertz, it is helpful to enable the internal bandwidth limitation of the digital-storage oscilloscope. Limiting the bandwidth has the advantage that the strong RF field from the UHF RFID reader is suppressed, which furthermore increases the quality of the measurements.

4.2 Far-Field Measurements

Near-field probes are no longer suitable for far-field measurements, rather, electromagnetic antennas are required. The UHF RFID tags that have been examined in this work operate at a carrier frequency of about 868 MHz. Since our far-field measurements have concentrated on detecting data-dependent emanation of UHF RFID tags around their carrier frequency, no special broadband antenna is necessary. A self-made dipole antenna shown in Figure 4 whose length is tuned to the carrier frequency is sufficient. The length of a dipole antenna for a carrier frequency of 868 MHz is about 17 cm [1]. While near-field measurements require an additional preamplifier in order to obtain acceptable signal amplitudes, far-field measurements do not. A spectrum analyzer with special band-pass filters can be used to transform the 868 MHz signal to baseband which allows to reduce the required sampling rate of the digital-storage oscilloscope. A reduced sampling rate results in measurements that consume less storage space on the computer and can be analyzed in a faster way.

5 Side-Channel Analysis of UHF RFID Tags

For analyzing the susceptibility of UHF RFID tags to side-channel analysis (SCA) we have used differential electromagnetic analysis (DEMA) attacks which have originally emerged from differential power analysis (DPA) attacks [9]. Both attacks are a powerful instrument to reveal secrets from crypto devices. The difference between DEMA attacks and DPA attacks is the method in which measurements are acquired. For DEMA attacks, measurements of the electromagnetic field that is emanated by crypto devices are required. DPA attacks use power traces that are obtained by directly measuring the power consumption of crypto devices. Both attacks have the advantage that only a simple model of the analyzed crypto device is necessary and that even very noisy measurements can be used [15].

Before starting a DEMA attack an appropriate operation needs to be selected that is suitable for revealing data dependencies. The UHF RFID tags which we have examined are EPC Generation 2 tags. For those tags, it has turned out that the *Write* command is a useful operation to detect data dependencies. The *Write* command as it is defined in the EPC Generation 2 standard [2] allows to write a 2-byte value to the non-volatile memory of a UHF RFID tag. Since the

2-byte value is a freely selectable parameter of the *Write* command, the 2-byte value can be used as chosen input data of the DEMA attack.

By using the measurement setup and the measurement-acquisition strategy described in Section 4 we have obtained various electromagnetic traces. The electromagnetic traces are recorded while the examined UHF RFID tag executes a *Write* command with a chosen 2-byte value. Thereby, always the same memory location of the UHF RFID tag is used. This memory location is initialized with the value zero before a new chosen 2-byte value is written. Initializing the memory location has the purpose to bring the UHF RFID tag always to the same initial state.

After recording the electromagnetic traces, a hypothetical model is used to map the chosen 2-byte values to hypothetical values that try to predict the electromagnetic emanation of the UHF RFID tag. There exist various hypothetical models like the Hamming-weight model or the Hamming-distance model which are not explained here in more detail (see [15]). Taking the hypothetical values from all 2-byte values that have been used to obtain the electromagnetic traces results in a hypothesis that is assumed to be correct. Additionally, another several hundred hypotheses are created that are assumed to be wrong. Wrong hypotheses are determined by applying the hypothetical model to randomly chosen values that are different from the 2-byte values that have been used to obtain the electromagnetic traces.

Having all the hypotheses allows to compare them with the electromagnetic traces that have been recorded previously. Comparison is done with the help of statistical methods. A well known statistical method for DEMA attacks and DPA attacks which we have used is the correlation coefficient. The correlation coefficient shows the linear dependency between different values [15]. The higher the absolute value of the correlation coefficient the higher is the linear dependency between the values that are compared. Based on the correlation coefficient, a correlation trace can be computed for each hypothesis. For the side-channel analysis, we call a DEMA attack successful if the comparison between the electromagnetic traces and the hypothesis that is assumed to be correct leads to significant peaks in the corresponding correlation trace.

5.1   Side-Channel Analysis of the UHF Tag Prototype

The UHF tag prototype we have built and used for side-channel analysis operates semi passively and contains a microcontroller. Compared to a conventional passive UHF RFID tag, the power consumption of the deployed microcontroller is much higher. For any fixed hardware architecture, higher power consumption brings along higher electromagnetic emanation.

Results of Near-Field Measurements. Main part of the electromagnetic field that is emanated by the UHF tag prototype's microcontroller is located in the frequency range of some hundreds of megahertz. Since the strong RF signal of the UHF RFID reader is located around 868 MHz, the RF signal can be easily

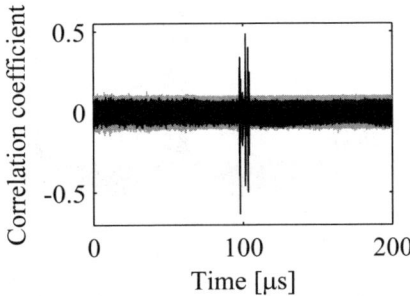 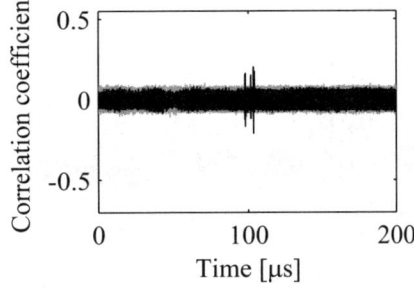

Fig. 5. Result of the DEMA attack on the UHF tag prototype by doing low-pass filtering directly on the digital-storage oscilloscope

Fig. 6. Result of the DEMA attack on the UHF tag prototype by doing low-pass filtering via software in an additional preprocessing step

suppressed by applying a low-pass filter. There are two possible ways for low-pass filtering: directly with the help of the digital-storage oscilloscope during the measurement acquisition, or via software in an additional preprocessing step before performing the DEMA attack.

Suppressing the strong RF signal by using the digital-storage oscilloscope results in electromagnetic traces with smaller amplitudes. As a consequence, a higher input sensitivity can be selected at the digital-storage oscilloscope which increases the accuracy of the measurements. Figure 5 shows the result of a successful DEMA attack on the UHF tag prototype in the near-field during the execution of a *Write* command. Thereby, recording 1000 individual electromagnetic traces has lead to a maximum absolute value of 0.63 for the correlation trace of the correct hypothesis. Low-pass filtering has been directly done on the digital-storage oscilloscope during measurement acquisition. For comparison, Figure 6 shows the result of the same DEMA attack by doing low-pass filtering of the electromagnetic traces in software. In this case, the maximum absolute value of the correlation trace reduces to about 0.21.

Results of Far-Field Measurements. Besides analyzing the emanation of the UHF tag prototype in the near field, we have also done analysis work in the far field. As mentioned in Section 4.2, we have concentrated on measuring the emanation of UHF RFID tags around the carrier frequency of the RF signal of about 868 MHz during our far-field measurements. With our measurement strategy we could not detect any data dependent emanation of the UHF tag prototype in the far field.

5.2   Side-Channel Analysis of Passive UHF RFID Tags

In contrast to our UHF tag prototype, passive UHF RFID tags have a power consumption of only some microwatts. With our measurement equipment we have not been able to directly measure the electromagnetic field that is emanated by

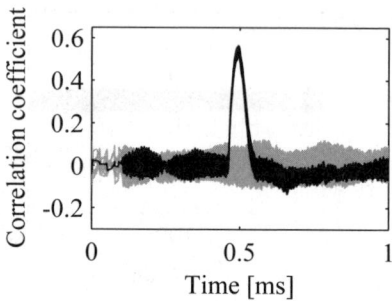

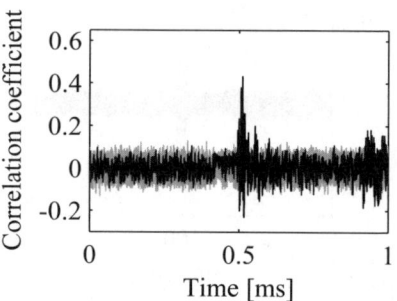

Fig. 7. Result of the DEMA attack on a passive UHF RFID tag in the near field

Fig. 8. Result of the DEMA attack on a passive UHF RFID tag from a different tag vendor in the near field

the microchip of a passive UHF RFID tag. Therefore, we have used an indirect effect named *parasitic backscatter* to detect data-dependent emanation of the microchip. Passive UHF RFID tags deploy backscatter modulation to transmit data to the UHF RFID reader. As described in [12], the power consumption of passive UHF RFID tags modulates the backscatter which results in *parasitic backscatter*. The most important observation is that the backscatter of a passive UHF RFID tag can be detected via a simple dipole antenna within several meters.

Results of Near-Field Measurements. Using the *parasitic backscatter* of passive UHF RFID tags in the near field has allowed to perform DEMA attacks successfully. When using a near-field probe, its placement toward the passive UHF RFID tag that is examined is an important factor for the success of the DEMA attack. Favorable placement of the near-field probe stronger attenuates the RF field that is emitted by the antenna of the UHF RFID reader. The stronger the RF field is attenuated the less measurements are necessary for a successful DEMA attack.

In this way we have been able to perform successful DEMA attacks by measuring less than 100 electromagnetic traces. Figure 7 shows the result of a DEMA attack on a passive UHF RFID tag by using 1000 measurements. In order to ensure that this is not a phenomenon of a specific tag vendor, we have tested passive UHF RFID tags from various tag vendors. Figure 8 shows the result of the same DEMA attack by using a passive UHF RFID tag from a different tag vendor. Although the two correlation traces in Figure 7 and Figure 8 are quite different, both illustrate that there is a strong data dependency.

Results of Far-Field Measurements. For the passive UHF RFID tags we have done the same measurements in the far field than for our UHF tag prototype. In contrast to the UHF tag prototype, the passive UHF RFID tags we have examined show data dependent emanation also in the far field. Thereby,

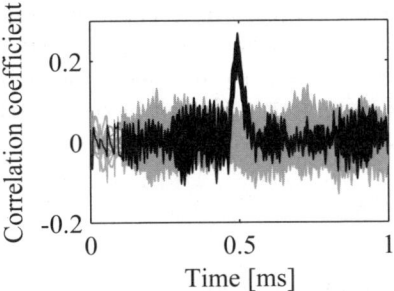
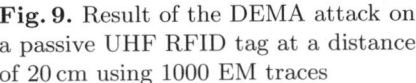
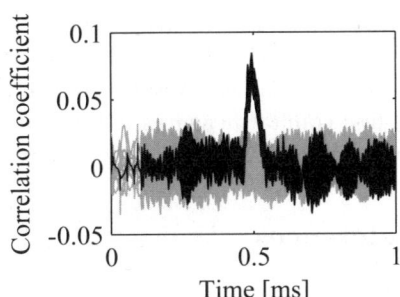

Fig. 9. Result of the DEMA attack on a passive UHF RFID tag at a distance of 20 cm using 1000 EM traces

Fig. 10. Result of the DEMA attack on a passive UHF RFID tag at a distance of 1 m using 10000 EM traces

we have analyzed the electromagnetic field with a self-made dipole antenna at various distances of the passive UHF RFID tags, starting from 20 cm up to 1 m.

All our DEMA attacks in the far field of the passive UHF RFID tags have been successful, even at a distance of 1 m. Figure 9 shows the result of a DEMA attack on a passive UHF RFID tag at a distance of 20 cm using 1000 measurements. Regardless of the distance, the peaks in the resulting correlation traces always look similar. For comparison, Figure 10 shows the correlation traces of the same passive UHF RFID tag at a distance of 1 m. The difference when the distance increases is the maximum absolute value of the correlation coefficient. Figure 9 shows a maximum absolute value of the correlation trace of 0.27 which decreases to 0.08 in Figure 10. As a consequence, the number of measurements must be increased to clearly identify the data dependency at greater distances. The correlation traces in Figure 10 have been obtained by using 10000 measurements.

6 Conclusion

In this work we have shown the susceptibility of UHF RFID tags to DEMA attacks. We have analyzed a self-made UHF tag prototype and commercially-available passive UHF RFID tags from various tag vendors. Whereas the UHF tag prototype that operates semi passively shows only data-dependent emanation in the near field, passive UHF RFID tags show data-dependent emanation in the far field too. We have performed successful DEMA attacks in the far field of passive UHF RFID tags at distances up to 1 m. However, increasing the number of acquired measurements should allow to realize successful DEMA attacks at greater distances as well.

Current RFID tags do not use cryptographic protection and furthermore store no secret that could be the aim of such attacks. Hence, this work has no practical relevance for current RFID products. Nevertheless, it was our goal to investigate the side-channel leakage and to determine the susceptibility of future UHF RFID tags to this class of attacks. Our results clearly show that once cryptographic functionality should be added to UHF RFID tags, countermeasures against SCA need to be applied as well. Analyzing our results we come to the conclusion that ad-hoc countermeasures will not suffice, but a complete re-design of the RFID tag's architecture will be necessary to protect effectively from SCA.

References

[1] Finkenzeller, K.: RFID-Handbook, 2nd edn. Carl Hanser Verlag München (2003)
[2] International Organization for Standardization: ISO/IEC 18000-6C: Air Interface for Radio-Frequenc Identification (RFID) Devices Operating in the 860 MHz to 960 MHz Industrial, Scientific, and Medical (ISM) Band used in Item Managment Applications. ISO/IEC (2006)
[3] Garfinkel, S., Rosenberg, B.: RFID: Applications, Security, and Privacy. Addison-Wesley Professional, Reading (2005)
[4] Aigner, M.: Seven reasons for application of standardized crypto functionality on low cost tags. EU RFID Forum (2007)
[5] Bailey, D., Juels, A.: Shoehorning Security into the EPC Standard. In: De Prisco, R., Yung, M. (eds.) SCN 2006. LNCS, vol. 4116, pp. 303–320. Springer, Heidelberg (2006)
[6] Feldhofer, M., Dominikus, S., Wolkerstorfer, J.: Strong Authentication for RFID Systems Using the AES Algorithm. In: Joye, M., Quisquater, J.-J. (eds.) CHES 2004. LNCS, vol. 3156, pp. 357–370. Springer, Heidelberg (2004)
[7] Yu, Y., Yang, Y., Fan, Y., Min, H.: Security Scheme for RFID Tag. Auto-ID Labs Fudan University, White Paper (2006)
[8] Kocher, P.C., Jaffe, J., Jun, B.: Differential Power Analysis. In: Wiener, M.J. (ed.) CRYPTO 1999. LNCS, vol. 1666, pp. 388–397. Springer, Heidelberg (1999)
[9] Gandolfi, K., Mourtel, C., Olivier, F.: Electromagnetic Analysis: Concrete Results. In: Koç, Ç.K., Naccache, D., Paar, C. (eds.) CHES 2001. LNCS, vol. 2162, pp. 251–261. Springer, Heidelberg (2001)
[10] Mangard, S.: Exploiting Radiated Emissions – EM Attacks on Cryptographic ICs. In: Ostermann, T., Lackner, C. (eds.) Austrochip 2003, Proceedings, Linz, Austria, October 1, 2003, pp. 13–16 (2003) (ISBN 3-200-00021-X)
[11] Hutter, M., Feldhofer, M., Mangard, S.: Power and EM Attacks on Passive 13.56 MHz RFID Devices. In: Paillier, P., Verbauwhede, I. (eds.) CHES 2007. LNCS, vol. 4727, pp. 320–333. Springer, Heidelberg (2007)
[12] Oren, Y., Shamir, A.: Remote password extraction from RFID tags. IEEE Transactions on Computers 56(9), 1292–1296 (2007)
[13] Zhu, Z.: RFID Analog Front End Design Tutorial (version 0.0). Auto-ID Labs University of Adelaide (2004)
[14] Karthaus, U., Fischer, M.: Fully Integrated Passive UHF RFID Transponder IC With 16.7-µW Minimum RF Input Power. IEEE Journal of Solid-State Circuits, 1602–1608 (2003)
[15] Mangard, S., Oswald, E., Popp, T.: Power Analysis Attacks – Revealing the Secrets of Smart Cards. Springer, Heidelberg (2007) (ISBN 978-0-387-30857-9)

Online/Offline Signature Schemes for Devices with Limited Computing Capabilities

Ping Yu[1] and Stephen R. Tate[2]

[1] Department of Computer Science and Engineering, University of North Texas,
Denton, TX 76203
pingyu@unt.edu
[2] Department of Computer Science, University of North Carolina at Greensboro,
Greensboro, NC 27402
srtate@uncg.edu

Abstract. We propose a family of three efficient digital signature schemes, which are proved secure under the strong RSA assumption without requiring a random oracle. The new signature schemes can operate in an online/offline manner, doing most of their work in the offline precomputation phase. The online phase of even the least efficient variant is very fast, requiring only a single non-modular multiplication of a short (160-bit) value by a longer (1022-bit) value. Online/offline signatures are useful in settings in which signatures need to be produced with few operations, either when there is a large volume of requests or if the device performing the signature is not computationally powerful. Our schemes have extremely low computation cost so are particularly suitable for devices with limited computing capabilities such as smart cards or mobile devices.

This paper provides three specific contributions. First, we propose our basic online/offline signature scheme, which could be viewed as the online/offline extension of the Camenisch-Lysyanskaya (CL) signature scheme. Compared to using the general Shamir-Tauman technique for converting the CL signature scheme into one that operates in an online/offline fashion, our direct adaptation has the same online efficiency, while having advantages of a more efficient offline phase, simpler key management that only requires one keypair, and a shorter signature. In addition, when used as a traditional one-phase signature our basic scheme is more efficient than the Camenisch-Lysyanskaya scheme, due to our operation restructuring. While this first scheme has advantages over using the Shamir-Tauman/Camenisch-Lysyanskaya construction, we describe two additional techniques that further improve efficiency of both online and offline phases. Our first improvement uses computations over a small subgroup of Z_n^* to reduce the size of the required computations. Our second improvement uses division intractable hash functions to remove the requirement of generating random primes for use in this class of signature schemes. As we present these three schemes, each one is more efficient than the previous one, but requires increasingly strong complexity assumptions.

Keywords: Digital Signature, Strong RSA Assumption, Standard Model, Online/Offline Signing, Devices with Limited Computing Capabilities.

T. Malkin (Ed.): CT-RSA 2008, LNCS 4964, pp. 301–317, 2008.

1   Introduction

The digital signature concept is a fundamental cryptographic primitive in modern cryptography. In such schemes, a signer prepares a keypair which includes a signing key and a verification key. The signing key is kept secret by the signer while the verification key is public for potential verifiers. The signer generates a string by signing a message using his signing key. This string is called the signer's signature on this particular message. Later a verifier can check the validity of a signature on a message using the signer's verification key.

The idea of digital signatures was first proposed by Diffie and Hellman [1]. Since then, numerous constructions have been proposed in the literature based on different security assumptions. Many schemes are based on the well-known RSA assumption and a variant known as the strong RSA assumption, including PSS [2] and the Cramer-Shoup scheme [3]. Other schemes are based on variants of the discrete logarithm or computational/decisional Diffie-Hellman assumption, including ElGamal signatures [4], Schnorr signatures [5], and many others.

As a fundamental cryptographic primitive, it is important to clarify the exact requirements for a secure digital signature scheme. In 1988, Goldwasser et al. defined a security notion for signature schemes, which is called existential unforgeability under adaptive chosen message attacks [6]. Since then, this notion has been widely used to judge whether a digital signature scheme is strong enough to be deployed in a real application. Many digital signature schemes (e.g., Fiat-Shamir, Schnorr, ElGamal, PSS) can be proved secure under this requirement in the random oracle model, which was first proposed by Fiat and Shamir [7], and formalized by Bellare and Rogaway [8]. In the random oracle model, a cryptographic hash function is abstracted as a random function that can be accessed by all participants in the protocol, including adversaries. However, Canetti et al. constructed a scheme that can be proved secure in the random oracle model, while any real implementation will result in an insecure construction [9]. Therefore, security proofs in the random oracle model do not necessarily imply security in the standard model, so a proof of security in the random oracle model can only be treated as a heuristic argument that a scheme is secure.

Gennaro, Halevi and Rabin [10] and Cramer and Shoup [3], in 1999 and 2000, respectively, independently proposed practical digital signature schemes secure under adaptive chosen message attacks under the strong RSA assumption in the standard model. Before these two constructions, available schemes secure under an adaptive chosen message attack in the standard model were not practical for real applications [6,11,12]. Based on the ideas of Cramer and Shoup, Camenisch and Lysyanskaya [13], Zhu [14,15], and Fischlin [16] all proposed schemes with similar structure based on the strong RSA assumption. In 2005, Groth extended these results to work over a small subgroup of Z_n^*, improving the efficiency of signature generation [17].

1.1   Online/Offline Signature Schemes

The notion of online/offline signatures was first introduced by Even, Goldreich, and Micali in 1989 [18]. An online/offline signature scheme does most of its work in an offline precomputation phase, before the message is known. The online phase, which is performed after the message to be signed is known, should be very efficient and can be completed quickly. Online/offline signing is important for applications when signatures need to be produced quickly. For instance, consider a stock broker's server that has "bursty" requests that need to be signed, where there are periods of low activity and infrequent bursts of rapid transaction requests (e.g., immediately after financial updates or news releases). Another example comes from the area of mobile computing, in which a mobile device with limited computing capabilities needs to authenticate itself by quickly producing a valid signature on a challenge from a server, but can precompute offline results at low speed and power in preparation for authentication requests. In still another example, the authenticating device could be a smart card, with a very weak processor, but which can be loaded with precomputed results of the offline phase from a more powerful device. In these scenarios, using precomputation can enable quick signature generation.

In addition to proposing the online/offline signature scenario, Even et al. proposed a generic method to convert any signature scheme into an online/offline one [18]. However, their method is not efficient enough to be practical. A little over a decade later, in 2001, Shamir and Tauman proposed another generic method to achieve online/offline signing, which is quite efficient [19] (we will discuss the efficiency further in Sect. 1.2). Their method is based on a new type of hash function called a trapdoor hash function, which was proposed by Krawczyk and Rabin [20], and allows the use of a "hash-sign-switch" paradigm. A trapdoor hash function uses a keypair with a public "hash key" HK and a private "trapdoor key" TK, and the hash for message m is produced by the function $h(HK, r, m)$, where r is a supplemental random input. For a secure trapdoor hash function, given HK, r, m, and alternative message m', it is infeasible to find an r' such that $h(HK, r, m) = h(HK, r', m')$; however, given TK it is easy to find such an r'. Thus, the hash-sign-switch paradigm is to first use any signature algorithm to sign $h(HK, r, m)$ for a random message m, and then when the real message m' is known (in the online phase) the signer simply computes the r' so that the hash remains the same and the precomputed signature is valid. Note that the new signature must contain this r' value, expanding the size of the signature, and that the signer must manage a keypair (HK, TK) for the trapdoor hash in addition to the keypair for the basic signature scheme.

In addition to these general online/offline constructions, a signature scheme due to Schnorr [5] operated in a two-phase manner, and was efficient enough for smartcards and other limited devices. Schnorr constructed a three-round identification scheme over a small prime-order subgroup of Z_p^*, which was then converted into a signature scheme using the Fiat-Shamir heuristic [7]. Schnorr's scheme is an efficient online/offline construction in which the signer needs about 160 modular multiplications for the offline phase and one modular multiplication for the online phase[1]. However, Schnorr's scheme can only be proved secure in the random oracle model due to the reliance on the Fiat-Shamir technique, and in fact Goldwasser and Kalai have recently published a result that casts doubt on the general applicability of the Fiat-Shamir technique [21].

---

[1] In the parameters used in the original system, the signer needs only 140 modular multiplications. However, 160 could be considered more appropriate due to the advance of computing technology in the past 20 years.

1.2   Our Results and Comparison with Previous Work

In this paper we propose a family of three efficient digital signature schemes, which we call OOSIG1, OOSIG2, and OOSIG3 ("OOSIG" is for online/offline signature). These signature schemes are progressively more efficient but rely on increasingly strong assumptions. All security proofs are in the standard model, without requiring a random oracle. We summarize the relevant features here:

OOSIG1: This is our basic online/offline signature scheme, and is proved secure under the strong RSA assumption. The scheme has the same verification protocol as the Camenisch-Lysyanskaya signature scheme [13], could be viewed as the online/offline extension of their scheme. We note that even when not applied in the online/offline setting, our scheme provides benefits and efficiency improvements over the Camenisch-Lysyanskaya scheme, saving several operations. Comparing the cost of the online phase of our algorithm with the online phase when using the Shamir-Tauman construction[2], our online phase requires a single non-modular multiplication of a 160-bit value by a 1022-bit value as well as one or two additions, whereas the Shamir-Tauman construction requires the modular reduction of a 1184-bit value by a 1024-bit modulus as well as one or two additions, which is essentially the same cost. Note that Shamir and Tauman relate this cost to a full-length modular multiplication and estimate that it is roughly 0.1 modular multiplication [19], and while this applies to our OOSIG1 algorithm as well we caution that this is not general complexity result and requires several assumptions about the concrete sizes of the values and the algorithms used.

An additional benefit of our OOSIG1 algorithm is that since OOSIG1 is a direct online/offline signature scheme, we avoid the use of trapdoor hash functions and hence do not need to manage the second keypair (although our private signing key is slightly larger than that in the original Camenisch-Lysyanskaya scheme), and do not have to include the random commitment value r' in the signature. As a result, OOSIG1 has simpler key management and shorter signatures than the Shamir-Tauman technique combined with the Camenisch-Lysyanskaya signature. Furthermore, the use of the trapdoor hash requires an additional complex operation (a modular powering with an exponent larger than the modulus) which is not required by our more direct approach, so our offline phase is significantly faster. In cases where the offline phase is run during idle or lightly loaded time on an embedded or other small device, this could be significant.

OOSIG2: In this scheme, we use computations over a small subgroup of Z_n^* to further improve the efficiency of our basic signature scheme while retaining the online/offline capability, and prove this scheme secure under the strong RSA subgroup assumption which was proposed by Groth [17].

OOSIG3: In this scheme, we remove one of the troubling requirements of this class of signature schemes — the necessity of generating a prime number for each signature (in the offline phase of our OOSIG1 and OOSIG2 variants). Instead, we use a hash function that satisfies certain properties, and so this scheme is proved secure under

---

[2] When we refer to the Shamir-Tauman construction in this paper, we always assume the use of the third and most efficient trapdoor hash family given in their paper [19].

the assumption that this hash function is division intractable [10] as well as the strong RSA subgroup assumption.

OOSIG3 is very efficient, and the assumptions seem reasonable. OOSIG3 needs only about 200 modular multiplications for the offline phase, which is only slightly more costly than the offline phase of Schnorr's scheme. The online phase requires only a few non-modular multiplications in which both operands are relatively short (under 200 bits), and hence is extremely efficient. To our knowledge, this is the first highly efficient online/offline signature scheme which has security based on variants of the strong RSA assumption in the standard model.

The rest of the paper is organized as follows. Sect. 2 reviews some cryptographic notations and definitions. Sect. 3 presents our basic scheme, which can be viewed as an online/offline extension of the Camenisch-Lysyanskaya signature scheme. In Sect. 4, we discuss ways to improve the efficiency of the basic scheme by using small subgroups of Z_n^*, leading to two new schemes for devices with limited computing capabilities. Finally, we give the conclusions in Sect. 5.

2   Preliminaries

This section reviews some notations and definitions which are used throughout the paper.

Definition 1 (Special RSA Modulus). *An RSA modulus $n = pq$ is called special if $p = 2p' + 1$ and $q = 2q' + 1$ where p' and q' also are prime numbers.*

Definition 2 (Quadratic Residue Group QR_n). *Let Z_n^* be the multiplicative group modulo n, which contains all positive integers less than n and relatively prime to n. An element $x \in Z_n^*$ is called a* quadratic residue *if there exists an $a \in Z_n^*$ such that $a^2 \equiv x \bmod n$. The set of all quadratic residues of Z_n^* forms a cyclic subgroup of Z_n^*, which we denote by QR_n. If n is the product of two distinct primes, then $|QR_n| = \frac{1}{4}|Z_n^*|$.*

Recently, Groth investigated cryptography over a small subgroup of Z_n^* [17]. Two of our proposed schemes are also constructed using this special kind of group, so we present the definition as used by Groth here.

Definition 3 (Small Subgroup G of Z_n^*). *Let $n = pq$ such that $p = 2p'r_p + 1$ and $q = 2q'r_q + 1$, where p, p', q, q' are all prime. There is a unique cyclic subgroup G of Z_n^* of order $p'q'$. For the purpose of efficient cryptographic construction, the order of G, i.e., $p'q'$, is chosen small and kept private. Let g be a random generator of G, and we call (n, g) an RSA subgroup pair.*

A hash function is a function mapping arbitrary strings of finite length to binary strings of fixed length. For cryptographic purposes, the most basic property that a hash function should satisfy is the collision-intractability property defined by Damgard [22]. One of our schemes in this paper will use another property for a hash function called "division intractability," which was introduced by Gennaro et al. [10]. Informally, a hash function is division intractable if it is infeasible to find distinct inputs for this hash function such that the hash value of one input divides the product of hash values of all other inputs.

Definition 4 (Division Intractability [10]). *A hashing family* $\mathcal{H}$ *is division intractable if it is infeasible to find distinct inputs* $X_1, \ldots, X_n, Y$ *such that* $h(Y)$ *divides the product of the* $h(X_i)$*'s.*

Formally, for every probabilistic polynomial time algorithm $\mathcal{A}$, *there exists a negligible function negl() such that*

$$Pr_{h \in \mathcal{H}_k} \left[\begin{array}{l} \mathcal{A}(h) = \langle X_1, \ldots, X_n, Y \rangle \\ s.t.\ Y \neq X_i \ for\ i = 1 \ldots n, \\ and\ h(Y)\ divides\ \prod_{i=1}^{n} h(X_i) \end{array} \right] = negl(k).$$

Now we introduce the strongest notion of a secure signature scheme, existential unforgeability under adaptive chosen message attacks, which was proposed by Goldwasser, Micali and Rivest [6]. The definition we give here is from Gennaro et al. [10].

Definition 5 (Secure Signatures [10]). *A signature scheme* $S = \langle Gen, Sig, Ver \rangle$ *is existentially unforgeable under an adaptive chosen message attack if it is infeasible for a forger who only knows the public key to produce a valid (message, signature) pair, even after obtaining polynomially many signatures on messages of its choice from the signer.*

Formally, for every probabilistic polynomial time forger algorithm $\mathcal{F}$, *there exists a negligible function* $negl()$ *such that*

$$Pr \left[\begin{array}{l} \langle pk, sk \rangle \leftarrow Gen(1^k); \\ for\ i = 1 \ldots n \\ \quad M_i \leftarrow \mathcal{F}(pk, M_1, \sigma_1, \ldots, M_{i-1}, \sigma_{i-1});\ \sigma_i \leftarrow Sig(sk, M_i); \\ \langle M, \sigma \rangle \leftarrow \mathcal{F}(pk, M_1, \sigma_1, \ldots, M_n, \sigma_n), \\ M \neq M_i\ for\ i = 1 \ldots n,\ and\ Ver(pk, M, \sigma) = accept \end{array} \right] = negl(k).$$

The security of the signature schemes presented in this paper is based on a well-accepted cryptographic assumption called the strong RSA assumption, which was first proposed by Baric and Pfitzmann [23] and Fujisaki and Okamoto [24].

Assumption 1 (Strong RSA Assumption). *Let* n *be an RSA modulus. The* Flexible RSA Problem *is the problem of taking a random element* $u \in Z_n^*$ *and finding a pair* (v, e) *such that* $e > 1$ *and* $v^e = u \mod n$. *The* strong RSA assumption *says that no probabilistic polynomial time algorithm can solve the flexible RSA problem for random inputs with non-negligible probability.*

3   The Basic Signature Scheme

In 2002, Camenisch and Lysyanskaya proposed a signature scheme secure in the standard model under the strong RSA assumption [13]. The Camenisch-Lysyanskaya signature scheme produces a triple (v, e, s) as a signature, where e and s are chosen randomly, and v is computed from these values, the message, and the private key. Our basic scheme could be viewed as an online/offline extension of the Camenisch-Lysyanskaya scheme — it produces the same triple (v, e, s) for the signature, and we will show that the distribution of triples from our scheme is statistically indistinguishable from the distribution of triples from the Camenisch-Lysyanskaya scheme. The key difference is that we randomly select v and e, and compute s, which requires a new key generation and signing algorithm. The benefit is that all but a small part of the computation for s can be done without knowing the message to be signed, so can be done in an offline phase. Since our signing algorithm produces the same signatures as in the Camenisch-Lysyanskaya scheme, our verification algorithm is the same as in the Camenisch-Lysyanskaya scheme.

3.1 Signature Scheme OOSIG1

In this section, we define our first online/offline signature scheme, called OOSIG1.

Public System Parameters. Let k be the security parameter, and define the following lengths: l_m is the length of the message to be signed, with the restriction that $l_m < k - 2$. l is a security parameter that controls the statistical closeness of distributions, and should be at least polynomial in k (in practice $l = 160$ is sufficient). For convenience, we also define some lengths based on these parameters: $l_e \geq l_m + 2$ is the length of an exponent in the signature algorithm, $l_n = 2k$ is the length of the public modulus, and $l_s = l_n + l_m + l$ is the length of another exponent used in the signing algorithm.

Key Generation. On input 1^k, pick two k-bit safe primes p and q (so that $p = 2p' + 1$, and $q = 2q' + 1$, where p' and q' are also prime), and let $n = pq$. Select b as a random generator of QR_n. Select $\alpha, \beta \in_R [0, p'q')$ and compute $a = b^\alpha \mod n$ and $c = b^\beta \mod n$. Let $K = \lfloor 2^{l_s}/p'q' \rfloor$. Output public key (n, a, b, c), and private key $(p'q', \alpha, \beta, K)$.

Signing Algorithm. The signing procedure includes two phases.
OFFLINE PHASE: The signer picks a random $\gamma \in_R [0, p'q')$, a random l_e-bit prime number e, and a random $k' \in_R [0, K)$, and then computes

$$v = b^\gamma \mod n, \quad \lambda = k'p'q' + \gamma e - \beta \mod Kp'q'.$$

ONLINE PHASE: When a message $m \in [0, 2^{l_m})$ appears, the signer computes

$$s = \lambda - \alpha m \mod Kp'q'.$$

Note that while this is stated as a modular operation, the ranges of the values ensure that an adjustment to keep the value in range is only needed with negligible probability, and even then this is accomplished with a single addition. The signature is (v, e, s) for the message m.

Verification Algorithm. To verify that (v, e, s) is a signature on message m, check that e's length is l_e, and

$$v^e \equiv a^m b^s c \mod n. \tag{1}$$

It can be verified that a valid signature can always pass the verification algorithm. Since these operations are being performed in QR_n, we consider operations in the exponent modulo $p'q'$, and get

$$s \equiv \gamma e - \beta - \alpha m \pmod{p'q'},$$

and so

$$a^m b^s c \equiv b^{\alpha m + (\gamma e - \beta - \alpha m) + \beta} \equiv b^{\gamma e} \equiv v^e \mod n.$$

The salient characteristic for the signing algorithm is its online/offline mechanism. Most of the computation can be done before the appearance of a message, and the online phase only needs a single multiplication (where one of the values is short) and a subtraction.

3.2 Comparison with the Camenisch-Lysyanskaya Scheme

OOSIG1 produces signatures that are indistinguishable from those of the Camenisch-Lysyanskaya scheme. To see this, consider (1). In their scheme, a, b, c are randomly chosen from QR_n, and v is calculated as

$$v = (a^m b^s c)^{e^{-1}} \mod n, \tag{2}$$

where $s \in_R [0, 2^{l_s})$. In OOSIG1, a, b, c are also random elements of QR_n, but s is calculated; however, in the following lemma we show that s in OOSIG1 is statistically indistinguishable from $[0, 2^{l_s})$.

Lemma 1. *Let $K = \lfloor 2^{l_s} / p'q' \rfloor$, where K is superpolynomial in the security parameter k. Let e be a value that is relatively prime to $p'q'$, α and β be constants in $[0, p'q')$, and m be a constant in $[0, 2^{l_m})$. Let $k' \in_R [0, K)$ and $\gamma \in_R [0, p'q')$. If we define $s = (k'p'q' + \gamma e - \beta - \alpha m) \mod Kp'q'$, then s is statistically indistinguishable from uniform over $[0, 2^{l_s})$.*

Proof. First, we prove that s is uniformly distributed over $[0, Kp'q')$. For any $x \in [0, Kp'q')$, since e is relatively prime to $p'q'$ there exists exactly one pair (k', γ) such that $x = (k'p'q' + \gamma e - \beta - \alpha m) \mod Kp'q'$. Therefore there is a one-to-one mapping between pairs (k', γ) and values in $[0, Kp'q')$, and since pairs (k', γ) are chosen uniformly, the resulting distribution of s over $[0, Kp'q')$ is uniform.

Next, we prove that the uniform distribution over $[0, Kp'q')$ is statistically indistinguishable from uniform over $[0, 2^{l_s})$. Let $Pr_D(x)$ denote the probability of x in distribution D. Then the distance between two distributions D_1 and D_2 is

$$dist(D_1, D_2) = \frac{1}{2} \sum_x |Pr_{D_1}(x) - Pr_{D_2}(x)|,$$

and note that for any two distributions $dist(D_1, D_2) \le 1$. Two distributions D_1 and D_2 are *statistically indistinguishable* if $dist(D_1, D_2)$ is negligible.

Distribution D_1 is uniform over $[0, Kp'q')$, and distribution D_2 is uniform over $[0, 2^{l_s})$. Doing the basic algebra, we get

$$dist(D_1, D_2) = \frac{1}{2} \sum_x |Pr_{D_1}(x) - Pr_{D_2}(x)|$$

$$= \frac{1}{2} [(\frac{1}{Kp'q'} - \frac{1}{2^{l_s}}) Kp'q' + \frac{1}{2^{l_s}} (2^{l_s} - Kp'q')]$$

$$= 1 - \frac{Kp'q'}{2^{l_s}}$$

$$= 1 - \frac{\lfloor 2^{l_s}/p'q' \rfloor p'q'}{2^{l_s}}$$

$$< 1 - \frac{((2^{l_s} - p'q')/p'q')p'q'}{2^{l_s}}$$

$$= 1 - (1 - \frac{p'q'}{2^{l_s}})$$

$$= \frac{p'q'}{2^{l_s}} .$$

Thus, the distance between D_1 and D_2 is less than $\frac{p'q'}{2^{l_s}}$, and since $\frac{2^{l_s}}{p'q'}$ is superpolynomial in the security parameter k, this distance is negligible. So, the uniform distribution over $[0, Kp'q')$ is statistically indistinguishable from uniform over $[0, 2^{l_s})$.

Therefore, the distribution of s is statistically indistinguishable from uniform over $[0, 2^{l_s})$.   □

As a consequence of Lemma 1, the view of an attacker with respect to our scheme is statistically indistinguishable from the view of an attacker with respect to the Camenisch-Lysyanskaya scheme, which gives the following theorem.

Theorem 1. *OOSIG1 is existentially unforgeable under an adaptive chosen message attack, assuming the strong RSA assumption.*

For concrete parameters, we use the recommended parameter settings from the Camenisch-Lysyanskaya scheme, with $k = 512$, so n is 1024 bits long. l_m can be chosen as 160, and messages longer than 160 bits can first be sent through a collision-resistant hash function (e.g., SHA-1) to produce a 160-bit message digest, which is then signed. As stated earlier, $l = 160$ is sufficient to ensure the statistical closeness of the signature's actual distribution to the simulated distribution in the proof of the scheme [13], so $l_s = 1024 + 160 + 160 = 1344$. For this setting of parameters, the cost of the signing algorithm is about 1022 modular multiplications and the generation of a 162-bit prime number in the offline phase, and one multiplication in the online phase. Our algorithm avoids multiplication related to s in the Camenisch-Lysyanskaya scheme, which is about 1344 modular multiplications. Furthermore, note that OOSIG1 does not require computation of the multiplicative inverse of e as required by the Camenisch-Lysyanskaya scheme (see (2)), so has advantages even when not used in the online/offline mode.

The verification algorithm requires about (1344 + 162 + 160) modular multiplications. However this can be expedited by providing precomputed inverses of a and b and then verifying

$$v^e(a^{-1})^m(b^{-1})^s \equiv c \mod n,$$

since multi-base exponentiation can be done only slightly slower than single-base exponentiation (Algorithm 15.2 in [25]). Therefore, the verification needs only slightly more than 1344 modular multiplications.

4   Further Efficiency Improvements

The basic scheme can accommodate most application scenarios when online/offline signing is needed. However, it is possible to improve the efficiency further, particularly in the offline phase, which could be useful in situations such as mobile devices precomputing values in the offline phase during idle time. The main costs of the offline phase are due to an exponentiation taking 1022 modular multiplications and the generation of a 162-bit prime e. In this section, we reduce both of these costs: the exponentiation cost is reduced by working over a small subgroup of Z_n^*, and we remove the requirement of generating a prime e by using a division intractable hash function. These improvements also make the online phase more efficient: the online phase of OOSIG1 requires the multiplication of a 1022-bit number by a 160-bit number, and the algorithms of this section reduce the size of the first number to 200 bits.

Using computations over a small subgroup of Z_n^* in this class of signature schemes was first investigated by Groth [17]. We incorporate these ideas into our basic scheme OOSIG1, reducing the bit length of $p'q'$ from the 1022 bits suggested for OOSIG1. Following Groth's suggestions, we set the bit length of $p'q'$ to 200, which reduces the number of modular multiplications for calculating v from 1022 to 200. The security of these new schemes is based on Groth's variant of the strong RSA assumption, called strong RSA subgroup assumption over this small subgroup of Z_n^* — we give this definition here, with terminology slightly cleaned up from the original paper [17].

Assumption 2 (Strong RSA Subgroup Assumption). *Let K be a key generation algorithm that produces an RSA subgroup pair (n, g). The flexible RSA subgroup problem is to find $u, w \in Z_n^*$ and $d, e > 1$ such that $g = uw^e$ mod n and $u^d = 1$ mod n. The strong RSA subgroup assumption for this key generation algorithm states that it is infeasible to solve the flexible RSA subgroup problem with non-negligible probability for inputs generated by K.*

In Sect. 4.1 we give our first subgroup-based signature scheme, using an idea of Groth's in order to reduce the cost of finding a prime e: we pick a much smaller prime number e, and a security parameter t such that $e^t \geq 2^{l_m + 2}$. However, with a reduced range for e, this method could lead to multiple selections of the same prime number. Therefore, the signature scheme should pick e in an incremental way, and always remember the last prime number used. This way, the signature becomes a stateful construction. In a later section (Sect. 4.2) we will introduce another method which completely circumvents the cost of prime number generation, and keeps the scheme as a stateless one. The technique is based on the smoothness property of a random integer and the division intractability of a hash function, which have been introduced by Gennaro et al. [10], and further investigated by other authors [26,27]. The security of both signature schemes will be proven in Sect. 4.3.

4.1   OOSIG2: A Stateful Signature Scheme

In this section, we present our second online/offline signature scheme, called OOSIG2.

Public System Parameters. The main parameters are similar to the OOSIG1 scheme, but with an additional parameter t chosen and l_e reduced subject to $t \times l_e \geq l_m + 2$. We also define a length $l_{p'q'}$ which determines the size of the subgroup used, and use this length to define $l_s = l_{p'q'} + l_m + l$.

Key Generation. On input 1^k, pick two k-bit primes p and q as in Definition 3 (so $p = 2p'r_p + 1$, and $q = 2q'r_q + 1$, where p' and q' are also prime, each with length $l_{p'q'}/2$), and let $n = pq$. Let G be the unique subgroup of Z_n^* of order $p'q'$, and select a random generator b of G. Select $\alpha, \beta \in_R [0, p'q')$ such that $a = b^\alpha \mod n$, and $c = b^\beta \mod n$. Let $K = \lfloor 2^{l_s}/p'q' \rfloor$. Output public key (n, a, b, c), and private key $(p'q', \alpha, \beta, K)$.

Signing Algorithm. The signing procedure includes two phases.
OFFLINE PHASE: Pick a random $\gamma \in_R [0, p'q')$, the next unused prime number e with length l_e, and a random $k' \in_R [0, K)$. Compute

$$v = b^\gamma \mod n, \quad \lambda = k'p'q' + \gamma \times e^t - \beta \mod Kp'q'.$$

ONLINE PHASE: For $m \in [0, 2^{l_m})$, compute

$$s = \lambda - \alpha \times m \mod Kp'q'.$$

The signature is (v, e, s).

Verification Algorithm. To verify that (v, e, s) is a signature on message m, check that e's length is l_e, and

$$v^{e^t} \equiv a^m b^s c \mod n.$$

A concrete example would be $k = 512$ so that $l_n = 1024$. We further set up other system parameters as $l_{p'q'} = 200$, $l_m = 160$, $l_e = 28$, $t = 6$, $l = 120$, and $l_s = 200 + 160 + 120 = 480$. The whole signing cost now is about 200 modular multiplications, and finding the next prime number with bit length of 28, which is significantly easier than finding a 162-bit prime number.

4.2 OOSIG3: A Stateless Signature Scheme

In this section we show how to avoid using prime numbers explicitly for the exponent e. In signature schemes OOSIG1 and OOSIG2, there were two important system requirements: that $e^t \geq 2^{l_m+2}$, and e should not be chosen more than once. If we can somehow generate a random integer that always has a prime factor greater than m, we don't have to use a prime number explicitly. In order to accomplish this, we bring in results from Gennaro et al. [10] in a different signature scheme, where the authors introduce the notion of a division intractable hash function and also prove some important properties regarding the smoothness of random integers. In particular, they show that a k-bit random integer (for sufficiently large k) has at least one prime factor greater than $2^{2\sqrt{k}}$ with overwhelming probability. For example, suppose $k = 1024$, then this prime factor is greater than 2^{64}. Unfortunately, this bound is too small compared to our message size which we set $l_m = 160$. To overcome this obstacle, we can split the m into three pieces as $m = m_1||m_2||m_3$ where "||" represents string concatenation, and the bit length of each sub-message is shorter than 63. For simplicity of notation in this section, we split the message into three pieces as just described, but clearly this generalizes to other numbers of pieces. This technique has also been used in the Camenisch-Lysyanskaya scheme for block message signing, and in Fischlin's scheme for reducing the size of e. Therefore, we have the following stateless scheme.

Public System Parameters. The parameters are similar to the OOSIG2 scheme, but with $l_s = l_{p'q'} + l_m/3 + l$. We define a new length l_h to be the length of message digests produced by a division intractable hash function $h : \{0, 1\}^* \rightarrow \{0, 1\}^{l_h}$, with the requirement that $2\sqrt{l_h} \geq l_m/3 + 2$.

Key Generation. On input 1^k, pick two k-bit primes p, q such that $p = 2p'r_p + 1$, and $q = 2q'r_q + 1$, where p' and q' are also prime. Let $n = pq$, and let G be the unique subgroup of Z_n^* of order $p'q'$. Select a random generator b of G, select $\alpha_1, \alpha_2, \alpha_3, \beta \in_R [0, p'q')$, and define

$$a_1 = b^{\alpha_1} \bmod n, \quad a_2 = b^{\alpha_2} \bmod n, \quad a_3 = b^{\alpha_3} \bmod n, \quad c = b^{\beta} \bmod n.$$

Finally, let $K = \lfloor 2^{l_s}/p'q' \rfloor$. The public key is (n, a_1, a_2, a_3, b, c), while the private key is $(p'q', \alpha_1, \alpha_2, \alpha_3, \beta, K)$.

Signing Algorithm. The signing procedure includes two phases.
OFFLINE PHASE: Pick a random $\gamma \in_R [0, p'q')$, a random $r \in_R [0, 2^{l_r})$, and a random $k' \in_R [0, K)$. Compute

$$v = b^{\gamma} \bmod n, \quad \lambda = k'p'q' + \gamma \times h(r) - \beta \bmod Kp'q'.$$

ONLINE PHASE: For $m \in [0, 2^{l_m})$, break m into pieces such that $m = m_1 \| m_2 \| m_3$ and the length of each piece is at most $\lceil l_m/3 \rceil$ bits. Compute

$$s = \lambda - \alpha_1 \times m_1 - \alpha_2 \times m_2 - \alpha_3 \times m_3 \bmod Kp'q'.$$

The signature is (v, r, s).

Verification Algorithm. To verify that (v, r, s) is a signature on message m, check that r's length is l_r, and

$$v^{h(r)} \equiv a_1^{m_1} a_2^{m_2} a_3^{m_3} b^s c \bmod n.$$

This new scheme is extremely efficient for the signer. Given parameters $l_n = 1024$, $l_{p'q'} = 200$, $l_h = 1024$, $l_r = 256$, $l_m = 180$, $l = 120$, and $l_s = 200 + 180/3 + 120 = 380$, the offline signing cost is about 200 modular multiplications, while the online signing needs three multiplications with small numbers.

4.3 Security of OOSIG2 and OOSIG3

Our construction is similar to the schemes by Camenisch and Lysyanskaya [13], Zhu [14,15], Fischlin [16], and Groth [17], except none of these schemes operate in the online/offline paradigm, and none except our OOSIG3 scheme make use of a division intractable hash function. However, due to the similarities, our security proof is similar to those in the previous schemes. The proof we give in this section is specifically for the OOSIG3 scheme, and the proof for OOSIG2 is simply a relaxation of the proof for OOSIG3 where we are guaranteed that the exponent e is a prime number, rather than relying on probabilistic results regarding large factors of random integers.

To prove the security of our scheme we use a multiple generator version of the strong RSA subgroup assumption: Given an appropriate modulus n so that Z_n^* has a subgroup G of size $p'q'$, let $g_1, \ldots, g_k$ be random generators of this subgroup. The problem is to find values $(y, e, e_1, \ldots, e_k)$ such that $y^e = g_1^{e_1} \ldots g_k^{e_k} \mod n$. The following lemma (due to Groth [17]) shows that under the strong RSA subgroup assumption, a probabilistic polynomial time algorithm $\mathcal{A}$ can only reliably find solutions to this problem that have a restricted form.

Lemma 2. *Let $n, g_1, \ldots, g_k$ be defined as above. If a probabilistic polynomial time algorithm $\mathcal{A}$ produces $(y, e, e_1, \ldots, e_k)$ such that $y^e \equiv g_1^{e_1} \ldots g_k^{e_k} \pmod{n}$, then with high probability either $e = e_1 = e_2 = \ldots = e_k = 0$, or $e|e_1, \ldots, e|e_k$ and $y = u \prod_{i=1}^{k} g_i^{e_i/e} \mod n$, where $u^e = 1 \mod n$.*

The next lemma addresses the smoothness of a random integer. The proof can be found in the proof of Lemma 6 presented by Gennaro et al. [10].

Lemma 3. *Let e be a random k-bit integer. The probability of e being $2^{2\sqrt{k}}$-smooth (i.e., all e's prime factors are no larger than $2^{2\sqrt{k}}$) is no larger than $2^{-2\sqrt{k}}$. In other words, the probability of e having at least one prime factor larger than $2^{2\sqrt{k}}$ is at least $1 - 2^{-2\sqrt{k}}$.*

The following lemma is used directly in the security proof for our stateless digital signature scheme — note that the condition on w is met for sufficiently large k whenever w is polynomial in k. The division intractability property of a hash function is based on this lemma: when a hash function outputs k-bit random integers for arbitrary inputs, it is intractable to find an input whose hash value can divide the product of other hash values. Due to the importance of this lemma, we re-write its proof to facilitate understanding of the subsequent proof. The original proof is presented by Gennaro et al. [10].

Lemma 4. *Let $e_1, e_2, \cdots, e_w$ be random k-bit integers, where $w < 2^{0.5\sqrt{k}}$. Let j be a randomly chosen index from $[1, w]$, and define $E = (\prod_{i=1}^{w} e_i)/e_j$. Then the probability that e_j divides E is less than $2^{-\sqrt{k}}$.*

Proof. We denote by smooth the event that e_j is $2^{2\sqrt{k}}$-smooth. From Lemma 3, we know that $Pr[\text{smooth}] \leq 2^{-2\sqrt{k}}$.

Consider the case in which e_j is not $2^{2\sqrt{k}}$-smooth. Then e_j has at least one prime factor $p > 2^{2\sqrt{k}}$, so $Pr[e_j \text{ divides } E]$ is bounded by the probability that at least one of the e_i ($i \neq j$) is divisible by p. Since the e_i's are chosen uniformly, the probability that any specific e_i is divisible by p is at most $1/p < 2^{-2\sqrt{k}}$. Then, the probability that there exists an e_i which is divisible by p is at most $w \times 2^{-2\sqrt{k}}$, and based on the bound on w given in the lemma we get $w \times 2^{-2\sqrt{k}} < 2^{-1.5\sqrt{k}}$. Therefore, $Pr[p \text{ divides } E | \neg\text{smooth}] < 2^{-1.5\sqrt{k}}$, and since p is a prime factor of e_j, we get $Pr[e_j \text{ divides } E | \neg\text{smooth}] < 2^{-1.5\sqrt{k}}$.

Therefore, the probability that e_j divides E is at most

$$Pr[\text{smooth}] + Pr[e_j \text{ divides } E|\neg\text{smooth}] < 2^{-2\sqrt{k}} + 2^{-1.5\sqrt{k}} < 2^{-\sqrt{k}},$$

which completes the proof. $\qquad\qquad\qquad\qquad\qquad\qquad\qquad\qquad\qquad\qquad\qquad\quad$ $\square$

Given these lemmas, we can now prove the security of OOSIG3.

Theorem 2. *Under the strong RSA subgroup assumption, OOSIG3 is existentially un-forgeable under an adaptive chosen message attack.*

Proof. Suppose there exists a probabilistic polynomial time forgery algorithm $\mathcal{F}$, which can launch an adaptive chosen message attack on OOSIG3 and output a valid signature which has not been produced by the signing algorithm. Then we can construct a probabilistic polynomial time algorithm $\mathcal{A}$ for the multiple generator version of the strong RSA subgroup problem, defined immediately before Lemma 2. $\mathcal{A}$ takes a random input (n, g_1, g_2, g_3), with $g_1, g_2, g_3 \in_R G$, and uses $\mathcal{F}$ as a subroutine. In the following proof, all exponentiations are done modulo n.

Since G has order $p'q'$, where p' and q' are prime, the probability that g_1, g_2, g_3 are generators of G is $(p' - 1)(q' - 1)/p'q'$, which is an overwhelming probability. So, in the sequel, we assume g_1, g_2, g_3 are generators of G.

Suppose $\mathcal{F}$ asks w signature queries for messages $m_1, \ldots, m_w$, obtaining signatures $(v_1, r_1, s_1), \ldots, (v_w, r_w, s_w)$ before forging a valid signature (v, r, s) for a message m. We can define three types of forgeries.

Type I: For all $1 \leq i \leq w, r \neq r_i$.
Type II: For some $1 \leq i \leq w, r = r_i, v = v_i$.
Type III: For some $1 \leq i \leq w, r = r_i, v \neq v_i$.

Any forgery algorithm which succeeds in producing forgeries must produce forgeries of at least one of these types with non-negligible probability. Next, we show how to construct three different algorithms for $\mathcal{A}$ such that if $\mathcal{F}$ succeeds in producing a forgery of a particular type, then the corresponding $\mathcal{A}$ will succeed in solving the multiple generator version of the strong RSA subgroup problem. In all three cases we show that such an $\mathcal{A}$ is impossible, so we conclude that no successful forger $\mathcal{F}$ can exist.

Type I: For all $1 \leq i \leq w, r \neq r_i$. $\mathcal{A}$ works as follows: choose according to the signature algorithm distinct l_r-bit integers $r_1, \ldots, r_w$. Set $E = \prod_{i=1}^{w} h(r_i)$. $\mathcal{A}$ selects $t_1, t_2 \in_R [0, 2^{l_s})$, and sets $a_1 = g_1^E, a_2 = a_1^{t_1}, a_3 = a_1^{t_2}, b = g_2^E, c = g_3^E$. $\mathcal{A}$ gives (n, a_1, a_2, a_3, b, c) to the forger $\mathcal{F}$. $\mathcal{A}$ can answer the forger $\mathcal{F}$'s signature query $m_i = m_{i1}||m_{i2}||m_{i3}$ by choosing $s_i \in_R [0, 2^{l_s})$ and computing

$$v_i = g_1^{(m_{i1}+t_1 m_{i2}+t_2 m_{i3})E/h(r_i)} g_2^{s_i E/h(r_i)} g_3^{E/h(r_i)} = (a_1^{m_{i1}} a_2^{m_{i2}} a_3^{m_{i3}} b^{s_i} c)^{h(r_i)^{-1}}.$$

$\mathcal{A}$ gives $\mathcal{F}$ the signature (v_i, r_i, s_i) for message m_i, which is statistically indistinguishable from the signature produced by OOSIG3.

Consider that the forger $\mathcal{F}$ outputs (v, r, s) for a message m, so we have

$$v^{h(r)} = a_1^{m_1} a_2^{m_2} a_3^{m_3} b^s c = g_1^{(m_1+t_1 m_2+t_2 m_3)E} g_2^{sE} g_3^E.$$

By Lemma 2, $h(r)$ must divide E, but since E is the product of hash values this would violate the division intractability of the hash function h. Therefore, the $\mathcal{F}$ cannot produce a Type I forgery with non-negligible probability.

Type II: For some $1 \leq i \leq w, r = r_i, v = v_i$. $\mathcal{A}$ follows the same method as in Type I to prepare E, a_1, a_2, a_3, b, c, and answers the forger $\mathcal{F}$'s signature queries.

Consider now that the forger's signature is (v_i, r_i, s) on message m. We have

$$a_1^{m_{i1}} a_2^{m_{i2}} a_2^{m_{i3}} b^{s_i} = a_1^{m_1} a_2^{m_2} a_3^{m_3} b^s,$$

and so $a_1^{m_{i1}-m_1} a_2^{m_{i2}-m_2} a_3^{m_{i3}-m_3} b^{s_i-s} = 1$, which gives

$$g_1^{E((m_{i1}-m_1)+t_1(m_{i2}-m_2)+t_2(m_{i3}-m_3))} g_2^{E(s_i-s)} = 1 = y^0$$

for any non-zero y. Since $m_i \neq m$, $(m_1 - m_{i1}) + t_1(m_2 - m_{i2}) + t_2(m_3 - m_{i3}) \neq 0$. However, this is infeasible by Lemma 2 under the strong RSA subgroup assumption.

Type III: For some $1 \leq i \leq w, r = r_i, v \neq v_i$. $\mathcal{A}$ guesses the forger $\mathcal{F}$ will make the forgery by reusing r_i. $\mathcal{A}$ prepares E as in Type I, and picks at random an l_s-bit long t, and $(l_s - l_m/3 - l/2)$-bit long t_1, t_2, t_3, t_4. Then set up

$$b = g_2^{E/h(r_i)}, \ a_1 = b^{t_1}, \ a_2 = b^{t_2}, \ a_3 = b^{t_3}, \ c = b^{h(r_i)t_4 - t}.$$

$\mathcal{A}$ can answer all queries $j \neq i$ as in Type I. For query i, $\mathcal{A}$ computes $s_i = t - t_1 m_{i1} - t_2 m_{i2} - t_3 m_{i3}$, $v_i = b^{t_4}$ such that (v_i, r_i, s_i) is also a valid signature. Due to length restriction over t, t_1, t_2, and t_3, the distribution of s_i is statistically indistinguishable from the uniform distribution over $[0, 2^{l_s})$, which is in turn indistinguishable from the distribution of signatures produced by OOSIG3.

Consider now that the forgery $\mathcal{F}$ outputs a new signature (v, r_i, s) on message m. That is, $v^{h(r_i)} = a_1^{m_1} a_2^{m_2} a_3^{m_3} b^s c$. We can obtain

$$(v/v_i)^{h(r_i)} = g_2^{((m_1-m_{i1})t_1 + (m_2-m_{i2})t_2 + (m_3-m_{i3})t_3 + (s-s_i))E/h(r_i)}.$$

We can assume that $h(r_i)$ has a prime factor $\pi > 2^{2\sqrt{l_h}}$ and that $h(r_i)$ divides $((m_1 - m_{i1})t_1 + (m_2 - m_{i2})t_2 + (m_3 - m_{i3})t_3 + (s - s_i))E/h(r_i)$ — these assumptions fail with negligible probability, due to Lemma 3 and Lemma 2, respectively. Furthermore, by Lemma 4, the probability that $E/h(r_i)$ contains π as a factor is negligible, so if $\mathcal{F}$ succeeds with non-negligible probability, it must be the case that

$$((m_1 - m_{i1})t_1 + (m_2 - m_{i2})t_2 + (m_3 - m_{i3})t_3 + (s - s_i)) \tag{3}$$

is divisible by π with non-negligible probability.

Let $t_1 = x_1 p' q' + t_1'$, and note that the forger's view is independent of x_1. Therefore, if the forger succeeds for this value of t_1 it must also succeed for a random $\widehat{t_1} = \widehat{x_1} p' q' + t_1'$ with $\widehat{x_1} \neq x_1$. Thus, π must divide (3) when t_1 is replaced by $\widehat{t_1}$, and so must divide the difference of these two values, leading to the requirement that π divides $(m_1 - m_{i1})(t_1 - \widehat{t_1}) = (m_1 - m_{i1})(x_1 - \widehat{x_1})p' q'$. However, since $2\sqrt{l_h} \geq l_m/3 + 2$, π cannot divide $m_1 - m_{i1}$. Furthermore, the probability that a random factor π divides $p' q'$ is negligible, and since x_1 and $\widehat{x_1}$ are chosen randomly the probability that π divides $(x_1 - \widehat{x_1})$ is also negligible. As we have exhausted all possibilities for $\mathcal{F}$ to succeed with a Type III forgery, we conclude that Type III forgeries are infeasible. $\qquad\square$

5 Conclusions

In this paper we have presented a family of three new efficient digital signature schemes, which are proved secure under the strong RSA assumption or its variant (the strong RSA subgroup assumption) in the standard model. All our schemes works in a two-phase offline/online model, so after some offline precomputation that is independent of the message to be signed, the online phase is highly efficient (one non-modular multiplication). Our schemes are particularly suitable for the application scenarios introduced in Sect. 1.

Our first scheme OOSIG1 is proved secure in the standard model without any special assumptions. It greatly improves current research results on provable signature schemes such as the Cramer-Shoup scheme, the Camenisch-Lysyanskaya's scheme, etc. However, it still has a shortcoming that a 162-bit random prime number is needed to generate a signature. This might not be a problem for powerful devices such as personal computer, but it does affect the applicability of the scheme on embedded devices.

Our OOSIG2 and OOSIG3 schemes are designed for embedded devices. These two schemes work in a small subgroup of Z_n^*, which greatly reduces computational overhead related to modular multiplication. OOSIG2 uses a technique from Groth to reduce the cost of random prime number generation. OOSIG3 completely excludes a prime number generation by using a division-intractable hash function. OOSIG2 and OOSIG3 were proved secure in the standard model.

Our OOSIG3 has some particular contributions. Compared to the well-known digital signature scheme for smart cards by Schnorr, our scheme has comparable computational requirements and is provably secure in the standard model, while Schnorr's scheme relies on the Fiat-Shamir heuristic and thus can only be demonstrated secure in the random oracle model. To the best of our knowledge, OOSIG3 is the most efficient online/offline signature scheme provably secure in the standard model, and suitable for embedded devices.

References

1. Diffie, W., Hellman, M.E.: New directions in cryptography. IEEE Transactions on Information Theory 11, 644–654 (1976)
2. Bellare, M., Rogaway, P.: The exact security of digital signatures — how to sign with RSA and Rabin. In: Maurer, U.M. (ed.) EUROCRYPT 1996. LNCS, vol. 1070, pp. 399–416. Springer, Heidelberg (1996)
3. Cramer, R., Shoup, V.: Signatures schemes based on the strong RSA assumption. In: ACM Transaction on Information and System Security, pp. 161–185 (2000)
4. El Gamal, T.: A public key cryptosystem and a signature scheme based on discrete logarithms. In: Blakely, G.R., Chaum, D. (eds.) CRYPTO 1984. LNCS, vol. 196, pp. 10–18. Springer, Heidelberg (1985)
5. Schnorr, C.: Efficient signature generation for smart cards. Journal of Cryptology 4(3), 161–174 (1991)
6. Goldwasser, S., Micali, S., Rivest, R.: A digital signature scheme secure against adaptive chosen-message attacks. SIAM J. Computing 17, 281–308 (1988)
7. Fiat, A., Shamir, A.: How to prove yourself: practical solutions to identification and signature problems. In: Odlyzko, A.M. (ed.) CRYPTO 1986. LNCS, vol. 263, pp. 186–194. Springer, Heidelberg (1987)

8. Bellare, M., Rogaway, P.: Random oracles are practical: A paradigm for designing efficient protocols. In: First ACM Conference on Computer and Communication Security, pp. 62–73 (1993)

9. Canetti, R., Goldreich, O., Halevi, S.: The random oracle model, revisited. In: 30th Annual ACM Symposium on Theory of Computing, pp. 209–218 (1998)

10. Gennaro, R., Halevi, S., Rabin, T.: Secure hash-and-sign signatures without the random oracle. In: Stern, J. (ed.) EUROCRYPT 1999. LNCS, vol. 1592, pp. 123–139. Springer, Heidelberg (1999)

11. Cramer, R., Damgard, I.: New generation of secure and practical RSA-based signatures. In: Koblitz, N. (ed.) CRYPTO 1996. LNCS, vol. 1109, pp. 173–185. Springer, Heidelberg (1996)

12. Dwork, C., Naor, M.: An efficient existentially unforgeable signature scheme and its applications. J. Cryptology 11(3), 187–208 (1988)

13. Camenisch, J., Lysyanskaya, A.: A signature scheme with efficient protocols. In: Cimato, S., Galdi, C., Persiano, G. (eds.) SCN 2002. LNCS, vol. 2576, pp. 268–289. Springer, Heidelberg (2003)

14. Zhu, H.: New digital signature scheme attaining immunity to adaptive chosen-message attack. Chinese Journal of Electronic 10(4), 484–486 (2001)

15. Zhu, H.: A formal proof of Zhu's signature scheme (2003),
    http://eprint.iacr.org/

16. Fischlin, M.: The Cramer-Shoup strong-RSA signature scheme revisited. In: Desmedt, Y.G. (ed.) PKC 2003. LNCS, vol. 2567, pp. 116–129. Springer, Heidelberg (2002)

17. Groth, J.: Cryptography in subgroups of Z_n^*. In: Kilian, J. (ed.) TCC 2005. LNCS, vol. 3378, pp. 50–65. Springer, Heidelberg (2005)

18. Even, S., Goldreich, O., Micali, S.: On-line/off-line digital signatures. In: Brassard, G. (ed.) CRYPTO 1989. LNCS, vol. 435, pp. 263–275. Springer, Heidelberg (1990)

19. Shamir, A., Tauman, Y.: Improved online/offline signature schemes. In: Kilian, J. (ed.) CRYPTO 2001. LNCS, vol. 2139, pp. 355–367. Springer, Heidelberg (2001)

20. Krawczyk, H., Rabin, T.: Chameleon signatures. In: Symposium on Network and Distributed Systems Security – NDSS 2000, pp. 143–154 (2000)

21. Goldwasser, S., Kalai, Y.T.: On the (in)security of Fiat-Shamir paradigm. In: Proceedings of the 44th Annual IEEE Symposium on Foundations of Computer Science — FOCS 2003, pp. 102–114 (2003)

22. Damgard, I.: Collision free hash functions and public key signature schemes. In: Price, W.L., Chaum, D. (eds.) EUROCRYPT 1987. LNCS, vol. 304, pp. 203–216. Springer, Heidelberg (1988)

23. Baric, N., Pfitzmann, B.: Collision-free accumulators and fail-stop signature schemes without trees. In: Fumy, W. (ed.) EUROCRYPT 1997. LNCS, vol. 1233, pp. 480–494. Springer, Heidelberg (1997)

24. Fujisaki, E., Okamoto, T.: Statistical zero knowledge protocols to prove modular polynomial relations. In: Kaliski Jr., B.S. (ed.) CRYPTO 1997. LNCS, vol. 1294, pp. 16–30. Springer, Heidelberg (1997)

25. Mao, W.: Modern Cryptography: Theory & Practice. Prentice-Hall, Englewood Cliffs (2004)

26. Coron, J.S., Naccache, D.: Security analysis of the Gennaro-Halevi-Rabin signature scheme. In: Preneel, B. (ed.) EUROCRYPT 2000. LNCS, vol. 1807, pp. 91–101. Springer, Heidelberg (2000)

27. Kurosawa, K., Schmidt-Samoa, K.: New online/offline signature schemes without random oracles. In: Yung, M., Dodis, Y., Kiayias, A., Malkin, T.G. (eds.) PKC 2006. LNCS, vol. 3958, pp. 330–346. Springer, Heidelberg (2006)

RFID Security: Tradeoffs between Security and Efficiency

Ivan Damgård and Michael Østergaard Pedersen

University of Aarhus
{ivan,michael}@daimi.au.dk

Abstract. We propose a model and definition for anonymous (group) identification that is well suited for RFID systems. This is based on the definition of Juels and Weis of strong privacy for RFID tags, where we add requirements for completeness and soundness. We also propose a weaker and more realistic definition of privacy. For the case where tags hold independent keys, we prove a conjecture by Juels and Weis, namely in a strongly private and sound RFID system using only symmetric cryptography, a reader must access virtually all keys in the system when reading a tag. It was already known from work by Molnar, Soppera and Wagner that when keys are dependent, the reader only needs to access a logarithmic number of keys, but at a cost in terms of privacy: For that system, privacy is lost if an adversary corrupts just a single tag. We propose protocols offering a new range of tradeoffs between security and efficiency. For instance, the number of keys accessed by a reader to read a tag can be significantly smaller than the number of tags while retaining soundness and privacy, as long as we assume suitable limitations on the adversary.

1 Introduction

RFID tags are small wireless devices that react to electromagnetic fields generated by an RFID reader; they can emit some prestored information and can also do computation. The computing power one can assume an RFID tag to have, however, is severely limited in many applications by requirements for extremely low price tags. RFID technology holds great promise in many scenarios, but can also lead to serious privacy problems, for instance because it becomes possible to track the behavior and whereabouts of people carrying tagged items.

Several research works have proposed protocols for addressing the privacy problem in RFID systems. However, until recently, not much work has addressed formal definitions of security for RFID systems. In [9], Juels and Weis propose a definition of what they call "strong privacy" (based on earlier work by Avoine [2]). Strong privacy is indeed a strong notion, primarily because the adversary is given a lot of power: He can corrupt any number of tags (but not the reader) and read their contents, he can eavesdrop and schedule the tag/reader communication any way he wants, and he can himself select the tags whose privacy he wants to break. In independent work, Burmester, Le and Medeiros

T. Malkin (Ed.): CT-RSA 2008, LNCS 4964, pp. 318–332, 2008.
© Springer-Verlag Berlin Heidelberg 2008

propose a security definition based on Canetti's Universal Composability framework [6] and they also propose a protocol secure in their model [14].

The work of Juels and Weis only addresses privacy, that is, making sure that the communication of a tag does not allow an external adversary to determine the identity of the tag. Of course, another natural requirement is that a reader should be able to determine whether the tag it reads is valid and not fabricated by an adversary, for instance. Indeed, if this was not required, tags could just return random information all the time or just not reply at all. This would trivially be private, but would of course lead to a useless system.

In this paper, we propose an extension to the strong privacy definition so one can also require completeness and soundness, with the intuitive meaning that the reader accepts valid tags and valid tags only. More specifically, soundness in the weakest sense means that we assume the adversary cannot corrupt tags, and when the reader accepts an instance of the read protocol, an (uncorrupted) tag has been involved in that instance at some point. So in this weak flavor, it is not required that the reader knows which tag it has been talking to. We also suggest a stronger version where corruptions are allowed and the reader must output the identity of the (honest) tag that was involved.

The concept of strongly private and sound systems is closely related to existing concepts for anonymous identification schemes, such as identity escrow schemes [11] or group signature schemes [1,7,10]. They are not the same, however: Our system model is designed to model RFID systems, and where identity escrow and group signature schemes are by definition public-key techniques, we want to cover techniques based on secret-key algorithms only.

The most important privacy issue regarding RFID tags is the issue of being able to systematically track individuals as they carry RFID enabled goods from the supermarket, embedded in the their clothes, etc. In this scenario, it is reasonable to assume that the adversary cannot himself choose the tags he wants to track. Strong privacy is therefore more than we need in this scenario, so we introduce a weaker, but more suitable, definition called *benign-selection privacy*.

Juels and Weis suggest a system that satisfies their definition, building on earlier work by Weis, Sarma and Rivest [15]. In this scheme, each tag is given an independently chosen key, and the reader must search exhaustively through all keys every time a tag is read. This of course does not scale well, but Juels and Weis conjecture that this is, in a certain sense, unavoidable: In strongly private systems that use only symmetric cryptography, and where tags are independently keyed, the reader must access all, or at least a large fraction of the keys in the system. Here, we prove this conjecture. We need to assume that the system is complete and sound, but this is of course a natural requirement and is necessary anyway to exclude degenerate cases, such as when tags only send random information.

The limitation to symmetric cryptography is clearly necessary: With public-key technology, a tag could send its identity encrypted under the reader's public-key, and then prove its identity using some shared-key technique, for instance. This does not require the reader to look at any information that is not related to the relevant tag. There has in fact been recent work in the direction of implementing public-key on very small devices [5], but even if public-key enabled RFID tags are only slightly more expensive than symmetric-key only tags, this will still inhibit the use of public-key technology in large scenarios that require millions of tags, in order to maximize profit. Therefore we believe the question of what can be done with symmetric techniques is of interest, both theoretically and in practice.

The limitation to schemes with independent keys is not surprising. It follows from work by Molnar, Soppera and Wagner [12] that when dependent keys are allowed, we can have a system where the reader only needs to look at a logarithmic (in the number of tags) number of keys. This comes at the price that strong privacy only holds if the adversary is "radio-only", i.e., he does not corrupt any tags. If the adversary corrupts even a single tag, strong privacy is lost, and benign-selection privacy is lost with large probability. This makes it natural to ask if there are alternative solutions where we can get some amount of privacy with a larger number of corruptions without going back to systems where the reader does exhaustive search over all keys.

In this paper, we first argue that for a wide range of RFID systems, there has to be a tradeoff between the efficiency of the reader and the resources we can allow the adversary to have. We then propose a class of protocols offering a new range of tradeoffs between security and efficiency. For instance, the number of keys accessed by a reader to read a tag can be significantly smaller than the number of tags while retaining soundness and privacy, as long as we assume suitable limitations on the adversary.

2   Model and Definition

Juels and Weis define *strong privacy* for RFID systems using a model of which we give a summary here, for details refer to [9].

The system consists of tags $T_i, i = 1..n$ and a reader $\mathcal{R}$. For simplicity, we assume that there is only one reader. Tags can receive SETKEY messages which will cause the tag to reveal its secret key, and the caller may then send a new key to the tag. This can be used to initialize the system and also models an attacker corrupting a tag to learns its key. A tag may receive a (TAGINIT, *sid*) message (where *sid* is a session id), which is used in the start of a session. The tag will forget any previous value of *sid*, so a tag may only run a single session at a time. Finally, the tag may respond to a protocol message c_i, called a challenge in [9], by a response r_i. A protocol may consist of several rounds of challenges and responses.

A Reader may receive READERINIT messages, causing it to generate a fresh session identifier *sid* and a first protocol message c_0 to be sent to a tag. It may also receive pairs of the form (sid, r_i). It will then return either a new message c_{i+1} to be sent to the tag or *Accept* or *Reject*. In [9], a reader, if it returns *Accept*, is not required to say which tag it thinks it has been talking to. We assume here that it may also return the identity of a tag. The reader keeps an internal log of all challenge and response pairs for each session id that is active, and decides based on this whether to accept or reject. A reader may be involved in several sessions simultaneously, but its behavior in a session only depends on messages it receives in that session and the fixed key material it holds.

We allow the adversary $\mathcal{A}$ to schedule all messages as it wants, and generate its own messages. The adversary is parameterized as follows: r is the number of READERINIT messages it generates, s is the number of computational steps and t is the number of TAGINIT messages it generates. Finally, k is a cryptographic security parameter. Juels and Weis do not treat the number of SETKEY messages, i.e., the number of corrupted tags, as a separate parameter, but simply say it has to be at most $n-2$. As we shall see, however, the number of corrupted tags is a very important parameter, so we will define u to be the number of tags corrupted by the adversary. A summary of these parameters can be found in Figure 1. Note that this model also captures an adversary that passively listens to a session between reader and tag, namely he starts a session with the reader and one with the tag and simply relays messages between the parties.

k: security parameter n: number of tags in the RFID system $\mathcal{S}$
r: number of READERINIT messages allowed s: number of computational steps allowed
t: number of TAGINIT messages allowed u: number of SETKEY messages allowed

Fig. 1. Description of parameters

The system is initially setup by running a probabilistic key generation algorithm $\text{GEN}(1^k)$ which produces a set of keys $key_1, ..., key_n$ to be assigned to the tags. Of course, $\mathcal{A}$ does not know these keys initially.

Let $\mathcal{S} = (\text{GEN}, \mathcal{R}, \{\mathcal{T}_i\})$ denote an RFID system. Strong privacy is defined via an experiment called $\text{Exp}_{\mathcal{A},\mathcal{S}}^{priv}[k, n, r, s, t]$. Here, we run the system where the adversary may corrupt tags, initiate sessions, etc., observing the limitations put on him. This ends by the adversary selecting two uncorrupted tags, called $\mathcal{T}_0^*, \mathcal{T}_1^*$. He is then given oracle access to $\mathcal{T}_b^*$ where b is a random bit. He may now again corrupt other tags and initiate sessions, and must finally guess the value of b. However, we have to assume that in this last phase, when using the reader to interact with $\mathcal{T}_b^*$, he only learns whether the reader outputs accept or reject and not the identity found by the reader. Otherwise, he could just let the reader identify $\mathcal{T}_b^*$. The system is said to be (r, s, t)-private if any (r, s, t)-adversary's advantage over $1/2$ in guessing b is negligible as a function of k. We propose here to define also (r, s, t, u)-privacy, which is the same, except that the adversary may only corrupt at most u tags. However, for some systems, the advantage that can be achieved depends not only k, but on all the parameters, and does not tend to 0 as we increase k, if other parameters are constant. We will therefore use a variant of strong privacy here:

Definition 1. *Strong $(k, r, s, t, u, n, \epsilon)$-privacy is defined via the experiment* $\text{Exp}_{\mathcal{A},\mathcal{S}}^{priv}[k, n, r, s, t, u]$ *which is the same as Juels and Weis' except that the*

adversary can only corrupt up to u tags. We say that the system is strongly
$(k, r, s, t, u, n, \epsilon)$-*private if any adversary observing the limitations in the exper-*
iment has advantage at most ϵ.

Experiment $\mathbf{Exp}_{A,S}^{priv}[k, n, r, s, t, u]$ **Setup:**

1. $\text{GEN}(1^k) \rightarrow (key_0, ..., key_n)$
2. Initialize $\mathcal{R}$ with $(key_0, ..., key_n)$
3. Set each $\mathcal{T}_i$'s key to key_i with a SETKEY call

Phase 1 (Learning):

4. $\mathcal{A}$ may do the following in any interleaved order:
   (a) Make READERINIT calls, without exceeding r overall calls
   (b) Make TAGINIT calls, without exceeding t overall calls
   (c) Make SETKEY calls, without exceeding u overall calls
   (d) Communicate and compute without exceeding s overall steps

Phase 2 (Challenge):

5. $\mathcal{A}$ selects two tags $\mathcal{T}_i$ and $\mathcal{T}_j$ to which it did *not* send SETKEY messages
6. Let $\mathcal{T}_0^* = \mathcal{T}_i$ and $\mathcal{T}_1^* = \mathcal{T}_j$ and remove both of these from the current tag set
7. Choose a random bit $b \in \{0, 1\}$ and provide $\mathcal{A}$ access to $\mathcal{T}_b^*$
8. $\mathcal{A}$ may do the following in any interleaved order:
   (a) Make READERINIT calls, without exceeding r overall calls
   (b) Make TAGINIT calls, without exceeding t overall calls
   (c) Make SETKEY calls, without exceeding u overall calls to any tag in the current tag set
   (d) Communicate and compute without exceeding s overall steps
9. $\mathcal{A}$ outputs a guess bit b'

$\mathcal{A}$ succeeds if $b = b'$

As Juels and Weis note in [9], strong privacy may be too strong a notion for many real world applications. In particular, the adversary can freely choose the target tags he wants to be challenged on. He may not have that much power in real life, where the choice may be forced on him by the environment he operates in. One may try to model this by having the target tags be chosen from some distribution independently of the adversary – this ideas is already present in the work of Avoine [2]. But it is very difficult to single out a distribution that realistically models the environment. We therefore propose a new model called *benign-selection privacy* where we allow any distribution as long as it only selects uncorrupted tags.

Definition 2. *Benign-selection privacy is defined via an experiment called* $\mathbf{Exp}_{A,S,\mathcal{D}}^{bspriv}[k, n, r, s, t, u]$ *which is the same as* $\mathbf{Exp}_{A,S}^{priv}[k, n, r, s, t, u]$, *except that the adversary does not select the two tags* $\mathcal{T}_0^*, \mathcal{T}_1^*$. *Instead they are chosen at random from distribution D among all uncorrupted tags. We think of D as a probabilistic algorithm that only gets the set of corrupted tags as input, and outputs* *the index of the target tags, i.e., the choice is uncorrelated to adversarial activity other than corruptions. We say that the system is $(k, r, s, t, u, n, \epsilon)$-private with benign D-selection if any adversary observing the limitations in the experiment has advantage at most ϵ.*

In the following, it will often be cumbersome and unnecessarily complicated to specify s, the number of computational steps, exactly. We will often replace s by a $poly(k)$, meaning that the statement involved holds for any adversary that uses time polynomial in k.

It is natural to expect a system as described here to also have the properties that valid tags are accepted, and that the adversary cannot impersonate a tag unless he corrupts it. This aspect was not treated in [9] (but was also not the main goal there). We propose to define this as follows:

Completeness. Assume that at the end of session sid the internal log of the reader $\mathcal{R}$ for that session contains pairs (c_j, r_j) where all r_j's were generated by an honest tag in correct order. Completeness means that $\mathcal{R}$ outputs Accept with probability 1 for any such session.

Strong Soundness. Consider the following experiment similar to the privacy experiment of Juels and Weis:

Experiment $\mathbf{Exp}_{\mathcal{A},\mathcal{S}}^{sound}[k, n, r, s, t, u]$:

Setup:
1. $\textsc{Gen}(1^k) \rightarrow (key_0, ..., key_n)$
2. Initialize $\mathcal{R}$ with $(key_0, ..., key_n)$
3. Set each $\mathcal{T}_i$'s key to key_i with a $\textsc{SetKey}$ call

Attack:
4. $\mathcal{A}$ may do the following in any interleaved order:
   (a) Make $\textsc{ReaderInit}$ calls, without exceeding r overall calls
   (b) Make $\textsc{TagInit}$ calls, without exceeding t overall calls
   (c) Make $\textsc{SetKey}$ calls, without exceeding u overall calls
   (d) Communicate and compute without exceeding s overall steps

Let E be the event that occurs if $\mathcal{R}$ at some point outputs $(Accept, i)$ at the end of session sid where $\mathcal{T}_i$ is not corrupted, yet $\mathcal{R}$'s internal entry for sid only contains pairs (c_j, r_j) where r_j was not sent by T_i as a response to c_j, i.e., $\mathcal{T}_i$ has not been involved in the session. We say that the system provides strong (r, s, t, u)-soundness if the probability that E occurs is negligible in k.

Weak Soundness. Weak (r, s, t)-soundness is defined by the same experiment as above, except that $\mathcal{R}$ now only has to output Accept or Reject at the end of a session, $\mathcal{A}$ is not allowed to corrupt tags, and the error event E is now defined to be that $\mathcal{R}$ outputs Accept, and yet no tag has been involved in the session.

3   Independent Keys

As mentioned earlier, our goal in this section is to prove the speculation by Juels and Weis: In any strongly private, complete and sound RFID system, the reader must access a key for every tag, or at least a large fraction of them, when reading a tag. This can only be expected to hold, however, when keys for different tags are independently chosen, and the system "only" uses symmetric cryptography. If public-key cryptography was allowed, a tag could first encrypt its identity under the reader's public-key, and then show possession of some secret that is shared between reader and this tag only.

To prove something, we need to formalize the constraints on the system. For the independence of keys, this is easy, we simply assume that each tag T_i gets a key K_i chosen independently from all other keys by a key generation algorithm G_i, i.e., $K_i \leftarrow G_i(1^k)$ where k is the security parameter. As for the constraint that "symmetric cryptography and nothing else is used", we will give the system access to a pseudorandom function, $\phi.(\cdot)$, and we will assume that every key K_i in the system is used only as a key to this function, i.e., tag T_i or reader use $\phi_{K_i}(\cdot)$ as a black box. This means that we can equivalently give tags and reader oracle access to $\phi_{K_i}(\cdot)$ for any key they need to use. Therefore, when in the following we say that "the reader accesses a key", this means it calls the oracle that holds that key.

Now, to model that the pseudorandom functions are the essential cryptographic resource used, we will simply assume that the keys $\{K_i\}$ held by the reader and tags are the only secret data in the system. More precisely, we think of the reader's algorithm as an interactive Turing machine that takes no private input, but may make oracle calls to $\phi_{K_i}(\cdot)$ for any K_i. Similarly, a tag may only call its own pseudorandom function, whereas the adversary may only call $\phi_{K_i}(\cdot)$ if he has corrupted T_i. We will say that such a system is *essentially symmetric*.

Note that an essentially symmetric system is not prevented from using public-key, or using secret-key techniques in a non-blackbox way – the reader could try to do a Diffie-Hellman key exchange with a tag, for instance, or generate a key for a pseudorandom function and use this key in any way it wants. Nevertheless, the constraints we have defined are sufficient to show what we are after. To get better intuition for why this is the case, one may note that, while the reader is free to generate a public encryption key and send it to a tag, the tag cannot immediately verify that the key comes from the reader and not the adversary. Thus it would not be secure to send the tag's id encrypted under the public key.

The first lemma formalizes the straightforward intuition that if keys are independent, a reader cannot determine if it is talking to a valid tag unless it accesses the key for that tag. More formally:

Lemma 1. *Consider an RFID system that is complete, weakly $(1, poly(k), 0)$-sound, and uses independent keys. Consider a session between reader and a tag where the adversary does not modify the traffic. In any such session, the algorithm executed by the reader when reading a tag T_i will access K_i, except with negligible probability.*

Proof. We consider all probabilities as taken over the choice of keys and the random coins used by tag and reader in the session. Let E be the event that the reader does not access ϕ_{K_i}. By completeness, the reader should accept with probability 1, so the probability that the reader accepts and E occurs equals $Pr(E)$. Assume for contradiction that $Pr(E)$ is non-negligible. Then an adversary could fabricate his own tag T_i' with a key K_i' generated by G_i, and start a session between this tag and the reader, while simply following the protocol. Now by independence of keys, as long as E occurs, conversations with T_i' and T_i are perfectly indistinguishable. Hence, the reader accepts with probability at least $Pr(E)$, which contradicts weak soundness.   □

The next theorem uses the observation that in an essentially symmetric system, the *only* difference between the honest reader and an adversary is that the reader has access to all keys, while the adversary initially does not. He can, however, corrupt tags and get access to (some of) the keys. He can therefore potentially run the same algorithm that the reader uses when reading a tag.

Theorem 1. *Assume an essentially symmetric RFID system is complete and weakly $(1, poly(k), 0)$-sound. Assume also that the reader algorithm accesses at most αn of the keys, for a constant $\alpha < 1/2$. Such a system cannot have strong $(k, 0, poly(k), 1, \alpha n, n, 1/2 - \alpha)$-privacy*

Proof. We describe an adversary that will break strong privacy for any system that is complete and weakly sound and where only αn oracles are accessed. The adversary picks uniformly a pair of tags T_i, T_j, and uses these two as the challenge pair (T_0^*, T_1^*) from the strong privacy definition. It then gets oracle access to T_b^*, where $b = 0$ or 1 and should try to guess which of the two it is talking to. To do this, it executes the read protocol with T_b^*, and while doing so, it emulates the reader's algorithm. Whenever the reader algorithm wants to access K_t, the adversary corrupts T_t, and may now call the pseudorandom function with key K_t. This goes on until the reader algorithm wants to access K_t where $t = i$ or j. In this case the adversary outputs 0 if $t = i$ and 1 otherwise and then stops.

   To analyze the probability that this adversary has success, suppose, for instance, that $b = 0$. Since our adversary follows the protocol when talking to T_b^*, we can apply Lemma 1 to conclude that the reader will access K_i when talking to T_b^* with probability essentially 1. On the other hand, the probability that it will not access K_j is greater than $1 - \alpha$ because only αn keys are accessed (one of which is K_i), and given i, j is uniform over all values different from i. It follows that the adversary's guess is correct with probability $1 - \alpha$ which is a constant greater than $1/2$ and hence we contradict strong privacy.   □

Note that since the adversary we construct in the proof selects target tags uniformly, this same argument also shows that a system as specified in the theorem cannot even have benign D-selection privacy where D is the uniform distribution.

   One might use some form of pre-computation to perform key lookups more efficiently. For example Avoine, Dysli, and Oechslin [4,3] propose to use Hellman tables [8] in the protocol of Ohkubo, Suzuki and, Kinoshita, to reduce key lookup time to $O(n^{2/3})$ at the cost of using an additional $O(n^{2/3})$ space [13]. Since the construction of the table requires accessing all keys in the system, methods using Hellman tables do not immediately contradict the lower bound. We can, however, argue that such methods cannot provide both soundness and privacy: To initialize such a table one must predict all possible outputs from the tag, which in turn means that the tag can only have a fixed number of outputs, m. Juels and Weis show how to break strong privacy for such a scheme, simply by querying a tag m times, and use the reader to distinguish it from another tag that has been queried less than m times [9]. Note that the reader can only accept having the same conversation once with a tag, otherwise a simply replay attack could break the soundness.

4   Correlated Keys

We have shown in the previous section that if we want strong privacy and tags have independent keys, the reader has to access least half of the keys in the worst case. This obviously does not scale well, so we now look at how much privacy and soundness we will loose in return for efficiency if we allow the keys to be correlated.

It was already known from the work of Molnar, Soppera and Wagner that using correlated keys, one can obtain the property that the reader only needs to access a logarithmic number of keys [12]. Unfortunately, this comes at the price that strong privacy is lost already if the adversary corrupts a single tag. This is due to the fact that the system works with a pair of keys (K_0, K_1), where half the tags hold K_0, the other half hold K_1 - as well as many other keys, arranged in a tree structure, which is not important here, however. Corrupting a single tag tells the adversary one of the keys, say K_0. The protocol is such that one can easily extract from the responses tags give, a part that is computed only from K_0 or K_1. This gives the adversary a way to compute from the responses of an uncorrupted tag which of the two keys it holds. Since half the tags hold K_0, 2 sessions with random chosen tags will locate two tags holding different keys with probability $1/2$. Clearly, using two such tags as the target in the privacy experiment, the adversary can identify with certainty which tag he talks to. It is not even private with benign selection, no matter which distribution is used: the distribution is by definition independent of which keys are held by uncorrupted tags, so we again have that the target tags hold different keys with probability $1/2$. Of course, an error probability of $1/2$ is too large in practice.

This makes it natural to ask if we can get privacy with a larger number of corruptions without going back to systems where the reader does exhaustive search over all keys.

4.1   A Necessary Tradeoff

First, it is useful to observe that in the kind of systems we look at here, some tradeoff between efficiency of the reader and privacy is unavoidable: suppose the key generation algorithm works by generating independently a number of keys, and then assigning to each tag a subset of these keys. The system we propose below, as well as the systems proposed by Molnar, Soppera and Wagner, and by Juels and Weis, are all of this type.

Let K be one of the keys used. We will say that K is *efficiently decidable* if there is an efficient algorithm that, when given K and a session between a tag $\mathcal{T}$ and the reader, can decide whether $\mathcal{T}$ holds K or not. For instance, it may be that the tag, if indeed it holds K, computes a particular part of its response only from K. One can then from K compute what the tag should say if it knows K and compare to what it actually said. In the systems from [9,12], all keys are efficiently decidable.

An efficiently decidable key can be used by the reader towards identifying the tag it is reading, because it can tell whether the tag is in the set of tags that know K or in the complement. However, such a key can also be used by the adversary, who may learn K by corrupting a tag, and can now also distinguish tags that know K from those who do not. Clearly, if the adversary can locate two tags, of which one holds K and the other doesn't, then he can break strong privacy. Let $p(K)$ be the number of tags that hold the key K. The case where $p(K) = n/2$ is the case where the reader gets maximal information from knowing K, namely one bit of information on the identity of the tag. Unfortunately, this is also the optimal case for the adversary, since interactions with a constant number of tags will be sufficient to locate two target tags that can be used to break the privacy.

One may treat this problem either by letting every part of the tag response depend on several keys, or make sure that $p(K)$ is small for every efficiently decidable key K. Both approaches make life harder for the adversary as well as for the reader. We give below an example of the second approach.

4.2 A Tradeoff Construction

Our construction depends on two parameter, v, c. Typically, v will be quite large, say $v = n^d$ for some constant $d < 1$, while c may be something small, say constant or logarithmic in n. We will assume that we have a pseudorandom function $\phi.(\cdot)$. It is straightforward to construct such functions from a cryptographic hash function by simply hashing the key together with the input, this is provably secure in the random oracle model. Other constructions based on, e.g. AES are also possible.

The key generation involves generating c lists of keys to the pseudorandom function ϕ, $K^j = (k_1^j, k_2^j, ..., k_j^j)$ for $j = 1..c$.

We assign to each tag $\mathcal{T}_i$ a random string $str_i = (s_{i,1}, ..., s_{i,c}) \in Z_v^c$, c keys $(k_{s_{i,1}}^1, ..., k_{s_{i,c}}^c)$, and a key k_i that is unique to $\mathcal{T}_i$ (see Figure 2). The probability that two tags will be assigned the same string is at most n^2/v^c, we assume v, c are chosen such that this is negligible. Let n_T, n_R be nonces chosen by tag, respectively reader, such that these values do not repeat. Then the protocol between the tag $\mathcal{T}_i$ and reader is:

1. $\mathcal{R}_i \longrightarrow \mathcal{T}_i$: n_R
2. $\mathcal{R} \longleftarrow \mathcal{T}_i$: n_T, $\phi_{k_{s_{i,j}}}(n_T, n_R)$, for $j = 1, .., c$, and $\phi_{k_i}(n_T, n_R)$. The intuition is that the first c values allow the reader to identify the tag, while the final value proves that the tag is who it claims to be.

$$K^1 = k_1^1, \boxed{k_2^1}, k_3^1, k_4^1, \ldots, k_v^1$$

$$K^2 = \boxed{k_1^2}, k_2^2, k_3^2, k_4^2, \ldots, k_v^2$$

$$K^3 = k_1^3, k_2^3, \boxed{k_3^3}, k_4^3, \ldots, k_v^3$$

$$\ldots$$

$$K^c = k_1^c, \boxed{k_2^c}, k_3^c, k_4^c, \ldots, k_v^c \quad \boxed{k_i}$$

Fig. 2. Example: Keys assigned to a tag T_i with string $str_i = (2, 1, 3, \ldots, 2)$

For the j'th pseudorandom function value received, $j = 1 \ldots c\}$, the reader searches through the v keys in K^j and checks if one of these will generate the value received, i.e., for each $k \in K^j$ one checks if $\phi_k(n_T, n_R) = \phi_{k_{s_{i,j}}}(n_T, n_R)$. If this is not the case, reject and stop. Otherwise note the index of the key. The indices noted form a string $(s_1, .., s_c)$. If this string matches the string assigned to some tag T_i, and the final pseudorandom value received is equal to $\phi_{k_i}(n_T, n_R)$, output (accept, i). Else output reject.

To show security of the system, we first go to the *independent oracles model*, i.e., we replace each call to ϕ using key k by a call to a random oracle O_k, using independent oracles for different keys. The adversary can only call an oracle O_k if he corrupts a tag that holds k.

It is straightforward to see that if we model the hash function used in the proposed construction of ϕ by a random oracle, then an adversary playing the privacy or soundness game is exactly working in the oracle model just described. For this reason and for simplicity, we will analyze the system in this model. At the cost of a more complicated proof, it is also possible to argue security based only on pseudo-randomness of ϕ, i.e., without using the random oracle model.

The first result on our system shows that, without loss of generality, we may consider only adversaries who do no talk to the reader:

Lemma 2. *In both the privacy and soundness games, sessions that the adversary initiates with the reader can be simulated without access to the reader, but with access to those oracles that the adversary can access. The simulation is perfect, except with probability negligible in k.*

Proof. We describe an algorithm for simulating the sessions in question: In any session, the reader first sends a nonce n_R, this can be simulated by simply following the reader's algorithm for selecting nonces. The message that the adversary returns must consist of a nonce n_T and $c + 1$ values $r_1, ..., r_c, s$. Note that the reader checks these values against oracle outputs generated from the fresh input n_R, n_T, and that we may assume that oracle answers are sufficiently long so they cannot be guessed except with negligible probability. For these reasons, the adversary can only hope to have the reader accept if he generated each of the $c + 1$ response values by either using an oracle he has direct access to, or by starting a session with an uncorrupted tag and using (part of) the tag's response.

If this is not the case, we can return *reject* to the adversary: in real life the reader will reject such a response except with negligible probability. But if the adversary has indeed generated the entire response by calling oracles (directly or indirectly), we know the identity k' of the oracle the generated the last value in the response. If the call to oracle $O_{k'}$ was made by an uncorrupted tag $\mathcal{T}_j$, this has to be because that tag received n_R as a challenge and therefore produced a correct response for nonces n_R, n_T. If we see that the adversary forwards this response to the reader, we return $(accept, j)$ as the real reader would have done. If the adversary has replaced any of the first c values with other oracle responses, we return *reject*, which is correct except with negligible probability.

The only remaining possibility is that it was the adversary who called $O_{k'}$. This means he must have corrupted the tag $\mathcal{T}_i$ giving access to this oracle, and so he also has access to to the other c oracles that this tag possesses. Therefore, having generated the message sent to the reader, we can check whether this is a correct response from $\mathcal{T}_i$. If this is not the case, we return reject to the adversary. Otherwise, we return $(accept, i)$.   □

The following lemma turns out to be essential for privacy:

Lemma 3. *Consider an adversary that does not start any session with the reader. Let M be the set of oracles that the adversary gains access to during the privacy game. Let E be the event that the following condition is satisfied after the game: the adversary has started at least one session with some uncorrupted tag $\mathcal{T}$, and one of the oracles assigned to $\mathcal{T}$ is in M. In the privacy game, by convention, the adversary selecting the two target tags counts as starting a session with both tags. Let t' be the number of different tags the adversary talks to during the game. The probability that E occurs is at most*

$$\frac{ct'u}{v} + \frac{ct'u}{v - u}.$$

Proof. Suppose we are at some point in the game where E has not occurred yet. This means that for all uncorrupted tags the adversary has talked to, he knows that they only have oracles he has no access to, but due to the randomness of the oracles, he has no information on their identity.

The adversary may now start a session with a new tag he did not talk to before, or corrupt a tag. For each of these moves, we bound the probability that E will occur after the move:

Start new session: Since the adversary has not previously talked to the tag $\mathcal{T}_i$, given what he knows, str_i is uniform. We can therefore model what goes on as follows: look at one of the c positions in str_i, and let $x \in Z_v$ be the number in this position. Now, x is uniform over v possibilities, and the adversary has success, if x happens to be one of the $\leq u$ values corresponding to oracles he can access. So the adversary has success in one position with probability at most u/v, and therefore has success in any position with probability at most $\frac{cu}{v}$

Corrupt new tag: For the previously uncorrupted tag $\mathcal{T}_i$, consider again x, the number at some position in str_i. Then given what the adversary knows, before he corrupts $\mathcal{T}_i$, x is uniform over at least $v - u$ possibilities, if the adversary talked to $\mathcal{T}_i$ before, he knows x does not match any of the $\leq u$ possibilities he knows from already corrupted tags. The adversary hopes x will hit one of the $\leq t'$ possibilities for tags he talked to, so the probability of success is at most $t'/(v - u)$ for one position and $\frac{ct'}{v-u}$ for all positions.

Finally, since there are at most t' respectively u steps that could cause the first respectively second kind of event, the lemma follows.   □

We are now ready to prove security of our construction.

Theorem 2. *For the RFID system described above, we have that if the hash function used in the construction is modeled by a random oracle, then the system is $(poly(k), poly(k), poly(k), n)$-strongly sound, and is strongly $(k, r, poly(k), t, u, n, \epsilon)$-private, where*

$$\epsilon = \frac{ct'u}{v} + \frac{ct'u}{v - u} + negl(k)$$

and where $negl(k)$ is a negligible function of k.

Proof. Completeness is obvious from the fact that the strings assigned to tags are unique except with negligible probability.

For soundness, recall that the adversary wins the soundness game if a session is generated where the reader outputs $(accept, i)$, but the (uncorrupted) tag $\mathcal{T}_i$ did not participate. Since the input nonces are fresh and oracles answers cannot be guessed in advance except with negligible probability, the oracle O_{k_i} must have been called to generated the last part of the response. But this is impossible since $\mathcal{T}_i$ did not participate and the adversary does not have access to O_{k_i} as long as $\mathcal{T}_i$ is uncorrupted.

Finally, for privacy, note that by Lemma 2, any adversary $\mathcal{A}$ playing the privacy game can be replaced by a new adversary $\mathcal{A}'$, who does not start sessions with the reader, and such that the advantage of $\mathcal{A}'$ is smaller than that of $\mathcal{A}$ by at most a negligible amount. This, together with Lemma 3 immediately implies the privacy result.   □

Finally, we show that the adversary's advantage in the benign selection privacy game is much smaller:

Theorem 3. *Our system is $(k, r, poly(k), t, u, n, \epsilon)$-private with benign D-selection for any D, and where $\epsilon = 2cu/v + negl(k)$*

Proof. As above, we can assume that the adversary does not talk to the reader, at the cost of adding a negligible amount to the advantage. Now consider the situation when the target tags are chosen. For each of the c positions in the strings assigned to tags, the adversary can access at most u of the v oracles assigned to this position. Hence, when an uncorrupted tag is chosen, no matter how this is done, the probability that its oracle for this position is known to the adversary is u/v since "names" of tags are assigned uniformly and independently. Since the two target tags hold a total of $2c$ oracles that could be used to distinguish them, the probability that at least one of them is known to the adversary is at most $2cu/v$. On the other hand, if the adversary has no oracles in common with the target tags, he cannot distinguish them at all.   □

5   Efficiency

The interest in this result is that it shows a possibility for a new tradeoff between security and efficiency for large systems, where the adversary can be expected to only corrupt or talk to a number of the tags that is very small compared to the total number of tags in the system. More precisely, for parameter values such that $r, t', u << v << n$, but still $n^2 < v^c$. However, for particular values of r, t', u and c, v and hence n must very large to make the privacy advantage be small. This has to do with the fact that we are asking for strong privacy and this is a very strong demand. Below, we show that the systems performs much better under the privacy definition with benign selection. On the practical side, note that the reader needs to look at only cv keys which can be much smaller than n. Also, each tag only has to hold $c + 1$ keys. Although the total number of keys in the system is greater than n, this does not mean that the reader has to store this many keys – they can be generated pseudorandomly from a single key when they are needed.

Let us look at a concrete example of parameters in the benign selection model for any distribution. Suppose we choose $v = 2^{16}$ and $c = 4$. Then we can accommodate over 33 million tags, say $n = 2^{25}$, and each tag only needs to store 5 keys. If the adversary can corrupt 100 tags, the above says that his chance of distinguishing two tags that are chosen for him is at most $1/100$. Note that even if the adversary is lucky with one pair of tags, his chance against another pair is still only $1/100$, so we think this can be quite reasonable in practice. In other words, even though a probability of $1/100$ is not negligible in the usual sense, this is not necessary, if the "bad event" does not imply a complete break of the system. With these parameters, the reader must search through at most 2^{18} keys to identify a tag, which is clearly better than 2^{25}, which was needed to get strong privacy. We can even increase n without increasing the number of keys to search through, as long as we keep the probability that two tags will be assigned the same key n^2/v^c reasonably small.

6   Conclusion

We have proposed a new definition of security for RFID systems, incorporating both strong privacy, soundness and completeness, and also a weaker but more realistic variant of privacy, with benign selection. We have shown that in sound, complete and essentially symmetric RFID system where tags are independently keyed, the reader must access at least half of all keys when reading a tag, or privacy is violated. Finally, we have proposed a new RFID system based on symmetric cryptography offering a tradeoff between reader efficiency and privacy.

References

1. Ateniese, G., Camenisch, J., Joye, M., Tsudik, G.: A practical and provably secure coalition-resistant group signature scheme. In: Bellare, M. (ed.) CRYPTO 2000. LNCS, vol. 1880, pp. 255–270. Springer, Heidelberg (2000)
2. Avoine, G.: Adversarial model for radio frequency identification. Cryptology ePrint Archive, Report 2005/049 (2005)
3. Avoine, G., Dysli, E., Oechslin, P.: Reducing time complexity in rfid systems. In: Preneel, B., Tavares, S. (eds.) SAC 2005. LNCS, vol. 3897, pp. 291–306. Springer, Heidelberg (2006)
4. Avoine, G., Oechslin, P.: A scalable and provably secure hash-based rfid protocol. In: Stajano, F., Thomas, R. (eds.) PerSec 2005, vol. 00, pp. 110–114. IEEE Computer Society Press, Los Alamitos (2005)
5. Batina, L., Guajardo, J., Kerins, T., Mentens, N., Tuyls, P., Verbauwhede, I.: An elliptic curve processor suitable for RFID-tags. Cryptology ePrint Archive, Report 2006/227 (2006)
6. Burmester, M., van Le, T., de Medeiros, B.: Provably secure ubiquitous systems: Universally composable rfid authentication protocols. Cryptology ePrint Archive, Report 2006/131 (2006)
7. Camenisch, J., Lysyanskaya, A.: Signature schemes and anonymous credentials from bilinear maps. In: Franklin, M. (ed.) CRYPTO 2004. LNCS, vol. 3152, pp. 56–72. Springer, Heidelberg (2004)
8. Hellman, M.E.: A cryptanalytic time-memory tradeoff. IEEE Transactions on Information Theory 26(6), 401–406 (1980)
9. Juels, A., Weis, S.A.: Defining strong privacy for rfid. In: PERCOMW 2007, vol. 1462, pp. 342–347. IEEE Computer Society Press, Los Alamitos (2007)
10. Kiayias, A., Yung, M.: Group signatures with efficient concurrent join. In: Cramer, R.J.F. (ed.) EUROCRYPT 2005. LNCS, vol. 3494, pp. 198–214. Springer, Heidelberg (2005)
11. Kilian, J., Petrank, E.: Identity escrow. In: Krawczyk, H. (ed.) CRYPTO 1998. LNCS, vol. 1462, pp. 169–185. Springer, Heidelberg (1998)
12. Molnar, D., Soppera, A., Wagner, D.: A scalable, delegatable pseudonym protocol enabling ownership transfer of rfid tags. Cryptology ePrint Archive, Report 2005/315 (2005)
13. Ohkubo, M., Suzuki, K., Kinoshita, S.: Efficient hash-chain based rfid privacy protection scheme. In: Ubicomp, Privacy Workshop: Current Status and Future Directions (2004)
14. de Medeiros, B., van Le, T., Burmester, M.: Universally composable and forward secure rfid authentication and key exchange. Cryptology ePrint Archive, Report 2006/448 (2006)
15. Weis, S.A., Sarma, S.E., Rivest, R.L., Engels, D.W.: Security and privacy aspects of low-cost radio frequency identification systems. In: Hutter, D., Müller, G., Stephan, W., Ullmann, M. (eds.) Security in Pervasive Computing. LNCS, vol. 2802, pp. 201–212. Springer, Heidelberg (2004)

Program Obfuscation and One-Time Programs

Shafi Goldwasser[1,2]

[1] RSA Professor in EECS, Massachusetts Institute of Technology
[2] Weizmann Institute of Science

Abstract. Program obfuscation is the process of taking a program as an input and modifying it so that the resulting program has the same I/O behavior as the input program but otherwise looks 'garbled' to the entity that runs it, even if this entity is adversarial and has full access to the program. Intuitively, by looking garbled to an adversarial entity, we mean that it should be impossible to understand the internal working of the program, or more generally to compute anything that cannot be computed by seeing only the legitimate outputs of the program on inputs of choice.

Traditionally, program obfuscation has been regarded as a software-based technique to curb the use of programs in commercial contexts such as preventing illegal re-distribution of copyrighted information. Here the obfuscation process is aimed at preventing 'reverse engineering' that would subvert the curbs and restrictions that were embedded into the original program. Another domain in which program obfuscation is considered imperative is within the on-line gaming industry, where in order to maintain a fair and consistent gaming environment which will keep gamers coming, one must ensure that hackers cannot modify the games so as to gain an unfair advantage. Also, as more and more web sites deliver Javascript source code to be run locally on browsers, programmers are naturally interested in obfuscating their source code in order to make it hard for competitors to learn how it works.

The design of program obfuscators (or at least attempts at it) has been standard fare in practice. However, in spite of the large effort dedicated to develop program obfuscators, these efforts have been successful only in the very short run. Indeed, the general belief in the industry has remained very skeptic regarding the viability of obfuscation methods, as expressed in the following recent quote:

This feeling seems to be supported by theoretical impossibility results that assert that several strong (albeit natural) formulations of obfuscation are impossible. That is, there is no generic mechanism that can successfully obfuscate large classes of programs.

Yet, even more recent theoretical results have pointed out a way in which, in spite of these generic impossibility results, the basic concept of program obfuscation is obtainable in many settings. One setting on which we will elaborate is of one-time programs: programs that can be executed only a restricted and pre-specified number of times. Naturally, these programs cannot be achieved using software alone. We show how to build them using 'simple' and 'universal' secure hardware components.

One-time programs serve many of the same purposes of program obfuscation, the obvious one being software protection. However, the

T. Malkin (Ed.): CT-RSA 2008, LNCS 4964, pp. 333–334, 2008.

applications of one-time programs go well beyond those of obfuscation, since one-time programs can only be executed once (or more generally, a limited number of times) while obfuscated programs have no such bounds. For example, one-time programs lead naturally to electronic cash or token schemes and to "one-time proofs", proofs that can only be verified once and then become useless and unconvincing. We show how to use a classical witness and simple secure hardware to efficiently construct such "one-time proofs" for any NP statement.

In this talk we will survey all of these exciting developments.

Efficient Two-Party Password-Based Key Exchange Protocols in the UC Framework

Michel Abdalla[1], Dario Catalano[2], Céline Chevalier[1], and David Pointcheval[1]

[1] École Normale Supérieure, LIENS-CNRS-INRIA, Paris, France
[2] Università di Catania, Italy

Abstract. Most of the existing password-based authenticated key exchange protocols have proofs either in the indistinguishability-based security model of Bellare, Pointcheval, and Rogaway (BPR) or in the simulation-based of Boyko, MacKenzie, and Patel (BMP). Though these models provide a security level that is sufficient for most applications, they fail to consider some realistic scenarios such as participants running the protocol with different but possibly related passwords. To overcome these deficiencies, Canetti *et al.* proposed a new security model in the universal composability (UC) framework which makes no assumption on the distribution on passwords used by the protocol participants. They also proposed a new protocol, but, unfortunately, the latter is not as efficient as some of the existing protocols in BPR and BMP models. In this paper, we investigate whether some of the existing protocols that were proven secure in BPR and BMP models can also be proven secure in the new UC model and we answer this question in the affirmative. More precisely, we show that the protocol by Bresson, Chevassut, and Pointcheval (BCP) in CCS 2003 is also secure in the new UC model. The proof of security relies in the random-oracle and ideal-cipher models and works even in the presence of adaptive adversaries, capable of corrupting players at any time and learning their internal states.

1 Introduction

Password-based authenticated key exchange (PAKE) protocols allow users to securely establish a common key over an insecure channel only using a low-entropy, human-memorizable, secret key called a password. Since PAKE protocols do not require complex public-key infrastructure (PKI) or trusted hardware capable of storing high-entropy keys, they have become quite popular since being introduced by Bellovin and Merritt [3].

Due to the low entropy of passwords, PAKE protocols are subject to dictionary attacks in which the adversary tries to break the security of the scheme by trying all values for the password in the small set of the possible values (i.e., the dictionary). Unfortunately, these attacks can be quite damaging since the attacker has a non-negligible probability of succeeding. To address this problem, one should invalidate or block the use of a password whenever a certain number of failed attempts occurs. However, this is only effective in the case of *online*

T. Malkin (Ed.): CT-RSA 2008, LNCS 4964, pp. 335–351, 2008.

dictionary attacks in which the adversary must be present and interact with the system in order to be able to verify whether its guess is correct. Thus, the goal of PAKE protocol is restrict the adversary to *online* dictionary attacks only. In other words, *off-line* dictionary attacks, in which the adversary verifies if a password guess is correct without interacting with the system, should not be possible in a PAKE protocol.

SECURITY MODELS. Even though the notion of password-based authentication dates back to the seminal work by Bellovin and Merritt [3], it took several years for the first formal security models to appear in the literature [5,4]. In [5], Bellare, Pointcheval, and Rogaway (BPR) proposed an indistinguishability-based security model extending the framework of Bellare and Rogaway [7,8] while, in [4], Boyko, MacKenzie, and Patel (BMP) proposed a simulation-based security model based on the framework of Shoup [18]. In both cases, the level of security provided by the models is quite reasonable and sufficient for most applications and it captures the intuition given above in which the success of an adversary in breaking the security of a scheme should be limited to its online attempts.

Unfortunately, as pointed out by Canetti *et al.* [10], the BPR and BMP security models are not as general or as strong as they could be and they fail to consider some realistic scenarios such as participants running the protocol with different but possibly related passwords. To overcome these deficiencies, Canetti *et al.* [10] proposed a new security model for PAKE schemes in the universal composability (UC) framework [9] which makes no assumption on the distribution on passwords used by the protocol participants. Their model was later extended to the verifier-based scenario by Gentry *et al.* [13].

In addition to the new security model, Canetti *et al.* [10] also proposed a new protocol based on the PAKE schemes by Katz, Ostrovsky, and Yung [15] and by Gennaro and Lindell schemes [12] and proved it secure in the new model against static adversaries based on standard computational assumptions. Unfortunately, the new protocol is not as efficient as some of the existing protocols in BPR and BMP models (e.g., [2,1,15,17]), an issue that can significantly limit its applicability. Given this limitation, one natural question to ask is whether some of the more efficient protocols that were proven secure in BPR and BMP models can also be proven secure in the model of Canetti *et al.* [10]. In this paper, we answer this question in the affirmative by showing that the protocol by Bresson, Chevassut, and Pointcheval (BCP) [2] is also secure in the model of Canetti *et al.* [10]. We view this as the main contribution of our paper.

In addition to proving the security of the BCP protocol in the model of Canetti *et al.* [10], another contribution of our paper is to show that their protocol remains secure even against adaptive adversaries, capable of corrupting adversaries at any time and learning their internal states. Despite this being first time that such a strong security level is achieved in the password-based scenario, we do not consider this result very surprising given the use of the random-oracle and ideal-cipher models in the security proof.

ORGANIZATION. In Section 2, we extend the ideal functionality of PAKE protocols to include client authentication, which not only ensures the parties that nobody else knows the common secret, but also that they actually share the same secret. As in [10], passwords are chosen by the environment who then hands them to the parties as input. This is the strongest security model, since it does not assume any distribution on passwords. Furthermore, it allows the environment to even make players run the protocol with different (possibly related) passwords. For example, this models a user mistyping a password. As in [10], we also provide the adversary with a Test-Password query to model the vulnerability of the passwords (whose entropy may be low). This models the case in which the adversary tries to impersonate a player by guessing its password. If the guess is correct (which may happen with non-negligible probability), the adversary should succeed in its impersonation.

Next, in Section 3, we recall the password-based protocol of [2] and prove it secure in the new extended model, even against adaptive adversaries which can perform strong corruptions at any time. The proof is given in Section 4. As we mentioned above, this is the first time that such a strong security level is achieved in the password-based scenario: adaptive and strong corruptions in the UC framework.

In the appendix, we also provide ideal functionalities for the ideal-cipher and the random-oracle models [6].

2   Definition of Security

Notations. We denote by k the security parameter. An event is said to be negligible if it happens with probability that is less than the inverse of any polynomial in k. If G is a finite set, $x \xleftarrow{R} G$ indicates the process of selecting x uniformly and at random in G (thus we implicitly assume that G can be sampled efficiently).

The UC Framework. Throughout this paper we assume basic familiarity with the universal composability framework. Here we provide a brief overview of the framework. The interested reader is referred to [9] for complete details. In a nutshell, security in the UC framework is defined in terms of an ideal functionality $\mathcal{F}$, which is basically a trusted party that interacts with a set of players to compute some given function f. In particular, the players hand their input to $\mathcal{F}$ which computes f on the received inputs and gives back to each player the appropriate output. Thus, in this idealized setting, security is inherently guaranteed, as any adversary, controlling some of the parties, can only learn (and possibly modify) the data of corrupted players. In order to prove that a candidate protocol π realizes the ideal functionality, one considers an environment $\mathcal{Z}$, which is allowed to provide inputs to all the participants and that aims to distinguish the case where it receives the outputs produced from a real execution of the protocol (involving all the parties and an adversary $\mathcal{A}$, controlling some of the parties and the communication among them), from the case where it receives outputs obtained from an ideal execution of the protocol (involving only dummy parties interacting with $\mathcal{F}$ and an ideal adversary $\mathcal{S}$ also interacting with $\mathcal{F}$). Then we say that π realizes the functionality $\mathcal{F}$ if for every (polynomially bounded) $\mathcal{A}$, there exists a (polynomially bounded) S such that no (polynomially bounded) Z can distinguish a real execution of the protocol from an ideal one with a significant advantage. In particular, the universal composability theorem assures us that π continues to behave like the ideal functionality even if it is executed in an arbitrary network environment.

SESSION ID'S AND PLAYER'S IDs. In the UC framework there may be many copies of the ideal functionality running in parallel. Each one of such copies is supposed to have a unique session identifier (SID). Every time a message has to be sent to a specific copy of $\mathcal{F}$, such a message should contain the SID of the copy it is intended for. Following [10], we decided to make things simple and to assume that each protocol that realizes $\mathcal{F}$ expects to receive inputs that already contain the appropriate SID. See [10] for further details about this. Moreover we assume that every player starts a new session of the protocol with input (NewSession, sid, P_i, P_j, pw, role), where P_i is the identity of the player, pw his or her password, P_j the identity of the player with whom he or she intends to share a session key and role being either client or server.

UC WITH JOINT STATE. The original UC theorem allows to analyze the security of a system viewed as a single unit, but it says nothing if different protocols share some amount of state and randomness (such as a common reference string, for instance). Thus for the application we have in mind, the UC theorem cannot be used as it is, since different sessions of the protocol share the same random oracles and the same ideal cipher.

In [11] Canetti and Rabin introduced the notion of universal composability with joint state. Informally, they put forward a new composition operation that allows different protocols to have some common state, while preserving security. Very informally, this is done by defining a multisession extension $\hat{\mathcal{F}}$ of $\mathcal{F}$, which basically runs multiple executions of $\mathcal{F}$. Each copy of $\mathcal{F}$ is identified by means of a sub-session id (SSID). This means that, if $\hat{\mathcal{F}}$ receives a message m with SSID $ssid$ it hands m to the copy of $\mathcal{F}$ having SSID $ssid$. If no such copy exists, $\hat{\mathcal{F}}$ invokes a new one on the spot. Notice that, whenever $\hat{\mathcal{F}}$ is executed, the calling protocol has to specify both the SID (i.e. the usual session id, as in any ideal functionality) *and* the SSID.

Adaptive Adversaries. In this paper, we will consider protocols that are secure against adaptive adversaries, i.e. adversaries that are allowed to arbitrarily corrupt players at any moment during the execution of the protocol. The adversary corrupts a player by getting complete access to its internal memory. Note that at the end of an execution of the protocol, the adversary recovers nothing, as if the internal state has been completely erased. In a real execution of the protocol this is modeled by letting the adversary $\mathcal{A}$ obtain the password and the internal state of the corrupted player. Moreover, the adversary can arbitrarily modify the player's strategy. In an ideal execution of the protocol, the simulator S gets the player's password and has to simulate its internal state, in a way that remains consistent to what already provided to the environment.

The Random Oracle and the Ideal Cipher. For lack of space, a description of these functionalities is given in Appendix A.

The Password-Based Key-Exchange Functionality With Client Authentication. In this section, we motivate and present our formulation of an ideal functionality for password-based key exchange with client authentication (see Figure 1). The starting point for our approach is the definition for universally composable password-based key exchange with no authentication [10]. Our aim is to define a functionality that achieves the same effect, except that we also incorporate the authentication of the client. Mutual authentication would have been easier to model. However, client-authentication is usually enough in most cases and often results in more efficient protocols.

First notice that the functionality is not in charge of providing the password(s) to the participants (the client Alice and the server Bob). Rather we let the environment do this. As already pointed out in [10], such an approach allows to model, for example, the case where some users may use the same password for different protocols and, more generally, the case where password(s) are chosen according to some arbitrary distribution (i.e. not necessarily the uniform one). Moreover, notice that allowing the environment to choose the password(s) guarantees forward secrecy, basically for free.

The queries NewSession and TestPwd are dealt with in the same manner as in [10], but we introduce the client authentication in the way the functionality answers the NewKey queries. In the definition of $\mathcal{F}_{pwKE}^{CA}$, the server receives an error if the players don't meet all the conditions to receive the same, randomly-chosen key. We could have chosen to send to the server a pair consisting of a key chosen independently from that of the client and a flag warning the server that the protocol has failed, but we preferred to keep the functionality as straightforward as possible.

CLIENT AUTHENTICATION. The first reason why the initial functionality didn't achieve this property is that we had to deal with the order of the queries NewKey. More precisely, if the server asks the first query, it is impossible to answer it, because we don't know what is going to happen to the client afterwards: If the session was fresh for both players and the server was the only one to have received his key, the client's session could possibly become compromised or interrupted after the server had received his key, whereas the functionality should have been able to determine whether or not the server should receive a key or an error message. We solved this issue by making it mandatory for the adversary to ask the query for the server after the corresponding query for the client. This is not a strong restriction, since this situation frequently happens in real protocols, and in particular in the one that we are studying: the server has to accept the client before generating the session key.

Thus, if the adversary asks for the key of a client, everything is as before, except that we also provide a flag ready for the session. The aim of this flag is to help determine, when the adversary asks for the key of the server, that the corresponding client has already got her key.

On the other hand, if the adversary asks for the key of a server, the server is given an error message in the easy failure cases (interrupted or compromised sessions, corrupted players – if the passwords are different in the two latter cases). If the session is fresh and the corresponding client hasn't yet received her key, we simply postpone the query of the adversary until the client has received her key. In the latter case, when the client has received her key, the server is given the same key if they have the same password and an error message otherwise. We finally obtain the following definition, which remains trivially secure and correct.

3   Our Scheme

3.1   Description of the Protocol

The protocol presented in Figure 2 is based on that of [2], with two slight differences: In the standard model using the security definition of Bellare *et al.* [5], the session identifier is obtained at the end of the program execution as the concatenation of the random values sent by the players; in particular, it is unique. In contrast, in the model of universal composability [9], these identifiers are uniquely determined in advance, before the beginning of the protocol. Thus, this difference must be taken care of in the definition of the protocol. Another difference has been made, in order to match the definition of the functionality: in case of a failure, the server receives an error message, this feature guaranteeing the client authentication.

3.2   Security Theorem

We consider here the Theorem of Universal Composability in its joint-state version. Let $\widehat{\mathcal{F}}_{pwKE}^{CA}$ be the multi-session extension of $\mathcal{F}_{pwKE}^{CA}$ and let $\mathcal{F}_{RO}$ and $\mathcal{F}_{IC}$ be the ideal functionalities that provide a random oracle and an ideal cipher to all parties. Note that only these two functionalities belong to the joint state.

Theorem 1. *The above protocol securely realizes $\widehat{\mathcal{F}}_{pwKE}^{CA}$ in the $(\mathcal{F}_{RO}, \mathcal{F}_{IC})$-hybrid model, in the presence of adaptive adversaries.*

4   Proof of Theorem 1

4.1   Description of the Proof

In order to show that the protocol UC-realizes the functionality $\mathcal{F}_{pwKE}^{CA}$, we need to show that for all environments and all adversaries, we can construct a simulator such that the interactions, from the one hand between the environment, the players (say, Alice and Bob) and the adversary (the real world), and from the other hand between the environment, the ideal functionality and the simulator (the ideal world), are indistinguishable for the environment.

In this proof, we incrementally define a sequence of games starting with the real execution of the protocol and ending up with game $\mathbf{G_6}$, which we prove to be indistinguishable from the ideal experiment.

- $\mathcal{F}_{pwKE}^{CA}$ owns a list L initially empty of values of the form (P_i, P_j, pw).
- **Upon receiving a query (NewSession, $ssid$, P_i, P_j, pw, role) from P_i:**
  - Send (NewSession, $ssid, P_i, P_j$, role) to $\mathcal{S}$.
  - If this is the first NewSession query, or if it is the second NewSession query and there is a record $(P_j, P_i, pw', \text{role}) \in L$, then record $(P_i, P_j, pw, \text{role})$ in L and mark this record fresh.
- **Upon receiving a query (TestPwd, $ssid$, P_i, pw') from the adversary $\mathcal{S}$:**
  If there exists a record of the form $(P_i, P_j, pw, \text{role}) \in L$ which is fresh, then do:
  - If $pw = pw'$, mark the record compromised and reply to $\mathcal{S}$ with "correct guess".
  - If $pw \neq pw'$, mark the record interrupted and reply to $\mathcal{S}$ with "wrong guess".
- **Upon receiving a query (NewKey, $ssid, P_i, sk$) from $\mathcal{S}$, where $|sk| = k$:**
  If there is a record of the form $(P_i, P_j, pw, \text{role}) \in L$, and this is the first NewKey query for P_i, then:
  If role=client:
  - If the session is compromised, or if one of the two players P_i or P_j is corrupted, then send $(ssid, sk)$ to P_i, record $(P_i, P_j, pw, \text{client}, \text{completed})$ in L, as well as $(ssid, P_i, pw, sk, \text{client}, \text{status}, \text{ready})$ (with status being the status of the session at that moment).
  - Else, if the session is fresh or interrupted, choose a random key sk' whose length is k and send $(ssid, sk')$ to P_i. Record $(P_i, P_j, pw, \text{client}, \text{completed})$ in L, as well as $(ssid, P_i, pw, sk', \text{client}, \text{status}, \text{ready})$ where status stands for fresh or interrupted;
  If role=server:
  - If the session is compromised, if one of the two players P_i or P_j is corrupted, and if there are two records of the form $(P_i, P_j, pw, \text{server})$ and $(P_j, P_i, pw, \text{client})$, set $s = sk$. Otherwise, if the session is fresh and there exists any recorded element of the form $(ssid, P_j, pw', sk', \text{client}, \text{fresh}, \text{ready})$, set $s = sk'$.
    * If $pw = pw'$, send $(ssid, s)$ to P_i record $(P_i, P_j, pw, \text{server}, \text{completed})$ in L, as well as $(ssid, P_i, pw, s, \text{server}, \text{status})$.
    * If $pw \neq pw'$, send $(ssid, \text{error})$ to P_i, record $(P_i, P_j, pw, \text{server}, \text{completed})$ in L, as well as $(ssid, P_i, pw, \text{server}, \text{error}, \text{status})$.
  - If the session is fresh and there doesn't exist any recorded element of the form $(ssid, P_j, pw', sk', \text{client}, \text{fresh}, \text{ready})$, then do not do anything;
  - If the session is interrupted, then send $(ssid, \text{error})$ to player P_i, and record in L $(P_i, P_j, pw, \text{server}, \text{completed})$ and $(sid, P_i, pw, \text{server}, \text{error}, \text{interrupted})$.

Fig. 1. Functionality $\mathcal{F}_{pwKE}^{CA}$: it is parametrized by a security parameter k. It interacts with an adversary $\mathcal{S}$ and a set of parties $P_1, \ldots, P_n$.

Since we have to deal with adaptive corruptions, we consider different cases according to the number of corruptions that have occurred up to now. $\mathbf{G}_0$ is the real world. In $\mathbf{G}_1$, we start by explaining how $\mathcal{S}$ simulates the ideal cipher and the random oracle. Then, in $\mathbf{G}_2$, we get rid of such a situation in which the adversary wins by chance. The passive case, in which no corruption occurs before the end of the protocol, is dealt with in $\mathbf{G}_3$. Next, we completely explain

Client U		Server S
$x \overset{R}{\leftarrow} [\![\, 1\,;\, q-1\,]\!]$		$y \overset{R}{\leftarrow} [\![\, 1\,;\, q-1\,]\!]$
$(U1)\ X \leftarrow g^x$	$\xrightarrow{\ U,X\ }$	
		$(S2)\ Y \leftarrow g^y$
		$Y^* \leftarrow \mathcal{E}_{ssid\|pw}(Y)$
	$\xleftarrow{\ S,Y^*\ }$	$K_S \leftarrow X^y$
$(U3)\ Y = \mathcal{D}_{ssid\|pw}(Y^*)$		
$K_U \leftarrow Y^x$		
$Auth \leftarrow \mathcal{H}_1(ssid\|U\|S\|X\|Y\|K_U)$		
$sk_U \leftarrow \mathcal{H}_0(ssid\|U\|S\|X\|Y\|K_U)$		
completed	$\xrightarrow{\ Auth\ }$	
		$(S4)$
		if $(Auth = \mathcal{H}_1(ssid\|U\|S\|X\|Y\|K_S))$
		then $sk_S \leftarrow \mathcal{H}_0(ssid\|U\|S\|X\|Y\|K_S)$
		completed
		else error

Fig. 2. Client-authenticated two-party password-based key exchange

the simulation of the client in $\mathbf{G}_4$, whatever corruption may occur. As for the server, we divide it into two steps: We first show in $\mathbf{G}_5$ how to simulate the last step of the protocol, and then we simulate it from the beginning in $\mathbf{G}_6$. $\mathbf{G}_7$ sums up the situation, and is shown to be indistinguishable from the ideal world.

Note that these games are sequential and built on each other. When we say that a game consider a specific case, one has to understand that in all other cases, the simulation is dealt with as described in the former game.

We first describe two hybrid queries that are going to be used in the games. The GoodPwd query checks whether the password of a certain player is the one we have in mind or not. The SamePwd query checks if the players share the same password, without disclosing it. In some games the simulator has actually full access to the players. In such a case, a GoodPwd (or a SamePwd) can easily be implemented by simply letting the simulator look at the passwords. When the players are entirely simulated, $\mathcal{S}$ will replace the queries above with a TestPwd and with a NewKey, respectively.

We say that a flow is *oracle-generated* if it was sent by an honest player and arrives without any alteration to the player it was meant to. We say it is *non-oracle-generated* otherwise, that is either if it was sent by an honest player and modified by the adversary, or if it was sent by a corrupted player or a player impersonated by the adversary.

4.2   Proof of Indistinguishability

Game G_0:   Real Game. G_0 is the real game in the random-oracle and ideal-cipher models.

Game G_1:   Simulation of the oracles. Here we modify the previous game by simulating the hash and the encryption/decryption oracles, in a quite natural and usual way.

For the ideal cipher, we allow the simulator to maintain a list Λ_ε of entries (queries, responses) of length $q_\varepsilon + q_D$. Such a list is used by $\mathcal{S}$ to be able to provide answers which are consistent with the following requirements. First, if the simulator receives twice the same question for the same password, it has to give twice the same answer. Second, the simulator should make sure that the simulated scheme (for each password) is actually a permutation. Third, in order to help the simulator to later extract the password used in the encryption of Y^* in the first flow, there should not be two entries (question, answer) with identical ciphertext, but different passwords. More precisely, Λ_ε is actually composed of two sublists: $\Lambda_\varepsilon = \{(ssid, pw, Y, \alpha, \mathcal{E}, Y^*)\} \cup \{(ssid, pw, Y, \alpha, \mathcal{D}, Y^*)\}$. The first (resp. second) sublist is used to indicate that the element Y (resp. Y^*) has been encrypted ("$\mathcal{E}$") (resp. decrypted ("$\mathcal{D}$")) to produce the ciphertext Y^* (resp. Y) via a symmetric encryption algorithm that uses the key $ssid\|pw$. The role of α will be explained below. The simulator manages the list through the following rules:

- For an encryption query $\mathcal{E}_{ssid\|pw}(Y)$ such that $(ssid, pw, Y, *, *, Y^*)$ appears in Λ_ε, the answer is Y^*. Otherwise, choose a random element $Y^* \in G^* = G \setminus \{1\}$. If a record $(*, *, *, *, *, Y^*)$ already belongs to the list Λ_ε, then abort, else add $(ssid, pw, Y, \perp, \mathcal{E}, Y^*)$ to the list.
- For a decryption query $\mathcal{D}_{ssid\|pw}(Y^*)$ such that $(*, pw, Y, *, *, Y^*)$ appears in Λ_ε, the answer is Y. Otherwise, choose a random element $\varphi \in \mathbb{Z}_q^*$ and evaluate the answer $Y = g^\varphi$. If $(*, *, Y, *, *, *)$ already belongs to the list Λ_ε, abort, else add $(ssid, pw, Y, \varphi, \mathcal{D}, Y^*)$ to the list.

The two abort-cases will be useful later in the proof: when one sees a ciphertext Y^*, it cannot have been obtained as the encryption with two different passwords, but a unique one.

In addition, the simulator maintains a list $\Lambda_{\mathcal{H}}$ of length q_h. This list is used to properly manage the queries for the random oracles $\mathcal{H}_0$ and $\mathcal{H}_1$. In particular, the simulator updates $\Lambda_{\mathcal{H}}$ using the following general rule (n stands for 0 or 1).

- For a hash query $\mathcal{H}_n(q)$ such that (n, q, r) appears in $\Lambda_{\mathcal{H}}$, the answer is r. Otherwise, choose a random $r \in \{0, 1\}^{\ell_{\mathcal{H}_n}}$. If $(n, *, r)$ already belongs to the list $\Lambda_{\mathcal{H}}$, abort, else add (n, q, r) to the list.

Due to the birthday paradox, G_1 is indistinguishable from the real game G_0.

Game G_2:   Case where the adversary wins by chance. This game is almost the same as the previous one. The only difference is that we allow the simulator to abort if the adversary manages to guess $Auth$ without having asked a corresponding query to the oracle. This happens with negligible probability so that $\mathbf{G}_2$ and $\mathbf{G}_1$ are indistinguishable.

Game $\mathbf{G}_3$:   Passive Case: No Corruption Before Step 4. In this game, we deal with the passive case in which no corruption occurs before step 4. We give the simulator some partial control on the players involved in the protocol. In particular, we assume that the simulator is given oracle access to each player, for the first three rounds of the protocol. Then in $S4$, if no corruption occurred, we require $\mathcal{S}$ to completely simulate their behavior. More precisely, during this game, we consider two cases. If no corruption occurred before $S4$, we require $\mathcal{S}$ to simulate the execution of the protocol on behalf of the two players. If, on the other hand, some party has already been corrupted before starting $S4$, the simulator does nothing. Notice that, in any case, we still allow $\mathcal{S}$ to know the passwords of both players.

If at the beginning of $S4$, the two players are still honest and all the flows were oracle-generated, the simulator asks a SamePwd query. Notice that, since we are assuming that $\mathcal{S}$ knows both passwords, this boils down to verify that both passwords are actually the same.

Now we distinguish two cases. If the two passwords are the same, $\mathcal{S}$ chooses a random key K (in the key space) and "gives" K to all players. Otherwise, $\mathcal{S}$ chooses a random key and gives it to the client whereas the server just receives an error message.

Notice that, if the two players have the same password, such a strategy makes this game indistinguishable with respect to previous one. If, conversely, the players do not have the same passwords, an execution of the protocol in this game is indistinguishable from a real execution except for the risk of collision, which is negligible. This is because, if the two players do not share the same passwords, the server will end-up computing a different $Auth$, thus getting an error message, with all but negligible probability. Hence $\mathbf{G}_3$ and $\mathbf{G}_2$ are indistinguishable.

Game $\mathbf{G}_4$:   Simulation of the Client From the Beginning of the Protocol. In this game, we let $\mathcal{S}$ simulate the non-corrupted client from the beginning of the protocol, but we don't allow him to have access to her password anymore. The simulation is done as follows. In $S1$, the client chooses a random x and sends the corresponding X to the server. In $S3$, if she is still honest, then she doesn't ask a decryption query for Y^*.

If all flows were oracle-generated, then she computes $Auth$ with the oracle $\mathcal{H}_1'$ private to the simulator: $Auth = \mathcal{H}_1'(ssid\|U\|S\|X\|Y^*)$ instead of $\mathcal{H}_1$. A problem can occur if the server gets corrupted, as we describe it more formally later on.

Otherwise, if the flow received by the client is not oracle-generated, we face two different cases:

- If the server was corrupted sooner in the protocol, the simulator knows his password, or if the Y^* sent by the adversary in S2 has been obtained via an encryption query, then the simulator recovers his password too (with the help of the encryption list). Then, when receiving Y^*, the client asks a GoodPwd query for the functionality. If it is a correct guess, then S uses $\mathcal{H}_1$ for the client, otherwise it uses its private oracle $\mathcal{H}_1'$: $Auth = \mathcal{H}_1'(ssid\|U\|S\|X\|Y^*)$.

- If the adversary has not obtained Y^* via an encryption query, there is a negligible chance that it knows the corresponding y and the client also uses $\mathcal{H}_1'$ in this case. The event AskH can then make the game to abort (we will bound its probability later on; simply note that it is negligible and related to the CDH):

  AskH: $\mathcal{A}$ queries one of the oracles $\mathcal{H}_0$ or $\mathcal{H}_1$ on $ssid\|U\|S\|X\|Y\|K_U$ or $ssid\|U\|S\|X\|Y\|K_S$, ie the common value of $ssid\|U\|S\|X\|Y\|CDH(X,Y)$

We now show how to simulate the second part of U3 (the computation of sk_U). We need to separate the cases in which the client remains honest, and those in which she gets corrupted.

- If the client remains honest, she is given sk_U by a query to $\mathcal{H}_0'$ if $Auth$ was obtained by a query to $\mathcal{H}_1'$ and no corruption occurred, and by a query to $\mathcal{H}_0$ if $Auth$ was obtained by a query to $\mathcal{H}_1$ or if $Auth$ was obtained by a query to $\mathcal{H}_1'$ and there was a corruption afterwards.
- If she is corrupted during U3, $\mathcal{A}$ is given her internal state: the simulator already knows x and learns her password; it is thus able to compute a correct Y. S then recomputes $Auth$ by a query to $\mathcal{H}_1$ (there is no need that this query gives the same value as the value previously computed by the query to $\mathcal{H}_1'$ since $Auth$ has not been published) and the client is given sk by a query to $\mathcal{H}_0$.

If the two players are honest at the beginning of S4 and all the flows were oracle-generated, there will be no problem as in the former game we prevented the server from computing $Auth$. If the server gets corrupted after $Auth$ has been sent, and if the passwords are the same, the simulator reprograms the oracles such that on the one hand $\mathcal{H}_1(ssid\|U\|S\|X\|Y\|K_U) = \mathcal{H}_1'(ssid\|U\|S\|X\|Y^*)$ and on the other hand $\mathcal{H}_0(ssid\|U\|S\|X\|Y\|K_U) = \mathcal{H}_0'(ssid\|U\|S\|X\|Y^*)$. This programming will only fail if this query to $\mathcal{H}_1$ or $\mathcal{H}_0$ has already been asked before the corruption, in which case the event AskH has happened.

Finally, if the client is being corrupted, S does the same reprogramming.

Thus, omitting the events AskH, which probability will be computed later on, the games $\mathbf{G_4}$ and $\mathbf{G_3}$ are indistinguishable.

Game $\mathbf{G_5}$: Simulation of the Server in the Last Step of the Protocol. In this game, we let S simulate the non-corrupted server in step S4. More precisely, during this game, we consider two cases. If no corruption occurred before S4 and all the flows were oracle-generated, the behavior of S was described in $\mathbf{G_3}$. If, on the other hand, the client has already been corrupted before starting S4, or if a flow was non-oracle-generated, the simulation is done as follows.

If the client is either corrupted or impersonated by the adversary who has decrypted Y^* to obtain the Y sent in $Auth$, then the server recovers the password used (by the corruption or by the decryption list) and he verifies the Diffie-Hellman sent by the client. If it is correct, then the simulator asks a GoodPwd query for the server (otherwise, the latter is given an error message). If the password is correct, then the server is given the same key as the client; otherwise, he is given an error message.

If the client is impersonated by the adversary who has sent anything else, we abort the game. This happens only if it has guessed Y by chance, which happens with negligible probability.

Finally, if the server is corrupted during S4, the adversary is given y and Y. More precisely, the simulator recovers the password of the server and gives something consistent with the lists to $\mathcal{A}$. Thus, $\mathbf{G}_5$ and $\mathbf{G}_4$ are indistinguishable.

Game $\mathbf{G}_6$: Simulation of the Server from the Beginning of the Protocol. In this game, we let $\mathcal{S}$ simulate the non-corrupted players from the beginning of the protocol. We have already seen how $\mathcal{S}$ simulates the client. The simulation, for a non-corrupted server, is done as follows.

In S2, the server sends a random Y^* (chosen without asking the encryption oracle). If he gets corrupted, the simulator recovers his password, and can then provide the adversary with adequate y and Y with the help of the encryption and decryption lists. The simulation of S4 has already been described.

$\mathbf{G}_6$ is indistinguishable from $\mathbf{G}_5$, since if the two players remain honest until the end of the game, they have the same key depending on their passwords and nothing else in $\mathbf{G}_3$. And the case in which one of the two gets corrupted has been dealt with in the two former games, and the execution doesn't depend on the value of Y^*, recalling that the encryption is $G \rightarrow G$ such that there is always a plaintext corresponding to a ciphertext.

Game $\mathbf{G}_7$: Summary of the Simulation and Replacement of the Hybrid Queries. Here we modify the previous game by replacing the hybrid queries GoodPwd and SamePwd with their ideal versions. If a session aborts or terminates, then $\mathcal{S}$ reports it to $\mathcal{A}$.

Figure 3 sums up the simulation until this point and describes completely the behavior of the simulator. At the beginning of a step of the protocol, the player is assumed to be honest (otherwise we don't have to simulate him or her), and he or she can get corrupted at the end of this step. We assume that U3 (1) has to be executed before both U3 (2) and U3 (3). But the two last can be executed in either order. For simplicity, we assume later on that the order is respected.

We show that $\mathbf{G}_7$ is indistinguishable from the ideal game by first recalling the only difference between $\mathbf{G}_6$ and $\mathbf{G}_7$: the GoodPwd queries are replaced by TestPwd queries to the functionality and the SamePwd by NewKey ones. Say that the players have matching sessions if they share the same *ssid*, have two opposite roles (client and server) and agree on the values of X and Y^*.

First, if the two players remain honest until the end of the game, they will obtain a random key, both in $\mathbf{G}_7$ and IWE (the ideal game), as there are no TestPwd queries and the sessions remain fresh.

We need to show that a honest client will receive the same key as a honest server in $\mathbf{G}_7$ if and only if it happens in IWE. We first deal with the case of client and server with matching sessions. If they have the same password in $\mathbf{G}_7$,

	Client	Server	Simulation
U1	honest	honest	random x, $X = g^x$
		adversary	
	gets corrupted	honest	reveal x to $\mathcal{A}$
		adversary	
S2	honest	honest	random Y^*
	adversary		
	honest	gets corrupted	learn pw compute y and Y via decryption query reveal X, y, Y to $\mathcal{A}$
	adversary		
U3 (1)	honest	honest	no decryption query on Y^*
		adversary	
	gets corrupted	honest	learn pw compute y and Y via decryption query reveal x, X, Y to $\mathcal{A}$
		adversary	
U3 (2)	honest	honest	use $\mathcal{H}_1'$ for $Auth$
		adversary	GoodPwd(pw) false, use $\mathcal{H}_1'$
			GoodPwd(pw) correct, use $\mathcal{H}_1$
			if pw unknown, abort
	gets corrupted	honest	learn pw compute y and Y via decryption query reveal x, X, Y to $\mathcal{A}$
		adversary	
U3 (3)	honest	honest	use $\mathcal{H}_0'$ for $Auth$
		adversary	GoodPwd(pw) false, use $\mathcal{H}_0'$
			GoodPwd(pw) correct, use $\mathcal{H}_0$
S4	honest	honest	if SamePwd correct, then same key sk
			if SamePwd incorrect, then error message
	adversary		if pw unknown, then abort
			if pw known, DH false, then error
			if pw known, DH correct, GoodPwd(pw) correct, then same key
			if pw known, DH correct, GoodPwd(pw) false, then error

Fig. 3. Simulation and adaptive corruptions

they will receive the same key: if they are honest, their key is given to them from $\mathbf{G}_3$; if the client is honest with a corrupted server, they will receive their key from $\mathbf{G}_4$; and if the client is corrupted, they will receive it from $\mathbf{G}_5$.

In IWE, the functionality will receive two NewSession queries with the same password. If both players are honest, it will not receive any TestPwd query, so that the key will be the same for both of them. And if one is corrupted and a TestPwd query is done (and correct, since they have the same password), then they will also have the same key, chosen by the adversary.

If they don't have the same password in $\mathbf{G_7}$, the server will always be given an error. In IWE, this is simply the definition of the functionality.

We now deal with the case of client and server with no matching sessions. It is clear that in $\mathbf{G_7}$ the session keys of a client and a server in such a case will be independent because they are not set in any of the games. In IWE, the only way that they receive matching keys is that the functionality receives two NewSession queries with the same passwords, and $\mathcal{S}$ sends NewKey queries for these sessions without having sent any TestPwd queries. But if the two sessions do not have a matching conversation, they must differ in either X, Y^* or $\mathcal{Auth}$. The probability that they share the same pair (X, Y^*) is bounded by q_ε^2/q and thus negligible, q_ε being the number of encryption queries to the oracle.

If the client is corrupted until the end of the game, then in $\mathbf{G_7}$, the server recovers the password and uses it in a TestPwd query to the functionality. If it is incorrect, he is given an error message, and if it is correct, he is given the same key as the client (which was chosen by the simulator). This is exactly the behavior of the functionality in IWE.

If the server gets corrupted, we still have a TestPwd query concerning the client in $\mathbf{G_7}$. If the password is correct, the simulator chooses the key, otherwise it is the adversary. The same thing happens in IWE.

4.3   Simulating Executions Via the CDH Problem

As in [2], we compute the probability of event AskH with the help of a reduction to the CDH problem, given one CDH instance (A, B). More precisely, AskH means that there exists one session in which we replaced the random oracles $\mathcal{H}_0$ or $\mathcal{H}_1$ by $\mathcal{H}_0'$ or $\mathcal{H}_1'$ respectively and $\mathcal{A}$ asks the corresponding hash query. We thus choose at random one session, denoted by $ssid$, and we inject the CDH instance in this specific session. With probability $1/q_s$ we have chosen the right session. In this specific session $ssid$, we maintain a list Λ_B, and

- the client sets $X = A$;
- the server still chooses Y^* at random, but the behavior of the decryption is modified on this specific input Y^*, whatever the key is, but only for this session $ssid$: choose a random element $\beta \in \mathbb{Z}_q^*$ and compute $Y = Bg^\beta$, and store (β, Y) in the list Λ_B, as well as the usual tuple in Λ_ε. If Y already belongs in this list, one aborts as before.

Note that this only affects the critical session $ssid$ and doesn't change anything else. Contrary to the earlier simulation, we do not know the values of x and φ, but they are not needed since the values of K_U and K_S are no longer required to compute the authenticator and the session key: the event AskH raised for this session (X, Y) means that the adversary has queried the random oracles $\mathcal{H}_0$ or $\mathcal{H}_1$ on $U\|S\|X\|Y\|Z$, where $Z = CDH(X, Y)$. By choosing randomly in the list $\Lambda_{\mathcal{H}}$, we obtain this Diffie-Hellman triple with probability $1/q_h$, where q_h is the number of hash queries. We can then simply look into the list Λ_B for the values β such that $Y = Bg^\beta$: $CDH(X, Y) = CDH(A, Bg^\beta) = CDH(A, B)A^\beta$.

Note however that in case of corruption, we may need to reveal internal states, with x and φ: If the corruption happens before the end of U3, with the publication of $\mathcal{A}uth$, there is no problem since the random oracles will not be replaced by the private oracles, and then the guess for the session was not correct, which contradicts the assumption of good choice. If the corruption happens after the end of U3, with the publication of $\mathcal{A}uth$, there is no problem either:

- the corruption of the client does not reveal any internal state, since she has completed her execution;
- the corruption of the server leads to a "reprogramming" of the public oracles that immediately raises the event AskH if the query had already been asked. We can thus stop our simulation, and extract the Diffie-Hellman value from the list $\Lambda_{\mathcal{H}}$, without having to wait the end of the whole attack game.

Acknowledgments

This work was supported in part by the European Commission through the IST Program under Contract IST-2002-507932 ECRYPT, and by the French government through the PAMPA ANR project.

References

1. Pointcheval, D., Abdalla, M.: Simple password-based encrypted key exchange protocols. In: Menezes, A. (ed.) CT-RSA 2005. LNCS, vol. 3376, pp. 191–208. Springer, Heidelberg (2005)
2. Bresson, E., Chevassut, O., Pointcheval, D.: Security proofs for an efficient password-based key exchange. In: ACM CCS 2003, October, 2003, pp. 241–250. ACM Press, New York (2003)
3. Bellovin, S.M., Merritt, M.: Encrypted key exchange: Password-based protocols secure against dictionary attacks. In: 1992 IEEE Symposium on Security and Privacy, May 1992, pp. 72–84. IEEE Computer Society Press, Los Alamitos (1992)
4. Boyko, V., MacKenzie, P.D., Patel, S.: Provably secure password-authenticated key exchange using Diffie-Hellman. In: Preneel, B. (ed.) EUROCRYPT 2000. LNCS, vol. 1807, pp. 156–171. Springer, Heidelberg (2000)
5. Bellare, M., Pointcheval, D., Rogaway, P.: Authenticated key exchange secure against dictionary attacks. In: Preneel, B. (ed.) EUROCRYPT 2000. LNCS, vol. 1807, pp. 139–155. Springer, Heidelberg (2000)
6. Bellare, M., Rogaway, P.: Random oracles are practical: A paradigm for designing efficient protocols. In: ACM CCS 1993, November 1993, pp. 62–73. ACM Press, New York (1993)
7. Bellare, M., Rogaway, P.: Entity authentication and key distribution. In: Stinson, D.R. (ed.) CRYPTO 1993. LNCS, vol. 773, pp. 232–249. Springer, Heidelberg (1994)

8. Bellare, M., Rogaway, P.: Provably secure session key distribution — the three party case. In: 28th ACM STOC, May 1996, pp. 57–66. ACM Press, New York (1996)

9. Canetti, R.: Universally composable security: A new paradigm for cryptographic protocols. In: 42nd FOCS, October 2001, pp. 136–145. IEEE Computer Society Press, Los Alamitos (2001)

10. Canetti, R., Halevi, S., Katz, J., Lindell, Y., MacKenzie, P.D.: Universally composable password-based key exchange. In: Cramer, R.J.F. (ed.) EUROCRYPT 2005. LNCS, vol. 3494, pp. 404–421. Springer, Heidelberg (2005)

11. Canetti, R., Rabin, T.: Universal Composition with Joint State. In: Boneh, D. (ed.) CRYPTO 2003. LNCS, vol. 2729, pp. 265–281. Springer, Heidelberg (2003)

12. Gennaro, R., Lindell, Y.: A framework for password-based authenticated key exchange. In: Biham, E. (ed.) EUROCRYPT 2003. LNCS, vol. 2656, pp. 524–543. Springer, Heidelberg (2003)

13. Gentry, C., MacKenzie, P., Ramzan, Z.: A method for making password-based key exchange resilient to server compromise. In: Dwork, C. (ed.) CRYPTO 2006. LNCS, vol. 4117, pp. 142–159. Springer, Heidelberg (2006)

14. Hofheinz, D., Müller-Quade, J.: Universally composable commitments using random oracles. In: Naor, M. (ed.) TCC 2004. LNCS, vol. 2951, pp. 58–76. Springer, Heidelberg (2004)

15. Katz, J., Ostrovsky, R., Yung, M.: Efficient password-authenticated key exchange using human-memorable passwords. In: Pfitzmann, B. (ed.) EUROCRYPT 2001. LNCS, vol. 2045, pp. 475–494. Springer, Heidelberg (2001)

16. Liskov, M., Rivest, R.L., Wagner, D.: Tweakable block ciphers. In: Yung, M. (ed.) CRYPTO 2002. LNCS, vol. 2442, pp. 31–46. Springer, Heidelberg (2002)

17. MacKenzie, P.D.: The PAK suite: Protocols for password-authenticated key exchange. Contributions to IEEE P1363.2 (2002)

18. Shoup, V.: On formal models for secure key exchange. Technical Report RZ 3120, IBM (1999)

A   The Random Oracle and the Ideal Cipher

In [10], Canetti *et al.* show that there doesn't exist any protocol that UC-emulates $\mathcal{F}_{pwKE}$ in the plain model (i.e. without additional setup assumptions). Here we show how to securely realize a similar functionality without setup assumption but working in the random oracle and ideal cipher models instead.

RANDOM ORACLES. The random oracle functionality was already defined by Hofheinz and Müller-Quade in [14]. We present it again in Figure 4 for completeness. It is clear that the random oracle model UC-emulates this functionality.

IDEAL CIPHER [16]. An ideal cipher is a block cipher that takes a plaintext or a ciphertext as input. We describe the ideal cipher functionality $\mathcal{F}_{IC}$ in Figure 5. Notice that the ideal cipher model UC-emulates this functionality. Note that this functionality characterizes a perfectly random permutation, by ensuring injectivity for each query simulation.

The functionality $\mathcal{F}_{RO}$ proceeds as follows, running on security parameter k, with parties $P_1, \ldots, P_n$ and an adversary $\mathcal{S}$:

- $\mathcal{F}_{RO}$ keeps a list L (which is initially empty) of pairs of bitstrings.
- Upon receiving a value (sid, m) (with $m \in \{0, 1\}^*$) from some party P_i or from $\mathcal{S}$, do:
  - If there is a pair $(m, \tilde{h})$ for some $\tilde{h} \in \{0, 1\}^k$ in the list L, set $h := \tilde{h}$.
  - If there is no such pair, choose uniformly $h \in \{0, 1\}^k$ and store the pair $(m, h) \in L$.

  Once h is set, reply to the activating machine (i.e., either P_i or $\mathcal{S}$) with (sid, h).

Fig. 4. Functionality $\mathcal{F}_{RO}$

The functionality $\mathcal{F}_{IC}$ takes as input the security parameter k, and interacts with an adversary $\mathcal{S}$ and with a set of (dummy) parties $P_1, \ldots, P_n$ by means of these queries:

- $\mathcal{F}_{IC}$ keeps a (initially empty) list L containing 3−tuples of bitstrings and a number of (initially empty) sets $C_{key,sid}, M_{key,sid}$.
- **Upon receiving a query (sid, ENC, key, m) (with $m \in \{0, 1\}^k$) from some party P_i or $\mathcal{S}$, do:**
  - If there is a 3−tuple $(key, m, \tilde{c})$ for some $\tilde{c} \in \{0, 1\}^k$ in the list L, set $c := \tilde{c}$.
  - If there is no such record, choose uniformly c in $\{0, 1\}^k - C_{key,sid}$ which is the set consisting of ciphertexts not already used with key and sid. Next, it stores the 3−tuple $(key, m, c) \in L$ and sets $C_{key,sid} \leftarrow C_{key,sid} \cup \{c\}$.

  Once c is set, reply to the activating machine with (sid, c).
- **Upon receiving a query (sid, DEC, key, c) (with $c \in \{0, 1\}^k$) from some party P_i or $\mathcal{S}$, do:**
  - If there is a 3−tuple $(key, \tilde{m}, c)$ for some $\tilde{m} \in \{0, 1\}^k$ in L, set $m := \tilde{m}$.
  - If there is no such record, choose uniformly m in $\{0, 1\}^k - M_{key,sid}$ which is the set consisting of plaintexts not already used with key and sid. Next, it stores the 3−tuple $(key, m, c) \in L$ and sets $M_{key,sid} \leftarrow M_{key,sid} \cup \{m\}$.

  Once m is set, reply to the activating machine with (sid, m).

Fig. 5. Functionality $\mathcal{F}_{IC}$

Beyond Secret Handshakes:
Affiliation-Hiding Authenticated Key Exchange

Stanisław Jarecki, Jihye Kim, and Gene Tsudik

Computer Science Department
University of California, Irvine
{stasio, jihyek, gts}@ics.uci.edu

Abstract. Public key based authentication and key exchange protocols are not usually designed with privacy in mind and thus involve cleartext exchanges of identities and certificates before actual authentication. In contrast, an Affiliation-Hiding Authentication Protocol, also called a *Secret Handshake*, allows two parties with certificates issued by the same organization to authenticate each other in a *private* way. Namely, one party can prove to the other that it has a valid organizational certificate, yet this proof hides the identity of the issuing organization unless the other party also has a valid certificate from the same organization.

We consider a very strong notion of Secret Handshakes, namely *Affiliation-Hiding Authenticated Key Exchange* protocols (AH-AKE), which guarantee security under arbitrary composition of protocol sessions, including man-in-the-middle attacks. The contribution of our paper is three-fold: First, we extend existing notions of AH-AKE security to Perfect Forward Secrecy (PFS), which guarantees session security even if its participants are later corrupted or any other sessions are compromised. Second, in parallel to PFS security, we specify the exact level of privacy protection, which we call *Linkable Affiliation-Hiding* (LAH), that an AH-AKE protocol can provide in the face of player corruptions and session compromises. Third, we show an AH-AKE protocol that achieves both PFS and LAH properties, under the RSA assumption in ROM, at minimal costs of 3 communication rounds and two (multi)exponentiations per player.

Keywords: secret handshakes, authenticated key exchange, privacy.

1 Introduction

Affiliation-Hiding Authentication protocols, also known as *Secret Handshakes* (SH) [BDS+03], allow two members of the same group to authenticate each other in a way that hides their affiliation from all others. For example, two FBI agents, Alice and Bob, want to discover and communicate with other agents, but they don't want to reveal their affiliations to non-agents. Since the environment is potentially hostile, Alice wants to authenticate herself to Bob *only if* Bob is another FBI agent, and vice versa for Bob. Affiliation-hiding authentication scheme ensures that if only one of the two is a genuine agent, the other

T. Malkin (Ed.): CT-RSA 2008, LNCS 4964, pp. 352–369, 2008.

(the impostor) learns nothing about the counterpart's affiliation. More generally, a non-member adversary who stages an active attack against group members (even playing a man in the middle) should not determine whether any of the parties he interacts with is a member of the targeted group.

Affiliation-hiding authentication schemes were introduced as *Secret Handshakes* by Balfanz et al. [BDS+03], together with a construction based on the security of the Bilinear Diffie-Hellman (BDH) problem in a group with a bilinear map. Subsequently, Castelluccia, et al. [CJT04] constructed a more efficient scheme secure under the Computational Diffie Hellman (CDH) assumption. (Recently [Ver05] proposed an RSA-based SH scheme, but the scheme fails to provide affiliation-hiding.[1]) However, the schemes of [BDS+03] and [CJT04] are only entity authentication schemes, and not authenticated key agreements (AKE's). Secondly, these papers consider a weak notion of (affiliation-hiding) authentication scheme, which looks only at security of isolated protocol instances. In particular, their model excludes man in the middle attacks.

This restricted notion of affiliation-hiding authentication was strengthened to *Affiliation-Hiding Authenticated Key Exchange* protocols (AH-AKE) in [JKT07] and [JL07]. In [JKT07] this notion was defined for *group* key agreement protocols, which generalize two-party AKE's. In [JL07] the notion was strengthened to AKE's which are both affiliation-hiding and *unlinkable* (see a more on unlinkability below), but this in particular implies a two-party affiliation-hiding AKE protocol. The notion of affiliation-hiding authenticated key exchange strengthens the notion of (affiliation-hiding) entity authentication considered in [BDS+03, CJT04] in two ways: First, an authenticated key agreement is a more useful protocol tool because it outputs an authenticated key which can be used for any secure communication task, including entity authentication. Second, the AH-AKE notion of [JKT07, JL07] satisfies the standard *security* requirements demanded of AKE protocols, as formalized by Bellare, Canetti, Krawczyk [BCK98, CK01] and Shoup [Sho99] (but without perfect forward secrecy). Essentially, each protocol session remains secure even if all protocol sessions are arbitrarily scheduled by the adversary, and even if the adversary compromises

---

[1] Providing affiliation-privacy in RSA-based protocols requires extra care, because one must prevent any correlation of protocol messages with the RSA modulus n which is a part of the public key of a given group. The proposal for an RSA-based secret handshake scheme, [Ver05], based on the OSBE scheme of [LDB03], fails to achieve affiliation-hiding because it tries to prevent such correlation by obfuscating only the *size* of the protocol messages (and hence, in their intention, the RSA modulus), leaving intact other possibilities of correlation. In fact, instances of the protocol of [Ver05] can be correlated with the group public key by computing the Jacobian symbol of several protocol messages: If the protocol instance involves modulus n, the corresponding Jacobian symbols are related in predictable ways, thus providing a test whether the protocol session involves players affiliated with a group whose public key is n. Our RSA-based protocol fixes this problem by requiring a safe RSA modulus n, and making sure that all the exchanged messages are random elements in the group Z_n^*, before applying a simple method to masks the modulus size.

the key on any *other* protocol session. In particular, this implies security against a man in the middle attack.

However, [JKT07] and [JL07] consider only a simplified notion of *privacy* (i.e. of affiliation-hiding) for AH-AKE protocols, where the adversary can arbitrarily interleave protocol sessions, but it cannot compromise any of them. Consequently, it is unclear how *any* information about the agreed session keys affects the privacy protection offered by such schemes. Note that in most applications even a passive adversary learns whether two sessions have succeeded, and produced the same session key, and indeed this information reveals something about the affiliations of two interacting parties, namely that they are the same.

Efficiency-wise, the 2-party AH-AKE protocols implied by the (affiliation-hiding) *group* key agreement protocols of [JKT07] take three rounds, involve three exponentiations, and work under either RSA or CDH assumptions in ROM. The 2-party AH-AKE protocol implied by the affiliation-hiding *and unlinkable* AKE scheme of [JL07] works only assuming that the revocation lists of the two interacting players are no farther than some constant Δ apart, moreover the protocol requires $O(\Delta * \log n)$ exponentiations where n is the upper bound on the number of players affiliated with a single organization.

There are other results on secret handshakes [TX06, XY04], but none of them address security under arbitrary protocol composition.

Our Contributions. **First,** we strengthen the AH-AKE security notion of [JKT07] and [JL07] to include *Perfect Forward Secrecy* (PFS), which ensures that each session remains secure even if its participants are eventually corrupted, revealing all their long-term secrets to the adversary. Note that since the adversary in the AH-AKE models of [JKT07, JL07] is not allowed to compromise any group members future corruption of the long-term secrets of any of them may endanger the security of previous protocol sessions in which this group member was a participant. We upgrade the AH-AKE security definition to a more robust and useful notion by modeling corruptions in the security model.

Second, in parallel to PFS security, we formalize the exact level of privacy protection, which we call *Linkable Affiliation-Hiding* (LAH), that an AH-AKE protocol might provide in the face of player corruptions and session compromises. Intuitively, a linkably affiliation-hiding AKE protocol can reveal only as much information about affiliations of the participating parties as is revealed in the following idealized process: The process assigns a random "pseudonym" value to each certificate in the system. Denote a value assigned to certificate $\mathsf{cert}_i{}^{(j)}$ which user P_i holds for group G_j as $id_{i,j}$. Every time player P_i runs an AH-AKE protocol using the public key of G_j, the session reveals the pseudonym $id_{i,j}$. Every time an adversary compromises a session, it learns only if this session failed or succeeded, and consequently it learns whether the pseudonyms $id_{i,j}$ and $id_{i',j'}$ of the two players participating in this session correspond to the same group (and thus it learns whether P_i and P_j have a shared affiliation), because otherwise these sessions are not supposed to succeed. Finally, whenever some player P_i is corrupted, the process reveals, for each pseudonym $id_{i,j}$ of player P_i, which group G_j this pseudonym corresponds to. We stress that the AH-AKE privacy models of [JKT07, JL07] modeled only the information which the adversary learns from protocol messages, and not the information learned from subsequent compromises of session keys and/or corruptions of their participants.

Third, we show an optimal-cost AH-AKE protocol, which satisfies our PFS and LAH notions, in the Random Oracle Model (ROM) under the RSA assumption. The protocol takes 3 rounds, and it is an *implicitly authenticated* version of the Diffie-Hellman Key Exchange protocol. The cost of the protocol appears minimal because its computation costs are very similar to the cost of the unauthenticated Diffie-Hellman key exchange, namely one (off-line) exponentiation and one (on-line) multi-exponentiation per participant. Moreover, three rounds of interaction again matches the round complexity of non-private PFS AKE protocols.

Linkability Disclaimer. As the name suggests, the privacy guaranteed by our notion of *linkable affiliation-hiding* does not include unlinkability, and in this aspect the new notion is similar to the (weaker) notions of affiliation-hiding considered in [BDS+03, CJT04, JKT07]. Indeed, since the ideal process we use to define the LAH privacy property reveals the same pseudonym every time a player uses the same certificate in an AH-AKE session, the adversary can link two instances of the same player. Note, however, that these pseudonyms do not leak the *affiliation* of this player, except if a player is corrupted (which reveals the linkage between the corrupted player's pseudonyms and groups) or when a session is compromised (which can reveal whether or not the pseudonyms of the two players involved in this session correspond to the same group). We stress that even though unlinkable and affiliation-hiding schemes exist [TX06, JL07], they have severe limitations (synchronization in revocation lists, expensive operation), while affiliation-hiding protocols, as we show here, can be achieved at seemingly minimal expense, and thus it is important to understand the exact privacy guarantees offered by the "merely" affiliation-hiding (but linkable) authentication protocols.

Moreover, it is worth pointing out that while our security and privacy models assume that every user has only a single certificate from any group, in which case any two instances involving the same group member are linkable, there are various heuristics which can ameliorate this issue in practice. For example, heuristic unlinkability can be achieved by users rotating through a small set of certificates, by setting strict time limits on usage of each certificate, or by associating different certificates with different locations or aspects of user's activity.

Organization. In Section 2 we define AH-AKE protocols with perfect-forward secrecy (PFS) and linkable affiliation-hiding (LAH), and show that linkable affiliation-hiding implies perfect forward secrecy. In Section 3 we show an AH-AKE scheme which satisfies the LAH property (and hence the PFS property), based on the RSA assumption.

2   Affiliation-Hiding Authenticated Key Exchange

Entities. Our AH-AKE model is based on the existing models for standard (i.e. non affiliation-hiding) authenticated key exchange protocols, e.g. [BPR00, CK01] in the setting of a Public Key Infrastructure (PKI). One difference between our model and the standard PKI setting is that in the standard setting it is assumed that certificates, which in many applications contain information about owners' *affiliation*, are publicly available. Since AH-AKE protocols aim to protect affiliation privacy of the participants, in our model all certificates are private. Moreover, the standard PKI model involves a *certification hierarchy*, where the integrity of the association between entities and their public keys is vouched by a chain of certificates all leading to a commonly trusted CA. In this paper we consider only a restricted version of this general PKI model with a "flat" certification structure, where certification hierarchies and chains are not allowed. In our model there are only "top" CA's, and entities certified by such CA's; there are no intermediate CA's and no delegation of certificates. (Indeed, it remains an open question how to enable affiliation-hiding AKE's with efficient support for general certificate chains.)

An AH-AKE *scheme* operates in an environment that includes a set of *users* $\mathcal{U}$ and a set of *groups* $\mathcal{G}$. Each group is administered by a CA responsible for creating the group, admitting entities as members and revoking membership. We assume upper bounds m and n, respectively, on the total number of groups and the number of members in any given group, i.e., $|\mathcal{G}| \leq m$ and $|\mathcal{U}| \leq n$. We assume that each user can be a member of many groups. We denote the fact that user $U \in \mathcal{U}$ is a member of group $G \in \mathcal{G}$ as $U \prec G$.

AH-AKE Protocol. The main part of an AH-AKE scheme is an AH-AKE *protocol*, which is executed by any pair of users. Player U_i, participating in an instance of the AH-AKE protocol executes the protocol instructions on inputs a public key of some group $G \in \mathcal{G}$ s.t. $U_i \prec G$, and U_i's certificate of membership in G. The purpose of the AH-AKE protocol is for a pair of players to establish an authenticated shared secret key as long as (1) both run the protocol on the public key of the same group G, and (2) it holds that $U_i \prec G$ and $U_j \prec G$. To avoid any misunderstanding, we stress that such protocol does not in general imply an efficient solution for an (affiliation-hiding) *group discovery* problem, where each player starts a protocol on a *set* of certificates, and the protocol succeeds, for example, as long as the sets contributed by the two players have a non-empty union. In contrast, our AH-AKE schemes are most practical in scenarios where each user is a member of at most one group. However, we stress that if a user is a member of many groups, this would affect execution efficiency, but it would not affect the *security* and the *affiliation-hiding* of our schemes. While the *protocols* we give are efficient only if each player is always a member of at most a few groups, the *security definitions* stated below without loss of generality assume that each user is a member of every group.

Public Information and Network Assumptions. We assume that all groups $G \in \mathcal{G}$ are publicly known. In particular the public keys of all CA's and the certificate revocation lists they maintain are public. Before any group can be created, a common security parameter must be publicly chosen, and a public Setup procedure is executed on that parameter. The Setup procedure creates common cryptographic parameters which are used as inputs in all subsequent protocols, but it does not need to be executed by a trusted authority: For example, it can be executed by one of the CA's, and the other CA's can verify the validity of its outputs. We assume that communication between the users and the CA's, i.e. the certificate issuance process and the CRL retrieval, are conducted over anonymous and authenticated channels. For example, a user might communicate with the CA, e.g., while retrieving the most recent CRL for its group, over an anonymous channel such as TOR [DMS04]. Alternatively, the CRL's of all groups can be combined and stored at some highly-available site where they can be either retrieved in bulk (if small) or via a Private Information Retrieval (PIR) protocol, e.g., [CKGS98].

We assume that all communication within the AH-AKE protocol takes place over an unauthenticated channel. In our model, the adversary is assumed to have full control of the underlying network: it sees the messages sent by each participant in a given round, and decides which messages will be *delivered* to each participant in that round. The adversary can delete, modify or substitute any message and it can choose to deliver different messages to different participants.

AH-AKE Syntax. We define an AH-AKE scheme as a collection of the following algorithms:

- Setup: on input of security parameter κ, it generates public parameters params.
- KGen: executed by the group CA, on input params, it outputs the group public key $\mathcal{PK}$ and the corresponding secret key $\mathcal{SK}$ for this group, and an empty certificate revocation list $\mathcal{CRL}$. We denote the group corresponding to the public key $\mathcal{PK}$ as Group($\mathcal{PK}$).
- Add: executed by the CA of group G, on input $\mathcal{SK}$ and $U \in \mathcal{U}$, it adds U to G by generating a certificate for U, denoted cert. If cert is issued under a public key $\mathcal{PK}$, we say that cert $\in$ Certs($\mathcal{PK}$).
- Revoke: executed by the group CA, on input $U \in \mathcal{U}$, it retrieves the corresponding certificate cert issued for U, and revokes it in the $\mathcal{CRL}$ object (e.g. by adding a new entry to a list). We denote this as cert $\in$ RevokedCerts($\mathcal{CRL}$).
- Handshake: This is an interactive protocol executed by two users, e.g. U_i and U_j. The inputs of user U_i is a tuple (cert$_i$, $\mathcal{PK}_i$, $\mathcal{CRL}_i$, role$_i$), where $\mathcal{PK}_i$ is the public key of the group with whose members U_i wants to establish an authenticated connection, $\mathcal{CRL}_i$ is U_i's current CRL for this group, cert$_i$ is U_i's certificate in that group, and role$_i \in \{\text{init, resp}\}$. The inputs of user U_j is the corresponding tuple (cert$_j$, $\mathcal{PK}_j$, $\mathcal{CRL}_j$, role$_j$). Either party can reject, or output an authenticated secret key, K_i or K_j.

Instances and Session IDs. As in the work on standard AKE's, e.g. [BPR00] or [CK01], our model allows for multiple executions of the AH-AKE protocol scheduled in an arbitrary way. Player $U_i \in \mathcal{U}$ can have many *instances*, involved in distinct concurrent executions of AH-AKE protocol. We denote s-th instance of player U as Π_U^s. Each player instance can either reject, or accept and output a key. We say that an instance Π_U^s runs a *protocol session*, and we use *player* *instance* and *protocol session* interchangeably, denoting both as Π_U^s. When referring to a specific user U_i we use Π_i^s as a short-hand for $\Pi_{U_i}^s$. Each instance Π_i^s keeps a state variable, sid_i^s called *session id*, which is always set in our protocols as a concatenation of all public inputs and all the messages sent and received by instance Π_i^s. (This value is defined only for a session which has completed.)

Matching Sessions, Partnered Sessions, and Correctness of AH-AKE's. The intended execution of the secret handshake scheme is to allow two players running two *matching* instances to establish an authenticated (and secret) key K. Two instances Π_i^s and Π_j^t care called **matching** if the respective inputs used on these sessions satisfy the following conditions: $\mathcal{PK}_i^s = \mathcal{PK}_j^t$, $\mathsf{cert}_i^s \in \mathsf{Certs}(\mathcal{PK}_i^s)$, $\mathsf{cert}_j^t \in \mathsf{Certs}(\mathcal{PK}_j^t)$, $\mathsf{cert}_i^s \notin \mathsf{RevokedCerts}(CRL_j^t)$, $\mathsf{cert}_j^t \notin \mathsf{RevokedCerts}(CRL_i^s)$, and $\mathsf{role}_i^s \neq \mathsf{role}_j^t$. We call two protocol instances **partnered** to denote instances which communicate without adversary's interference. Namely, we say that two instances Π_i^s and Π_j^s are partnered if $\mathsf{sid}_i^s = \mathsf{sid}_j^t$, because this condition implies a complete agreement among these two instances with regard to the set of messages sent and delivered between them.

Finally, we say that an AH-AKE scheme is **correct** if, assuming that all keys, certificates and CRL's are generated by following the Setup, KGen, Add and Revoke procedures, for every instances Π_i^s and Π_j^t it holds that if these two instances are *matching and partnered* then they output the same key $K_i^s = K_j^t$.

Security with Perfect Forward Secrecy. We model the security of an AH-AKE scheme similarly to the way security has been defined for general AKE schemes (e.g. [BPR00, CK01]). Namely, we define security via a game between a challenger $\mathcal{C}$ playing the part of a network of m groups and n users, and an adversary $\mathcal{A}$ who starts any number of sessions between these users, and who is in complete control of the network over which they communicate, who is allowed to reveal any number of agreed-upon keys *and* corrupt any number of players, and yet he cannot distinguish from random a key of *any* un-revealed session of some currently uncorrupted player. This is modeled in a standard way, by having an adversary test some session executed by a currently uncorrupted player, which also has not been revealed before (also, the adversary is barred from revealing a session which is partnered with the tested session), at which point a random coin-toss b determines if the adversary sees a key computed on this session or a random bitstring of the same length. The attacker can continue starting and revealing sessions (except of the tested session or the session which is partnered with it) and corrupting players (including the players on the tested session), and finally he outputs his guess b' as to bit b, i.e. as to whether the tested key was real or random.

Formally, security is defined via an interaction of an adversarial algorithm $\mathcal{A}$ and a challenger $\mathcal{C}$ on common inputs (κ, n, m). The interaction starts with $\mathcal{C}$ generating params via Setup(κ), and initializing m groups $G_1, ..., G_m$, by running the KGen(params) algorithm m times. $\mathcal{C}$ initializes all members in these groups, by running the Add$(\mathcal{SK}_j)$ algorithm, for each $\mathcal{SK}_j$, for n times per each key. This way, $\mathcal{C}$ generates n certificates for every group, so that all n users can be members of all m groups. The adversary $\mathcal{A}$ gets the public keys $\mathcal{PK}_1, ..., \mathcal{PK}_m$. It then chooses a subset $\mathsf{Rev} \subseteq \mathcal{U}$ of initially corrupted players and receives the set of certificates $\{\mathsf{cert}_i{}^{(j)}\}_{U_i \in \mathsf{Rev}, j \in [1..m]}$. For each group G in $\mathcal{G}$, the challenger runs the Revoke algorithm to revoke all corrupted members $U \in \mathsf{Rev}$, and outputs the resulting CRL's for each group, $\mathcal{CRL}_1, ..., \mathcal{CRL}_m$.

After this initialization, $\mathcal{A}$ schedules any number of $\mathsf{Handshake}$ protocols, arbitrarily manipulates their messages, requests the keys on any number of the (accepting) sessions, and corrupts any number of additional players, all of which can be modeled by $\mathcal{A}$ issuing any number of the commands listed below. Finally, $\mathcal{A}$ stops and outputs a single bit b'. The commands the adversary can issue, and the way the challenger $\mathcal{C}$ responds to them, are listed below. In all commands we assume that $U \in \mathcal{U} \setminus \mathsf{Rev}$.

- $\mathsf{Start}(U, G, s, \mathsf{role})$: If $U \in \mathcal{U} \setminus \mathsf{Rev}$ and $G \in \mathcal{G}$ and s was not used already in another Start query on the same user, the challenger retrieves key $\mathcal{PK}$ for group G, certificate $cert$ issued to player U for group G, and the $\mathcal{CRL}$ corresponding to this group, initiates instance Π_U^s, and follows the $\mathsf{Handshake}$ protocol on behalf of user U on inputs ($\mathsf{cert}, \mathcal{PK}, \mathcal{CRL}, \mathsf{role}$), forwarding any message generated by U to $\mathcal{A}$. The challenger keeps the state of all initiated instances Π_U^s. We denote the group G upon which Π_U^s is initiated as $\mathsf{Group}(\Pi_U^s)$.

- $\mathsf{Send}(U, s, \mathcal{M})$: If instance Π_U^s has been initiated and is waiting for a message, $\mathcal{C}$ delivers M to this instance, and forwards to $\mathcal{A}$ any message Π_U^s generates in response. If Π_U^s outputs a key, $\mathcal{C}$ stores it with the session state.

- $\mathsf{Reveal}(U, s)$: If instance Π_U^s has been initiated and has output a session key K, $\mathcal{C}$ delivers this key to $\mathcal{A}$. If the session has either not completed yet or has rejected, $\mathcal{C}$ sends a null value to $\mathcal{A}$. The challenger does not respond if $\mathcal{A}$ queries Reveal on instance Π_U^s for which $\mathcal{A}$ previously issued a Test query (see below) *or* s.t. Π_U^s *matches* and is *partnered* with some $\Pi_{U'}^{s'}$ for which $\mathcal{A}$ issued a Test query.

- $\mathsf{Test}(U, s)$: This query is allowed only once, at any time during the adversary's execution. If Π_U^s is *fresh* (see below), and has output some key K, then $\mathcal{C}$ responds depending on its private input bit b. If $b = 1$ then $\mathcal{C}$ sends to $\mathcal{A}$ the key K. If $b = 0$ then $\mathcal{C}$ sends to $\mathcal{A}$ a random κ-bit long value K'. If the session does not exist, has failed, or is still active, the challenger ignores this Test command.

- $\mathsf{Corrupt}(U)$: This query models adaptive corruptions. The challenger adds U to the revocation list Rev, and replies with all long-term secrets $\mathsf{cert}^{(1)}$, ..., $\mathsf{cert}^{(m)}$ of U. In particular, the adversary can query $\mathsf{Corrupt}(U)$ even if it has *previously* issued $\mathsf{Test}(U, s)$ for some s.[2]

---

[2] On the other hand, $\mathcal{C}$ does refuse the $\mathsf{Test}(U, s)$ query if U is corrupted. (Compare with the notion of freshness.) This reflects the simple fact that we can protect security of a protocol run even if one or both of the two players involved is corrupted in the future, but it makes no sense to ask for security of sessions in which one of the participants is already corrupted.

Freshness. Following [BPR00], we define a notion of *freshness* appropriate for modeling perfect forward secrecy. An instance Π_U^s is *fresh* unless, for any session $\Pi_{U'}^{s'}$ which *matches* and is *partnered* with Π_U^s, the adversary has issued any of the following queries: Reveal(U, s), Reveal(U', s'), Corrupt(U), or Corrupt(U').

Definition 1. *Denote $\mathcal{A}$'s output in the above interaction with $\mathcal{C}$ on bit b and (κ, n, m) as $\mathcal{A}^{\mathcal{C}(b)}(\kappa, n, m)$. Define the adversary's advantage as follows (the probability goes over the randomness of $\mathcal{A}$ and $\mathcal{C}$):*

$$Adv_{\mathcal{A}}^{\text{sec}}(\kappa, n, m) = \left| \Pr[1 \leftarrow \mathcal{A}^{\mathcal{C}(1)}(\kappa, n, m)] - \Pr[1 \leftarrow \mathcal{A}^{\mathcal{C}(0)}(\kappa, n, m)] \right|$$

We call an AH-AKE scheme secure with perfect forward secrecy, *or* PFS, *if for any efficient probabilistic adversary $\mathcal{A}$, for parameters n an m polynomially related to κ, $Adv_{\mathcal{A}}^{\text{sec}}(\kappa, n, m)$ is negligible in κ.*

Linkable Affiliation-Hiding. We define the affiliation-hiding property, similarly to security, using a game between an adversary and a challenger $\mathcal{C}^{\text{ah}}$. However, the adversary's goal in the affiliation-hiding game is not to violate semantic security of some session key but to learn about the participants' affiliation by corrupting players, and by learning whether certain sessions were successful. Note that by corrupting players the adversary learns their affiliations, and that by revealing whether two partnered sessions are successful the adversary learns that these two players belong to the same group. We model the property of the attacker's *inability to learn anything above this information*, by comparing two executions of the adversary: One where the challenger follows the protocols faithfully on behalf of all honest participants, and the other where the adversary interacts with a *simulator*. The simulator attempts to follow adversary's instructions on behalf of honest users, except that it is never told the groups for which the (scheduled by the adversary) Handshake protocol instances are executed, i.e., if the adversary issues a Start(U, G, s, role) query, the simulator gets only an identifier id which is uniquely but arbitrarily assigned to the pair $(U, G) \in \mathcal{U} \times \mathcal{G}$.

Consequently, these inputs are also the only thing that the adversary can possibly learn from the messages produced by this simulator. In other words, the simulated protocol messages can reveal only whether or not two sessions involve *the same* (user,group) pair. However, the adversary does not learn which group it is, nor can he decide if two instances of two different users belong to the same group. Note that we allow the adversary to be able to *link* instances which involve the same (user,group) pair because the simulator gets the same id for such instances. Indeed, all AH-AKE schemes we propose in this paper are linkable in this sense.

Formally, we model the affiliation-hiding property using an interactive algorithm $\mathcal{SIM}$, function F indexed by the public parameters params of the scheme, and the following game between adversary $\mathcal{A}$ and challenger $\mathcal{C}^{\text{ah}}$, on inputs κ, n, m: $\mathcal{C}^{\text{ah}}$ runs Setup$(\kappa) \rightarrow$ params, KGen(params) $\rightarrow (\mathcal{PK}_j, \mathcal{SK}_j)$, for $j \in [1..m]$, and Add$(\mathcal{SK}_j) \rightarrow \text{cert}_i^{(j)}$, for $(i, j) \in [1..n] \times [1..m]$, and gives

$\{\mathcal{PK}_j\}_{j\in[1..m]}$ to $\mathcal{A}$. After this initialization, $\mathcal{A}$ can issue any number of queries of the form $\mathsf{Start}(U, G, i, \mathsf{role})$, $\mathsf{Send}(U, s, \mathcal{M})$, $\mathsf{Reveal}(U, s)$, and $\mathsf{Corrupt}(U)$ to $\mathcal{C}^{\mathsf{ah}}$, as in the security game (except there's no Test query). The challenger $\mathcal{C}^{\mathsf{ah}}$ runs on an additional input of bit b, and it responds to $\mathcal{A}$'s commands depending on whether $b = 0$ or 1. If $b = 1$, $\mathcal{C}^{\mathsf{ah}}$ responds to all $\mathcal{A}$'s commands by following the corresponding protocol on behalf of the honest users. If $b = 0$ then $\mathcal{C}^{\mathsf{ah}}$ replies to $\mathcal{A}$'s commands using an *ideal affiliation-hiding* process and a *simulator*, an interactive machine $\mathcal{SIM}$ running on input params, as follows:

- (1) On Start and Send, $\mathcal{C}^{\mathsf{ah}}$ replies with messages output by $\mathcal{SIM}$, which instead of $\mathsf{Start}(U_i, G_j, s, \mathsf{role})$ and $\mathsf{Send}(U_i, s, \mathcal{M})$ gets inputs $\mathsf{Start}(id_i^j, s, \mathsf{role})$ and $\mathsf{Send}(id_i^j, s, \mathcal{M})$, respectively, where $id_i^j = F_{\mathsf{params}}(\mathsf{cert}_i^{(j)})$.
- (2) On $\mathsf{Corrupt}(U_i)$, $\mathcal{C}^{\mathsf{ah}}$ gives to $\mathcal{A}$ all the long-term secrets of U_i, i.e. $\{\mathsf{cert}_i^{(j)}\}_{j\in[1..m]}$.
- (3) On $\mathsf{Reveal}(U_i, s)$, $\mathcal{C}^{\mathsf{ah}}$ returns value $\bar{K}_i^s$ chosen as follows. If (a) Π_i^s is *matched* with some session Π_j^t (note that $\mathcal{C}^{\mathsf{ah}}$ knows this), (b) all the messages between Π_i^s and Π_j^t up to this point were correctly exchanged, (c) Π_i^s has received all messages needed to complete the protocol, and (d) $\bar{K}_i^s$ is not yet set, then $\mathcal{C}^{\mathsf{ah}}$ picks $\bar{K}_i^s$ at random in $\{0,1\}^\kappa$, sets $\bar{K}_j^t \leftarrow \bar{K}_i^s$, and returns $\bar{K}_i^s$ to $\mathcal{A}$. If (a),(b),(c) holds but not (d), i.e. if $\bar{K}_i^s$ is already set then $\mathcal{C}^{\mathsf{ah}}$ returns this $\bar{K}_i^s$ to $\mathcal{A}$. In every other case $\mathcal{C}^{\mathsf{ah}}$ returns $\bar{K}_i^s = \bot$. (In particular, if two sessions are matching and partnered, their keys will be the same.)

Remark. Note that if an adversary $\mathcal{A}$ exchanges all messages between Π_i^s and Π_j^t, and then reveals whether or not Π_i^s established a key, then $\mathcal{A}$ learns whether or not sessions Π_i^s and Π_i^t are matching, and hence learns that these sessions relate to the same group. This is unavoidable, since session is supposed to be successful only if it is partnered with a matching one, but the way $\mathcal{C}^{\mathsf{ah}}$ replies to Reveal queries implies that this is the only information divulged by revealing a session key. In particular, the adversary does not learn *which* group these two protocol instances share.

Definition 2. *Denote the output of adversary $\mathcal{A}$ in the above interaction with $\mathcal{C}^{\mathsf{ah}}$ on inputs (κ, n, m), $\mathcal{C}^{\mathsf{ah}}$'s private input b, and $\mathcal{C}^{\mathsf{ah}}$'s access to procedure $\mathcal{SIM}$ and function F, as $\mathcal{A}^{\mathcal{C}^{\mathsf{ah}}(b), \mathcal{SIM}, F}(\kappa, n, m)$. Define $\mathcal{A}$'s advantage as follows, where the probabilities are taken over the randomness of $\mathcal{A}$, $\mathcal{C}^{\mathsf{ah}}$, and $\mathcal{SIM}$:*

$$Adv_{\mathcal{A}, \mathcal{SIM}, F}^{\mathsf{ah}}(\kappa, n, m) =$$

$$\left| \Pr[1 \leftarrow \mathcal{A}^{\mathcal{C}^{\mathsf{ah}}(1), \mathcal{SIM}, F}(\kappa, n, m)] - \Pr[1 \leftarrow \mathcal{A}^{\mathcal{C}^{\mathsf{ah}}(0), \mathcal{SIM}, F}(\kappa, n, m)] \right|$$

We call an AH-AKE scheme linkably affiliation-hiding, *or* LAH, *if there exists an efficiently computable F and an efficient probabilistic algorithm $\mathcal{SIM}$ s.t.*

1. *For any efficient probabilistic algorithm $\mathcal{A}$ and any n an m polynomially related to κ, the adversarial advantage $Adv_{\mathcal{A}, \mathcal{SIM}, F}^{\mathsf{ah}}(\kappa, n, m)$ is a negligible function of κ.*

2. *There is a negligible function ϵ s.t. for any params output by Setup(κ), and any two keys pairs $(\mathcal{PK}_0, \mathcal{SK}_0)$ and $(\mathcal{PK}_1, \mathcal{SK}_1)$ output by KGen(params), the statistical distance between distribution D_0 and D_1 is bounded by $\epsilon(\kappa)$, where $D_b = \{F(\text{cert}) \mid \text{cert} \leftarrow \text{Add}(\mathcal{SK}_b)\}$.*

Remark. Intuitively, requirement (2) implies that $F(\text{cert}_i{}^{(j)})$ reveals no information about the group G_j that issued $\text{cert}_i{}^{(j)}$. Therefore, by requirement (1), the only information that $\mathcal{A}$ learns when attacking a LAH scheme, is a "pseudonym" $id_{i,j} = F(\text{cert}_i{}^{(j)})$ which corresponds to user U_i and group G_j, but which does not leak what group this pseudonym corresponds to.

Linkable Affiliation-Hiding Implies PFS Security. By a simple hybrid argument we can show that LAH implies PFS. Intuitively, this is because the affiliation-hiding game compares the view of the real execution with a "fully-random" view, where all messages and keys are chosen by the challenger and a simulator, whereas the security game compares the real view with a view modified so that only the key of the tested session is chosen at random. It's not difficult to see that a significant difference between the views in the second pair implies a significant difference between the views in the first pair. The exact security in this reduction decreases by a small constant factor.

Lemma 1. *If AH-AKE scheme is Linkably Affiliation-Hiding (def. 2) then it is Secure with Perfect Forward Secrecy (def. 1).*

Proof. Let $\mathcal{A}$ be an (adaptive) adversary which attacks the security game. The construction of an adaptive adversary $\mathcal{A}'$ which attacks the affiliation-hiding game is trivial: $\mathcal{A}'$ forwards the messages from its challenger $\mathcal{C}^{ah}$ to $\mathcal{A}$, and it similarly forwards all the commands from $\mathcal{A}$ to $\mathcal{C}^{ah}$, except the Test(U, s) command for which $\mathcal{A}'$ issues Reveal(U, s) to $\mathcal{C}^{ah}$, and returns $\mathcal{C}^{ah}$'s response to $\mathcal{A}$. When $\mathcal{A}$ stops and outputs a bit b', $\mathcal{A}'$ returns the same bit.

Let p-sc$_b$ denote the probability that $\mathcal{A}$ outputs 1 on the interaction defined as in the security game with the challenger $\mathcal{C}(b)$, and p-ah$_b$ be the probability that $\mathcal{A}'$ outputs 1 when interacting with $\mathcal{C}^{ah}(b)$. Note that p-ah$_1$ = p-sc$_1$ because in both cases this is the interaction of $\mathcal{A}$ with the real protocol. Let $p \approx p'$ denote that $|p - p'|$ is a negligible function of the security parameter. Then by the assumption that the scheme is linkably affiliation-hiding, p-ah$_0 \approx$ p-ah$_1 =$ p-sc$_1$. We will argue that p-sc$_0 \approx$ p-sc$_1$ as well. Let $p(b_1, b_2, b_3)$ denote the probability $\mathcal{A}$ outputs 1 on interaction with the challenger which computes all messages and keys as in the real protocol, except that (1) if $b_1 = 0$ then all players' messages are computed as $\mathcal{C}^{ah}$ on $b = 0$, i.e. via the simulator $\mathcal{SIM}$, and (2) if $b_2 = 0$ then all the revealed keys are chosen independently at random for every session (unless Π_i^s matches some partnered session $\Pi_{i'}^{s'}$, in which case $K_i^s = K_{i'}^{s'}$), again as in the procedure for $\mathcal{C}^{ah}$ on $b = 0$, and (3) if $b_3 = 0$ then also the key of the tested session is chosen at random. Using this notation we have p-ah$_0 = p(0, 0, 0)$, p-sc$_0 = p(1, 1, 0)$, p-ah$_1 =$ p-sc$_1 = p(1, 1, 1)$. Our assumption is that p-ah$_0 = p(0, 0, 0) \approx p(1, 1, 1) =$ p-ah$_1$, and so if we show that $p(0, 0, 0) \approx p(1, 1, 0)$, this will imply that p-sc$_0 = p(1, 1, 0) \approx p(1, 1, 1) =$ p-sc$_1$.

Now, if $p(0,0,0) \not\approx p(1,0,0)$ then by a trivial reduction which substitutes the revealed and tested session keys with independently chosen random keys (except for partnered matching sessions, as above), it would follow that $p(0,0,0) \not\approx p(1,1,1)$. By contradiction, we get $p(0,0,0) \approx p(1,0,0)$. By a similar argument on just the revealed keys, if $p(0,0,0) \not\approx p(1,1,0)$ then we'd have $p(0,0,0) \not\approx p(1,0,0)$, and hence it follows that $p(0,0,0) \approx p(1,1,0)$ as needed.

3   PFS-Secure Affiliation-Hiding AKE Based on RSA

- Setup: Given security parameter κ, we define κ' as the smallest integer s.t. the RSA assumption holds on $(2\kappa')$-long composites with security parameter κ. We also define a hash function $H_1 : \{0,1\}^* \to \{0,1\}^\kappa$.

- KGen: Generate a $2\kappa'$-bit safe RSA modulus $n = pq$, where $p = 2p' + 1$, $q = 2q'+1$, and p, q, p', q' are primes. Pick a random element g s.t. g generates a maximum subgroup in Z_n^*, i.e. $ord(g) = 2p'q'$, and s.t. $-1 \notin \langle g \rangle$. (This holds for about half of the elements in Z_n^*, and it is easily tested.) Note that in this case $Z_n^* \equiv \langle -1 \rangle \times \langle g \rangle$. Therefore, in particular, if $x \leftarrow Z_{2p'q'}$ and $b \leftarrow \{0,1\}$ then $(-1)^b g^x$ is distributed uniformly in Z_n^*. RSA exponents (e, d) are chosen in the standard way, as a small prime e and $d = e^{-1} \pmod{\phi(n)}$. The secret key is (p, q, d) and public key is (n, g, e). Key generation also fixes a hash function $H_n : \{0,1\}^* \to Z_n$, specific to the group modulus n.[3]

- Add: To add user U to the group, the manager picks a random string $id \leftarrow \{0,1\}^\kappa$ and computes a (full-domain hash) RSA signature on id, $\sigma = h^d \pmod n$, where $h = H_n(id)$. U's certificate is cert $= (id, \sigma)$.

- Revoke: To remove user U from the group, the manager appends string id to the group $\mathcal{CRL}$, where (σ, id) is U's certificate in this group.

- Handshake: This is an AKE protocol for users U_A and U_B of the honest players, where player U_A's inputs a tuple $(\text{cert}_A, \mathcal{PK}_A, \mathcal{CRL}_A, \text{init})$ and U_B's inputs $(\text{cert}_B, \mathcal{PK}_B, \mathcal{CRL}_B, \text{resp})$ s.t. $\text{cert}_A = (id_A, \sigma_A)$ is U_A's certificate for the public key $\mathcal{PK}_A = (n_A, e_A, g_A)$, i.e. $\text{cert}_A \in \text{Certs}(\mathcal{PK}_A)$, $\mathcal{CRL}_A$ is the (hopefully recent) CRL for group $\text{Group}(\mathcal{PK}_A)$, and similarly $\text{cert}_B = (id_B, \sigma_B)$, $\mathcal{PK}_B = (n_B, e_B, g_B)$, and $\mathcal{CRL}_B$ are defined for U_B. The protocol is in Figure 1.

To verify correctness, observe that $z_A = ((\theta_A)^e (h_A)^{-1})^2 = g^{2ex_A}$ and $z_B = g^{2ex_B}$, and therefore both $r_A = (z_B)^{x_A} = g^{2ex_Ax_B}$ and $r_B = (z_A)^{x_B} = g^{2ex_Ax_B}$. Here is the key reason why this protocol hides the modulus n either player uses: First, θ'_A is uniform in $Z_{n_A}^*$. Second, note that if θ'_A is (statistically) uniform in

---

[3] Selecting separate hash function H_n for every group is done purely for notational convenience. A family of hash functions $H_n : \{0,1\}^* \to Z_n$ s.t. each H_n is statistically close to a random function with range Z_n, can be easily implemented in the random oracle model with a single hash function with range $2^{2\kappa'+\kappa}$. E.g., $H_n(m) = H(n, m) \bmod n$.

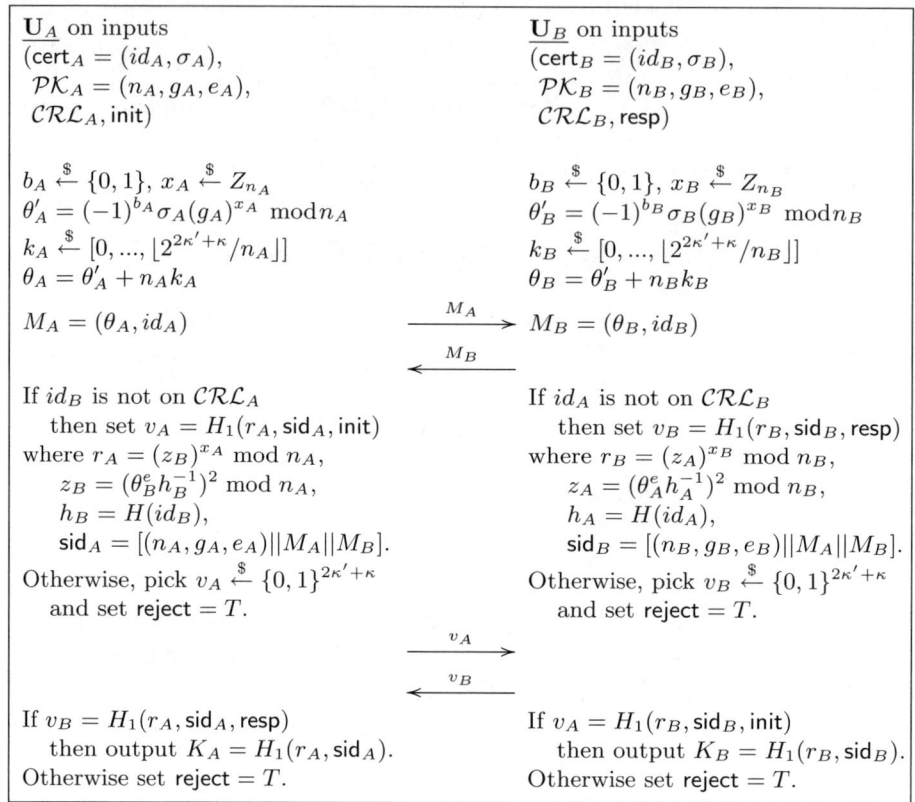

U_A on inputs
$(\text{cert}_A = (id_A, \sigma_A),$
$\quad \mathcal{PK}_A = (n_A, g_A, e_A),$
$\quad \mathcal{CRL}_A, \text{init})$

$b_A \xleftarrow{\$} \{0, 1\}, \, x_A \xleftarrow{\$} Z_{n_A}$
$\theta'_A = (-1)^{b_A} \sigma_A (g_A)^{x_A} \bmod n_A$
$k_A \xleftarrow{\$} [0, ..., \lfloor 2^{2\kappa'+\kappa}/n_A \rfloor]$
$\theta_A = \theta'_A + n_A k_A$

$M_A = (\theta_A, id_A) \qquad \xrightarrow{\quad M_A \quad}$

$\xleftarrow{\quad M_B \quad}$

U_B on inputs
$(\text{cert}_B = (id_B, \sigma_B),$
$\quad \mathcal{PK}_B = (n_B, g_B, e_B),$
$\quad \mathcal{CRL}_B, \text{resp})$

$b_B \xleftarrow{\$} \{0, 1\}, \, x_B \xleftarrow{\$} Z_{n_B}$
$\theta'_B = (-1)^{b_B} \sigma_B (g_B)^{x_B} \bmod n_B$
$k_B \xleftarrow{\$} [0, ..., \lfloor 2^{2\kappa'+\kappa}/n_B \rfloor]$
$\theta_B = \theta'_B + n_B k_B$

$M_B = (\theta_B, id_B)$

If id_B is not on $\mathcal{CRL}_A$
   then set $v_A = H_1(r_A, \text{sid}_A, \text{init})$
where $r_A = (z_B)^{x_A} \bmod n_A$,
   $z_B = (\theta_B^e h_B^{-1})^2 \bmod n_A$,
   $h_B = H(id_B)$,
   $\text{sid}_A = [(n_A, g_A, e_A)||M_A||M_B]$.
Otherwise, pick $v_A \xleftarrow{\$} \{0, 1\}^{2\kappa'+\kappa}$
   and set $\text{reject} = T$.

If id_A is not on $\mathcal{CRL}_B$
   then set $v_B = H_1(r_B, \text{sid}_B, \text{resp})$
where $r_B = (z_A)^{x_B} \bmod n_B$,
   $z_A = (\theta_A^e h_A^{-1})^2 \bmod n_B$,
   $h_A = H(id_A)$,
   $\text{sid}_B = [(n_B, g_B, e_B)||M_A||M_B]$.
Otherwise, pick $v_B \xleftarrow{\$} \{0, 1\}^{2\kappa'+\kappa}$
   and set $\text{reject} = T$.

$\xrightarrow{\quad v_A \quad}$

$\xleftarrow{\quad v_B \quad}$

If $v_B = H_1(r_A, \text{sid}_A, \text{resp})$
   then output $K_A = H_1(r_A, \text{sid}_A)$.
Otherwise set $\text{reject} = T$.

If $v_A = H_1(r_B, \text{sid}_B, \text{init})$
   then output $K_B = H_1(r_B, \text{sid}_B)$.
Otherwise set $\text{reject} = T$.

Fig. 1. AH-AKE protocol based on the RSA assumption

Z_{n_A} then the distribution of $\theta_A = \theta'_A + k_A n_A$, where k_A is picked as above, is statistically close to $U_{2^{2\kappa'+\kappa}}$. An alternative way to hide the range of θ_A, which does not take the κ bandwidth overhead, follows the idea given by [BBDP01] for key-private version of RSA encryption. Namely, one can repeat picking θ'_A until $\theta'_A \in \{0, 1\}^{2\kappa'-1}$. However, the expected running time of such procedure is at most twice that of our procedure, and this alternative procedure could also be subject to timing attacks. Note that the overhead of κ bits we incur is small compared to $|\theta'_A| = |n_A| = 2\kappa'$.

Theorem 1. *Under the RSA assumption on safe RSA moduli (see definition 3 below), the above AH-AKE scheme is Secure with Perfect Forward Secrecy and it is Linkably Affiliation-Hiding, in the Random Oracle Model.*

Definition 3. *Let* S-RSA-IG(κ') *be an algorithm that outputs safe RSA instances, i.e. pairs* (n, e) *where* $n = pq$, e *is a small prime that satisfies* $gcd(e, \phi(n)) = 1$, *and* p, q *are randomly generated* κ'-*bit primes subject to the constraint that* $p = 2p' + 1$, $q = 2q' + 1$ *for prime* $p', q', p' \neq q'$. *We say that the*

RSA problem is (ϵ, t)-hard on $2\kappa'$-bit safe RSA moduli, if for every algorithm $\mathcal{A}$ that runs in time t we have

$$\Pr[(n, e) \leftarrow \text{S-RSA-IG}(\kappa'), g \leftarrow \mathbb{Z}_n^* : \mathcal{A}(n, e, g) = z \ s.t. \ z^e = g \bmod n] \leq \epsilon.$$

Proof. By lemma 1, we only need to argue LAH, i.e. we need to show that $\mathcal{A}$'s view of the interaction with the challenger $\mathcal{C}^{\text{ah}}$ on bit $b = 1$ is indistinguishable from the view of the interaction with $\mathcal{C}^{\text{ah}}$ on bit $b = 0$. Let Game0 represent a real execution, i.e., interaction of $\mathcal{A}$ with challenger $\mathcal{C}^{\text{ah}}$ on bit $b = 1$, while Game2 represents a simulation, i.e., interaction of $\mathcal{A}$ with challenger $\mathcal{C}^{\text{ah}}$ on bit $b = 0$. Thus, our goal is to demonstrate that $\mathcal{A}$'s view in Game0 is indistinguishable from $\mathcal{A}$'s view in Game2. Consider Game1, which is like Game0, except that it stops if there is ever a collision in sid values of any two instances. Since θ_i sent by U_i is indistinguishable from $(2\kappa' + \kappa)$-bit string (see below), probability that there is a collision in polynomially many executions is negligible. Therefore Game0 and Game1 are indistinguishable.

Simulation. To describe Game2, i.e. the simulation, we need to define the simulator $\mathcal{SIM}$ and a function family F. Note that params $= (\kappa, \kappa')$ and that certificates cert are pairs of the form (id, σ) where $id \in \{0, 1\}^\kappa$. We will set $F_{(\kappa, \kappa')}(\text{cert}) = id$. Note that this function satisfies requirement (2) in the LAH definition because the id part of any certificate cert is a random κ-bit string independent of the group's key. Now we describe the simulator $\mathcal{SIM}$, and at the same time we recall how $\mathcal{C}^{\text{ah}}$ interacts with $\mathcal{A}$ using this simulator and function F. Note that the simulator is only involved in the Start and Send queries. Since the protocols of the initiator and the responder are symmetric, below we only describe the initiator's part.

- First $\mathcal{C}^{\text{ah}}$ initializes all the groups and all the users in these groups using Setup, KGen, and Add algorithms as in the real protocol, and gives all the group public keys to $\mathcal{A}$.
- On Start$(U_i, G_j, s, \text{init})$ from $\mathcal{A}$, $\mathcal{C}^{\text{ah}}$ runs $\mathcal{SIM}(id_{i,j}, s, \text{init})$, where $id_{i,j} = F_{(\kappa, \kappa')}(\text{cert}_i^{(j)})$, and responds with $\mathcal{SIM}$'s output, which is a pair $(\theta_i^s, id_{i,j})$ where $\theta_i^s \leftarrow \{0, 1\}^{2\kappa' + \kappa}$.
- On Send$(U_i, s, \mathcal{M})$ from $\mathcal{A}$, $\mathcal{C}^{\text{ah}}$ runs $\mathcal{SIM}(id_{i,j}, s)$ where $id_{i,j}$ is the id corresponding to Π_i^s and responds with $\mathcal{SIM}$'s output. The simulator, regardless of $\mathcal{M}$, returns v_i^s where $v_i^s \leftarrow \{0, 1\}^{2\kappa' + \kappa}$.
- On Corrupt(U_i), $\mathcal{C}^{\text{ah}}$ gives to $\mathcal{A}$ the set $\{\text{cert}_i^{(j)}\}_{j \in [1..m]}$.
- On Reveal(U_i, s), $\mathcal{C}^{\text{ah}}$ returns value $\bar{K}_i^s$ chosen as described in the LAH definition. Briefly, it's a random κ-bit string if the sessions are matched, partnered, and all messages were exchanged up to this point, and otherwise it's a $\perp$ symbol. The only exception is that $\bar{K}_j^t = \bar{K}_i^s$ on the two matched sessions on which all messages were exchanged properly.

Let HQuery be an event that $\mathcal{A}$ ever queries H_1 on arguments (r_i^s, sid_i^s), or $(r_i^s, \text{sid}_i^s, \text{init})$, or $(r_i^s, \text{sid}_i^s, \text{resp})$, for any Π_i^s that $\mathcal{A}$ starts, where r_i^s is defined via the combination of the message (θ_i^s, id_i, s) which instance Π_i^s of an honest player

U_i sent on that session, and message $M = (\hat{\theta}, \hat{id}, s)$ which $\mathcal{A}$ sent to Π_i^s in his
Send(U_i, s, M) command, as follows:

$$r_i^s = (\hat{z})^{x_i^s} \bmod n \text{ where } \hat{z} = g^{2e\hat{x}} = (\hat{\theta})^{2e}(\hat{h})^{-2} \text{ and } z_i^s = g^{2ex_i^s} = (\theta_i^s)^{2e}(h_i)^{-2} \tag{1}$$

where $\hat{h} = H_n(\hat{id})$ and $h_i = H_n(id_i)$. In other words, HQuery is an event that $\mathcal{A}$
computes (and enters into hash function H_1) the key-material r_i^s for *any* instance
Π_i^s run by an honest player.

Claim 1. Unless HQuery happens, $\mathcal{A}$'s view of the interaction with the challenger
in Game1 is indistinguishable from $\mathcal{A}$'s view in Game2.

Note that if HQuery does not happen then all the challenge/response values v_i^s
and keys K_i^s that $\mathcal{A}$ gets in Game1 are distributed the same as in the real protocol, i.e. as independently chosen random κ-bit strings. Moreover, all the messages
(θ_i^s, id_i) the adversary sees are also statistically close to the corresponding values in the real execution. The reason is that in both cases, the simulation and
the execution, each value θ_i^s is distributed statistically close to a uniform bitstring of length $2\kappa' + \kappa$, and it is independent of id_i (and σ_i). Note that since
$Z_n^* \equiv \langle -1 \rangle \times \langle g \rangle$ value $\theta' = (-1)^b g^x \sigma \pmod{n}$ is random in Z_n^* if b is a random
bit and if x is random in $Z_{2p'q'}$. Since $n - (4p'q')$ is on the order of $\sqrt{n}$, which is
negligible compared to n, the distribution of θ' for x chosen in Z_n is still statistically close to uniform in Z_n^*. Similarly, since there are only $O(\sqrt{n})$ elements in
Z_n/Z_n^*, this random variable is also statistically close to uniform in Z_n. Finally,
for any $s > n2^\kappa$, value $\theta = \theta' + k * n$ (over integers) for random θ' in Z_n and
$k \leftarrow [0, ..., \lfloor s/n \rfloor]$, is statistically indistinguishable from random in Z_s.

Claim 2. If event HQuery happens with non-negligible probability on input
(r_i^s, sid_i^s), or $(r_i^s, \text{sid}_i^s, \text{init})$, or $(r_i^s, \text{sid}_i^s, \text{resp})$, for any Π_i^s, then $\mathcal{A}$ can be used to
break the RSA assumption.

We divide the adversary into three types classes, depending on the message $M = (\hat{id}, \hat{\theta})$ which is involved in computation of r_i^s. Type I adversary makes HQuery
s.t. the related $\hat{id}$ and $\hat{\theta}$ are rerouted from another honest player's instance. Type
II adversary makes HQuery s.t. the related $\hat{id}$ is created by the adversary. Type
III adversary makes HQuery s.t. the related $\hat{id}$ is rerouted but $\hat{\theta}_j$ is created by
the adversary.

  We describe each reduction algorithm using a modified challenger algorithm
called F-$\mathcal{C}^{ah}$. Let G^* be the group s.t. the probability that $\mathcal{A}$ queries H_1 on r_i^s
corresponding to Π_i^s where $\text{Group}(\Pi_i^s) = G^*$, is at least $1/m$. For each type of
adversaries, F-$\mathcal{C}^{ah}$ takes an RSA public key (n, e) of one of the groups denoted
by G^* as an input and picks the private/public keys for all the remaining groups.
The F-$\mathcal{C}^{ah}$ issues users' certificates for all groups except of G^* correctly as in
the real execution. However, for users in G^* the modified challenger will need
to simulate the signatures on each id_i by setting $H_n(id_i)$ as $a^e \bmod n$ for some
random value a. This way F-$\mathcal{C}^{ah}$ can present the certificate of player id_i in G^*

as a. (The exact way that values a are chosen is described in adversary type I-III below.) The modified challenger can fail if $\mathcal{A}$ has made a query to H_n on the randomly chosen id_i value, for any i, but this happens with a negligible probability of at most $nq_H/2^\kappa$, and otherwise the certificates are distributed as in the execution. In each case, F-$\mathcal{C}^{ah}$ responds to Send commands as in Simulation above, additionally storing $[j, \Pi_i^s, \mathsf{sid}_i^s]$ in table denoted T_{H_1}, which is used by a reduction algorithm every time $\mathcal{A}$ makes a query to H_1 (see below).

Type I Adversary

- **Setup and Initialization.** On the RSA challenge (n, e, z), F-$\mathcal{C}^{ah}$ sets the public key of G^* as (n, e, g) where $g = h^{\alpha e^2}$ for $h \xleftarrow{\$} Z_n^*$ and $\alpha \xleftarrow{\$} Z_n$. Note that given a safe RSA modulus n, with probability about $1/2$ we have that $Z_n^* \equiv \langle -1 \rangle \times \langle g \rangle$. The rest part of initialization is the same as in the real protocol.
- **Hash queries to H_n and H_1.** F-$\mathcal{C}^{ah}$ sets $H_n(id_i) = (-1)^{d_i}/g^{a_i e} \pmod{n}$ for random $(d_i, a_i) \in Z_2 \times Z_n$ for each U_i. For the queries to H_1, F-$\mathcal{C}^{ah}$ simply passes these queries to H_1. However, for each query (r, sid) and $(r, \mathsf{sid}, \mathsf{role})$ to H_1, F-$\mathcal{C}^{ah}$ also tries to solve the RSA challenge as we describe below.
- **Corrupt queries.** F-$\mathcal{C}^{ah}$ responds to $\mathsf{Corrupt}(U_i)$ with $(id_i, (-1)^{d_i}/g^{a_i})$.
- **Start queries.** On $\mathsf{Start}(U_i, G^*, s, \mathsf{init})$ from $\mathcal{A}$, F-$\mathcal{C}^{ah}$ responds with the output of $\mathcal{SIM}(id_i, s, \mathsf{init})$, where $id_i = F_{(\kappa, \kappa')}(\mathsf{cert}_i^{(*)})$. On inputs (id_i, s, init), $\mathcal{SIM}$ returns (θ_i^s, id_i): $\mathcal{SIM}$ sets θ_i^s as either $hg^{c_i^s}(-1)^{b_i^s} \bmod n$ or $zg^{c_i^s}(-1)^{b_i^s} \bmod n$, plus the random kn shift, with probability $1/2$ each, for random $(c_i^s, b_i^s) \in Z_n \times Z_2$. Notice that all these values are distributed indistinguishably from the distribution produced by the real execution.
- **Reduction algorithm from HQuery event.** With probability of at least $\epsilon/16m$, for the Π_i^s and Π_j^t instances involved in $\mathcal{A}$'s query on r_i^s, we have $\theta_i^s = hg^{c_i^s}(-1)^{b_i^s} \bmod n$ and $\theta_j^t = zg^{c_j^t}(-1)^{b_j^t} \bmod n$. We replace RSA challenge z by h^k for some unknown k. Then it's easy to see that if $(\hat{\theta}, \hat{id}) = (\theta_j^t, id_j)$ then in equation (1) we get $z_i^s = g^{2ex_i^s} = g^{2e(1+\alpha e^2(a_i+c_i^s))(\alpha e^2)^{-1}}$ and $\hat{z} = g^{2e(k+\alpha e^2(a_j+c_j^t))(\alpha e^2)^{-1}}$. Therefore,

$$r_i^s = g^{2e[k(\alpha e^2)^{-1}+a_j+c_j^t][(\alpha e^2)^{-1}+a_i+c_i^s]}$$

Since F-$\mathcal{C}^{ah}$ knows α, e, a_j, c_j^t, a_i and c_i^s, F-$\mathcal{C}^{ah}$ can extract $g^{2k(\alpha^2 e^3)^{-1}}$, from which F-$\mathcal{C}^{ah}$ can compute $z^{2d\alpha^{-1}}$ since $g = h^{\alpha e^2}$ and $z = h^k$. Thus, z^{2d} can be extracted. Since $gcd(2, e) = 1$, therefore, computing z^{2d} leads to computing z^d. This reduction algorithm is executed whenever $\mathcal{A}$ makes a query (r, sid) (or $(r, \mathsf{sid}, \mathsf{role})$) to H_1 for each entry $[j, \Pi_i^s, \mathsf{sid}_i^s]$ in table T_{H_1} s.t. $\mathsf{sid}_i^s = \mathsf{sid}$. F-$\mathcal{C}^{ah}$ can verify which entry is related to the given value r, since after computing $w = z^{2d}$ as above F-$\mathcal{C}^{ah}$ can test if $w^e = z^2$.

Type II Adversary

- **Setup and Initialization.** On the RSA challenge (n, e, z), F-$\mathcal{C}^{ah}$ sets the public key of G^* as (n, e, g) where $g = \alpha^e$ for $\alpha \leftarrow Z_n^*$. (Note that a random

g in Z_n^* matches that chosen by a real key generation with probability about $1/2$). The rest part of initialization is the same as in the real protocol.

- **Hash queries to H_n and H_1.** F-$\mathcal{C}^{\text{ah}}$ answers a query to H_n on x depending on the source of x. Namely, F-$\mathcal{C}^{\text{ah}}$ responds with $H_n(x) = a_x^e/g \bmod n$ for a randomly chosen $a_x \in Z_n^*$ if x corresponds to some id of an honest player generated by the simulator. Let $H_n(id_i) = a_i^e g^{-1}$. If x does not match with any ids created by the simulator, F-$\mathcal{C}^{\text{ah}}$ replies with $H_n(x) = a_x^e/z \bmod n$ for a randomly chosen $a_x \in Z_n^*$. For the queries to H_1, F-$\mathcal{C}^{\text{ah}}$ simply passes these queries to H_1. However, for each query (r, sid) and $(r, \text{sid}, \text{role})$ to H_1, F-$\mathcal{C}^{\text{ah}}$ also tries to solve the RSA challenge as we describe below.
- **Corrupt queries.** F-$\mathcal{C}^{\text{ah}}$ responds to $\text{Corrupt}(U_i)$ with $(id_i, a_i/\alpha)$.
- **Start queries.** On $\text{Start}(U_i, G^*, s, \text{init})$ from $\mathcal{A}$, F-$\mathcal{C}^{\text{ah}}$ responds with the output of $\mathcal{SIM}(id_i, s, \text{init})$, where $id_i = F_{(\kappa,\kappa')}(\text{cert}_i^{(*)})$. On inputs (id_i, s, init), $\mathcal{SIM}$ returns (θ_i^s, id_i) where $\theta_i^s = (-1)^{b_i^s} a_i g^{\gamma_i^s}$ plus the random kn shift, for random $(\gamma_i^s, b_i^s) \in Z_n \times Z_2$.
- **Reduction algorithm on HQuery event.** With probability of (almost) at least $\epsilon/4m$, the query r corresponds to session Π_i^s on which $\mathcal{A}$ sends $\hat{id} \neq id_j$ for an honest U_j, i.e., $r = r_i^s = (\hat{z})^{x_i^s}$. Since $\theta_i^s = (-1)^{b_i^s}(H(id_i))^d g^{d+\gamma_i^s}$ and $\hat{z} = (\hat{\theta}/a)^{2e} z^2$ where $H_n(\hat{id}) = a^e z^{-1}$, we get $z_i^s = g^{2ex_i^s} = g^{2e(d+\gamma_i^s)}$ from eq. (1). Therefore $(\hat{z})^{x_i^s} = (\hat{\theta}/a)^{2e(d+\gamma_i)} z^{2(d+\gamma_i)} = (\hat{\theta}/a)^{(2+2e\gamma_i)} z^{2\gamma_i} z^{2d}$ and F-$\mathcal{C}^{\text{ah}}$ can extract z^{2d}. Since $gcd(2, e) = 1$, F-$\mathcal{C}^{\text{ah}}$ can compute z^d from $(\hat{z})^{x_i^s}$.

Type III Adversary

In the last case, we assume that with probability at least $\epsilon/4m$ the *first such* $(\hat{z})^{x_i^s}$ which makes event HQuery true corresponds to session Π_i^s on which $\mathcal{A}$ sends $\hat{id} = id_j$ for some session Π_j^t matching Π_i^s for some *currently uncorrupted* U_j, but $\hat{\theta} \neq \theta_j^t$. We show that F-$\mathcal{C}^{\text{ah}}$ can solve the RSA problem in this case. The view that it will present to $\mathcal{A}$ will match what $\mathcal{A}$ expects *until* the above query $(\hat{z})^{x_i^s}$, i.e. HQuery, is done, in which case F-$\mathcal{C}^{\text{ah}}$ will solve the RSA problem. Note that it is unimportant whether F-$\mathcal{C}^{\text{ah}}$ can continue presenting $\mathcal{A}$ with the correct view afterwards. Let U_j be a player s.t. the probability that this query is done *and* that it involves $\hat{id} = id_j$ is at least $\epsilon/4mn$. Setup, Start, and the reduction algorithm are the same as in Type II, and here are the remaining queries:

- **Hash queries to H_n and H_1.** F-$\mathcal{C}^{\text{ah}}$ responds with $H_n(x) = a^e/g$ for a randomly chosen $a \in Z_n^*$ if x corresponds to some id of an honest player $U_i \neq U_j$. If x match with the id of U_j, F-$\mathcal{C}^{\text{ah}}$ replies with $H_n(x) = a^e/z \bmod n$ for a randomly chosen $a \in Z_n^*$. Let $H_n(id_i) = a_i^e g^{-1}$.
- **Corrupt queries.** F-$\mathcal{C}^{\text{ah}}$ responds to $\text{Corrupt}(U_i)$ with $(id_i, a_i/\alpha)$ if $U_i \neq U_j$. On $\text{Corrupt}(U_j)$, F-$\mathcal{C}^{\text{ah}}$ stops. As we argued above, it does not matter that this reduction cannot open the state of player U_j with a valid-looking signature on id_j, since at the time $\mathcal{A}$ makes the crucial r_i^s query the player U_j must be still uncorrupted.

References

[BBDP01]   Bellare, M., Boldyreva, A., Desai, A., Pointcheval, D.: Key-privacy in public-key encryption. In: Boyd, C. (ed.) ASIACRYPT 2001. LNCS, vol. 2248, pp. 566–582. Springer, Heidelberg (2001)

[BCK98]   Bellare, M., Canetti, R., Krawczyk, H.: A modular approach to the design and analysis of authentication and key exchange protocols (extended abstract). In: STOC, pp. 419–428 (1998)

[BDS+03]   Balfanz, D., Durfee, G., Shankar, N., Smetters, D.K., Staddon, J., Wong, H.-C.: Secret handshakes from pairing-based key agreements. In: IEEE Symposium on Security and Privacy, pp. 180–196. IEEE Computer Society Press, Los Alamitos (2003)

[BPR00]   Bellare, M., Pointcheval, D., Rogaway, P.: Authenticated Key Exchange Secure against Dictionary Attacks. In: Preneel, B. (ed.) EUROCRYPT 2000. LNCS, vol. 1807, pp. 139–155. Springer, Heidelberg (2000)

[CJT04]   Jarecki, S., Tsudik, G., Castelluccia, C.: Secret handshakes from ca-oblivious encryption. In: Lee, P.J. (ed.) ASIACRYPT 2004. LNCS, vol. 3329, pp. 293–307. Springer, Heidelberg (2004)

[CK01]   Canetti, R., Krawczyk, H.: Analysis of key-exchange protocols and their use for building secure channels. In: Pfitzmann, B. (ed.) EUROCRYPT 2001. LNCS, vol. 2045, pp. 453–474. Springer, Heidelberg (2001)

[CKGS98]   Chor, B., Kushilevitz, E., Goldreich, O., Sudan, M.: Private information retrieval. J. ACM 45(6), 965–981 (1998)

[DMS04]   Dingledine, R., Mathewson, N., Syverson, P.F.: Tor: The second-generation onion router. In: USENIX Security Symposium, pp. 303–320 (2004)

[JKT07]   Jarecki, S., Kim, J., Tsudik, G.: Group secret handshakes or affiliation-hiding authenticated group key agreement. In: Abe, M. (ed.) CT-RSA 2007. LNCS, vol. 4377, pp. 287–308. Springer, Heidelberg (2006)

[JL07]   Jarecki, S., Liu, X.: Unlinkable secret handshakes and key-private group key management schemes. In: Katz, J., Yung, M. (eds.) ACNS 2007. LNCS, vol. 4521, pp. 270–287. Springer, Heidelberg (2007)

[LDB03]   Li, N., Du, W., Boneh, D.: Oblivious signature-based envelope. In: PODC, pp. 182–189 (2003)

[Sho99]   Shoup, V.: On formal models for secure key exchange. IBM Research Report RZ 3120 (1999)

[TX06]   Tsudik, G., Xu, S.: A flexible framework for secret handshakes. In: Danezis, G., Golle, P. (eds.) PET 2006. LNCS, vol. 4258, pp. 295–315. Springer, Heidelberg (2006)

[Ver05]   Vergnaud, D.: Rsa-based secret handshakes. In: Ytrehus, Ø. (ed.) WCC 2005. LNCS, vol. 3969, pp. 252–274. Springer, Heidelberg (2006)

[XY04]   Xu, S., Yung, M.: k-anonymous secret handshakes with reusable credentials. In: Atluri, V., Pfitzmann, B., McDaniel, P.D. (eds.) ACM Conference on Computer and Communications Security, pp. 158–167. ACM Press, New York (2004)

Improving the Efficiency of Impossible Differential Cryptanalysis of Reduced Camellia and MISTY1

Jiqiang Lu[1,*], Jongsung Kim[2,**], Nathan Keller[3,***], and Orr Dunkelman[4,†]

[1] Information Security Group, Royal Holloway, University of London
Egham, Surrey TW20 0EX, UK
lvjiqiang@hotmail.com
[2] Center for Information Security Technologies (CIST), Korea University
Anam Dong, Sungbuk Gu, Seoul, Korea
joshep@cist.korea.ac.kr
[3]Einstein Institute of Mathematics, Hebrew University
Jerusalem 91904, Israel
nkeller@math.huji.ac.il
[4]ESAT/SCD-COSIC, Katholieke Universiteit Leuven
Kasteelpark Arenberg 10, B-3001 Leuven-Heverlee, Belgium
orr.dunkelman@esat.kuleuven.be

Abstract. We observe that when conducting an impossible differential cryptanalysis on Camellia and MISTY1, their round structures allow us to partially determine whether a candidate pair is useful by guessing only a small fraction of the unknown required subkey bits of a relevant round at a time, instead of guessing all of them at once. Taking advantage of the early abort technique, we improve a previous impossible differential attack on 6-round MISTY1 without the FL functions, and present impossible differential cryptanalysis of 11-round Camellia-128 without the FL functions, 13-round Camellia-192 without the FL functions and 14-round Camellia-256 without the FL functions. The presented results are better than any previously published cryptanalytic results on Camellia and MISTY1 without the FL functions.

Keywords: Block cipher, Camellia, MISTY1, Impossible differential cryptanalysis.

* This author as well as his work was supported by a British Chevening / Royal Holloway Scholarship and the European Commission under contract IST-2002-507932 (ECRYPT).

** This author was supported by the MIC (Ministry of Information and Communication), Korea, under the ITRC (Information Technology Research Center) support program supervised by the IITA (Institute of Information Technology Advancement) (IITA-2006-(C1090-0603-0025)).

*** This author was supported by the Adams fellowship.

† This work was supported in part by the Concerted Research Action (GOA) Ambiorics 2005/11 of the Flemish Government and by the IAP Programme P6/26 BCRYPT of the Belgian State (Belgian Science Policy).

T. Malkin (Ed.): CT-RSA 2008, LNCS 4964, pp. 370–386, 2008.

1   Introduction

Camellia [1] is a 128-bit Feistel block cipher with a user key length of 128, 192 or 256 bits, and MISTY1 [19] is a 64-bit Feistel block cipher with a 128-bit user key. Both Camellia and MISTY1 were selected to be CRYPTREC [6] e-government recommended ciphers in 2002 and in the NESSIE [20] block cipher portfolio in 2003, and were adopted as ISO [10] international standards in 2005. Since Camellia and MISTY1 are increasingly being used in many real-life cryptographic applications, it is essential to continuing to investigate their security against different cryptanalytic techniques. For simplicity, we denote by Camellia-128/192/256 the three versions of Camellia that use 128, 192 and 256 key bits, respectively.

Many cryptanalytic results on Camellia and MISTY1 have been published so far [2,7,8,13,14,15,22,23,24,25,26,27]. In summary, in terms of the numbers of attacked rounds, the best cryptanalytic results on Camellia without the FL functions are the truncated differential cryptanalysis [11] on 8-round Camellia-128 [16], the impossible differential cryptanalysis on 12-round Camellia-192 [26], and the linear [18] and impossible differential cryptanalysis on 12-round Camellia-256 [22,26]; the best cryptanalytic result on MISTY1 without the FL functions is the impossible differential cryptanalysis on 6 rounds [14].

Impossible differential cryptanalysis [3,12], as a special case of differential cryptanalysis [5], uses one or more differentials with a zero probability, called impossible differentials, which are usually built in a miss-in-the-middle manner [4]. In the impossible differential attacks on Camellia and MISTY1 described in [14,26], the general approach is to guess all the unknown required subkey bits of a relevant round to partially decrypt (or encrypt) a candidate pair through the round function; finally one checks whether the pair could produce the expected difference just before (respectively after) the round.

In this paper, we observe that due to the round structures of Camellia and MISTY1, we can partially check whether a candidate pair could produce the expected difference by guessing only a small fraction of the unknown required subkey bits at a time, and do a series of partial checks by guessing other fractions of the unknown required subkey bits, instead of guessing all the unknown required subkey bits at once. Since some unuseful pairs can be discarded before the next guess for a different fraction of the required round subkey bits, we can reduce the computational workload for an attack, and even more importantly, we may break more rounds of a cipher. A similar technique is used in differential cryptanalysis of DES [5], and is referred to as the early abort technique. Taking advantage of the early abort technique, we improve a previous impossible differential attack on 6-round MISTY1 without the FL functions, and present impossible differential cryptanalysis of 11-round Camellia-128 without the FL functions, 13-round Camellia-192 without the FL functions and 14-round Camellia-256 without the FL functions, following the work described in [14,26]. Table 1 summarises our main cryptanalytic results and the best previously published on Camellia and MISTY1.

Table 1. Summary of our main cryptanalytic results and the best previously published on Camellia and MISTY1

Cipher	Attack Type	Rounds	FL/FL^{-1}	Data	Time	Paper
Camellia-128	Truncated differential	8	none	$2^{83.6}$CP	$2^{55.6}$	[16]
(18 rounds)	Impossible differential	11	none	2^{118}CP	2^{126}MA&2^{118}	This
		11	none	2^{118}CP	2^{126}MA	This
Camellia-192	Boomerang attack	9	all	2^{124}ACPC	2^{170}	[22]
(24 rounds)	Impossible differential	12	none	2^{120}CP	2^{181}	[26]
		13	none	2^{119}CP	$2^{167.9}$	This
		13	none	2^{119}CP	$2^{169.4}$MA	This
Camellia-256	High-order differential	11	all	2^{93}CP	2^{256}	[8]
(24 rounds)	Linear cryptanalysis	12	none	2^{119}CP	2^{247}	[22]
	Impossible differential	12	none	2^{120}CP	2^{181}	[26]
		13	none	2^{120}CP	$2^{168.9}$	This
		13	none	2^{120}CP	$2^{170.4}$MA	This
		14	none	2^{120}CP	$2^{232.5}$	This
		14	none	2^{120}CP	2^{231}MA	This
MISTY1	Slicing attack	4	all	$2^{22.25}$CP	2^{45}	[15]
(8 rounds)	Integral cryptanalysis	5	most	2^{34}CP	2^{48}	[13]
	Impossible differential	6	none	2^{54}CP	2^{61}	[14]
		6	none	2^{39}CP	2^{106}	[14]
		6	none	2^{39}CP	2^{85}	This

CP: Chosen Plaintexts, ACPC: Adaptive Chosen Plaintexts and Ciphertexts,
Time unit: Encryptions, if otherwise stated explicitly, MA: Memory Accesses,
"none" means "no FL function", "all" means "all the FL functions",
"most" means "all the FL functions except those in the final swap layer"

The rest of the paper is organised as follows. In the next section, we briefly describe the Camellia and MISTY1 ciphers. In Section 3, we introduce the early abort technique in a general way. In Sections 4 and 5, we present our cryptanalytic results on Camellia and MISTY1, respectively. Section 6 concludes this paper.

2   Preliminaries

Throughout the paper, we denote the bit-wise exclusive OR (XOR) operation by $\oplus$, and bit string concatenation by $\|$.

2.1   The Camellia Block Cipher

Camellia [1] takes a 128-bit plaintext P as input, and has a total of N rounds, where N is 18 for Camellia-128, and 24 for Camellia-192/256. Its encryption procedure is as follows.

1. $L^0||R^0 = P \oplus (KW_1||KW_2)$
2. For $i = 1$ to N:
   if $i = 6$ or 12 (or 18 for Camellia-192/256),
   $$L'^i = \mathrm{F}(L^{i-1}, K_i) \oplus R^{i-1},\ R'^i = L^{i-1};$$
   $$L^i = \mathrm{FL}(L'^i, KI_{i/3-1}),\ R^i = \mathrm{FL}^{-1}(R'^i, KI_{i/3});$$
   else
   $$L^i = \mathrm{F}(L^{i-1}, K_i) \oplus R^{i-1},\ R^i = L^{i-1};$$
3. Ciphertext $C = (R^N \oplus KW_3)||(L^N \oplus KW_4)$,

where KW, K and KI are 64-bit round subkeys, L^i, R^i, L'^i and R'^i are 64 bits long, and the F function comprises a XOR operation, then an application of 8 parallel nonlinear 8×8-bit bijective S-boxes $s_1, s_2, \cdots, s_8$, and, finally, a linear P function. As we consider the version of Camellia without the FL functions, we omit the description of the two functions FL and FL^{-1}; we refer the reader to [1] for their specifications. The P function and its inverse P^{-1} are defined over $GF(2^8)^8 \to GF(2^8)^8$, as follows.

$$
\mathrm{P} = \begin{pmatrix}
1 & 0 & 1 & 1 & 0 & 1 & 1 & 1 \\
1 & 1 & 0 & 1 & 1 & 0 & 1 & 1 \\
1 & 1 & 1 & 0 & 1 & 1 & 0 & 1 \\
0 & 1 & 1 & 1 & 1 & 1 & 1 & 0 \\
1 & 1 & 0 & 0 & 0 & 1 & 1 & 1 \\
0 & 1 & 1 & 0 & 1 & 0 & 1 & 1 \\
0 & 0 & 1 & 1 & 1 & 1 & 0 & 1 \\
1 & 0 & 0 & 1 & 1 & 1 & 1 & 0
\end{pmatrix}, \quad
\mathrm{P}^{-1} = \begin{pmatrix}
0 & 1 & 1 & 1 & 0 & 1 & 1 & 1 \\
1 & 0 & 1 & 1 & 1 & 0 & 1 & 1 \\
1 & 1 & 0 & 1 & 1 & 1 & 0 & 1 \\
1 & 1 & 1 & 0 & 1 & 1 & 1 & 0 \\
1 & 1 & 0 & 0 & 1 & 0 & 1 & 1 \\
0 & 1 & 1 & 0 & 1 & 1 & 0 & 1 \\
0 & 0 & 1 & 1 & 1 & 1 & 1 & 0 \\
1 & 0 & 0 & 1 & 0 & 1 & 1 & 1
\end{pmatrix}.
$$

2.2 The MISTY1 Block Cipher

MISTY1 [19] takes a 64-bit plaintext P as input, and has a total of 8 rounds; the user key is 128 bits long. Its encryption procedure is as follows.

1. $P = L^0||R^0$, $KL = KL_1||KL_2||\cdots||KL_{10}$, $KI = KI_1||KI_2||\cdots||KI_8$, $KO = KO_1||KL_2||\cdots||KO_8$.
2. For $i = 1, 3, 5, 7$:
   $$R^i = \mathrm{FL}(L^{i-1}, KL_i),\ L^i = \mathrm{FL}(R^{i-1}, KL_{i+1}) \oplus \mathrm{FO}(R^i, KO_i, KI_i),$$
   $$L^{i+1} = R^i \oplus \mathrm{FO}(L^i, KO_{i+1}, KI_{i+1}),\ R^{i+1} = L^i.$$
3. Ciphertext $C = \mathrm{FL}(R^8, KL_{10})||\mathrm{FL}(L^8, KL_9)$,

where KL, KI and KO are round subkeys, and the FL function takes a 32-bit block X and a 32-bit subkey KL as inputs, and outputs a 32-bit block Y, computed as defined below.

1. $X = X_L||X_R$, $KL = KL_{iL}||KL_{iR}$.
2. $Y_R = (X_L \wedge KL_{iL}) \oplus X_R$, $Y_L = X_L \oplus (Y_R \vee KL_{iR})$.
3. $Y = Y_L||Y_R$.

The FO function takes as inputs a 32-bit block X and two 32-bit subkeys KO_i and KI_i, and outputs a 32-bit block Y, and is defined as follows.

1. $X = XL_0||XR_0$, $KO_i = KO_{i1}||KO_{i2}||KO_{i3}||KO_{i4}$, $KI_i = KI_{i1}||KI_{i2}||KI_{i3}$.
2. For $j = 1, 2, 3$:
   $$XR_j = \mathrm{FI}(XL_{j-1} \oplus KO_{ij}, KI_{ij}) \oplus XR_{j-1},\ XL_j = XR_{j-1}.$$

3. $Y = (XL_3 \oplus KO_{i4})||XR_3$.

In the above description, the FI function takes a 16-bit block X and a subkey KI_{ij} as inputs, and outputs a 16-bit block Y, computed as follows.

1. $X = XL_0(9 \text{ bits})||XR_0(7 \text{ bits})$, $KI_{ij} = KI_{ijL}(7 \text{ bits})||KI_{ijR}(9 \text{ bits})$,
2. $XL_1 = XR_0$, $XR_1 = S_9(XL_0) \oplus \text{Extnd}(XR_0)$,
3. $XL_2 = XR_1 \oplus KI_{ijR}$, $XR_2 = S_7(XL_1) \oplus \text{Trunc}(XR_1) \oplus KI_{ijL}$,
4. $XL_3 = XR_2$, $XR_3 = S_9(XL_2) \oplus \text{Extnd}(XR_2)$,
5. $Y = XL_3||XR_3$,

where S_9 is a 9×9-bit bijective S-box, S_7 is a 7×7-bit bijective S-box, the function Extnd extends from 7 bits to 9 bits by concatenating two zeros on the left side, and the function Trunc truncates two bits from the left side.

3   A General Description of the Early Abort Technique

Impossible differential cryptanalysis is based on one or more impossible differentials, written $\alpha \nrightarrow \beta$, and it usually treats a block cipher $E : \{0,1\}^n \times \{0,1\}^k \rightarrow \{0,1\}^n$ as a cascade of three sub-ciphers $E = E_b \circ E_0 \circ E_a$, where E_0 denotes the rounds for which $\alpha \nrightarrow \beta$ holds, E_a denotes a few rounds before E_0, and E_b denotes a few rounds after E_0. Given a guess for the subkeys used in E_a and E_b, if a plaintext pair produces a difference of α just after E_a, and its corresponding ciphertext pair produces a difference of β just before E_b, then this guess for the subkeys must be incorrect. Thus, given a sufficient number of matching plaintext/ciphertext pairs, we can find the correct subkey by discarding all the wrong guesses.

When checking if a plaintext pair produces a difference of α just after E_a (or its corresponding ciphertext pair produces a difference of β just before E_b), the general approach is to guess all the unknown bits of the relevant round subkey necessary to partially encrypt (respectively decrypt) the pair through the substitution and diffusion layers; finally, one can check whether the pair could produce an expected difference just after (respectively before) the round. To make matters more specific, consider a Feistel structure as in Camellia; as shown in Fig. 1, we assume that it has an nonlinear substitution consisting of m parallel S-boxes and a linear diffusion function P. For simplicity, we assume the round in Fig. 1 is just before E_0; that is to say, the attacker is looking for a pair with difference $(\Delta L_{i+1}||\Delta R_{i+1}) = \alpha$. According to previous attack procedures, due to the diffusion of the P function, the attacker will guess all the required unknown subkey bits (i.e. those corresponding to the active S-boxes) at a time, then encrypt the left halves of the pair through the substitution layer to get the difference just after the P function, and finally XOR it with the difference ΔR_i to check if it has the difference α after the round.

However, the round structure can allow us to partially determine whether a candidate pair could produce the expected difference α by guessing only a small fraction of the required round subkey bits at a time, instead of all of them

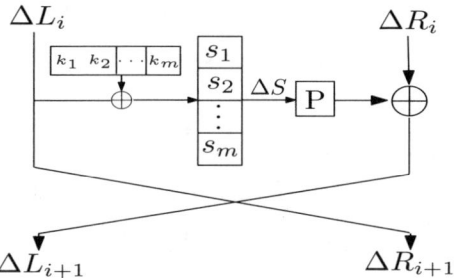

Fig. 1. A Feistel structure

simultaneously. More specifically, since we know the expected difference α and the intermediate values of the pair just before the round, we can compute the expected difference just before the P function, denoted by ΔS ($= P^{-1}(\Delta R_i \oplus \Delta L_{i+1})$), as the P function is usually linearly invertible. Only if the expected difference ΔS appears after the substitution layer could the pair produce the difference α after the round. Thus, in the following, we guess only those of the required unknown subkey bits corresponding to one (or more) active S-box, then encrypt the pair through the S-box, and finally check if it produces the corresponding partial difference in ΔS. If not, then the pair is not useful, and we can discard it immediately; otherwise, we guess another part of the required round subkey bits corresponding to another active S-box, and check the pair similarly. A pair is useful only if it could produce the partial difference out of the expected difference ΔS just before the P function, under every part of the required round subkey bits. Some unuseful pairs can be discarded before the next guess; by this observation we can reduce the computational workload of an attack, and even more significantly, we may break more rounds.

4   Impossible Differential Cryptanalysis of Reduced Camellia

As Camellia is byte-oriented, we represent the 128 bits of the (intermediate) state as 16 bytes; and we denote the l-th byte of a subkey K_i by $k_{i,l}$, ($1 \leq l \leq 8$). Let the question mark ? denote an unknown byte difference (two bytes marked with ? may be different).

In 2007, Wu et al. [26] presented an impossible differential attack on 12-round Camellia-192/256 without the FL functions, which is based on the following 8-round impossible differentials: $(0,0,0,0,0,0,0,0,a,0,0,0,0,0,0,0) \nrightarrow (h,0,0,0,0,0,0,0,0,0,0,0,0,0,0,0)$, where a and h are any two nonzero bytes. See Fig. 2 for more details, where the values of the forms $b_\times, c_\times, \cdots$, and $f_\times$ are all one byte long. A detailed explanation of these 8-round impossible differentials is given in [26].

In this section, we also consider the version of Camellia that excludes the FL (and FL^{-1}) functions. We present an impossible differential cryptanalysis on

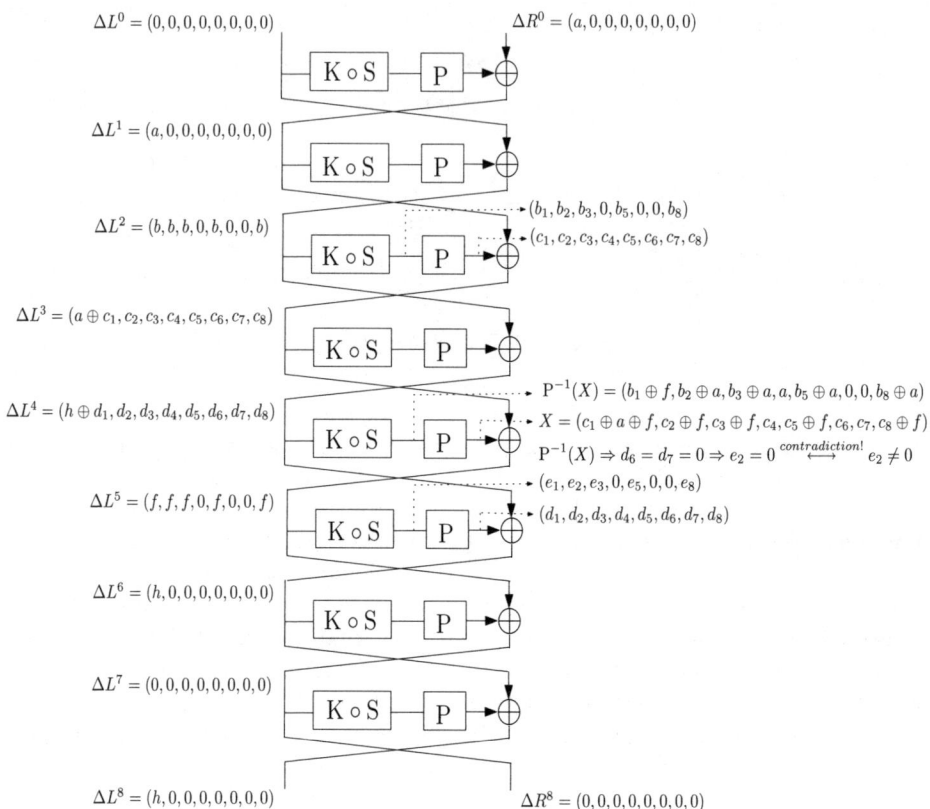

Fig. 2. 8-round impossible differentials of Camellia

14-round Camellia-256, 13-round Camellia-192 and 11-round Camellia-128, and finally give several extensions.

4.1   Attacking 14-Round Camellia-256 without the FL Functions

We attack Rounds 1 to 14, and use the 8-round impossible differentials in Rounds 4 to 11. As every S-box has a minimal nonzero differential probability of 2^{-7}, an output difference $(h, 0, 0, 0, 0, 0, 0, 0, 0, 0, 0, 0, 0, 0, 0, 0)$ of the 8-round impossible differentials propagates to at most 2^7 possible output differences $(g, g, g, 0, g, 0, 0, g, h, 0, 0, 0, 0, 0, 0, 0)$ after Round 12, where g is nonzero. Then, every $(g, g, g, 0, g, 0, 0, g, h, 0, 0, 0, 0, 0, 0, 0)$ propagates to at most $(2^7)^5$ possible output differences after Round 13. Hence, given the difference $(h, 0, 0, 0, 0, 0, 0, 0, 0, 0, 0, 0, 0, 0, 0, 0)$ just after Round 11, there are at most $(2^8 - 1) \cdot 2^7 \cdot (2^7)^5 \approx 2^{50}$ possible output differences after Round 13; we denote these possible differences by the set Δ_{13}. Every difference in Δ_{13} propagates to at most $(2^7)^8$ possible output differences after Round 14; therefore, given the difference $(h, 0, 0, 0, 0, 0, 0, 0, 0, 0, 0, 0, 0, 0, 0, 0)$ just after Round 11, there are at most $2^{50} \cdot 2^{56} = 2^{106}$ possible output differences after Round 14; we denote these possible differences by the set Δ_{14}.

We use the early abort technique in the first and last two rounds of the 14-round attack. We first give Property 1, as follows.

Property 1. *The following properties hold.*

1. *For a plaintext pair $(P_i = (L_i^0, R_i^0), P_j = (L_j^0, R_j^0))$, $P^{-1}(R_i^0 \oplus R_j^0 \oplus (u, u, u, 0, u, 0, 0, u))$ has a unique value in the first two bytes for every nonzero value of u (one byte long).*
2. *If a ciphertext pair (C_i, C_j) has an output difference $(\Delta L^{13} = L_i^{13} \oplus L_j^{13}, \Delta R^{13} = R_i^{13} \oplus R_j^{13})$ belonging to Δ_{13}, then the difference just after the S-box substitution layer of Round 13 must have the form $(?, ?, ?, 0, ?, 0, 0, ?)$, and there must be a h such that $P^{-1}(L_i^{13} \oplus L_j^{13} \oplus (h, 0, 0, 0, 0, 0, 0, 0))$ has the form $(?, ?, ?, 0, ?, 0, 0, ?)$. h has 255 possible values, but only one of them satisfies the above condition.*

Proof. The proof of Property 1-1 is follows. Suppose that there are two values u_1 and u_2 such that $P^{-1}(R_i^0 \oplus R_j^0 \oplus (u_1, u_1, u_1, 0, u_1, 0, 0, u_1)) \oplus P^{-1}(R_i^0 \oplus R_j^0 \oplus (u_2, u_2, u_2, 0, u_2, 0, 0, u_2)) = (0, 0, ?, ?, ?, ?, ?, ?)$, then we get $P^{-1}(u_1 \oplus u_2, u_1 \oplus u_2, u_1 \oplus u_2, 0, u_1 \oplus u_2, 0, 0, u_1 \oplus u_2) = (0, 0, ?, ?, ?, ?, ?, ?)$; by the P^{-1} function we know that the first byte should be $u_1 \oplus u_2$, meaning that $u_1 = u_2$.

The fore part of Property 1-2 is trivial; here we just prove the latter part of Property 1-2. Assume there are two different values h_1 and h_2 that satisfy the condition, then observe that $P^{-1}((h_1, 0, 0, 0, 0, 0, 0, 0) \oplus (h_2, 0, 0, 0, 0, 0, 0, 0))$ also has the form $(?, ?, ?, 0, ?, 0, 0, ?)$; note that the 4-th byte is 0; however, by the P^{-1} function we know that the 4-th byte should be $h_1 \oplus h_2 \neq 0$. This gives a contradiction. □

An impossible differential attack is generally conducted in the order of checking ciphertext pairs first and finally plaintext pairs in a chosen-plaintext attack scenario, or the reverse in a chosen-ciphertext attack scenario. However, it may be improved by using an optimised order, as shown by the 14-round Camellia-256 attack below.

The above analysis enables us to give the following procedure for attacking 14-round Camellia-256. Fig. 3 illustrates the attack.

1. Choose 2^8 structures: each structure contains a set of 2^{112} plaintexts $P_i = (L_i^0, R_i^0)$, with $L_i^0 = P(x_1, x_2, x_3, \alpha_4, x_5, \alpha_6, \alpha_7, x_8) \oplus (x, \beta_2, \beta_3, \beta_4, \beta_5, \beta_6, \beta_7, \beta_8)$ and $R_i^0 = (y_1, y_2, y_3, y_4, y_5, y_6, y_7, y_8)$, where the bytes with the forms $x_\times$ and $y_\times$ take all the possible values in $\{0, 1\}^8$, and the bytes with the forms $\alpha_\times$ and $\beta_\times$ are fixed to certain values in $\{0, 1\}^8$, $(i = 1, 2, \cdots, 2^{112})$. In a chosen-plaintext attack scenario, obtain all their ciphertexts; we denote them by $C_i = (L_i^{14}, R_i^{14})$, respectively. For different values of $(x_1, x_2, x_3, x_5, x_8, x, y_1, \cdots, y_8)$, the resultant 128-bit blocks are different; thus, there are $2^{112 \times 2}/2 = 2^{223}$ plaintext pairs (P_i, P_j) in a structure $(j = 1, 2, \cdots, 2^{112})$, so the 2^8 structures yield a total of 2^{231} ciphertext pairs. Keep only the pairs (C_i, C_j) with a difference belonging to Δ_{14}. The expected number of remaining pairs is about $2^{231} \cdot \frac{2^{106}}{2^{128}} = 2^{209}$.

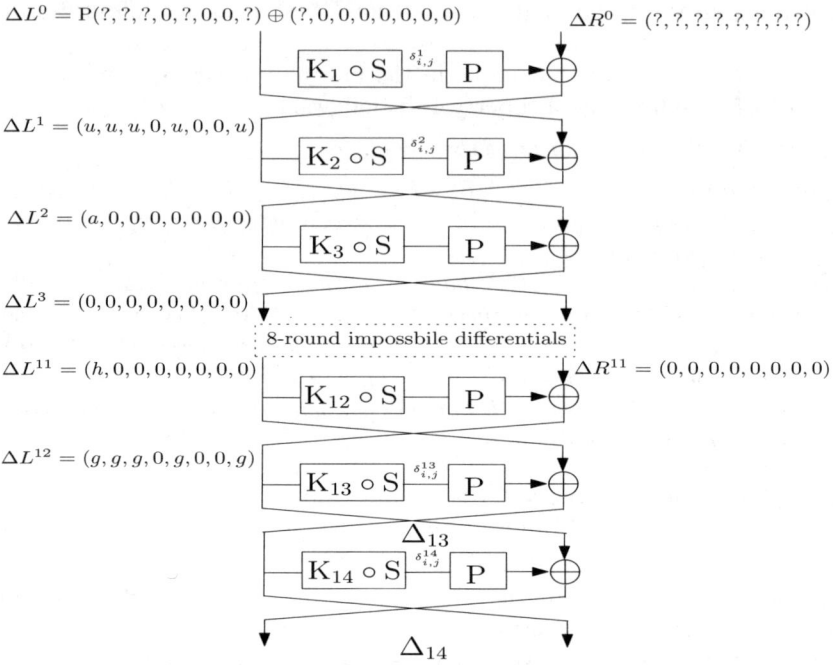

$\Delta L^0 = P(?, ?, ?, 0, ?, 0, 0, ?) \oplus (?, 0, 0, 0, 0, 0, 0, 0)$　　　　$\Delta R^0 = (?, ?, ?, ?, ?, ?, ?, ?)$

$\Delta L^1 = (u, u, u, 0, u, 0, 0, u)$

$\Delta L^2 = (a, 0, 0, 0, 0, 0, 0, 0)$

$\Delta L^3 = (0, 0, 0, 0, 0, 0, 0, 0)$

8-round impossbile differentials

$\Delta L^{11} = (h, 0, 0, 0, 0, 0, 0, 0)$　　　　$\Delta R^{11} = (0, 0, 0, 0, 0, 0, 0, 0)$

$\Delta L^{12} = (g, g, g, 0, g, 0, 0, g)$

Δ_{13}

Δ_{14}

Fig. 3. Impossible differential attack on 14-round Camellia-256

2. For every remaining plaintext pair (P_i, P_j), compute $P^{-1}(R_i^0 \oplus R_j^0 \oplus (u, u, u, 0, u, 0, 0, u))$ for all the 255 possible nonzero values of u; we denote the values by $\Delta_{i,j}^1$, respectively. Then, do as follows.
   (a) Guess the two bytes $(k_{1,1}, k_{1,2})$ of the subkey K_1. For every plaintext pair (P_i, P_j), partially encrypt the first two bytes of (L_i^0, L_j^0) through the s_1 and s_2 S-boxes, and check if they have a difference equal to any of the corresponding two-byte partial differences in $\Delta_{i,j}^1$. Keep only the qualified pairs. By Property 1-1 there is only one difference in $\Delta_{i,j}^1$ for a qualified pair, and we denote this difference from $\Delta_{i,j}^1$ by $\delta_{i,j}^1$. As there are 255 possible values in $\Delta_{i,j}^1$ for every pair, the expected number of remaining pairs is about $2^{209} \cdot \frac{255}{2^{16}} \approx 2^{201}$.
   (b) For $l = 3$ to 8:
       - Guess the byte $k_{1,l}$ of K_1;
       - For every remaining pair (P_i, P_j), partially encrypt the l-th byte of (L_i^0, L_j^0) through the s_l S-box, and check if they have a difference equal to the corresponding one-byte partial difference in $\delta_{i,j}^1$; keep only the qualified pairs. The difference $\delta_{i,j}^1$ is already fixed in Step 2-(a), so it is expected that a proportion of about $1 - 2^{-8}$ of the remaining pairs will be discarded after every iteration.
3. For every remaining plaintext pair (P_i, P_j), from Property 1-2 we similarly know that there is only one value of a such that $P^{-1}(L_i^0 \oplus L_j^0 \oplus$

$(a,0,0,0,0,0,0,0))$ has the form $(?,?,?,0,?,0,0,?)$; we denote by $\delta_{i,j}^2$ the value $P^{-1}(L_i^0 \oplus L_j^0 \oplus (a,0,0,0,0,0,0,0))$ with the form $(?,?,?,0,?,0,0,?)$. Then, for $l = 1, 2, 3, 5, 8$, do as follows.

- Guess the byte $k_{2,l}$ of the subkey K_2;
- For every remaining pair (P_i, P_j), partially encrypt the l-th byte of (L_i^1, L_j^1) through the s_l S-box, and check if they have a difference equal to the corresponding one-byte partial difference in $\delta_{i,j}^2$; keep only the qualified pairs. Similarly, it is expected that a proportion of about $1 - 2^{-8}$ of the remaining plaintext pairs will be discarded after every iteration.

Finally, for every remaining pair of plaintexts we can get the first bytes of their intermediate values just after Round 2.

4. Guess the byte $k_{3,1}$ of the subkey K_3. For every plaintext pair (P_i, P_j), partially encrypt the first bytes of (L_i^2, L_j^2) through the s_1 S-box of Round 3, and check if they have a difference equal to $L_{i,1}^1 \oplus L_{j,1}^1$. Keep only the qualified pairs. The expected number of remaining plaintext pairs is about $2^{113} \cdot 2^{-8} = 2^{105}$.

5. For every ciphertext pair (C_i, C_j) corresponding to a remaining plaintext pair (P_i, P_j), compute $P^{-1}(L_i^{14} \oplus L_j^{14} \oplus (g, g, g, 0, g, 0, 0, g))$ for all the 255 possible nonzero values of g; we denote the values by $\Delta_{i,j}^{14}$, respectively. Then, do as follows.

   (a) Guess the two bytes $(k_{14,1}, k_{14,2})$ of the subkey K_{14}. For every plaintext pair (C_i, C_j), partially encrypt the first two bytes of (R_i^{14}, R_j^{14}) through the s_1 and s_2 S-boxes, and check if they have a difference equal to any of the corresponding two-byte partial differences in $\Delta_{i,j}^{14}$. Keep only the qualified pairs. From Property 1-1 we can similarly get that there is only one difference in $\Delta_{i,j}^{14}$ for a qualified pair, and we denote this difference from $\Delta_{i,j}^{14}$ by $\delta_{i,j}^{14}$. As there are 255 possible values in $\Delta_{i,j}^{14}$ for every pair, the expected number of remaining pairs is about $2^{105} \cdot \frac{255}{2^{16}} \approx 2^{97}$.

   (b) For $l = 3$ to 8:
      - Guess the byte $k_{14,l}$ of K_{14};
      - For every remaining pair (C_i, C_j), partially encrypt the l-th byte of (R_i^{14}, R_j^{14}) through the s_l S-box, and check if they have a difference equal to the corresponding one-byte partial difference in $\delta_{i,j}^{14}$; keep only the qualified pairs. The difference $\delta_{i,j}^{14}$ is already fixed in Step 5-(a), so it is expected that a proportion of about $1 - 2^{-8}$ of the remaining pairs will be discarded after every iteration.

6. For every remaining ciphertext pair (C_i, C_j), by Property 1-2 there is only one value of h such that $P^{-1}(L_i^{13} \oplus L_j^{13} \oplus (h, 0, 0, 0, 0, 0, 0, 0))$ has the form $(?,?,?,0,?,0,0,?)$; we denote by $\delta_{i,j}^{13}$ the value $P^{-1}(L_i^{13} \oplus L_j^{13} \oplus (h, 0, 0, 0, 0, 0, 0, 0))$ with the form $(?,?,?,0,?,0,0,?)$. Then, for $l = 1, 2, 3, 5, 8$, do as follows.

   - Guess the byte $k_{13,l}$ of the subkey K_{13};
   - For every remaining pair (C_i, C_j), partially decrypt the l-th byte of (R_i^{13}, R_j^{13}) through the s_l S-box, and check if they have a difference equal to the corresponding one-byte difference in $\delta_{i,j}^{13}$; keep only the qualified

pairs. A proportion of about $1 - 2^{-7}$ of the remaining ciphertext pairs will be discarded after every iteration.

Finally, for every remaining pair of ciphertexts we can get the first bytes of their intermediate values just after Round 12.

7. Guess the byte $k_{12,1}$ of the subkey K_{12}. For every remaining ciphertext pair (C_i, C_j), compute $s_1(R_{i,1}^{12} \oplus k_{12,1})$ and $s_1(R_{j,1}^{12} \oplus k_{12,1})$, and check if they have a difference equal to $L_{i,1}^{12} \oplus L_{j,1}^{12}$. If there exists a ciphertext pair that passes this test, then discard this subkey guess, and try another; otherwise, for every subkey guess $(K_1, k_{2,1}, k_{2,2}, k_{2,3}, k_{2,5}, k_{2,8})$, exhaustively search for the remaining 152 key bits.

In Step 1, choosing the qualified pairs requires about $2^{120} \cdot 2^{106} = 2^{226}$ memory accesses in a simple implementation. Step 2 has a time complexity of about $2 \cdot 2^{209} \cdot 2^{16} \cdot \frac{1}{14} \cdot \frac{2}{8} + \sum_{i=0}^{5}(2 \cdot 2^{201-8 \cdot i} \cdot 2^{16+8 \cdot (i+1)} \cdot \frac{1}{14} \cdot \frac{1}{8}) \approx 2^{222.2}$ encryptions. Step 3 has a time complexity of about $\sum_{i=0}^{4}(2 \cdot 2^{153-8 \cdot i} \cdot 2^{64+8 \cdot (i+1)} \cdot \frac{1}{14} \cdot \frac{1}{8}) \approx 2^{221.5}$ encryptions. Step 4 has a time complexity of about $2 \cdot 2^{113} \cdot 2^{112} \cdot \frac{1}{14} \cdot \frac{1}{8} \approx 2^{219.2}$ encryptions. Step 5 has a time complexity of about $2 \cdot 2^{105} \cdot 2^{128} \cdot \frac{1}{14} \cdot \frac{2}{8} + \sum_{i=0}^{5}(2 \cdot 2^{97-8 \cdot i} \cdot 2^{128+8 \cdot (i+1)} \cdot \frac{1}{14} \cdot \frac{1}{8}) \approx 2^{230.2}$ decryptions. Step 6 has a time complexity of about $\sum_{i=0}^{4}(2 \cdot 2^{49-7 \cdot i} \cdot 2^{176+8 \cdot (i+1)} \cdot \frac{1}{14} \cdot \frac{1}{8}) \approx 2^{232.1}$ decryptions. In Step 7, the expected number of remaining subkey guesses is about $2^{224} \cdot (1 - 2^{-7})^{2^{14}} \approx 2^{39.7}$, meaning that $2^{191.7}$ trial encryptions are required to find the correct 256 key bits. Thus, Step 7 has a time complexity of about $2 \cdot 2^{224} \cdot [1 + (1 - 2^{-7}) + \cdots + (1 - 2^{-7})^{2^{14}}] \cdot \frac{1}{14} \cdot \frac{1}{8} + 2^{191.7} \approx 2^{225.2}$ encryptions.

Therefore, the attack has a total time complexity of about $2^{232.5}$ 14-round Camellia-256 computations.

Note that in the above attack we first check the plaintext pairs and finally ciphertext pairs. Using this order, we obtain an improvement of a factor of about 2^6 on the time complexity of that using the general order (i.e. checking ciphertext pairs first and finally plaintext pairs).

4.2   Attacking 13-Round Camellia-192 without the FL Functions

Using the 8-round impossible differentials we can break 13-round Camellia-192 without the FL Functions; the attack is basically the version of the above 14-round Camellia-256 attack when the last round is removed. The main difference is that in the last step we exhaustively search for the remaining 88 key bits for every subkey guess $(K_1, k_{2,1}, k_{2,2}, k_{2,3}, k_{2,5}, k_{2,8})$. After a similar analysis, we get that the 13-round Camellia-192 attack requires 2^{119} chosen plaintexts, and has a time complexity of $2^{167.9}$ 13-round Camellia-192 computations.

Note

1. Similarly, we can mount an attack on 13-round Camellia-256 without the FL functions, with a data complexity of 2^{120} chosen plaintexts and a time complexity of $2^{168.9}$ 13-round Camellia-256 computations.

2. As mentioned earlier, Wu et al. [26] presented an impossible differential cryptanalysis on 12-round Camellia-192 and Camellia-256 without the FL functions. The attack requires 2^{120} chosen plaintexts, and has a time complexity of 2^{181} Camellia-192/256 computations. However, it can be improved; the improved attack is basically the version of the above 14-round Camellia-256 attack when the last two rounds are removed. The improved attack on 12-round Camellia-192 requires 2^{119} chosen plaintexts, and has a time complexity of 2^{131} 12-round Camellia-192 computations; the improved attack on 12-round Camellia-256 requires 2^{120} chosen plaintexts, and has a time complexity of 2^{152} 12-round Camellia-256 computations.

4.3 Attacking 11-Round Camellia-128 without the FL Functions

To attack 11-round Camellia-128, we use the 8-round impossible differentials in Rounds 3 to 10, and use the early abort technique in the first round. We briefly describe the attack procedure as follows.

1. Choose 2^{30} structures: each structure contains a set of 2^{88} chosen plaintexts $P_i = (L_i^0, R_i^0)$, with $R_i^0 = P(x_1, x_2, x_3, \alpha_4, x_5, \alpha_6, \alpha_7, x_8) \oplus (x, \beta_2, \beta_3, \beta_4, \beta_5, \beta_6, \beta_7, \beta_8)$ and $L_i^0 = (y_1, y_2, y_3, \gamma_4, y_5, \gamma_6, \gamma_7, y_8)$, where the bytes with the forms $x_\times$ and $y_\times$ take all the possible values in $\{0,1\}^8$, and the bytes with the forms $\alpha_\times$, $\beta_\times$ and $\gamma_\times$ are fixed to certain values in $\{0,1\}^8$, $(i = 1, 2, \cdots, 2^{88})$. In a chosen-plaintext attack scenario, obtain their ciphertexts. Keep only the pairs such that $\Delta L^0 = (u, u, u, 0, u, 0, 0, u)$ and $(\Delta L^{11}, \Delta R^{11})$ belonging to the 2^{15} possible output differences after Round 11. The expected number of remaining plaintext pairs is about 2^{60}.
2. Conduct a step similar to Step 3 of the 14-round Camellia-256 attack presented in Section 4.1. This step has a time complexity of about $\sum_{i=0}^{4}(2 \cdot 2^{60-8 \cdot i} \cdot 2^{8 \cdot (i+1)} \cdot \frac{1}{11} \cdot \frac{1}{8}) \approx 2^{64.9}$ 11-round Camellia-128 computations.
3. Conduct a step similar to Step 4 of the 14-round Camellia-256 attack. This step has a time complexity of about $2 \cdot 2^{20} \cdot 2^{48} \cdot \frac{1}{11} \cdot \frac{1}{8} \approx 2^{62.5}$ 11-round Camellia-128 computations.
4. Conduct a step similar to Step 7 of the 14-round Camellia-256 attack; here, for every remaining guess for $(k_{1,1}, k_{1,2}, k_{1,3}, k_{1,5}, k_{1,8}, k_{2,1})$, exhaustively search for the remaining 80 key bits.

In Step 1, a structure yields about $\frac{2^{88 \times 2}}{2} \cdot \frac{255}{2^{40}} \approx 2^{143}$ plaintext pairs with $\Delta L^0 = (u, u, u, 0, u, 0, 0, u)$, so the 2^{30} structures yield a total of 2^{173} plaintext pairs with $\Delta L^0 = (u, u, u, 0, u, 0, 0, u)$, which generate $2^{173} \cdot \frac{2^{15}}{2^{128}} = 2^{60}$ useful pairs. To get the qualified pairs, we first store the ciphertexts into a hash table indexed by the 4-th, 6-th and 7-th bytes of L_i^{11}, the bytes from 2 to 8 of R_i^{11}, the XOR of the 1-st and 2-nd bytes of L_i^{11}, the XOR of the 1-st and 3-rd bytes of L_i^{11}, the XOR of the 1-st and 5-th bytes of L_i^{11} and the XOR of the 1-st and 8-th bytes of L_i^{11}; and then we choose the qualified pairs. Thus, it requires about $2^{118} \cdot 2^8 = 2^{126}$ memory accesses.

In Step 4, it is expected that about $2^{56} \cdot (1 - 2^{-7})^{2^{12}} \approx 2^{10}$ guesses for $(k_{1,1}, k_{1,2}, k_{1,3}, k_{1,5}, k_{1,8}, k_{2,1}, k_{11,1})$ remain; thus 2^{90} trial encryptions are required to find the 128 key bits. This step has a time complexity of about $2 \cdot 2^{56} \cdot [1 + (1 - 2^{-7}) + \cdots + (1 - 2^{-7})^{2^{12}}] \cdot \frac{1}{11} \cdot \frac{1}{8} + 2^{90} \approx 2^{90}$ 11-round Camellia-128 computations.

Therefore, the attack has a total time complexity of about 2^{118} 11-round Camellia-128 computations and 2^{126} memory accesses.

4.4   Extending the above Attacks

We next observe the following Property 2 for Camellia, which can be used to extend the attacks described in Sections 4.1 – 4.3.

Property 2. *Given an input difference and an output difference of a Camellia S-box, we can know the possible pairs of actual values input to the S-box. Every Camellia S-box has a differential probability of 2^{-6} or 2^{-7}, thus on average there is approximately only one ($\approx \frac{126}{255} \cdot 2 + \frac{1}{255} \cdot 4$) pair of actual values input to the S-box, given a randomly chosen pair of input and output differences.*

Property 2 suggests that, during the above attacks, we can pick up the pairs with the actual values equal to the XOR of the key guess and the possible inputs to the S-box, instead of partially encrypting or decrypting it through the S-box, which can be done by keeping a precomputation table storing the results. The resulting attacks using this way have a number of table lookups (i.e. memory accesses) comparable to the computational complexities of the above attacks.

All the attacks given above work in the following way: for a key guess, we try to find a plaintext pair such that an impossible differential holds for the pair under the key guess; thus the key guess is impossible, and can be discarded. Another way to conduct an impossible differential attack is that, for a plaintext or ciphertext pair, we can discard all the key guesses such that impossible differentials hold for the pair under these key guesses, by using Property 2. By this way, the 11-round Camellia-128 attack, the 13-round Camellia-192 attack, the 13-round Camellia-256 attack and the 14-round Camellia-256 attack have a time complexity of about 2^{126}, $2^{169.4}$, $2^{170.4}$ and 2^{231} memory accesses, respectively.

5   Impossible Differential Cryptanalysis of 6-Round MISTY1 without the FL Functions

In 2001, Kühn [14] presented an impossible differential cryptanalysis on 6-round MISTY1 (without the FL functions); the attack requires 2^{39} plaintexts, and has a time complexity of 2^{106} 6-round MISTY1 computations. Kühn also presented another impossible differential cryptanalysis on 6-round MISTY1, which requires more plaintexts but less computations. Both the attacks are based on the following generic 5-round impossible differentials for Feistel networks with bijective round structures: $(0, 0, \alpha_l, \alpha_r) \nrightarrow (0, 0, \alpha_l, \alpha_r)$, where $(\alpha_l, \alpha_r) \neq (0, 0)$.

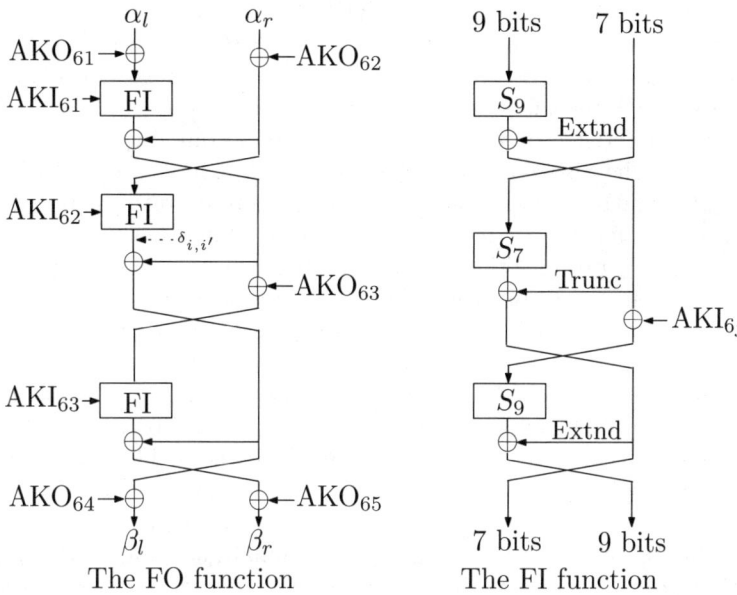

The FO function   The FI function

Fig. 4. Impossible differential attack on 6-round MISTY1

Kühn's attacks use a round structure equivalent to the original one, which is illustrated in Fig. 4; let $[KI_{6j}]_{15-9}$ denote the bits from 9 to 15 of KI_{6j}, $[KI_{6k}]_{8-0}$ denote the bits from 0 to 8 of KI_{6k}, and $KI'_{6j} = [KI_{6j}]_{15-9}\|00\|[KI_{6j}]_{15-9}$, the equivalent subkeys are as follows.

$$\text{AKO}_{6k} = KO_{6k}, \quad k = 1, 2.$$
$$\text{AKO}_{63} = KO_{62} \oplus KO_{63} \oplus KI'_{61}.$$
$$\text{AKO}_{64} = KO_{62} \oplus KO_{64} \oplus KI'_{61} \oplus KI'_{62}.$$
$$\text{AKO}_{65} = KO_{62} \oplus KI'_{61} \oplus KI'_{62} \oplus KI'_{63}.$$
$$\text{AKI}_{6k} = [KI_{6k}]_{8-0}, \quad k = 1, 2, 3.$$

MISTY1 has a nested Feistel structure, which is rather different from the "regular" one. Nevertheless, the MISTY1 round structure also allows us to use the early abort technique. As a result, we can improve the first attack due to Kühn, as follows.

1. Choose 2^7 structures: each structure contains 2^{32} plaintexts $P_i = (x, y, a_i, b_i)$, where x and y are 16-bit fixed constants, and a_i and b_i take all the possible 2^{16} values. Keep only the pairs $(P_i, P_{i'})$ with an output difference $(?, ?, \alpha_l, \alpha_r)$, where the question mark ? denotes an unknown difference of 16 bits long (two differences marked with ? may be different). The expected number of remaining ciphertext pairs is $2^7 \cdot \frac{2^{32 \times 2}}{2} \cdot 2^{-32} = 2^{38}$. (This step is exactly the same as that in Kühn's attack.)

2. Guess the 41 subkey bits $(\text{AKO}_{61}, \text{AKI}_{61}, \text{AKO}_{62})$ in Round 6. For every remaining ciphertext pair $(C_i, C_{i'})$, the 32-bit difference in the left side is known, say (β_l, β_r), (β_l and β_r are 16-bit long), so we can compute the difference just after the second FI in the FO by using $(\text{AKO}_{61}, \text{AKI}_{61})$; we denote the difference by $\delta_{i,i'}$, (see Fig. 4). As a consequence, using $\delta_{i,i'}$ we can compute the difference just after the S_7 S-box in the second FI by using AKO_{62}. On the other hand, we know the two inputs to this S-box S_7 for the pair, whose difference is the right 7 bits of α_r. Finally, keep the pair if the inputs to the S_7 produce the output difference obtained earlier. This imposes a 7-bit filtering condition; thus about $2^{38} \cdot 2^{-7} = 2^{31}$ pairs are expected to remain for every subkey guess. This step has a time complexity of about $2 \cdot 2^{38} \cdot 2^{41} \cdot \frac{1}{6} \cdot \frac{2}{3} \approx 2^{77}$ 6-round MISTY1 computations.

3. Guess the 9 subkey bits AKI_{62}. For a remaining pair $(C_i, C_{i'})$, with $\delta_{i,i'}$ we can compute the output difference of the second S_9 S-box in the second FI. Keep the pairs which produce these output differences. The expected number of remaining pairs is $2^{31} \cdot 2^{-9} = 2^{22}$. This step has a time complexity of about $2 \cdot 2^{31} \cdot 2^{50} \cdot \frac{1}{6} \cdot \frac{1}{3} \approx 2^{78}$ 6-round MISTY1 computations.

4. Guess the 16 subkey bits AKO_{63}. For a remaining pair, with (β_l, β_r) we can compute the difference just after the S_7 S-box of the third FI by using AKO_{63}. Keep the pairs which produce these output differences. The expected number of remaining pairs is $2^{22} \cdot 2^{-7} = 2^{15}$. This step has a time complexity of about $2 \cdot 2^{22} \cdot 2^{66} \cdot \frac{1}{6} \cdot \frac{1}{3} \approx 2^{85}$ 6-round MISTY1 computations.

5. Guess the 9 subkey bits AKI_{63}, and check whether or not there is a pair such that the difference just after the third FI is $\beta_l \oplus \beta_r$. If there is such a pair, the guess for $(\text{AKO}_{61}, \text{AKI}_{61}, \text{AKO}_{62}, \text{AKI}_{62}, \text{AKO}_{63}, \text{AKI}_{63})$ is impossible, discard it, and guess another. The expected number of remaining guesses for the 75 subkey bits is $2^{75} \cdot (1 - 2^{-9})^{2^{15}} \approx 2^{-17}$; thus we can assume it is the correct one. This step has a time complexity of about $2 \cdot 2^{75} \cdot [1 + (1 - 2^{-9}) + \cdots + (1 - 2^{-9})^{2^{15}}] \cdot \frac{1}{6} \cdot \frac{1}{3} \approx 2^{81}$ 6-round MISTY1 computations.

Therefore, this attack has a total time complexity of about 2^{85} 6-round MISTY1 computations, significantly lower than the complexity of 2^{106} for Kühn's attack.

6 Conclusions

In this paper, we observe that, when conducting an impossible differential cryptanalysis on Camellia and MISTY1, their round structures allow us to partially determine whether a candidate pair is useful by guessing only a small fraction of the unknown required subkey bits of a relevant round at a time, instead of all of them. This can reduce the computation complexity of an attack, and may allow us to break more rounds. Taking advantage of the early abort technique, we improve a previous impossible differential attack on 6-round MISTY1 without the FL functions, and present impossible differential cryptanalysis of 11-round Camellia-128 without the FL functions, 13-round Camellia-192 without the FL functions and 14-round Camellia-256 without the FL functions. The presented attacks are the best currently published cryptanalytic results on Camellia and MISTY1 without the FL functions.

Depending on the design of the round structure of a block cipher, the early abort technique can also be used to improve the efficiency of other cryptanalytic approaches, including differential cryptanalysis and its extensions. Its application to impossible differential cryptanalysis of AES [21] is investigated in [17].

Acknowledgments

The authors are very grateful to Jiqiang Lu's supervisor Prof. Chris Mitchell for his editorial comments and to the anonymous referees for their comments.

References

1. Aoki, K., Ichikawa, T., Kanda, M., Matsui, M., Moriai, S., Nakajima, J., Tokita, T.: *Camellia*: a 128-bit block cipher suitable for multiple platforms — design and analysis. In: Stinson, D.R., Tavares, S. (eds.) SAC 2000. LNCS, vol. 2012, pp. 39–56. Springer, Heidelberg (2001)
2. Babbage, S., Frisch, L.: On MISTY1 higher order differential cryptanalysis. In: Won, D. (ed.) ICISC 2000. LNCS, vol. 2015, pp. 22–36. Springer, Heidelberg (2001)
3. Biham, E., Biryukov, A., Shamir, A.: Cryptanalysis of Skipjack reduced to 31 rounds using impossible differentials. In: Stern, J. (ed.) EUROCRYPT 1999. LNCS, vol. 1592, pp. 12–23. Springer, Heidelberg (1999)
4. Biham, E., Biryukov, A., Shamir, A.: Miss in the middle attacks on IDEA and Khufu. In: Knudsen, L.R. (ed.) FSE 1999. LNCS, vol. 1636, pp. 124–138. Springer, Heidelberg (1999)
5. Biham, E., Shamir, A.: Differential cryptanalysis of the Data Encryption Standard. Springer, Heidelberg (1993)
6. CRYPTREC — Cryptography Research and Evaluatin Committees, report, Archive (2002), http://www.ipa.go.jp/security/enc/CRYPTREC/index-e.html
7. Duo, L., Li, C., Feng, K.: New observation on Camellia. In: Preneel, B., Tavares, S. (eds.) SAC 2005. LNCS, vol. 3897, pp. 51–64. Springer, Heidelberg (2006)
8. Hatano, Y., Sekine, H., Kaneko, T.: Higher order differential attack of Camellia(II). In: Nyberg, K., Heys, H.M. (eds.) SAC 2002. LNCS, vol. 2595, pp. 39–56. Springer, Heidelberg (2003)
9. He, Y., Qing, S.: Square attack on reduced Camellia cipher. In: Qing, S., Okamoto, T., Zhou, J. (eds.) ICICS 2001. LNCS, vol. 2229, pp. 238–245. Springer, Heidelberg (2001)
10. International Standardization of Organization (ISO), International Standard – ISO/IEC 18033-3, Information technology – Security techniques – Encryption algorithms – Part 3: Block ciphers (July 2005)
11. Knudsen, L.R.: Truncated and higher order differentials. In: Preneel, B. (ed.) FSE 1994, vol. 1008, pp. 196–211. Springer, Heidelberg (1995)
12. Knudsen, L.R.: DEAL — a 128-bit block cipher. Technical report, Department of Informatics, University of Bergen, Norway (1998)
13. Knudsen, L.R., Wagner, D.: Integral cryptanalysis. In: Daemen, J., Rijmen, V. (eds.) FSE 2002. LNCS, vol. 2365, pp. 112–127. Springer, Heidelberg (2002)

14. Kühn, U.: Cryptanalysis of reduced-round MISTY. In: Pfitzmann, B. (ed.) EURO-CRYPT 2001. LNCS, vol. 2045, pp. 325–339. Springer, Heidelberg (2001)
15. Kühn, U.: Improved cryptanalysis of MISTY1. In: Daemen, J., Rijmen, V. (eds.) FSE 2002. LNCS, vol. 2365, pp. 61–75. Springer, Heidelberg (2002)
16. Lee, S., Hong, S., Lee, S., Lim, J., Yoon, S.: Truncated differential cryptanalysis of Camellia. In: Kim, K.-c. (ed.) ICISC 2001. LNCS, vol. 2288, pp. 32–38. Springer, Heidelberg (2002)
17. Lu, J., Dunkelman, O., Keller, N., Kim, J.: Revisiting impossible differential cryptanalysis of AES. (manuscript, 2007)
18. Matsui, M.: Linear cryptanalysis method for DES cipher. In: Helleseth, T. (ed.) EUROCRYPT 1993. LNCS, vol. 765, pp. 386–397. Springer, Heidelberg (1994)
19. Matsui, M.: New block encryption algorithm MISTY. In: Biham, E. (ed.) FSE 1997. LNCS, vol. 1267, pp. 54–68. Springer, Heidelberg (1997)
20. NESSIE — New European Schemes for Signatures, Integrity, and Encryption, final report of European project IST-, -12324. Archive (1999), https://www.cosic.esat.kuleuven.be/nessie/Bookv015.pdf
21. NIST — National Institute of Standards and Technology, Advanced Encryption Standard (AES), FIPS-197 (2001)
22. Shirai, T.: Differential, linear, boomerang and rectangle cryptanalysis of reduced-Round Camellia. In: Proceedings of the Third NESSIE Workshop (2002)
23. Sugita, M., Kobara, K., Imai, H.: Security of reduced version of the block cipher Camellia against truncated and impossible differential cryptanalysis. In: Boyd, C. (ed.) ASIACRYPT 2001. LNCS, vol. 2248, pp. 193–207. Springer, Heidelberg (2001)
24. Tanaka, H., Hisamatsu, K., Kaneko, T.: Strength of MISTY1 without FL function for higher order differential attack. In: Fossorier, M.P.C., Imai, H., Lin, S., Poli, A. (eds.) AAECC 1999. LNCS, vol. 1719, pp. 221–230. Springer, Heidelberg (1999)
25. Wu, W., Feng, D., Chen, H.: Collision attack and pseudorandomness of reduced-round Camellia. In: Handschuh, H., Hasan, M.A. (eds.) SAC 2004. LNCS, vol. 3357, pp. 256–270. Springer, Heidelberg (2004)
26. Wu, W., Zhang, W., Feng, D.: Impossible differential cryptanalysis of reduced-round ARIA and Camellia. Journal of Computer Science and Technology 22(3), 449–456 (2007)
27. Yeom, Y., Park, S., Kim, I.: On the security of Camellia against the square attack. In: Daemen, J., Rijmen, V. (eds.) FSE 2002. LNCS, vol. 2365, pp. 89–99. Springer, Heidelberg (2002)
28. Yeom, Y., Park, S., Kim, I.: A study of integral type cryptanalysis on Camellia. In: Proceedings of the 2003 Symposium on Cryptography and Information Security, pp. 453–456 (2003)

Small Secret Key Attack on a Variant of RSA
(Due to Takagi)

Kouichi Itoh[1], Noboru Kunihiro[2], and Kaoru Kurosawa[3]

[1] Fujistu Laboratories
[2] University of Electro-Communications
[3] Ibaraki University, Japan
kito@labs.fujitsu.com, kunihiro@ice.uec.ac.jp, kurosawa@mx.ibaraki.ac.jp

Abstract. For a variant of RSA with modulus $N = p^r q$ and $ed \equiv 1$ mod $(p-1)(q-1)$, we show that d can be recovered if $d < N^{(2-\sqrt{2})/(r+1)}$. (Note that $\phi(N) \neq (p-1)(q-1)$.) Boneh-Durfee's result for the standard RSA is obtained as a special case for $r = 1$. Technically, we develop a method of a finding small root of a trivariate polynomial equation $f(x, y, z) = x(y-1)(z-1) + 1 = 0 \pmod{e}$ under the condition that $y^r z = N$. Our result cannot be obtained from the generic method of Jochemsz-May.

Keywords: lattice, LLL, trivariate polynomial, RSA.

1 Introduction

1.1 Background

Based on LLL algorithm [11], Coppersmith showed a polynomial time algorithm to find a small root of univariate modular polynomial equations [4]. Howgrave-Graham showed a variant which is easy to understand and apply [8]. This algorithm and its extensions to more than one variable have been used extensively to study the security of RSA and its variants.

For example, Coron and May showed a deterministic equivalence between factoring $N = pq$ and finding d from (N, e), where (N, e) is a public key and d is the private key [5]. Ernst et al. studied Partial Key Exposure Attacks on RSA [7]. Boneh, Durfee and Howgrave-Graham [3] showed an efficient algorithm for factoring $N = p^r q$ in time $O(p^{\frac{2}{r+1}})$. Recently, Jochemsz-May gave a generic method of lattice construction for multivariate polynomials and applied it to attacks on some variants of RSA [9].

Meanwhile, an important problem of RSA is to study the security of small d because the decryption or signature generation can be made faster if d is small. Wiener showed that d can be recovered from (e, N) in polynomial time if $d < N^{1/4}$ [17]. Boneh-Durfee improved the bound up to $d < N^{0.292}$ by using the technique of Coppersmith and Howgrave-Graham [2][1].

---

[1] They also studied the unbalanced case by extending their technique to a similar trivariate modular polynomial equation. Durfee-Nguyen studied a small d attack on another unbalanced RSA [6].

T. Malkin (Ed.): CT-RSA 2008, LNCS 4964, pp. 387–406, 2008.

Meanwhile, there are two variants of RSA with modulus $N = p^r q$. In the first variant, $ed \equiv 1 \bmod p^{r-1}(p-1)(q-1)$ while in the second variant, $ed \equiv 1 \bmod (p-1)(q-1)$. The first variant is more natural because $\phi(N) = p^{r-1}(p-1)(q-1)$, where ϕ is Euler's totient function.

For the first variant, Takagi [14,15] extended Wiener's attack up to $d < N^{1/2(r+1)}$. May showed two more efficient attacks by using the technique of Coppersmith and Howgrave-Graham [12].

For the second variant, Takagi observed that the decryption can be significantly faster [14,15]. Kunihiro and Kurosawa [10] proved the deterministic polynomial time equivalence between factoring and computing d.

1.2 Our Contributions

In this paper, we study the security of small d of the second variant of RSA, i.e., $ed \equiv 1 \bmod (p-1)(q-1)$. We show that d can be recovered from (e, N) if $d < N^{(2-\sqrt{2})/(r+1)}$. It is interesting to see that Boneh-Durfee's result for the standard RSA [2] is obtained as a special case for $r = 1$. Hence, our result is a natural generalization of Boneh-Durfee [2]. Table 1 summarizes previous results and our results. Let $d = N^\delta$. Technically, we develop a method of a finding small root

Table 1. Best Known Results and Our Contribution

	$N = pq$ $ed \equiv 1(\bmod \operatorname{lcm}(p-1,q-1))$	$N = p^r q$ $ed \equiv 1(\bmod\ p^{r-1}(p-1)(q-1))$	$N = p^r q$ $ed \equiv 1(\bmod\ (p-1)(q-1))$
Basic Attack	$\delta < 0.25$ Wiener [17]	$\delta < 1/2(r+1)$ Takagi [15]	$\delta < 1/2(r+1)$ Remark 4 in this paper
Advanced Attack	$\delta < 0.292$ Boneh-Durfee[2]	$\delta < \frac{r}{(r+1)^2}, \delta < \left(\frac{r-1}{r+1}\right)^2$ for $r \geq 2$ May [12]	$\delta < (2-\sqrt{2})/(r+1)$ Theorem 2 in this paper

of a trivariate polynomial equation $f(x, y, z) = x(y-1)(z-1) + 1 = 0 \pmod{e}$ under the condition such that $y^r z = N$ while Boneh-Durfee used a method of a finding small root of a bivariate polynomial equation $f(x, y) = 0 \pmod{e}$ [2]. In fact, we show how to derive a *good* lattice from the previous lattice for $h(x, y) = (y-1)(z-1)$ with $y^r z = N$ [10]. This is of independent interest.

As mentioned earlier, Jochemsz-May gave a generic method of lattice construction for multivariate polynomials [9]. However, we cannot obtain our result by using this method.

1.3 Organization

The rest of paper is organized as follows. In section 2, we explain preliminaries. First, we review LLL algorithm and Howgrave-Graham's Lemma. Then we explain Takagi's variant of RSA. In section 3, we introduce and show small d attacks on Takagi's RSA. In particular, we prove if $d < N^{(7-2\sqrt{7})/3(r+1)}$ then d can be recovered from e and N. In section 4, we use "Geometrically Progressive

Matrices" introduced in [2] to improve the bound of d to $d < N^{(2-\sqrt{2})/(r+1)}$. In section 5, we give some experimental results and we show the our attack is effective. Finally, section 6 gives some discussions.

2   Preliminaries

This section describes LLL algorithm, Howgrave-Graham's lemma and Takagi's variant of RSA.

2.1   Notations

For a vector b, $||b||$ denotes the Euclidean norm of b. For a trivariate polynomial $h(x,y,z) = \sum h_{ijk} x^i y^j z^k$, define

$$||h(x,y,z)|| = \sqrt{\sum h_{ijk}^2}.$$

That is, $||h(x,y,z)||$ denotes the Euclidean norm of the vector which consists of coefficients of $h(x,y,z)$.

2.2   LLL Algorithm and Howgrave-Graham's Lemma

Let $M = \{a_{ij}\}$ be a $w \times n$ matrix of integers. The rows of M generate a lattice L if row vectors of M are linearly independent. Rows form a basis of L. The lattice L is also represented as follows. Letting $a_i = (a_{i1}, a_{i2}, \ldots, a_{in})$, the lattice L spanned by $\langle a_1, \ldots, a_w \rangle$ consists of all integral linear combinations of $a_1, \ldots, a_w$, that is :

$$L = \left\{ \sum_{i=1}^{w} n_i a_i \mid n_i \in \mathbb{Z} \right\}. \tag{1}$$

LLL algorithm outputs a short vector in the lattice L. This algorithm works in a deterministic polynomial time.

Proposition 1 (LLL [11]). *Let $M = \{a_{ij}\}$ be a nonsingular $w \times n$ matrix of integers. The rows of M generates a lattice L. Given M, the LLL algorithm finds w vectors $b_1, \ldots, b_w \in L$ such that*

$$||b_i|| \leq 2^{(w+i-2)/4} (\det L)^{1/(w-i+1)}$$

in time polynomial in (w, B), where $B = \max \log_2 |a_{ij}|$. Especially, it is important for our application that

$$||b_1|| \leq 2^{(w-1)/4} (\det L)^{1/w} \quad and \quad ||b_2|| \leq 2^{w/4} (\det L)^{1/(w-1)}.$$

Lemma 1 (Howgrave-Graham [8]). *Let $h(x,y,z) \in \mathbb{Z}[x,y,z]$ be a polynomial, which is a sum of at most w monomials. Let m be an integer and X, Y and Z be some positive integers. Suppose that*

1. *$h(x_0, y_0, z_0) = 0 \bmod \phi^m$, where $|x_0| < X$, $|y_0| < Y$ and $|z_0| < Z$.*
2. *$||h(xX, yY, zZ)|| < \phi^m / \sqrt{w}$.*

Then $h(x_0, y_0, z_0) = 0$ holds over integers.

2.3   Takagi's Variant of RSA

Takagi proposed a variant of RSA such that $N = p^r q$ and $ed \equiv 1 \bmod (p-1)(q-1)$ and showed that a faster decryption algorithm can be obtained [14,15]. For example, for $r = 2$, it is 42% faster than the original RSA decryption algorithm. We should notice again that e and d are not set as $ed \equiv 1 \bmod p^{r-1}(p-1)(q-1)$ in Takagi's scheme.

Key Generation. Generate two distinct primes p and q. Let $N = p^r q$. Find e and d such that

$$ed \equiv 1 \bmod (p-1)(q-1). \tag{2}$$

Let $d_p = d \bmod p-1$ and $d_q = d \bmod q-1$. Then, e and N are the encryption keys and d_p, d_q, p, q are the decryption keys.

Encryption. For a plaintext $M \in \mathbb{Z}_N^*$, the ciphertext is computed as follows.

$$C = M^e \bmod N. \tag{3}$$

Decryption. Given a ciphertext C, do:
1. Compute $M_q = C^{d_q} \bmod q$, where $M_q = M \bmod q$.
2. Compute $M_p = C^{d_p} \bmod p$, where $M_p = M \bmod p$.
3. Find $M_p^{(r)}$ such that $M_p^{(r)} = M \bmod p^r$ by using Hensel lifting.
4. Compute M by applying Chinese remainder theorem to M_q and $M_p^{(r)}$.

Remark 1. Since e is less than $(p-1)(q-1)$, $e < N^{2/(r+1)}$.

3   Small d Attack on Takagi's Variant of RSA

3.1   Formulation: Small d Attack on Takagi's RSA

We say that (r, N, e, d) is a Takagi's RSA parameter if

$$N = p^r q, ed = 1 \bmod (p-1)(q-1),$$

where the bit length of p and q are the same. First, we will show that we can recover d if $d < N^{(7-2\sqrt{7})/3(r+1)}$ from the public information (r, N, e).

Since the bit length of p and q are the same, it holds that $p < 2N^{1/(r+1)}$ and $q < 2N^{1/(r+1)}$. Letting $e = N^\alpha$, we have $p, q < e^{1/\alpha(r+1)}$. Furthermore, let $d < N^\delta$.

Since $ed = 1 (\bmod (p-1)(q-1))$, there exists an integer k such that $ed - k(p-1)(q-1) = 1$. The range of k is given by

$$k = \frac{ed - 1}{(p-1)(q-1)} < \frac{2ed}{pq} < 2e^{1 + \frac{\delta}{\alpha} - \frac{2}{\alpha(r+1)}}. \tag{4}$$

We used the inequality: $(p-1)(q-1) > pq/2$. In a usual choice, α is set as $\alpha = 2/(r+1)$. In this case, we have

$$p, q < 2e^{1/2}, k < e^{\frac{r+1}{2}\delta}. \tag{5}$$

In Appendix B, we give the general solution for arbitrary α.

Key recovery attack on Takagi's variant of RSA is described as follows. Given a trivariate polynomial $f(x, y, z) = x(y-1)(z-1)+1$, find (x_0, y_0, z_0) satisfying

$$f(x_0, y_0, z_0) = 0 (\mathrm{mod}\, e), \tag{6}$$

and $y_0^r z_0 = N$, where $|x_0| < e^{\frac{r+1}{2}\delta}, |y_0|, |z_0| < 2e^{1/2}$. Our aim is to derive the condition of δ when the solution of the above problem is obtained.

In the next subsection, we show we can recover d if $\delta < (7 - 2\sqrt{7})/(3(r+1))$ from the public information (r, N, e) in polynomial time under the reasonable assumption.

3.2   Deriving the Bound of δ

Algorithm for key recovery on Takagi's RSA.

Step 1. Construct a set of polynomials $g_{[i,j,k,l]}(x, y, z)$ such that

$$g_{[i,j,k,l]}(x_0, y_0, z_0) \equiv 0 \bmod e^m.$$

Step 2. Apply LLL algorithm to the coefficient matrix of $\{g_{[i,j,k,l]}(xX, yY, zZ)\}$ to obtain $g_1(x, y, z)$ and $g_2(x, y, z)$, where $g_1(x, y, z)$ and $g_2(x, y, z)$ are non-zero integer combinations of $g_{[i,j,k,l]}(x, y, z)$ with small coefficients.

Step 3. Let $g_1'(x, y) = \mathrm{Resultant}_z(g_1(x, y, z), y^r z - N)$ and $g_2'(x, y) = \mathrm{Resultant}_z(g_2(x, y, z), y^r z - N)$. In this computation, the resultants never vanish.

Step 4. Compute the resultant of $g_1'(x, y)$ and $g_2'(x, y)$. If the resultant does not vanish, we obtain univariate polynomial $g_3(x)$.

Step 5. Solve the equation $g_3(x) = 0$ over integers.

As many similar papers [2,4,7,9], we cannot guarantee that $g_1'(x, y)$ and $g_2'(x, y)$ are algebraic independent polynomials. (i.e., Resultant is not vanished.) Hence, we make the following "heuristic" assumption.

Assumption 1. *The resultant computation for the polynomials $g_1'(x, y)$ and $g_2'(x, y)$ yields non-zero polynomials.*

Ernst et al. [7] and Jochemsz et al. [9] used the same assumption. So, we will also use the assumption. However, experiments are needed for specific cases to justify the assumption. In Section 5, we give some examples and cannot find any counter examples.

How to construct lattice. Remember that $f(x, y, z) = x(y-1)(z-1)+1$. Since it holds that $ed = 1 \bmod (p-1)(q-1)$, there exist an integer k such that $ed - k(p-1)(q-1) = 1$. Then, we have

$$k(p-1)(q-1) + 1 = 0 \bmod e, \tag{7}$$

which leads that $f(k, p, q) = 0 \bmod e$. Note that p and q satisfy that $N = p^r q$ and we know N. This setting of $f(x, y, z)$ is almost the same as the polynomial in cryptanalysis of unbalanced RSA in [2] and in Durfee-Nguyen's attack to RSA schemes with short secret exponent [6].

We need to construct a lattice to find the small roots of $f(x, y, z) = 0(\bmod e)$. In our analysis, we will replace each occurrence of $y^r z$ by N because $N = p^r q$ (based on Durfee-Nguyen technique [6]). How to construct it for our problem is not trivial.

For fixed a positive integer m, define

$$g_{[i,j,k,l]}(x, y, z) = x^i y^j z^k f(x, y, z)^l e^{m-l}. \tag{8}$$

It is easy to see that $g_{[i,j,k,l]}(k, p, q) = 0 \bmod e^m$ for any non-negative integers (i, j, k, l). This setting is general. However, we need to carefully set the order of indexes $[i, j, k, l]$.

In our construction, we will utilize a lattice construction for $h(y, z) = (y - 1)(z - 1)$ with constraint: $y^r z = N$, which is introduced by Kunihiro-Kurosawa [10]. To be self-contained, we show such a list in Appendix A. Since in their construction, Durfee-Nguyen technique is taken into account, we need not to consider it again in constructing that for $f(x, y, z)$.

We set the list of polynomials $G = \{g_{[i,j,k,l]}\}$ for $f(x, y, z) = x(y - 1)(z - 1)$ with constraint: $y^r z = N$ as follows.

> $G \leftarrow \emptyset.$
> for $u = 0, \cdots, m$, do;
>   for $i = 0, \cdots, u - 1$, do;
>   for $j = 0, 1$, do; append $g_{[u-i,j,0,i]}$ to G.
>   for $j = r - 1, \cdots, 1$, do; append $g_{[u-i,j,1,i]}$ to G.
>   for $j = 0, \cdots, s_u$, do ; append $g_{[0,j,0,u]}$ to G.
>   for $k = 1, \cdots, t_u$, do ;
>   for $j = r - 1, \cdots, 0$, do ; append $g_{[0,j,k,u]}$ to G.
> return G.

Remark 2. This construction is not trivial. Actually, in Jochemsz-May's lattice construction, monomials will appear in lexicographic order. Table 3 shows that the order of monomials in our lattice is not lexicographic.

Remark 3. To guarantee the coefficient matrix M is triangular, parameters s_u, t_u should be set as

$$s_0 \geq s_1 \geq \cdots \geq s_m, t_0 \geq t_1 \geq \cdots \geq t_m. \tag{9}$$

Actually, we will set $s_0 = s_1 = \cdots = s_m = t_0 = t_1 = \cdots = t_m = s$.

Remark 4. Suppose that we omit to append $g_{[0,j,0,u]}$ and $g_{[0,j,k,u]}$. This case corresponds to an extension of Wiener's attack on Takagi's variant of RSA. Then, if $\delta < 1/2(r + 1)$, d can be recovered.

We show a small example of the list G of $\{g_{ijkl}\}$ for $r = 2, m = 3, s_0 = s_1 = s_2 = s_3 = 2, t_0 = t_1 = t_2 = t_3 = 2$.

$G = \{g_{[0000]}, g_{[0100]}, g_{[0200]}, g_{[0110]}, g_{[0010]}, g_{[0120]}, g_{[0020]},$
$g_{[1000]}, g_{[1100]}, g_{[1110]}, g_{[0001]}, g_{[0101]}, g_{[0201]}, g_{[0111]}, g_{[0011]}, g_{[0121]}, g_{[0021]},$
$g_{[2000]}, g_{[2100]}, g_{[2110]}, g_{[1001]}, g_{[1101]}, g_{[1111]}, g_{[0002]}, g_{[0102]}, g_{[0202]}, g_{[0112]}, g_{[0012]}, g_{[0122]}, g_{[0022]},$
$g_{[3000]}, g_{[3100]}, g_{[3110]}, g_{[2001]}, g_{[2101]}, g_{[2111]}, g_{[1002]}, g_{[1102]}, g_{[1112]}, g_{[0003]}, g_{[0103]}, g_{[0203]}, g_{[0113]}, g_{[0013]}, g_{[0123]}, g_{[0023]}\}$

The 1st line of G corresponds to the list for $u = 0$. The 2nd, 3rd and 4th line also corresponds to that for $u = 1, 2, 3$.

We briefly show that $G = \{g_{[i,j,k,l]}(xX, yY, zZ)\}$ is a lower triangular. Note that in our setting, $i + l$ is always u. Consider $g_{[u-l,j,k,l]}(x, y, z)$.

$$g_{[u-l,j,k,l]}(x, y, z) = x^{u-l} y^j z^k (xh(y,z) + 1)^l e^{m-l}$$

$$= x^{u-l} y^j z^k \{x^l h(y,z)^l + \sum_{i=1}^{l} \binom{l}{i} x^{l-i} h(y,z)^i\} e^{m-l}$$

$$= x^u y^j z^k h(y,z)^l e^{m-l} + \sum_{i=1}^{l} \binom{l}{i} x^{u-i} y^j z^k h(y,z)^i e^{m-l}.$$

The number of the term including x^u is only one in $g_{[u-l,j,k,l]}$. The rest of all are terms with lower degree terms of x. Hence, they must appear in the former polynomial. To guarantee it, the parameters should be adequately chosen.

Table 2 shows a matrix M derived from G. For each $u = 0, 1, .., m$, $M^{(u)}$ consists of three sub-matrixes $M^{(u,0)}$, $M^{(u,1)}$ and $M^{(u,2)}$ illustlated in Figure 1. $M^{(u,0)}$ is $(r+1)u \times (r+1)u$ matrix, where the order of monomials is represented as $x^u z^i, x^u y^{i+1}, x^u y^{r-1} z^{i+1}, x^u y^{r-2} z^{i+1}, \ldots, x^u y z^{i+1}$ for $i = 0, 1, \ldots u-1$. Note that $M^{(u,0)}$ is empty when $u = 0$. $M^{(u,1)}$ is $(s_u + 1) \times (s_u + 1)$ matrix, where the order of monomials is represented as $x^u z^u, x^u y^{u+1}, x^u y^{u+2}, \ldots, x^u y^{u+s_u}$. $M^{(u,2)}$ is $rt_u \times rt_u$ matrix, where the order of monomials is represented as $x^u y^{r-1} z^{u+k}$, $x^u y^{r-2} z^{u+k}, \ldots, x^u z^{u+k}$ for $k = 1, \ldots t_u$. As an example, the explicit forms of $M^{(1)}$ is given in Table 3. The determinant of M can be easily calculated.

Determinant of Sub-matrix. Let $M^{(u)}$ be the sub-matrix corresponding to the list for u and monomials with x^u. The determinant of M is given by $\det M = \prod_{u=0}^{m} \det M^{(u)}$. Letting w_u be the dimension of sub-matrix $M^{(u)}$, the dimension of lattice is given by $w = \sum_{u=0}^{m} w_u$. Next, we compute $\det M^{(u)}$ and w_u for each $M^{(u)}$.

Simple analysis gives w_u and $\det M^{(u)}$ as follows. w_u is given by

$$w_u = (r+1)u + (1 + s_u) + rt_u = (r+1)u + rt_u + s_u + 1. \tag{10}$$

Table 2. Small example

	monomials with x^0	with x^1	with x^2	with x^3
the list for $u = 0$	$M^{(0)}$		0	
the list for $u = 1$	*	$M^{(1)}$		
the list for $u = 2$	*	*	$M^{(2)}$	
the list for $u = 3$	*	*	*	$M^{(3)}$

Table 3. Example of $M^{(1)}$ for $r = 2, m = 3, s = 2$

	x	xy	xyz	xz	xy^2	xy^3	xyz^2	xz^2	xyz^3	xz^3
$g_{[1000]}(xX, yY, zZ)$	Xe^3									
$g_{[1100]}(xX, yY, zZ)$	$*$	XYe^3								
$g_{[1110]}(xX, yY, zZ)$	$*$	$*$	$XYZe^3$							
$g_{[0001]}(xX, yY, zZ)$	$*$	$*$	$*$	XZe^2						
$g_{[0101]}(xX, yY, zZ)$	$*$	$*$	$*$	$*$	XY^2e^2					
$g_{[0201]}(xX, yY, zZ)$	$*$	$*$	$*$	$*$	$*$	XY^3e^2				
$g_{[0111]}(xX, yY, zZ)$	$*$	$*$	$*$	$*$	$*$	$*$	XYZ^2e^2			
$g_{[0011]}(xX, yY, zZ)$	$*$	$*$	$*$	$*$	$*$	$*$	$*$	XZ^2e^2		
$g_{[0121]}(xX, yY, zZ)$	$*$	$*$	$*$	$*$	$*$	$*$	$*$	$*$	XYZ^3e^2	
$g_{[0021]}(xX, yY, zZ)$	$*$	$*$	$*$	$*$	$*$	$*$	$*$	$*$	$*$	XZ^3e^2

$\det M^{(u)}$ can be described by

$$\det M^{(u)} = e^{\gamma_e} X^{\gamma_X} Y^{\gamma_Y} Z^{\gamma_Z}. \tag{11}$$

Each $\gamma_e, \gamma_X, \gamma_Y$ and γ_Z are given as follows.

$$\gamma_e = \sum_{i=0}^{u-1}(m - i)(r + 1) + (m - u)(1 + s_u + rt_u)$$
$$= w_u m - \frac{u}{2}((r + 1)u - (r - 1) + 2(s_u + rt_u)).$$

γ_X is given by

$$\gamma_X = uw_u.$$

γ_Y is given by

$$\gamma_Y = \frac{(u + s_u)(u + s_u + 1)}{2} + \frac{r(r - 1)(u + t_u)}{2}.$$

γ_Z is given by

$$\gamma_Z = \frac{r(u + t_u)(u + t_u + 1)}{2}.$$

Substituting $X = e^{\frac{r+1}{2}\delta}, Y = 2e^{1/2}, Z = 2e^{1/2}$ into Eq. (11), we have

$$\log_e \det M^{(u)} = mw_u - \frac{u}{2}((r + 1)u - (r - 1) + 2(s_u + rt_u)) + \frac{r + 1}{2}\delta uw_u$$
$$+ \frac{1}{2}\left(\frac{(u + s_u)(u + s_u + 1)}{2} + \frac{r(u + t_u)(u + t_u + r)}{2}\right).$$

By setting $s_u = t_u = s$ for any u, the logarithm of $\det M$ can be simplified as

$$\log_e \det M^{(u)} = mw_u - \frac{u}{2}((r + 1)u - (r - 1) + 2(r + 1)s) + \frac{r + 1}{2}\delta uw_u$$
$$+ \frac{(u + s)(u + s + 1)(r + 1)}{4} + \frac{r^2 - r}{4}(u + s). \tag{12}$$

The dimension of sub-lattice is given as $w_u = (r + 1)(u + s) + 1$.

The necessary condition to obtain two equations over integer from Howgrave-Graham's Lemma is given by

$$\det M < e^{m(w-1)}/\gamma, \tag{13}$$

where γ is a small term. By ignoring low order terms, it is enough to consider the inequality

$$\det M < e^{mw}. \tag{14}$$

The reason is Eq. (14) is approximated to Eq. (13) when m is large.

Derivation of the condition for δ. Eq. (14) leads to an inequality:

$$\log_e \det M = \sum_{u=0}^{m} \log \det M^{(u)} < m \sum_{u=0}^{m} w_u. \tag{15}$$

By ignoring low order terms, we carry out the computation. We have

$$m(r+1) \sum_{u=0}^{m} (u+s) - \frac{r+1}{2} \sum_{u=0}^{m} u(u+2s) + \frac{\delta(r+1)^2}{2} \sum_{u=0}^{m} u(u+s) \tag{16}$$
$$+ \frac{r+1}{4} \sum_{u=0}^{m} (u+s)^2 < m(r+1) \sum_{u=0}^{m} (u+s).$$

By straightforward computation, we have

$$3s^2 - 3(1 - (r+1)\delta)sm + (2(r+1)\delta - 1)m^2 < 0.$$

By setting $s = \frac{1-(r+1)\delta}{2} m$, we obtain an inequality:

$$3((r+1)\delta)^2 - 14(r+1)\delta + 7 > 0. \tag{17}$$

By solving the inequality, we have

$$\delta < \frac{7 - 2\sqrt{7}}{3(r+1)}. \tag{18}$$

$\square$

To summing up our claim, we have the following theorem.

Theorem 1. *Under assumption 1, we can recover d from the public information* (r, N, e) *in polynomial time if* $\delta < \dfrac{7 - 2\sqrt{7}}{3(r+1)}.$

Boneh-Durfee's result is obtained as a special case for $r = 1$.

Remark 5. In our theorems, Assumption 1 is crucial. Under Assumption 1, our theorems are valid. If one does not agree on Assumption 1, our theorems will not be "Theorem" but just "Claim".

In appendix B, we will show analysis for arbitrary e.

4 Improved Bound of δ

In this section, we improve the upper bound of δ. Bound of δ is derived from $\det M < e^{mw}$. As $\det M$ is a function of δ, bound of δ can be improved if the value of $\det M$ is reduced. $\det M$ is a product of element on the diagonal, so its value will be reduced if we can eliminate the large values of element on the diagonal. To do this, we eliminate some rows whose element on the diagonal is large. By this elimination, M will not be full-rank matrix, which is hard to evaluate the determinant. This problem is solved by using the "Geometrically Progressive Matrices (GPM)" introduced by Boneh-Durfee [2]. By applying their technique to our problem, we have the following theorem.

Theorem 2. *If* $\delta < \dfrac{2 - \sqrt{2}}{r + 1}$, *we can recover d from the public information in polynomial time.*

Proof. We set $s = (1 - (r+1)\delta)m$. We can prove that our matrix $M^{(u)}$ contains GPM with some parameter. Then we take only those $\{g_{[i,j,k,l]}\}$ of $M^{(u)}$ whose element on the diagonal is less than or equal to e^m. By eliminating rows whose element on the diagonal greater than e^m, we transform $M^{(u)}$ to $M'^{(u)}$ to calculate $\prod_{u=0,1,\ldots,m} det(M'^{(u)}) < e^{mw'}$, where w' is a number of rows of $M'^{(u)}$. By using this transformation, we can prove the improved upper bound of δ.

As the first step for calculating the determinant, we show the structure of $M^{(u)}$ in Figure 1. If we define $M^{(u,0)}, M^{(u,1)}, M^{(u,2)}, M'^{(u,0)}, M'^{(u,1)}$ and $M'^{(u,2)}$ as

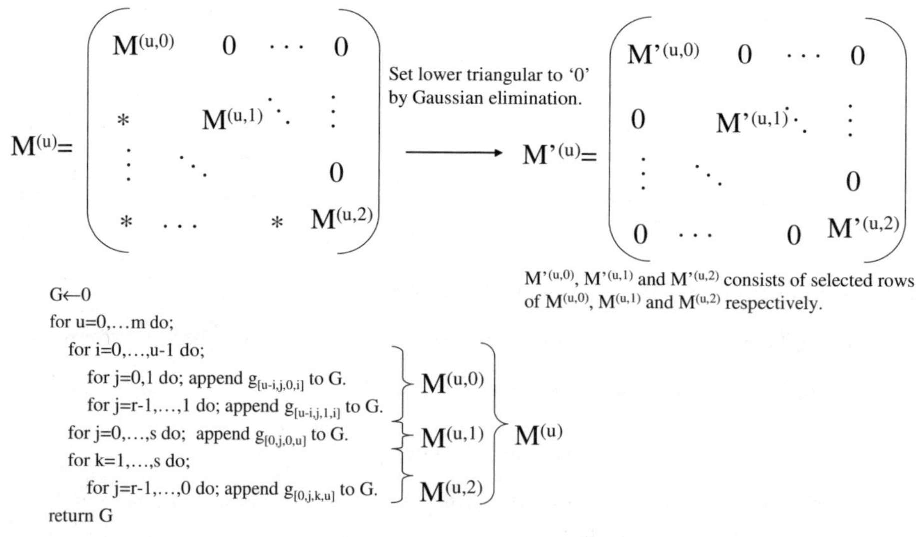

Fig. 1. Structure of $M^{(u)}$

Figure 1, we can see $det(M'^{(u)}) = det(M'^{(u,0)}) \times det(M'^{(u,1)}) \times det(M'^{(u,2)})$. In [10], these matrices are said to satisfy Lemma 2.

Lemma 2 (Kunihiro-Kurosawa [10]). *Coefficient matrix $M^{(u,0)}, M^{(u,1)}$ and $M^{(u,2)}$ for $\{g_{[i,j,k,l]} (xX, yY, zZ)\}$ and $X = e^{\frac{r+1}{2}\delta}, Y = Z = e^{1/2}$ satisfies following a),b) and c) respectively, where $E(M,i,j)$ represents the element of i-th row and j-th column of matrix M.*

a) Sequences of $E(M^{(u,0)}, i, i)$ for $i = 0, 1, \ldots, u \times (r+1) - 1$ are:

$e^{\frac{1}{2}((r+1)\delta u + 2m)}, e^{\frac{1}{2}((r+1)\delta u + 2m+1)}, e^{\frac{1}{2}((r+1)\delta u + 2m+r)}, \ldots, e^{\frac{1}{2}((r+1)\delta u + 2m+2)}, \ldots$

$e^{\frac{1}{2}((r+1)\delta u - k + 2m)}, e^{\frac{1}{2}((r+1)\delta u - k + 2m+1)}, e^{\frac{1}{2}((r+1)\delta u - k + 2m+r)}, \ldots, e^{\frac{1}{2}((r+1)\delta u - k + 2m+2)}, \ldots$

$e^{\frac{1}{2}((r+1)\delta u - u + 1 + 2m)}, e^{\frac{1}{2}((r+1)\delta u - u + 1 + 2m+1)}, e^{\frac{1}{2}((r+1)\delta u - u + 1 + 2m+r)}, \ldots, e^{\frac{1}{2}((r+1)\delta u - u + 1 + 2m+2)}$.

b) Sequences of $E(M^{(u,1)}, i, i)$ for $i = 0, 1, \ldots, s$ are:

$e^{\frac{1}{2}((r+1)\delta u + 2m - u)}, e^{\frac{1}{2}((r+1)\delta u + 2m - u + 1)}, \ldots, e^{\frac{1}{2}((r+1)\delta u + 2m - u + s)}$.

And $|E(M^{(u,1)}, i, j)| \leq K_b \times e^{\frac{r+1}{2r}(s+m)+u} \times E(M^{(u,1)}, i, i)$ is satisfied for $i > j$ and $K_b = u^{3u}$.

c) Sequences of $E(M^{(u,2)}, i, i)$ for $i = 0, 1, \ldots, s \times r - 1$ are:

$e^{\frac{1}{2}((r+1)\delta u + 2m - u + r)}, e^{\frac{1}{2}((r+1)\delta u + 2m - u + r - 1)}, \ldots, e^{\frac{1}{2}((r+1)\delta u + 2m - u + 1)}, \ldots$

$e^{\frac{1}{2}((r+1)\delta u + 2m - u + r - 1 + k)}, e^{\frac{1}{2}((r+1)\delta u + 2m - u + r - 2 + k)}, \ldots, e^{\frac{1}{2}((r+1)\delta u + 2m - u + k)}, \ldots$

$e^{\frac{1}{2}((r+1)\delta u + 2m - u + r - 1 + s)}, e^{\frac{1}{2}((r+1)\delta u + 2m - u + r - 2 + s)}, \ldots, e^{\frac{1}{2}((r+1)\delta u + 2m - u + s)}$.

And $|E(M^{(u,2)}, i, j)| \leq K_c \times e^{\frac{r+1}{2r}(m+r-1)+u} \times E(M^{(u,2)}, i, i)$ is satisfied for $i > j$ and $K_c = u^{3u}$.

In Lemma 2, equation of $E(i,i)$ is trivial from [10], but relation between $E(i,j)$ and $E(i,i)$ for $i > j$ is not trivial. We describe why this relationship holds in Appendix D.

We can see sequences of the exponents of $E(M^{(u,0)}, i, i)$ contains irregular pattern (i.e. $(r+1)\delta u - k + 2m, (r+1)\delta u - k + 2m + 1, (r+1)\delta u - k + 2m + r, (r+1)\delta u - k + 2m + r - 1, \ldots, (r+1)\delta u - k + 2m + 2)$, but that of $E(M^{(u,1)}, i, i)$ and $E(M^{(u,2)}, i, i)$ are regular pattern. So $M^{(u,1)}$ and $M^{(u,2)}$ satisfy the condition of GPM in Definition 1. We denote $M'^{(u,1)}$ and $M'^{(u,2)}$ as the matrix obtained by eliminating the rows of $M^{(u,1)}$ and $M^{(u,2)}$ whose element on the diagonal is greater than e^m. Determinant of $M'^{(u,1)}$ and $M'^{(u,2)}$ can be calculated by Theorem 3 proposed by Boneh-Durfee [2]. Hence we can calculate the determinant as $\prod_{u=0,1,\ldots,m} det(M'^{(u)}) = \prod_{u=0,1,\ldots,m}\{det(M^{(u,0)}) \times det(M'^{(u,1)}) \times det(M'^{(u,2)})\}$ to obtain the improved bound of δ.

Definition 1 (Geometrically Progressive Matrices (GPM)). *Let M be an $(a+1)b \times (a+1)b$ matrix, where rows and columns are divided into $a+1$ blocks and each block is $b \times b$ matrix. And $M(i, j, k, l)$ denotes the element of $(ib+j)$-th row and $(kb+l)$-th column of M. Let $C, D, c_0, c_1, c_2, c_3, c_4$ and β be real numbers with $C, D, \beta \geq 1$. A matrix M is said to be geometrically progressive matrices (GPM) with parameters $(C, D, c_0, c_1, c_2, c_3, c_4, \beta)$ if the following conditions i), ii), iii) and iv) hold for all $i, k = 0, \ldots, a$ and $j, l = 1, \ldots, b$.*

i) $|M(i, j, k, l)| \leq C \times D^{c_0 + c_1 i + c_2 j + c_3 k + c_4 l}$.

ii) $M(k, l, k, l) = D^{c_0 + c_1 i + c_2 j + c_3 k + c_4 l}$.

iii) $M(i, j, k, l) = 0$ *whenever* $i > k$ *or* $j > l$.
iv) $\beta c_1 + c_3 \geq 0$ *and* $\beta c_2 + c_4 \geq 0$.

Theorem 3. *Let M be an $(a+1)b \times (a+1)b$ GPM with parameters $(C, D, c_0, c_1, c_2, c_3, c_4, \beta)$ satisfying $C, D, \beta \geq 1$ and let B be real number. Define $S_B = \{(k, l) \in \{0, \ldots, a\} \times \{1, \ldots, b\} | M(k, l, k, l) \leq B\}$, and set $w' = |S_B|$. If L is a lattice defined by the rows $(k, l) \in S_B$ of M, then $\det(L) \leq ((a+1)b)^{w'/2}(1 + C)^{w'^2} \prod_{(k,l) \in S_B} M(k, l, k, l)$*

Lemma 3. *$M^{(u,1)}$ is $(s+1) \times (s+1)$ GPM with parameters $(u^{3u}, e, \frac{r+1}{2}\delta u + m - \frac{u}{2}, -(\frac{r+1}{2r}(m+s) + u), 0, \frac{r+1}{2r}(m+s) + u + 1, 0, 1)$.*

Lemma 4. *$M^{(u,2)}$ is $sr \times sr$ GPM with parameters $(u^{3u}, e, \frac{r+1}{2}\delta u + m - \frac{u}{2}, -(\frac{r+1}{2r}(m+r-1) + u)r, -(\frac{r+1}{2r}(m+r-1) + u), (\frac{r+1}{2r}(m+r-1) + u)r + 1, \frac{r+1}{2r}(m+r-1) + u - 1, 1 - \frac{2r}{(r+1)(m+r-1)+2ru})$.*

Proof. $M^{(u,1)}$ and $M^{(u,2)}$ satisfy iii) of Definition 1, because these are lower triangular matrix. It is easy to see that parameter $(c_0, c_1, c_2, c_3, c_4, \beta)$ in Lemma 3,4 satisfies iv). We can confirm parameter $(c_0, c_1, c_2, c_3, c_4, \beta)$ satisfies ii) from the sequences of $E(M^{(u,1)}, i, i)$ and $E(M^{(u,2)}, i, i)$ in Lemma 2. And these GPM parameters mean $E(M^{(u,1)}, i - v, i) \leq C \cdot e^{(\frac{r+1}{2r}(m+s)+u)v} \cdot E(M^{(u,1)}, i, i)$ and $E(M^{(u,2)}, i - v, i) \leq C \cdot e^{(\frac{r+1}{2r}(m+r-1)+u)v} \cdot E(M^{(u,2)}, i, i)$ must be satisfied for $v \geq 1$ and $C = u^{3u}$. We can see this conditions are satisfied by the relationship between $E(M^{(u,*)}, i, j)$ and $E(M^{(u,*)}, i, i)$ for $i > j$ in Lemma 2. Note that if m is large enough, β is approximated to 1.

Finally, we prove Theorem 2. Firstly, we calculate $e^{mw'}$. Let w_0, w'_1, w'_2 be a number of rows of $M^{(u,0)}, M'^{(u,1)}$ and $M'^{(u,2)}$ respectively. w'_1 for some u is obtained by counting the number of i satisfying $((r+1)\delta u + 2m - u + i)/2 < m(i = 0, 1, \ldots, s)$, that is, $i < u(1 - (r+1)\delta)$. w'_2 for some u is obtained by counting the i satisfying $((r+1)\delta u + 2m - u + i + v)/2 < m(i = 0, 1, \ldots, s, v = 0, 1, \ldots, r-1)$, that is, $i < u(1 - (r+1)\delta) - v$. By summing these numbers for $u = 0, 1, \ldots, m$, we obtain $w'_1 = (m^2 R)/2 + o(m^2)$ and $(m^2 Rr)/2 + o(m^2)$ where $R = (1 - (r+1)\delta)$. w_0 satisfies $w_0 = (m^2(r+1))/2 + o(m^2)$. From $w' = w_0 + w'_1 + w'_2 = ((r+1)(R+1)m^2)/2 + o(m^2)$, we get

$$e^{mw'} = e^{\frac{(r+1)(R+1)}{2}m^3 + o(m^3)}. \tag{19}$$

Secondly, we calculate $\det(M'^{(u,1)})$ and $\det(M'^{(u,2)})$ by using Theorem 3 in case $B = e^m$. Hence, $w'_1, w'_2, s, r, K_b, K_c$ are functions of δ and u, but not that of e, so that the term other than $\prod_{(k,l) \in S_{e^m}} M'^{(u,1)}(k, l, k, l)$ and $\prod_{(k,l) \in S_{e^m}} M'^{(u,2)}(k, l, k, l)$ are negligible in comparison with e^{m^3}. So we can calculate each determinant by $\prod_{(k,l) \in S_{e^m}} M'^{(u,1)}(k, l, k, l)$ and $\prod_{(k,l) \in S_{e^m}} M'^{(u,2)}(k, l, k, l)$. By using this idea, $\det(M'^{(u,1)})$ is calculated as follows.

$$\log_e(\det(M'^{(u,1)})) = \sum_{u=0}^{m} \sum_{i=0}^{\lfloor uR \rfloor} \frac{1}{2}((r+1)\delta u + 2m - u + i)$$

$$\leq \sum_{u=0}^{m} \sum_{i=0}^{uR} \frac{1}{2}(2m - uR + i)$$

$$= \frac{1}{2} \sum_{u=0}^{m} [2muR - \frac{1}{2}u^2 R^2] + o(m^3)$$

$$= \frac{1}{12}(-R^2 + 6R)m^3 + o(m^3). \tag{20}$$

$\det(M'^{(u,2)})$ is calculated as follows.

$$\log_e(\det(M'^{(u,2)})) = \sum_{u=0}^{m} \sum_{v=0}^{r-1} \sum_{i=0}^{\lfloor uR \rfloor} \frac{1}{2}((r+1)\delta u + 2m - u + v + i)$$

$$\leq \sum_{u=0}^{m} \sum_{v=0}^{r-1} \sum_{i=0}^{uR} \frac{1}{2}(2m - uR + v + i)$$

$$= \frac{1}{2} \sum_{u=0}^{m} \sum_{v=0}^{r-1} [uR(2m - uR + v) + \frac{1}{2}u^2 R^2] + o(m^3)$$

$$= \frac{1}{2} \sum_{u=0}^{m} \sum_{v=0}^{r-1} [2muR - \frac{1}{2}u^2 R^2] + o(m^3)$$

$$= \frac{r}{12}(-R^2 + 6R)m^3 + o(m^3). \tag{21}$$

$\det(M^{(u,0)})$ is calculated as follows.

$$\log_e(\det(M^{(u,0)})) = \sum_{u=0}^{m} \sum_{v=0}^{r} \sum_{i=0}^{u-1} \frac{1}{2}((r+1)\delta u + 2m + v - i)$$

$$= \frac{1}{2} \sum_{u=0}^{m} \sum_{v=0}^{r} [(r+1)\delta u^2 + 2mu + vu - \frac{1}{2}u^2] + o(m^3)$$

$$= \frac{1}{2} \sum_{u=0}^{m} \sum_{v=0}^{r} [2mu + u^2((r+1)\delta u - \frac{1}{2})] + o(m^3)$$

$$= \frac{1}{2} \sum_{u=0}^{m} \sum_{v=0}^{r} [2mu + u^2(\frac{1}{2} - R)] + o(m^3)$$

$$= \frac{r+1}{12}(7 - 2R)m^3 + o(m^3). \tag{22}$$

From (19), (20), (21) and (22), the condition for $\prod_{u=0,1,\ldots,m} \{ det(M^{(u,0)}) \times det(M'^{(u,1)}) \times det(M'^{(u,2)}) \} < e^{mw'}$ holds is equal to that $\frac{r+1}{12}(-R^2 + 4R + 7)m^3 + o(m^3) < \frac{r+1}{12}(6+6R)m^3 + o(m^3)$. It means $\frac{r+1}{12}(R^2 - 2R - 1)m^3 + o(m^3) > 0$, solved as $R > -1 + \sqrt{2}$. From $R = 1 - (r + 1)\delta$, this condition is equal to $1 - (r+1)\delta > -1 + \sqrt{2}$. So we finally obtain $\delta < \frac{2-\sqrt{2}}{r+1}$. Boneh-Durfee's result is obtained as a special case for $r = 1$.

5   Experiments

We performed experiments of our attacks described in section 3. We used Victor Shoup's NTL library [16] for performing LLL algorithm. The results are shown in Table 4. Our experiments were ran PC with Pentium-D 3.4GHz processor under Linux (Fedora core 6). In all cases, we succeeded to solve the resolution of described in section 3.2, and correctly obtained $y = p$ and $z = q$. Our experimental results show δ ("$\frac{r+1}{2}\delta$ (experiment)" column of Table 4) is smaller than that of theoretical upper bound ("$\frac{r+1}{2}\delta$ (upper bound for $m \to \infty$)" column of Table 4) as r gets bigger. This can be explained by following reason: If m goes to infinity, upper bound of $\frac{r+1}{2}\delta$ is 0.284, but if m is not large, upper bound of δ becomes smaller.

Table 4. Experimental Results

r	$log_2 N$	d	$\frac{r+1}{2}\delta$ (experiment)	$\frac{r+1}{2}\delta$ (upper bound) for $m \to \infty$	$\frac{r+1}{2}\delta$ (upper bound) for given m,r	m	s	dimension	running time	success rate
1	1024	281bits	0.275	0.284	0.220	8	2	117	9 hours.	17 out of 17
2	1026	128bits	0.188	0.284	0.172	8	2	171	56 hours.	2 out of 2
3	1024	61bits	0.120	0.284	0.124	8	2	225	70 hours.	2 out of 2

We theoretically analyzed the upper bound of δ for given m and r without ignoring the small term. (For further details, see Appendix C.) And we show the result in "$\frac{r+1}{2}\delta$ (upper bound for given m,r)" column of Table 4. Experimental results of δ can be equal or bigger or than this value, so this value can be used to estimate experiment results for small m.

6   Discussions

We studied small d attack on Takagi's variant of RSA. Our results are natural extension of Boneh-Durfee's. Boneh and Durfee described in their paper [2] "A bound of $d < N^{1-\frac{1}{\sqrt{2}}}$ cannot be the final answer. It is too unnatural. We believe the correct bound is $d < N^{1-\frac{1}{2}}$."

We also try to explain the final answer of our attack by using the idea of the counting argument. Under this idea, d is assumed to be large enough as far as the equation has the unique solution.

Our attack tries to solve the equation $x(y - 1)(z - 1) = 0 \pmod{\text{mode}}$, where $x < e^{\frac{r+1}{2}\delta}$ and $y, z < e^{\frac{1}{2}}$. This equation itself does not have the unique solution, because possible patterns of (x, y, z) is $e^{\frac{r+1}{2}\delta} \times e^{\frac{1}{2}} \times e^{\frac{1}{2}} = e^{1+\frac{r+1}{2}\delta}$, which is larger than the modulo e. But if we also consider the another equation $y^r z = N$, possible patterns of (x, y, z) will be reduced to $e^{\frac{r+1}{2}\delta} \times e^{\frac{1}{2}} = e^{\frac{1}{2}+\frac{r+1}{2}\delta}$ because z

is eliminated. By solving $\frac{1}{2} + \frac{r+1}{2}\delta < 1$, we can see these equations have unique solution for $\delta < \frac{1}{r+1}$. Therefore, $d < N^{\frac{1}{r+1}}$ could be the final answer of our attack.

Acknowledgements

We thank to anonymous referees for their helpful advices, especially for the final answer of d.

References

1. Blömer, J., May, A.: A Tool Kit for Finding Small Roots of Bivariate Polynomials over the Integers. In: Cramer, R.J.F. (ed.) EUROCRYPT 2005. LNCS, vol. 3494, pp. 251–267. Springer, Heidelberg (2005)
2. Boneh, D., Durfee, G.: Cryptanalysis of RSA with private key d less than $N^{0.292}$. IEEE Transactions on Information Theory 46(4), 1339 (2000) (Firstly appeared in Eurocrypt 1999)
3. Boneh, D., Durfee, G., Howgrave-Graham, N.: Factoring $N = p^r q$ for Large r. In: Wiener, M.J. (ed.) CRYPTO 1999. LNCS, vol. 1666, pp. 326–337. Springer, Heidelberg (1999)
4. Coppersmith, D.: Small Solutions to Polynomial Equations, and Low Exponent RSA Vulnerabilities. J. Cryptology 10(4), 233–260 (1997)
5. Coron, J.S., May, A.: Deterministic Polynomial Time Equivalence of Computing the RSA Secret Key and Factoring. Journal of Cryptology 20(1), 39–50 (2004) (IACR ePrint Archive: Report 2004/208 (2004))
6. Nguyên, P.Q., Durfee, G.: Cryptanalysis of the RSA Schemes with Short Secret Exponent from Asiacrypt 99. In: Okamoto, T. (ed.) ASIACRYPT 2000. LNCS, vol. 1976, pp. 14–29. Springer, Heidelberg (2000)
7. Ernst, M., Jochemsz, E., May, A., Weger, B.: Partial Key Exposure Attacks on RSA up to Full Size Exponents. In: Cramer, R.J.F. (ed.) EUROCRYPT 2005. LNCS, vol. 3494, pp. 371–386. Springer, Heidelberg (2005)
8. Howgrave-Graham, N.: Finding Small Roots of Univariate Modular Equations Revisited. In: IMA Int. Conf., pp. 131–142 (1997)
9. Jochemsz, E., May, A.: A Strategy for Finding Roots of Multivariate Polynomials with New Applications in Attacking RSA Variants. In: Lai, X., Chen, K. (eds.) ASIACRYPT 2006. LNCS, vol. 4284, pp. 267–282. Springer, Heidelberg (2006)
10. Kunihiro, N., Kurosawa, K.: Deterministic Polynomial Time Equivalence between Factoring and Key-Recovery Attack on Takagi. In: Okamoto, T., Wang, X. (eds.) PKC 2007. LNCS, vol. 4450, pp. 412–425. Springer, Heidelberg (2007)
11. Lenstra, A.K., Lenstra, H.W., Lovász, L.: Factoring polynomials with rational coefficients. Mathematische Annalen 261, 515–534 (1982)
12. May, A.: Secret Exponent Attacks on RSA-type Schemes with Moduli $N = p^r q$. In: Bao, F., Deng, R., Zhou, J. (eds.) PKC 2004. LNCS, vol. 2947, pp. 218–230. Springer, Heidelberg (2004)
13. Rivest, R., Shamir, A., Adleman, L.: A Method for Obtaining Digital Signatures and Public-Key Cryptosystems. Communications of the ACM 21(2), 120–126 (1978)

14. Takagi, T.: Fast RSA-Type Cryptosystem Modulo $p^k q$. In: Krawczyk, H. (ed.) CRYPTO 1998. LNCS, vol. 1462, pp. 318–326. Springer, Heidelberg (1998)
15. Takagi, T.: A Fast RSA-Type Public-Key Primitive Modulo $p^k q$ Using Hensel Lifting. IEICE Trans. Fundamentals 87(1), 94–101 (2004)
16. Shoup, V.: Number Theory Library (NTL), http://www.shoup.net/ntl/
17. Wiener, M.: Cryptanalysis of Short RSA Secret Exponents. IEEE Transactions on Information Theory 36, 553–558 (1990)

A   Lattice Construction for $h(y, z) = (y - 1)(z - 1)$ and Converting That for $xh(y, z) + 1$

We show the lattice for $h(y, z) = (y - 1)(z - 1) \equiv 0 \pmod{e}$ with constraint: $y^r z = N$. In construction, we need to take a care of Durfee-Nguyen technique. The construction of lattice for this type of bivariate polynomial is given in [10]. For a fixed positive integer u, define

$$h_{[j,k,l]}^{(u)}(y, z) = y^j z^k h(y, z)^l e^{u-l}. \tag{23}$$

The list of polynomials $G^{(u)} = (h_{[j,k,l]}^{(u)})$ as follows, where s and t will be determined later.

> $G \leftarrow \emptyset$.
> for $k = 0, \cdots, u - 1$, do;
>> append $h_{[0,0,k]}$ and $h_{[1,0,k]}^{(u)}$ into G in this order.
>> for $i = r - 1, \cdots, 1$, do; append $h_{[i,1,k]}^{(u)}$ to G.
> for $i = 0, \cdots, s$, do; append $h_{[i,0,u]}^{(u)}$ to G.
> for $j = 1, \cdots, t$, do;
>> for $i = r - 1, \cdots, 0$, do; append $h_{[i,j,u]}^{(u)}$ to G.
> return G.

We can easily verify that the coefficient matrix of $\{h_{[i,j,u]}^{(u)}(yY, zZ)\}$ is lower triangular.

Next, we show a transformation from lattice construction for $h(y, z)$ to that for $f(x, y, z) = xh(y, z) + 1$.

Suppose that the order of polynomials $h_{jkl}^{(u)}(y, z)$ and monomials are given for any u. Suppose that the list of $H_{(u)}$ is given as follows.

$$H_{(u)} = \{[j_1^{(u)}, k_1^{(u)}, l_1^{(u)}], [j_2^{(u)}, k_2^{(u)}, l_2^{(u)}], \cdots, [j_{w_u}^{(u)}, k_{w_u}^{(u)}, l_{w_u}^{(u)}]\},$$

where w_u is the length of list.

Then, we define G for $f(x, y, z)$ as follows.

> $G \leftarrow \emptyset$.
> for $u = 0, \cdots, m$, do;
>> for $i = 1, \cdots, w_u$, do;
>>> Append $[u - l_i^{(u)}, j_i^{(u)}, k_i^{(u)}, l_i^{(u)}]$
> return G.

B   Analysis for Arbitrary e

In this section, we analyze small d attack for arbitrary e. Given a trivariate polynomial $f(x, y, z) = x(y - 1)(z - 1) + 1$, find (x_0, y_0, z_0) satisfying

$$f(x_0, y_0, z_0) = 0 (\mathrm{mode}), \tag{24}$$

where $|x_0| < e^{1 + \frac{\delta}{\alpha} - \frac{2}{\alpha(r+1)}}$, $|y_0|, |z_0| < e^{\frac{1}{\alpha(r+1)}}$. Our aim is to obtain the condition of δ when the solution of the above problem is obtained.

Theorem 4. *If $\delta < \dfrac{7 - 2\sqrt{1 + 3\alpha(r + 1)}}{r + 1}$, we can recover d from the public information in polynomial time.*

Proof. We build the exact same lattice used in Section 3. Plugging

$$X = e^{1 + \frac{\delta}{\alpha} - \frac{2}{\alpha(r+1)}}, Y = Z = e^{\frac{1}{\alpha(r+1)}}, \tag{25}$$

we obtain

$$
\begin{aligned}
\log_e \det M^{(u)} = mw_u &- \frac{u}{2}((r + 1)u - (r - 1) + 2(r + 1)s) \\
&+ (1 + \frac{\delta}{\alpha} - \frac{2}{\alpha(r + 1)})uw_u \\
&+ \frac{1}{2\alpha(r+1)} \big((r+1)(u+s)(u+s+1)+(r^2-r)(u+s)\big).
\end{aligned} \tag{26}
$$

By the similar calculation as Section 3.2, we obtain

$$3s^2 - 3(1 - (r + 1)\delta)sm + (2(r + 1)\delta + \alpha(r + 1) - 3)m^2 < 0.$$

By setting $s = \frac{1-(r+1)\delta}{2}m$, we obtain

$$\big(3((r + 1)\delta)^2 - 14(r + 1)\delta + 15 - 4\alpha(r + 1)\big) < 0.$$

By solving the above inequality, we have

$$\delta < \frac{7 - 2\sqrt{1 + 3\alpha(r + 1)}}{3(r + 1)}. \tag{27}$$

$\square$

Boneh-Durfee's result is obtained as a special case for $r = 1$. Furthermore, for $\alpha = 2/(r + 1)$, we obtain the result of Theorem 1.

When $\alpha = \frac{15}{4(r+1)}$, the bound of δ is $\delta = 0$. Hence, this attack is ineffective when $e > N^{\frac{15}{4(r+1)}}$.

By similar analysis, we obtain the improved bound as follows.

Theorem 5. *If $\delta < \dfrac{2 - \sqrt{\alpha(r + 1)}}{r + 1}$ and $\delta < \alpha$, we can recover d from the public information in polynomial time.*

For $\alpha = 2/(r + 1)$, we obtain the result of Theorem 2.

C   Upper Bound of δ for Small m and r

If m goes to infinity, upper bound of δ satisfies $\frac{r+1}{2}\delta = 0.284$ as shown in the proof of section 3.2. But if m is not so large, $\frac{r+1}{2}\delta$ might be smaller than this value because term $o(m^3)$ can not be ignored. In this section, we analyze the upper bound of δ for smaller m and r by considering the term $o(m^3)$. By equation (12), we can see δ must satisfy

$$\sum_{u=0}^{m} \log_e \det M^{(u)} = \sum_{u=0}^{m} (mw_u - \frac{u}{2}((r+1)u - (r-1) + 2(r+1)s) + \frac{r+1}{2}\delta uw_u$$
$$+ \frac{(u+s)(u+s+1)(r+1)}{4} + \frac{r^2-r}{4}(u+s)) < mw. \quad (28)$$

We know sum of w_u for $u = 0, 1, \ldots, m$ is equal to w, so above relation is euqal to

$$\sum_{u=0}^{m} (-\frac{u}{2}((r+1)u - (r-1) + 2(r+1)s) + \frac{r+1}{2}\delta uw_u$$
$$+ \frac{(u+s)(u+s+1)(r+1)}{4} + \frac{r^2-r}{4}(u+s)) < 0, \quad (29)$$

where this equation is equal to

$$(6r+6)s^2$$
$$+ ((6\delta+6)r^2 + (12m\delta - 6m)r + 6m\delta - 6m + 6)s$$
$$+ ((4m^2 + 2m)\delta + 3m)r^2 + ((8m^2 + 10m)\delta - 2m^2 + 5m)r$$
$$+ (4m^2 + 8m)\delta - 2m^2 - 4m < 0. \quad (30)$$

Left hand side of this equation is minimized when $s = ((-r^2 - 2r - 1)m\delta + (r+1)m - (r^2+1))/(2r+2)$.

By plugging this value and performing tedious computation, above relation leads to

$$-((3r^4 + 12r^3 + 18r^2 + 12r + 3)m^2)\delta^2$$
$$-((-14r^3 - 42r^2 - 42r - 14)m^2 + (6r^4 + 8r^3 - 12r^2 - 24r - 10)m)\delta$$
$$-(7r^2 + 14r + 7)m^2 + (-12r^3 - 22r^2 - 8r + 2)m + 3r^4 + 6r^2 + 3 < 0. \quad (31)$$

If we plug m and r for some given value, we can solve this relation for δ with $\frac{r+1}{2}\delta < 0.284$ to obtain the exact value of δ. If $m = 8$, we get $\delta < 0.220$ for $r = 1$, $\delta < 0.115$ for $r = 2$ and $\delta < 0.062$ for $r = 3$, which lead to "$\frac{r+1}{2}\delta$ (upper bound for given m, r)" column of *Table* 4.

D   Complements of Lemma 2

In Lemma 2, we showed the relation between $E(M^{(u,*)}, \mu, \lambda)$ and $E(M^{(u,*)}, \mu, \mu)$ for $\mu > \lambda$, $M^{(u,1)}$ and $M^{(u,2)}$. These are

$$|E(M^{(u,1)}, \mu, \lambda)| \leq u^{3u} \times e^{\frac{r+1}{2r}(s+m)+u} \times E(M^{(u,1)}, \mu, \mu) \quad (32)$$

and

$$|E(M^{(u,2)}, \mu, \lambda)| \leq u^{3u} \times e^{\frac{r+1}{2r}(m+r-1)+u} \times E(M^{(u,2)}, \mu, \mu). \tag{33}$$

In this section, we explain why these relations hold.

For some $g_{[i,j,k,l]}(xX, yY, zZ)$, $E(M^{(u,*)}, \mu, \mu)$ is diagonal element and $E(M^{(u,*)}, \mu, \lambda)$ can be any non-diagonal element. So we prove (32) and (33) for maximum absolute value of $E(M^{(u,*)}, \mu, \lambda)$, this relation holds for any non-diagonal element.

By the definition of $g_{[i,j,k,l]}(x, y, z)$, we get

$$g_{[i,j,k,l]}(x, y, z) = e^{m-l}x^i y^j z^k \sum_{v_0=0}^{l} \binom{l}{v_0} x^{l-v_0}(y-1)^{v_0}(z-1)^{v_0}$$

$$= e^{m-l}x^i y^j z^k \sum_{v_0=0}^{l} \binom{l}{v_0} x^{l-v_0}(y-1)^{v_0} \left(\sum_{v_2=0}^{v_0} \binom{v_0}{v_2} z^{v_2}(-1)^{v_2-v_0} \right)$$

$$= e^{m-l}x^i y^j z^k \sum_{v_0=0}^{l} \binom{l}{v_0} x^{l-v_0} \left(\sum_{v_1=0}^{v_0} \binom{v_0}{v_1} y^{v_1}(-1)^{v_0-v_1} \right.$$

$$\left. \left(\sum_{v_2=0}^{v_0} \binom{v_0}{v_2} z^{v_2}(-1)^{v_0-v_2} \right) \right)$$

$$= e^{m-l} \sum_{v_0=0}^{l} \sum_{v_1=0}^{v_0} \sum_{v_2=0}^{v_0} \binom{l}{v_0}\binom{v_0}{v_1}\binom{v_0}{v_2}$$

$$x^{l-v_0+i} y^{v_1+j} z^{v_2+k}(-1)^{v_0-v_1}(-1)^{v_0-v_2}$$

So $g_{[i,j,k,l]}(xX, yY, zZ)$ is

$$g_{[i,j,k,l]}(xX, yY, zZ)$$

$$= e^{m-l} \sum_{v_0=0}^{l} \sum_{v_1=0}^{v_0} \sum_{v_2=0}^{v_0} \binom{l}{v_0}\binom{v_0}{v_1}\binom{v_0}{v_2}$$

$$X^{l-v_0+i} Y^{v_1+j} Z^{v_2+k} x^{l-v_0+i} y^{v_1+j} z^{v_2+k}(-1)^{v_0-v_1}(-1)^{v_0-v_2}.$$

If we consider $0 \leq v_1, v_2 \leq v_0 \leq l$, $i = 0$ and $l = u$ holds for $M^{(u,1)}$ and $M^{(u,2)}$, maximum absolute value of the coefficient is

$$u^{3u} \cdot e^{m-u} X^u Y^{j+u} Z^{k+u}.$$

By plugging $X = e^{\frac{r+1}{2}\delta}$ and $Y = Z = e^{\frac{1}{2}}$, we get

$$u^{3u} \cdot e^{\frac{1}{2}((r+1)\delta u+2m+j+k)}. \tag{34}$$

Above (34) represents the maximum coefficients without replacing $y^r z = N$. If this replacement occurs, (34) becomes $N^{\theta/r}$ times larger, where θ represents the maximum order of y. θ is maximum value of $j + u$. And if we consider j is ranged by $0 \leq j \leq s$ for $M^{(u,1)}$ and $0 \leq j \leq r - 1$ for $M^{(u,2)}$, $\theta = u + s$ holds for $M^{(u,1)}$, and $\theta = u + r - 1$ holds for $M^{(u,1)}$. Hence, maximum coefficients with replacing $y^r z = N = e^{\frac{r+1}{2}}$ is $u^{3u} \cdot e^{\frac{r+1}{2r}(u+s)} \cdot e^{\frac{1}{2}((r+1)\delta u + 2m + j + k)}$ for $M^{(u,1)}$ and $u^{3u} \cdot e^{\frac{r+1}{2r}(u+r-1)} \cdot e^{\frac{1}{2}((r+1)\delta u + 2m + j + k)}$ for $M^{(u,2)}$. Hence we get

$$|E(M^{(u,1)}, \mu, \lambda)| \leq u^{3u} \cdot e^{\frac{r+1}{2r}(u+s)} \cdot e^{\frac{1}{2}((r+1)\delta u + 2m + j + k)}$$

and

$$|E(M^{(u,2)}, \mu, \lambda)| \leq u^{3u} \cdot e^{\frac{r+1}{2r}(u+r-1)} \cdot e^{\frac{1}{2}((r+1)\delta u + 2m + j + k)}.$$

If we consider $E(M^{(u,*)}, \mu, \mu) = e^{\frac{1}{2}((r+1)\delta u + 2m + j + k - u)}$ holds for any j, k of $M^{(u,1)}$ and $M^{(u,2)}$, we finally get

$$|E(M^{(u,1)}, \mu, \lambda)| \leq u^{3u} e^{\frac{r+1}{2r}(u+s)+u} E(M^{(u,1)}, \mu, \mu)$$
$$\leq u^{3u} \cdot e^{\frac{r+1}{2r}(m+s)+u} \cdot E(M^{(u,1)}, \mu, \mu) \qquad (35)$$

and

$$|E(M^{(u,2)}, \mu, \lambda)| \leq u^{3u} e^{\frac{r+1}{2r}(u+r-1)+u} E(M^{(u,2)}, \mu, \mu)$$
$$\leq u^{3u} \cdot e^{\frac{r+1}{2r}(m+r-1)+u} \cdot E(M^{(u,2)}, \mu, \mu). \qquad (36)$$

Hence (32) and (33) are proved.

Super-Efficient Verification of Dynamic Outsourced Databases[*]

Michael T. Goodrich[1], Roberto Tamassia[2], and Nikos Triandopoulos[3]

[1] Dept. of Computer Science, UC Irvine, USA
goodrich@ics.uci.edu
[2] Dept. of Computer Science, Brown University, USA
rt@cs.brown.edu
[3] Dept. of Computer Science, University of Aarhus, Denmark
nikos@daimi.au.dk

Abstract. We develop new algorithmic and cryptographic techniques for authenticating the results of queries over databases that are outsourced to an untrusted responder. We depart from previous approaches by considering *super-efficient* answer verification, where answers to queries are validated in time asymptotically less that the time spent to produce them and using lightweight cryptographic operations. We achieve this property by adopting the decoupling of query answering and answer verification in a way designed for queries related to range search. Our techniques allow for efficient updates of the database and protect against replay attacks performed by the responder. One such technique uses an off-line audit mechanism: the data source and the user keep digests of the sequence of operations, yet are able to jointly audit the responder to determine if a replay attack has occurred since the last audit.

1 Introduction

Large databases are increasingly being outsourced to untrusted third parties (responders) and without some kind of verification mechanisms, users cannot trust the answers to queries. Thus, an important component of any outsourced database system is the security of its answer-verification process. Moreover, database outsourcing is typically realized for efficiency purposes in a distributed setting where clients are machines that have low computational power running applications that demand authentic responses of dynamic data at high rates. In this context, the cryptographic protocols for trustworthy answer verification

[*] Research supported in part by the U.S. National Science Foundation under grants IIS–0713403, IIS–0713046, CNS–0312760 and OCI–0724806, the Institute for Information Infrastructure Protection under an award from the Science and Technology Directorate at the U.S. Department of Homeland Security, and the Center for Algorithmic Game Theory at the University of Aarhus under an award from the Carlsberg Foundation. The views in this paper do not necessarily reflect the views of the sponsors.

T. Malkin (Ed.): CT-RSA 2008, LNCS 4964, pp. 407–424, 2008.

should incur small communication and computational overheads that ideally depend only on the answer size.

This paper studies protocols for authenticating the integrity of outsourced databases in ways that achieve high security and efficiency levels. Most database queries boil down to *one-dimensional range search* queries—asking to report those records having values of a certain field within a given interval—and most existing techniques for authenticating such queries have $O(\log n + t)$ communication and computational costs, where n is the total number of records in the database and t is the number of returned records. Instead, our goal is to design cryptographic techniques that allow *super-efficient* answer verification, that is, allow authentication of range search queries with only $O(t)$ associated costs, even when t is $o(\log n)$. Furthermore, we wish our protocols to involve lightweight cryptographic operations with, ideally, only $O(1)$ modular exponentiations performed during the answer-verification process.

Additionally, we seek authentication solutions that perform well even if the database evolves frequently over time. The main challenge in this context is that a malicious responder may perform a *replay attack*, i.e., provide verifiable (e.g., signed) but stale or currently invalid information (e.g., that was originated from the owner long in the past) to a client. But here is exactly where super-efficiency can hurt us, since we want to avoid a verification method that requires more than $O(t)$ work on the part of the client, and we want to avoid requiring the data owner to process (e.g., re-sign) all the records of the database with each update. Ideally, we would like a dynamic system that is super-efficient for the client, and immune to replay attacks launched by the responder, and that can process updates efficiently for the data owner and the responder.

Super-efficient verification is a theoretically interesting concept, since it advances the design of data authentication protocols by exploring the possibility of removing unnecessary computations at the verifier. But it is also a practically important property in database systems, since it provides trustworthy functionality in dynamic and highly distributed data dissemination models, where small mobile and computationally limited devices query continuously and at high rates data that is outsourced to untrusted, geographically dispersed, proxy machines.

Related Work. Extensive work exists on *authenticated data structures* [19, 25], which model secure data querying in adversarial environments, where data created by a trusted source becomes available to users through queries after it is replicated to an untrusted remote server. The general approach is to augment the data structures used by the source and the responder to support authentication protocols such that, along with an answer to a query, a cryptographic proof is provided to the user by the server that can be used to verify the authenticity of the answer. Research has mostly focused on hash-based authentication protocols, where extensions of Merkle's *hash tree* [16] are used for authenticating membership queries (e.g., [6, 11, 19, 26, 27]) or more general query types, such as basic operations on relational databases [9], pattern matching and orthogonal range searching [15], graph connectivity and geometric searching [13], XML queries [4, 8], and two-dimensional grid searching [1]. Many of these queries essentially boil down to one-dimensional range search queries. General authentication techniques have been also proposed for certain query classes, including read-write operations on memory cells [5], queries on static data that are modeled as search DAGs [15], and decomposable queries over sequences and iterative searches over catalogs [13]. These schemes are not super-efficient as they involve answer proofs and verification times that asymptotically equal the complexity of answering queries. In [26], for hash-based authentication of set-membership queries, it is showed that for a set of size n, all costs related to authentication are at least logarithmic in n in the worst case. Related work on consistency and privacy of committed databases appears in [6, 17, 22]. Authenticated dictionaries in the two-party model, where the source keeps minimal state to check the integrity of its outsourced data, appear in [10, 24]. Finally, in [12] it is showed how to use the RSA accumulator [7] to realize a dynamic authenticated dictionary that achieves constant (thus super-efficient) verification costs at the client.

There has also been a growing body of work on authenticating queries in outsourced databases. The model is essentially the one of authenticated data structures, but now the data sets are relational databases residing in external memory and are queries through SQL queries which are founded on one-dimensional range search. In [9, 13] range queries are supported with $O(\log n + t)$ authentication costs. In [21], cryptographic hashing and accumulators are used in the first hash-based super-efficient, but static, verification scheme that achieves $O(\log t)$ communication cost and $O(t)$ verification cost, whereas in [23], static hash trees, where each tree node is individually signed, are used to authenticate range queries, incurring cost of $O(t)$ signature verifications. In [20] signature aggregation is used to accelerate the verification of the (individually signed) answer records. Both schemes achieve super-efficiency, but not coupled with both efficient updates and replay-attack safety. Finally, in [14] authentication techniques based on B-trees and aggregated signatures are studied experimentally.

Table 1. A summary of how our results are qualitatively compared with existing work

	[5, 13]	[21, 18]	[20]	**this work**
super-efficient		•	•	•
dynamic	•		•	•
replay safe	•	n.a.		•

Our Contributions. We provide the first super-efficient authentication techniques for one-dimensional range searching (or queries based on it), that are both *dynamic* and *replay safe*. Our schemes can support fast query time for the untrusted responder, super-efficient verification for clients, and fast update time for the data source. Our main technique for achieving these properties involves the use of an optimal authentication structure (employed separately by the source and the responder) that divides a hash tree in a recursive fashion so that it has $O(\log^* n)$ "special" levels (i.e., a number proportional to the inverse of the tower-of-twos function). The database owner needs to authenticate only the hash values of tree nodes that lie on the special levels, which significantly speeds up data updates while also simplifying the means to achieve super-efficiency. Indeed, for all practical applications, there are only a constant number of special levels in our scheme. Table 1 summarizes the comparison of our work with the best existing methods for authentication of range searching in outsourced data.

To avoid the possibility of replay attacks, we provide two possible solutions. One solution involves the use of an RSA accumulator to allow clients to verify a single secure aggregation to check that the signed responses to a query are still valid even if some individual signatures are possibly quite old. We use a source-responder work trade-off to perform updates in $O(\sqrt{n})$ time, which is efficient for moderately large values of n. Our second solution provides a different trade-off, between the update cost at the source and responder and the immediacy in detecting a replay attack. We show how to build an off-line *auditing* mechanism to detect, and thereby deter, replay attacks through periodic audits of the responder. The key contribution here is that the auditor mechanism, based on an off-line memory-checking test introduced in [5], is implemented jointly but non-interactively by the source and the user and needs only store and process a constant-sized digest to check the responder (so that auditing is also a super-efficient computation), and that the responder cannot employ a replay attack without being caught by the auditing mechanism.

Section 2 describes our authentication model. Section 3 describes our approach for verifying answers to range queries by decoupling answer verification from query answering, and presents our core authentication structure designed to optimally support super-efficient verification. Section 4 describes a dynamic extension of our scheme that provides a trade-off in update and query costs, and Section 5 presents an augmentation of our scheme that realizes an efficient off-line auditing mechanism. Section 6 discusses extensions to support verification of other query types that are related to range searching, and also our final concluding remarks. Focused on one-dimensional range search and due to lack of space, this extended abstract omits some details of our design and proof techniques.

2   Authentication Model

We examine data authentication in the setting commonly used in today's Internet reality, where a database becomes available for queries at an intermediate entity that is distinct from the data owner and untrusted by the end user. In particular, we consider the following three-party data querying and authentication model. A data *source* S creates (and owns) a *dynamic data set* D, which evolves through update operations, and maintains an *authentication structure* for D, appropriately designed for a specific query type. Data set D is stored by a *responder* R who maintains the same authentication structure for D and answers queries issued by a *user* U. Along with an answer a to a query q, R provides U with a cryptographic proof p that is computed using the authentication structure of D; p is used by a verification process ran by U to check the validity of answer a subject to query q. On any update for D issued by the source, D and the authentication structure are appropriately updated by S and R.

The merits of this query model include scalability, decentralization and load-balance: by outsourcing D, S minimizes its operational costs by processing only data updates (e.g., it minimizes the time being on-line) and heavy query traffics of an unlimited population of users can be securely handled by one (or more) untrusted responders (e.g., proxy servers), *without* the need of creating or updating any trust relations, or installing any secure component at the server.

In this model, our goal is to design an authentication structure that allows trustworthy answer verification, that is, to check that the answer is as accurate as it would have been, had the answer come directly from S. To achieve this, we use the following general approach. Using a PKI, we assume that $\mathcal{U}$ knows the public key of S. The corresponding secret key is used by S in combination with some cryptographic primitives to produce one (or more) *authentication strings* (or digests) for data set D, which constitute short descriptions of D that capture structural information related to the type of queries of interest. Given any query q, $\mathcal{R}$ uses its authentication structure to produce a proof p for the answer a of q. On input a query-answer pair (q, a), a proof p, and the public key of S, $\mathcal{U}$ runs a verification algorithm that either accepts a as valid or rejects it as invalid: p securely relates a, q to (some of) the authentication string(s), which are authenticated by S using a signature scheme. We call the set of authentication and communication protocols and verification process, an *authentication scheme.*

We now describe the security requirement that any authentication scheme must satisfy. Security is captured as two individual requirements, modeling the desired property: for all queries the verification process should be trustworthy, accepting an answer-proof pair if and only if the returned answer is the correct answer to the query. First, we require *completeness*, which ensures that for any query the authentication structure generates a correct corresponding answer-proof pair that the verification algorithm accepts. Second, we require *soundness*, which ensures that if, given a query q, an answer-proof pair (a, p) is accepted by the verification algorithm, then a is the correct answer to q. With respect to this requirement, we assume the following threat model. The user $\mathcal{U}$ trusts only the source S, not the responder $\mathcal{R}$ which is modeled as an entity that is controlled by an adversary [1]. $\mathcal{R}$ can maliciously try to cheat, by providing an incorrect answer to a query and forging a false proof for this answer. Accordingly, the soundness requirement dictates that given any query issued by $\mathcal{U}$, no polynomial-time responder $\mathcal{R}$, having oracle access to the algorithm that the source runs to generate the authentication strings,[2] can come up with an answer-proof pair, such that the answer is incorrect, yet the verification algorithm accepts the answer as authentic. This definition implies safety against replay attacks.

In this work, we are interested in secure authentication schemes for verifying the results of range search queries that introduce low computational and communication overhead to the involved parties. In particular, we seek for

---

[1] We do not consider denial-of-service attacks but assume that $\mathcal{R}$ always participates in the communication protocol and interacts with S and $\mathcal{U}$.

[2] That is, $\mathcal{R}$ observes the authentication strings of D that are produced by S over time or selectively query for the authentication strings of specially chosen data sets.

authentication schemes that primarily incur low verification time, called *verification cost*. Other important secondary cost parameters are the *update cost* (for updating the authentication structure at $\mathcal{S}$ and $\mathcal{R}$ after updates) and the *query cost* (for producing the answer-proof pairs at $\mathcal{R}$ after queries), as well as the proof size. In the interest of super-efficient verification, we wish to design schemes that allow very fast answer verification, in time asymptotically less than the time needed for answer generation and tolerate reasonable trade-offs in the update and query costs or the update costs and the immediacy of replay-attack detections.

In our authentication schemes, we use standard cryptographic tools, such as collision-resistant hash functions and digital signatures, and the *dynamic RSA accumulator* [2, 3, 7]. Given a set X of size n, an accumulator can be used to incrementally and order-independently (through bivariate function $f(\cdot, \cdot)$) compute a constant-size *accumulation value* $A(X)$, with respect to which there exist (*i*) constant-size *witnesses* for all accumulated elements in X, and (*ii*) a constant-time computationally secure *verification test* that accepts witnesses of only elements existing in X. The RSA accumulator results by setting $f(a, x) = a^x$ mod N as the result of accumulating new element x in the current accumulation value a, where x is a prime number in the appropriate range and N is an RSA modulo, thus, $A(X) = a_0^{\prod_{x \in X} x \mod \phi(N)}$, where $\phi(\cdot)$ is Euler's function and a_0 an initial (public) value. Membership of x with witness w in set X is tested as $w^x = A(X)$, which is a secure test: under the strong RSA assumption [3, 7], it is computationally infeasible to find items that are not accumulated in the set and corresponding fake witnesses that pass the test. In [12], it is showed how to use this primitive in our three-party authentication model for optimally verifying set membership, that is, how to update elements' witnesses *without* the trapdoor information $\phi(N)$, using $O(\sqrt{n})$ modular operations and multiplications.

3   A New Super-Efficient Authentication Structure

In this section, we present a new authentication structure that allows super-efficient answer verification of one-dimensional range search queries and that is the key component of the authentication schemes presented in the next sections.

Let $D \triangleq \{(k_1, v_1), \ldots, (k_n, v_n)\}$ be a set of n key-value pairs (k, v), where each key k is a distinct element of a totally ordered universe K, where, for simplicity and without loss of generality, $k_1 < \ldots < k_n$ and $n = 2^d$. A one-dimensional *range search* query $q = [q_L, q_R]$ on D is an interval with $q_L, q_R \in K \cup \{-\infty, +\infty\}$, and maps to answer $A_q \triangleq \{(k, v) \in D : q_L \leq k \leq q_R\}$, the subset of D consisting of all pairs whose keys are in $[q_L, q_R]$. We assume that answer A_q can be computed by the responder $\mathcal{R}$ in $O(\log n + t)$ time, using some optimal technique (e.g., searching in a balanced range tree), where $n = |D|$ and $t = |A_q|$.

Our approach for achieving super-efficient verification of range searching is to decouple the authentication structure from the search data structure in order to authenticate a collection of certain relations defined over D. Let the *successor relation* $\sigma(X)$, defined over a totally ordered set X with n elements, be the set of size $n + 1$ that consists of all ordered pairs of consecutive elements in X, augmented with pairs $(-\infty, x_1)$ and $(x_n, +\infty)$, where x_1 and x_n are the smallest and largest elements of X, respectively (e.g., $\sigma(\{1, 5, 2\}) = \{(-\infty, 1)(1, 2)(2, 5)(5, +\infty)\}$). The successor relation of the keys of D is the essential information for verifying answers of range search queries on D.

Fact 1. *Let* $q = (q_L, q_R)$ *be a range search query on set* D *of key-value pairs,* K_D *be the set of keys in* D *and* $A_q = \{(k_{i_1}, v_{i_1}), \ldots, (k_{i_t}, v_{i_t})\}$, $k_{i_1} < \ldots < k_{i_t}$, *be a set of key-value pairs. Then* $A = A_q$ *if and only if there exist keys* $k_{i_0}, k_{i_{t+1}} \in K_D$ *such that: (1)* $\{(k_{i_0}, k_{i_1}), (k_{i_1}, k_{i_2}), \ldots, (k_{i_{t-1}}, k_{i_t}), (k_{i_t}, k_{i_{t+1}})\} \subseteq \sigma(K_D)$ *and* $A \subseteq D$; *and (2)* $k_{i_0} < q_L \leq k_{i_1}$ *and* $k_{i_t} \leq q_R < k_{i_{t+1}}$.

Indeed, keys $k_{i_0}, k_{i_{t+1}}$ correspond to the boundaries of the range interval, each one possibly coinciding with fictitious keys $-\infty$ or $+\infty$, with $(k_{i_0}, k_{i_{t+1}}) \in \sigma(K_D)$ if $A_q = \emptyset$. The first condition guarantees that the answer A consists of t consecutive key-value pairs of data set D, whereas the second that the query range is exactly covered by the answer range. Thus, in our formulation, answer correctness for range searching captures both inclusiveness (all returned pairs are in the query range) and completeness (all pairs in the query range are returned).

It follows that, if $A_q = \{(k_{i_1}, v_{i_1}), \ldots, (k_{i_t}, v_{i_t})\}$, $k_{i_1} < \ldots < k_{i_t}$, is the correct answer to query q, A_q can be authenticated by verifying (i) t pairs of the key-value relation, namely that $(k_{i_j}, v_{i_j}) \in D$, $1 \leq j \leq t$, (ii) $t + 1$ pairs of the successor relation on the keys, namely that $(k_{i_j}, k_{i_{j+1}}) \in K_D$, $0 \leq j \leq t$, where $k_{i_0} = -\infty$ if $k_{i_1} = k_1$, or $k_{i_0} = k_{i_1 - 1}$ otherwise, and, similarly, $k_{i_{t+1}} = +\infty$ if $k_{i_t} = k_n$, or $k_{i_{t+1}} = k_{i_t + 1}$ otherwise, and, finally, (iii) $t + 4$ inequalities (i.e., the ordering of these pairs and that $k_{i_0} < q_L \leq k_{i_1}$, $k_{i_t} \leq q_R < k_{i_{t+1}}$). Assuming uniquely defined representations for the key-value and successor relations, we denote by $\theta(q)$ the resulting set of $2t + 1$ pairs to be verified, i.e., $\theta(q) \triangleq \{(k_{i_1}, v_{i_1}), \ldots, (k_{i_t}, v_{i_t})\} \cup \{(k_{i_0}, k_{i_1}), \ldots, (k_{i_t}, k_{i_{t+1}})\}$.

By Fact 1, we have that the problem of authenticating any range search query q on a set D of size n is reduced to the problem of authenticating the membership of the relations of set $\theta(q)$ (of size $O(|A_q|) = O(t)$) in $D \cup \sigma(K_D)$ (the union of the key-value and successor relations defined by D, a set of size $O(n)$). We use this property to decouple the answer verification from the answer generation by designing an authentication structure that for any query q provides super-efficient verification of the corresponding special relations $\theta(q)$ over D. Our construction securely and compactly encodes and authenticates these special relations by associating, in a cryptographically sound manner, the answer A_q, a corresponding proof p and, overall, the relations in $\theta(q)$, with one or more authentication strings that are signed by the source. This structure is used both by $\mathcal{S}$, for computing and signing the authentication strings, and by $\mathcal{R}$, for producing the proofs that will allow $\mathcal{U}$ to verify the answer to queries.

Authentication Structure. Let D be the data set as before. Our authentication structure uses a hash tree built over D that essentially encodes the relations D and $\sigma(K_D)$. In particular, let h be a collision-resistant hash function. We build a balanced hash tree T of depth d, storing at the leaves from left to right the hash values $h_1, \ldots, h_n$ defined as follows, where $\|$ denotes string concatenation:

- $h_i \triangleq h(h(k_i) \| h(v_i) \| h(k_{i+1}))$, $i = 2, \ldots, n-1$, and
- $h_1 \triangleq h(h(-\infty) \| h(k_1) \| h(v_1) \| h(k_2))$, $h_n \triangleq h(h(k_n) \| h(v_n) \| h(+\infty))$.

Thus, the hash values at the leaves encode information about various relations: for $2 \le i \le n-1$, h_i is the digest of the key-value relation (k_i, v_i) and successor relation (k_i, k_{i+1}), h_1 is the digest of relations (k_1, v_1), $(-\infty, k_1)$ and (k_1, k_2), and h_n is the digest of relations (k_n, v_n), $(k_n, +\infty)$. Internal nodes in T store the hash of the concatenation of the hash values stored at their children. So, any node v in T stores a hash value h_v that is the digest of the key-value and successor relations R_v that are associated with the leaves of the subtree T_v of T rooted at v. For instance, a hash value stored at the parent u of two sibling leaf nodes j and $j+1$ is the digest of the set of relations $R_u = \{(k_j, v_j), (k_{j+1}, v_{j+1}), (k_j, k_{j+1}), (k_{j+1}, k_{j+2})\}$, whereas the hash value h_r of the root r of T is the digest of all relations $R_r = \sigma(D) \cup D$ defined in T.

As we know, in order to authenticate answer A_q, it suffices to authenticate set $\theta(q)$; consequently, due to the collision resistance property of function h and the fact that h_v is the digest of relations R_v associated with the leaves of tree T_v, it suffices to (1) authenticate any set $S_q = \{h_{v_1}, \ldots, h_{v_m}\}$ of hash values stored at tree nodes $v_1, \ldots, v_m$ such that the set of relations $R(S_q) = R_{v_1} \cup \ldots \cup R_{v_m}$ strictly contain set $\theta(q)$, and (2) provide, as proof p, the collection of hash values and relations that associates A_q with S_q in T. Indeed, S_q contains digests that serve as a cryptographic commitment of $\theta(q)$ (computational binding by the collision resistance of h), thus when, given the answer A_q and the proof p, the authenticated hashes in S_q can be recomputed, then one can be assured—subject to the underlying security assumptions of the cryptographic primitives—that A_q is correct, simply by checking the validity of A_q with the test of Fact 1. Set S_q is not uniquely defined but corresponds to a specific query q. Our goal is to define a fixed collection of *special hash values* S such that any query q can be super-efficiently verified by authenticating membership of set $S_q \subset S$ in S. In the simplest case, S can authenticate S by separately signing its hash values; A_q is verified at hashing and signing costs proportional to $|R(S_q)|$ and $|S_q|$ respectively.

Super-efficient Verification. An efficient approach is to set $S = h_r$, i.e., to use as special hash value for all queries the root hash h_r. Then, for any query q, A_q $|A_q| = t$, can be efficiently associated with h_r, by considering as proof the $O(\log t)$ subtrees of total size $O(t)$ that *exactly cover* the leaves in T that the relations in $\theta(q)$ are associated with, along with the paths connecting these subtrees to r through $O(\log n)$ other tree nodes. The total verification cost is $O(\log n + t)$, which is not super-efficient whenever $t = o(\log n)$ (e.g., $t = O(\log \log n)$ or t is constant). We improve the verification cost as follows.

Suppose that we only query for answers of size $t < \log n$ (see Figure 1). We define the set S_1 of special hash values to contain the hashes $h_1^1, \ldots, h_{m_1}^1$, $m_1 = n/\log n$, at level $\ell_1 = \log \log n$ of the hash tree. It is easy to see that any answer of size t is covered by the subtrees of at most two nodes at level

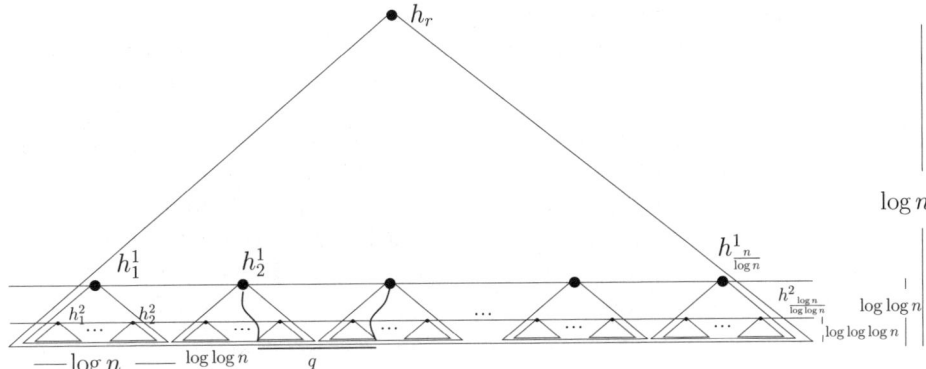

Fig. 1. Our new authentication structure. The set S of special hash values in the tree is defined recursively and consists of $\Theta(n)$ values residing at $\log^* n$ levels: h_r at level $\log n$, $\{h_1^1, ...\}$ at level $\log \log n$, $\{h_1^2, ...\}$ at level $\log \log \log n$, etc. Super-efficient verification is achieved: answer A_q of size at most $t = \log \log n$ to query q is verified by hashing along $O(\log t)$ nodes in the hash tree up to at most two special hash values and by optimally verifying that these hash values are indeed special, i.e., belong in S.

ℓ_1, thus can be verified at $O(\log \log n)$ cost and, if t is $o(\log n)$ and $\Omega(\log \log n)$ we have an improvement and optimal performance. To further improve the verification cost in the case where t is $o(\log \log n)$, we use the above technique to *recursively* define additional special hash values over the $n/\log n$ trees defined by the special hash values in S_1: we consider each one of the trees of size $\log n$ rooted at level ℓ_1 and apply the above technique, assuming that $t < \log \log n$. We define the set S_2 of special hash values to contain the hashes $h_1^2, \ldots, h_{m_2}^2$, $m_2 = m_1 \frac{\log n}{\log \log n}$, at level $\ell_2 = \log \ell_1 = \log \log \log n$ of the hash tree and answers of size t with $\log \log \log n < t < \log \log n$ can be authenticated *super-efficiently* at cost $O(\log \log \log n)$. We proceed as above: at the i-th step of the recursion we define the set S_i of special hash values, we stop before level $\log^* n$, effectively at the level 2 (or some other small constant) of T and set $S \triangleq h_r \cup S_1 \cup \ldots \cup S_{(\log^* n)-1}$ as the final set of special hash values, which is of $\Theta(n)$ size[3].

Our authentication structure lends itself to a first authentication scheme that achieves super-efficient verification: an answer of size t is verified with $O(\log t)$ hashing cost where $O(1)$ special hash values need be authenticated, essentially as being members of the set of special hash values S. In what follows, we consider the case where authentication in S is performed using a signatures scheme, i.e., each value in S is separately signed by the source $\mathcal{S}$. Updates on D are handled by appropriately updating the hash tree T (by hashing and restructuring T along a leaf-to-root path; see also Section 4), having $\mathcal{S}$ sign $O(\log^* n)$ updated special hashes. Replay attacks are eliminated by using *time-stamps* in the signed statements to check the freshness of a valid signature; this is a state-of-the-art solution (see, e.g., [13, 14, 19]) where time is partitioned into fixed and publicly

---

[3] In fact $|S| < n - 1$, thus, S has smaller size than the trivial solution of $S = T$.

known time-quanta, and verifiable signatures on digests are accepted only if their time-stamps belong in the current (at the time of verification), most recent, time-quantum. For *hash-based authentication*, i.e., in the most practical setting where only cryptographic hashing is used to produce the authentication strings, our authentication structure achieves *optimal performance* with respect to both the verification and the update costs. In particular, using the lower-bound framework of [26], we can show that in the worst case the source needs to authenticate a set S of $\Omega(n)$ special hash values in order to achieve verification costs that are independent of the size n of the database, and that, in this case, time-stamping and signature refreshing is an optimal technique against replay attacks. Thus:

Theorem 1. *There exists a super-efficient authentication scheme for range-search queries over a set of n key-value pairs with the following performance, where t denotes the number of pairs returned by a query: (i) a range query is answered in $O(\log n + t)$ time; (ii) the answer proof has size $O(\log t)$ and consists of two signatures, two keys, and $O(\log t)$ hash values; (iii) the answer to a range query is validated by performing $O(t)$ arithmetic computations, $O(t)$ hash operations, and $O(1)$ signature verifications; (iv) an update results in $O(\log n)$ hash operations (at both the source and the responder), $O(\log^* n)$ signature generations (at the source) and $O(n)$ signature renewals (at the source). This authentication scheme is secure with respect to data authentication, safe with respect to replay attacks, and optimal with respect to super-efficient verification in the hash-based data authentication model.*

4   Super-Efficient Dynamic Authentication Scheme

In this section, we propose an alternative technique that reduces the high update cost of the previous, optimal but less practical, hash-based authentication scheme to get the first super-efficient dynamic authentication scheme for range queries, which provides reasonable trade-offs between the update and query costs.

In Section 3 we constructed a hash tree for a set D of n key-value pairs that encodes information about the key-value and successor relations in D, and we defined a set S of $O(n)$ special hash values that are sufficient to support super-efficient answer verification, provided there is an optimal (in terms of verification) technique for authenticating set-membership queries. Recall that for any query, there are at most two special hash values, out of the total $O(n)$, that need to be verified as members of S, and note that only queries with positive answer need to be authenticated: a special hash value must be verified to be in set S.

We now describe our new authentication scheme. The main idea is to use a dynamic RSA accumulator for authenticating set membership queries for the set of special hash values S. This is performed as follows: the set S of special hash values is accumulated to accumulation value $\alpha = A(S)$ and α is signed by the source. Then, verifying that a special hash value belongs in S is performed in two steps, and still in optimal way ($O(1)$ verification cost): first, the hash value together with the membership witness are used to verify that the hash value was used by the accumulator in producing α and, second, the signature on α is verified. For security reasons, only the source knows the trapdoor information of the accumulator; the responder does not know the trapdoor. It follows that the verification is (as in the construction of the previous section) super-efficient.

Let us briefly describe the dynamization of the authentication structure, i.e., how updates on the data set can be handled. Assume for simplicity that only values are updated, that is, no keys are inserted or deleted in D. After any update of this type in D, we end up rehashing over a unique-per update operation leaf-to-root path in the hash tree. Thus $O(\log^* n)$ special hash values change and we need to remove the old special hash values from the accumulation α and add the new ones into this, i.e., to perform $O(\log^* n)$ element deletions and insertions in S and update $A(S)$. Inserting and deleting elements in an accumulator involves some computational cost for updating the new accumulation but also for updating the set-membership witnesses of all the elements. Suppose that the witnesses of the $O(n)$ accumulated special hash values are explicitly maintained in the source and the responder. In a highly dynamic setting updates can be of cost $O(n)$: the reason is that after any update all n membership witnesses must be updated. The problem of the high update cost becomes more challenging for deletions, especially under the necessary restriction that the responder cannot use the trapdoor information, but using the RSA accumulator and certain algorithmic techniques [12] we can achieve reasonable update and query costs. We can show:

Theorem 2. *There exists a dynamic super-efficient authentication scheme for range search queries over a set of n key-value pairs with the following performance, where t denotes the size of the returned answer: (i) a range query is answered in $O(\log n + t)$ time; (ii) the answer proof has size $O(\log t)$ and consists of one signature, two field elements, two keys and $O(\log t)$ hash values; (iii) the answer to a range query is validated by performing $O(t)$ arithmetic computations, $O(t)$ hash operations and $O(1)$ modular exponentiation and verifying $O(1)$ signatures; (iv) an update results in $O(\log n)$ hash operations (at both the source and the responder), $O(\sqrt{n} \log^* n)$ modular operations and $O(1)$ signature generations (at the source). This authentication structure is secure with respect to authentication and safe with respect to replay attacks.*

5   Detection and Elimination of Replay Attacks

In Section 3, we presented an authentication structure for range search queries that provides super-efficient answer verification, asymptotically optimally in the hash-based data authentication model. In this section, we propose a new scheme in our three-party authentication model $(\mathcal{S}, \mathcal{R}, \mathcal{U})$ that achieves efficient update costs at $\mathcal{S}$ and $\mathcal{R}$ (only logarithmic in the database size) and super-efficient verification costs at $\mathcal{U}$ (as before), but uses an alternative solution to the replay-attack problem. In particular, we slightly relax the security requirement with respect to the time when replay attacks are detected and replayed data is rejected. As before, invalid answers are *immediately* rejected by $\mathcal{U}$, but answers are checked to be consistent with the update history in an *off-line* fashion. We introduce a technique which implements an *auditing* mechanism and provides *delayed consistency checking* for detecting and effectively eliminating replay attacks. This mechanism augments the authentication scheme of Section 3, so that $\mathcal{U}$ can immediately check any received answer for correctness and at any later time check, in a batch, all received answers for freshness.

Delayed consistency checking is a useful property in application areas where the freshness of answers is not critical to be verified in real time. In many applications, risk management requires that invalid responses must be caught, but this determination does not always have to be immediate, as long as it is certain and sufficiently near-term. Indeed, such swift and sure justice is an ideal circumstance for risk management purposes. Additionally, delayed consistency checking is appropriate when consecutive queries occur sequentially in a short time window and share locality in risk management or equivalent trust relations.

In our auditing mechanism, the delayed consistency checking is performed by the user $\mathcal{U}$, collaboratively with the source $\mathcal{S}$ but without any direct interaction between the two, however. The auditing mechanism corresponds to securely, compactly and efficiently encoding a series of *transactions* with the responder $\mathcal{R}$, i.e., updates and queries over data set D issued by $\mathcal{S}$ and $\mathcal{U}$, respectively. In particular, $\mathcal{S}$ maintains an *update audit state* Σ_u, that encodes the history of updates, through information reported after update transactions with $\mathcal{R}$: for any update u performed on the data set D, an *update trail* T_u is provided to $\mathcal{S}$ by $\mathcal{R}$ that is used to update Σ_u through operation updU. Similarly, $\mathcal{U}$ maintains a *query audit state* Σ_q, that encodes the history of queries, through information reported after query transactions with $\mathcal{R}$: for any query q issued on D and returned answer-proof pair, a *query trail* T_q is provided to $\mathcal{U}$ by $\mathcal{R}$ that is used to update Σ_q through operation updQ. These trails correspond to "receipts" that the auditing mechanism collects (namely, the update and query trails that $\mathcal{S}$ and $\mathcal{U}$ receive). This series of updates of the states Σ_u and Σ_q corresponds to the *computation phase* of the auditing mechanism.

Verification of the consistency of the two transaction series (update and query) and, consequently, replay-attack detection are performed by $\mathcal{U}$ in the *audit phase*. At any point in time (predefined or decided instantly), $\mathcal{U}$ can invoke a request for checking the consistency of the reported transactions with the current set D that resides at $\mathcal{R}$. This is performed at $\mathcal{U}$ through operation audit, which receives as input the current audit query state Σ_q of $\mathcal{U}$ and the current audit update state Σ_u of $\mathcal{S}$, appropriately updated given the current data set D (provided to $\mathcal{S}$ by $\mathcal{R}$), and accepts or rejects its input, accordingly verifying the consistency of transactions. After an audit operation that accepts its input, the audit state remains unchanged and a new computation phase begins. If it rejects, the states are reset and the next computational phase starts for a new data set: in this case, the data source $\mathcal{S}$ is responsible for creating the new data set at $\mathcal{R}$. We call the triplet of algorithms (updU, updQ, audit) along with the protocols for formatting the trails an *auditing scheme*.

An auditing scheme (updU, updQ, audit) is secure if it satisfies the following property: operation audit accepts its input if and only if no malicious action has been performed by $\mathcal{R}$, i.e., all query-answer pairs verified by $\mathcal{U}$ are consistent with the update history of D and the states computed using operations updU, updQ. In particular, (updU, updQ, audit) is secure if the following conditions hold: *completeness*, dictating that all valid update and query transactions yield (through updU and updQ) audit states that when checked by audit with a valid (not corrupted by $\mathcal{R}$) data set D always result in accepting; and *soundness*, dictating that when audit accepts its inputs, then the audit states correspond to transactions of valid update/query operations subject to the current data set.

To detect and prevent replay attacks, we augment the authentication scheme of Section 3 with a secure auditing scheme (updU, updQ, audit) as follows. After updates, along with the update at $\mathcal{S}$ and $\mathcal{R}$ of the underlying authentication structure, $\mathcal{S}$ runs updU to update its update audit state, but now no signature refreshing is performed: only $O(\log^* n)$ hash values are signed by $\mathcal{S}$. After queries, along with the answer verification, $\mathcal{U}$ also runs updQ to update its query audit state. If $\mathcal{R}$ launches a replay attack at some point in time, it will be detected by $\mathcal{U}$ at the first audit phase occurring after the attack since, by the security property, audit will reject its input. So, a rejecting audit phase is equivalent to detecting a replay attack launched by $\mathcal{R}$, and a misbehaving $\mathcal{R}$ who performs replay attacks is always caught and exposed to its victim $\mathcal{U}$. Note that this technique provides only detection and cannot pinpoint which query-answer pairs were replayed.

To construct a secure auditing scheme, we use a simple cryptographic solution that is inspired from efficient and secure cryptographic mechanisms for off-line memory checking by Blum *et al.* [5]. In off-line memory checking, a trusted checker checks the correctness (or consistency) of an untrusted memory, where data is written in and read from the memory through operations load and store. The checker maintains some constant-size state and augments the data that is written into the untrusted memory with *time-stamps*, such that at any point in time, a check can be performed on the memory correctness. The idea is to use a cryptographic primitive A for generating and updating this state, as a short description of the memory history. A can produce short digests of large sets in an incremental fashion (i.e., elements are inserted in the set and the new digest is updated in $O(1)$ time without recomputing from scratch) and is used as follows. After any (augmented) load or store operation performed in the memory, a special encoding of the operation is created and securely enclosed in the state through A. In particular, two separate digests are maintained over two sets: a first set encodes the "load" history of the memory (i.e., reading operations); the second set encodes the "store" history of the memory (i.e., writing operations). An operation results in updating both sets (e.g., operation load(i) adds an item d_i in the "load" history and item d_i' with new time-stamp in the "store" history). The crucial property of the approach in [5] is that if the memory is correct, the encodings produce load and store digests that are the same when the check is performed. By choosing the cryptographic primitive A such that it is collision-resistant, meaning that its computationally infeasible to find distinct sets that produce the same digest, the memory checking problem is reduced to an equality testing problem (subject to an appropriate encoding for the operations in the memory). Primitives A for incrementally computing collision-resistant digests of sets exist (e.g., ϵ-biased hash functions in the original work [5]).

We next design an efficient secure auditing scheme that is based on the above checking technique. The challenge in applying this idea in our three-party model is to implement the checking functionality collaboratively by $\mathcal{S}$ and $\mathcal{U}$ without destroying super-efficiency at $\mathcal{U}$. We use the RSA accumulator as a collision-resistance primitive A for incrementally computing digests over sets and use $A(S)$ to denote the digest of set S. Thus, given $A(S)$ and a new element x not in S, $A(S \cup x)$ can be computed in $O(1)$ time; also, it is hard to find sets $S \neq S'$ such that $A(S) = A(S')$. We use A to define the audit states Σ_u and Σ_q stored by $\mathcal{S}$ and $\mathcal{U}$. The main idea is as follows. We view the set S of special values defined over our super-efficient authentication structure of Section 3 as an untrusted memory: memory locations correspond to the unique identifiers of the tree nodes (according to a fixed ordering, e.g., in-order tree traversal) and memory items correspond to the special hash values and their signatures.

Every transaction (update or query) uniquely defines a subset of special hash values in the tree: for updates, the hashes in the $O(\log^* n)$ special tree levels in the corresponding leave-to-root path; for queries, the two hashes of the lowest special tree level that exactly covers the answer. These two subsets of special hashes respectively define the update trail T_u and the query trail T_q that are returned by $\mathcal{R}$. For each tree node v in a subset, the tuple $(id_v, h_v, \sigma_v, t_v)$ is included in the corresponding trail. Here, id_v is the identifier of v, h_v the hash value, σ_v the corresponding signature and t_v the associated timestamp. Algorithms updU and updQ process these trails to update the audit states $\Sigma_u = (A_{u,l}, A_{u,s})$ and $\Sigma_q = (A_{q,l}, A_{q,s})$. Each audit state is a pair of values, one for "load" history, one for "store"; $A_{u,l}$, $A_{u,s}$ are integer values and $A_{q,l}$, $A_{q,s}$ are accumulations. The tuple of v is encoded (according to fixed way) to a unique string x_v (e.g., by applying an one-way hash function) and for each tuple in the trails the states are updated to $\Sigma'_u = (A'_{u,l}, A'_{u,s})$ and $\Sigma'_q = (A'_{q,l}, A'_{q,s})$, as follows: $A'_{u,l} = A_{u,l} \cdot e(x_v) \mod \phi(N)$, $A'_{u,s} = A_{u,s} \cdot e(x'_v) \mod \phi(N)$, $A'_{q,l} = A_{q,l}^{e(x_v)} \mod N$, $A'_{q,s} = A_{q,s}^{e(x'_v)} \mod N$, where $e(\cdot)$ is a function for computing prime representative values, N is the RSA modulo, and x'_v is encoding x_v but with a fresh time-stamp (monotonically increasing, synchronized for all parties) and possibly with a new identifier, hash value and signature (only for updates).

The audit phase is as follows. First $\mathcal{R}$ forwards the request for the audit to $\mathcal{S}$, along with a final audit trail that contains a tuple for each special node in set S (final reading of memory). $\mathcal{S}$ updates its update audit state (only the "load" part), signs the final Σ_u and forwards it to $\mathcal{U}$, through $\mathcal{R}$. Given $(A_{u,l}, A_{u,s})$, $(A_{q,l}, A_{q,s})$, audit (run at $\mathcal{U}$) accepts if and only if: $A_{q,l}^{A_{u,l}} \equiv A_{q,s}^{A_{u,s}} \mod N$.

Theorem 3. *There exists a hash-based, dynamic, super-efficient and audited authentication scheme for range search queries over a set of size n with the following performance, where t denotes the number of data items returned by a query: (i) a query is answered in $O(\log n + t)$ time; an update results in $O(\log n)$ hash operations (at both the source and the responder), $O(\log^* n)$ signature generations* *(at the source); (ii) the answer proof has size $O(\log t)$ and consists of two signatures, two keys and $O(\log t)$ hash values; (iii) the answer to a query is validated by performing $O(t)$ hash operations and verifying $O(1)$ signatures; (iv) the auditing scheme stores $O(1)$ audit state, performs $O(\log n)$ work per update (at the source) and $O(1)$ work per query (at the user) during the computation phase and performs $O(n)$ work (at the source) and $O(1)$ work (at the user) during the audit phase; (v) replay attacks performed by the responder are always detectable by the user at the audit phase.*

Proof. (Sketch.) The complexity for the queries and updates follow from Theorems 1 and 2, by observing that no signature refreshing is necessary after updates at $\mathcal{S}$. The update and query audit states are both of $O(1)$ size (a pair of values). At the computation phase, each update incurs $O(\log n)$ cost at $\mathcal{S}$ for updating the authentication structure (hashing along the update path and updating $O(\log^* n)$ signatures and the audit state with $O(\log^* n)$ exponent accumulations). Each query incurs $O(1)$ cost at $\mathcal{U}$ (at most two values are accumulated in the audit state). At the audit phase, the cost at $\mathcal{S}$ is $O(n)$, since $\mathcal{S}$ accumulates in the exponent all special hash values currently in the authentication structure; the cost at $\mathcal{U}$ is $O(1)$ as before. Security follows from the correctness of the checking mechanism of [5] and the collision-resistance property of the RSA accumulator. Recall that $\mathcal{R}$ does not know the trapdoor $\phi(N)$ of the accumulator. Regarding soundness, suppose that audit fails to detect a replay attack launched by $\mathcal{R}$. Either the provided by $\mathcal{R}$ update and query trails were correct or there existed one trail that was invalid. In the former case and given that the audit mechanism accepts, the memory checking technique is incorrect; in the latter case, there exist different sets S and S' that produce the same RSA-based accumulations $A(S) = A(S')$. We must conclude that either the $\mathcal{R}$ was able to compute the trapdoor $\phi(N)$ for the RSA modulo N (a task that is computationally equivalent to factoring N) or $\mathcal{R}$ was able, given $A_{q,l}^{A_{u,l}} \bmod N$, to compute (through the query trails that $\mathcal{R}$ provided $\mathcal{U}$ with, which are distinct from the update trails) values $A_{q,s}$ and $A_{u,s}$ such that $A_{q,l}^{A_{u,l}} \bmod N = A_{q,s}^{A_{u,s}} \bmod N$ (a task that is computationally infeasible under the strong RSA assumption).   □

6   Extensions and Concluding Remarks

Our authentication schemes are based on the authentication structure for range search queries of Section 3. Many other query types are related to range searching or consist of more complex search problems that eventually boil down to range searching. This suggests that our authentication schemes can be used as general design tools for achieving super-efficient authentication of other types of queries. Indeed, all that is needed is to consider a (different) hashing scheme over the data set D (computed along the hash tree), which should be appropriate for the target query type. Similar to the construction in Section 3, the hashing scheme over D should securely encode these relations that are sufficient for verifying the answers to the queries in consideration. Super-efficiency would then follow simply by authenticating at most two special hash values at the appropriate special level of the tree, depending on the exact range defined by the query.

We briefly discuss two types of queries that fall into this category. Consider the class of queries that ask for any *associative* function over some field of data records that lie in a query range. The canonical members of this class are *aggregate* queries, e.g., SUM, MAX, AVG. An appropriate hashing scheme for these queries would be constructed such that it encodes the information (relations) about ranges, the corresponding aggregation values and the neighboring data records. In particular, the hash tree node v defining subtree T_v stores a hash value that encodes information about the aggregation value a_v computed over the records that correspond to the leaves of T_v, the left-most and right-most records in T_v and, also, their predecessor and successor records (not in T_v), respectively. Using this hashing scheme, these queries can be authenticated by considering the (at most two) allocation nodes that correspond to the query range and lie in some special tree level and *without* applying any associate operation. Similarly, we can use our schemes for the class of *path property* queries that are studied in [13]—all related to range searching. Our hashing scheme of Section 3 and, accordingly, all of our authentication schemes can be extended to these classes of queries (aggregation and path-property queries).

In conclusion, in this paper we study data authentication in a setting where critical information is queried at high rates from dynamic outsourced databases that reside in untrusted sites. We propose a new approach for query authentication, where, by decoupling the answer-generation and answer-verification procedures, super-efficient answer verification is enabled, a theoretically interesting and practically important property. We design the first authentication schemes for range searching that achieve super-efficiency (answers of size t are verified in time $O(t)$, using only $O(1)$ modular exponentiations), allow for efficient updates on the database and eliminate the replay attacks from the database responder. To prevent replay attacks on old invalid data, we design an authentication protocol that implements an off-line auditing mechanism, which checks the consistency of a dynamic database and reliably reports malicious actions of the responder. Open problems include further improving the update costs of our authentication schemes and extending our auditing scheme in a multi-user setting.

References

[1] Atallah, M.J., Cho, Y., Kundu, A.: Efficient data authentication in an environment of untrusted third-party distributors. In: Proceedings of International Conference on Data Engineering (ICDE) (to appear, 2008)

[2] Barić, N., Pfitzmann, B.: Collision-free accumulators and fail-stop signature schemes without trees. In: Fumy, W. (ed.) EUROCRYPT 1997. LNCS, vol. 1233, pp. 480–494. Springer, Heidelberg (1997)

[3] Benaloh, J., de Mare, M.: One-way accumulators: A decentralized alternative to digital signatures. In: Proceedings of Advances in Cryptology — EUROCRYPT, pp. 274–285 (1994)

[4] Bertino, E., Carminati, B., Ferrari, E., Thuraisingham, B., Gupta, A.: Selective
     and authentic third-party distribution of XML documents. IEEE Transactions on
     Knowledge and Data Engineering 16(10), 1263–1278 (2004)
[5] Blum, M., Evans, W., Gemmell, P., Kannan, S., Naor, M.: Checking the correct-
     ness of memories. Algorithmica 12(2/3), 225–244 (1994)
[6] Buldas, A., Laud, P., Lipmaa, H.: Accountable certificate management using un-
     deniable attestations. In: Proceedings of ACM Conference on Computer and Com-
     munications Security, pp. 9–18. ACM Press, New York (2000)
[7] Camenisch, J., Lysyanskaya, A.: Dynamic accumulators and application to effi-
     cient revocation of anonymous credentials. In: Yung, M. (ed.) CRYPTO 2002.
     LNCS, vol. 2442, pp. 61–76. Springer, Heidelberg (2002)
[8] Devanbu, P., Gertz, M., Kwong, A., Martel, C., Nuckolls, G., Stubblebine, S.:
     Flexible authentication of XML documents. Journal of Computer Security 6, 841–
     864 (2004)
[9] Devanbu, P., Gertz, M., Martel, C., Stubblebine, S.G.: Authentic data publication
     over the Internet. Journal of Computer Security 11(3), 291–314 (2003)
[10] Di Battista, G., Palazzi, B.: Authenticated relational tables and authenticated
     skip lists. In: Proc. Working Conference on Data and Applications Security (DB-
     SEC), pp. 31–46 (2007)
[11] Gassko, I., Gemmell, P.S., MacKenzie, P.: Efficient and fresh certification. In:
     Imai, H., Zheng, Y. (eds.) PKC 2000. LNCS, vol. 1751, pp. 342–353. Springer,
     Heidelberg (2000)
[12] Goodrich, M.T., Tamassia, R., Hasic, J.: An efficient dynamic and distributed
     cryptographic accumulator. In: Chan, A.H., Gligor, V.D. (eds.) ISC 2002. LNCS,
     vol. 2433, pp. 372–388. Springer, Heidelberg (2002)
[13] Goodrich, M.T., Tamassia, R., Triandopoulos, N., Cohen, R.: Authenticated data
     structures for graph and geometric searching. In: Joye, M. (ed.) CT-RSA 2003.
     LNCS, vol. 2612, pp. 295–313. Springer, Heidelberg (2003)
[14] Li, F., Hadjieleftheriou, M., Kollios, G., Reyzin, L.: Dynamic authenticated index
     structures for outsourced databases. In: Proceedings of ACM SIGMOD Interna-
     tional Conference on Management of Data, pp. 121–132 (2006)
[15] Martel, C., Nuckolls, G., Devanbu, P., Gertz, M., Kwong, A., Stubblebine, S.G.: A
     general model for authenticated data structures. Algorithmica 39(1), 21–41 (2004)
[16] Merkle, R.C.: A certified digital signature. In: Brassard, G. (ed.) CRYPTO 1989.
     LNCS, vol. 435, pp. 218–238. Springer, Heidelberg (1990)
[17] Micali, S., Rabin, M., Kilian, J.: Zero-Knowledge sets. In: Proceedings of Sympo-
     sium of Foundations of Computer science (FOCS), pp. 80–91 (2003)
[18] Mykletun, E., Narasimha, M., Tsudik, G.: Authentication and integrity in out-
     sourced databases. In: Proceeding of Network and Distributed System Security
     (NDSS) (2004)
[19] Naor, M., Nissim, K.: Certificate revocation and certificate update. In: Proceed-
     ings 7th USENIX Security Symposium, pp. 217–228 (1998)
[20] Narasimha, M., Tsudik, G.: Authentication of outsourced databases using signa-
     ture aggregation and chaining. In: Proceedings of 11th International Conference
     on Database Systems for Advanced Applications, pp. 420–436 (2006)
[21] Nuckolls, G.: Verified query results from hybrid authentication trees. In: Proceed-
     ings of Data and Applications Security (DBSec), pp. 84–98 (2005)
[22] Ostrovsky, R., Rackoff, C., Smith, A.: Efficient consistency proofs for generalized
     queries on a committed database. In: Díaz, J., Karhumäki, J., Lepistö, A., San-
     nella, D. (eds.) ICALP 2004. LNCS, vol. 3142, pp. 1041–1053. Springer, Heidelberg
     (2004)

[23] Pang, H., Jain, A., Ramamritham, K., Tan, K.-L.: Verifying completeness of relational query results in data publishing. In: Proceedings of ACM SIGMOD Int. Conference on Management of data, pp. 407–418 (2005)
[24] Papamanthou, C., Tamassia, R.: Time and space efficient algorithms for two party authenticated data structures. In: Qing, S., et al. (eds.) ICICS 2007. LNCS, vol. 4861, pp. 1–15. Springer, Heidelberg (2007)
[25] Tamassia, R.: Authenticated data structures. In: Proceedings of European Symposium on Algorithms, pp. 2–5 (2003)
[26] Tamassia, R., Triandopoulos, N.: Computational bounds on hierarchical data processing with applications to information security. In: Caires, L., Italiano, G.F., Monteiro, L., Palamidessi, C., Yung, M. (eds.) ICALP 2005. LNCS, vol. 3580, pp. 153–165. Springer, Heidelberg (2005)
[27] Tamassia, R., Triandopoulos, N.: Efficient content authentication in peer-to-peer networks. In: Katz, J., Yung, M. (eds.) ACNS 2007. LNCS, vol. 4521, pp. 354–372. Springer, Heidelberg (2007)

A Latency-Free Election Scheme

Kristian Gjøsteen

Department of Mathematical Sciences
Norwegian University of Science and Technology
`kristian.gjosteen@math.ntnu.no`

Abstract. We motivate and describe the problem of finding protocols for multiparty computations that only use a single broadcast round per computation (latency-free computations). We show that solutions exists for one multiparty computation problem, that of elections, and more generally, addition in certain groups. The protocol construction is based on an interesting pseudo-random function family with a novel property.

1 Introduction

Consider a small cluster of stars, separated from each other by distances of between one and four light-years. Being in a civilised part of the universe, the cluster has a general assembly for discussing questions of importance for the cluster. Due to the inconvenience of gathering for this assembly, it has been decided that twice every decade, they shall have a vote on whether to convene the assembly or not.

The communication channel can easily be established with radio telescopes broadcasting a signal to every planet, but there is general agreement about a need for privacy, so some kind of secure election scheme must be used.

Most secure election schemes without a central authority (see for example [2]) that operate over a broadcast channel require two rounds of communication, typically one round to publish encrypted votes and one round to decrypt the result. Unfortunately, due to the speed of light, each extra round will require up to four years to complete, which is clearly too much. Therefore we need an election scheme that can produce a result in a single round.

A less frivolous but more realistic example is the case where an election protocol is used as a subprotocol. If the subprotocol must be repeated multiple times, minimising the number of rounds used for the election protocol will be important.

An election scheme is just one example of a multiparty computation problem. In general, we can consider an environment where a group of users want to perform some type of multiparty computation many times. If they are communicating over high-latency channels, there is a clear incentive to minimise the number of rounds. That raises the natural question: Can we do multiparty computations with just a single round of communication? Of course, some kind of setup will always be necessary, but can we hope to do one computation per additional round, after the setup?

T. Malkin (Ed.): CT-RSA 2008, LNCS 4964, pp. 425–436, 2008.

Sometimes it is possible to interleave independent protocol runs, to get essentially one computation result per round. But if we consider situations where the multiparty computations may happen in parallel, interleaving becomes impossible. The notion of self-tallying elections [9] allow the actual election to happen in one round, but requires one round of precomputation where the communication complexity depends on the number of election rounds.

We propose a solution for one multiparty computation problem (elections) where, after some initial setup, every communication round performs one multiparty group operation, and all communication rounds are essentially independent (any number of new rounds can start before the already started rounds complete). The communication complexity of the initial setup is independent of the number of election rounds.

Franklin and Yung [6] investigated how the communication complexity of multiparty computations could be reduced by performing computations in parallel. This work was in an information-theoretic setting. Our solution does in a certain sense show how to achieve optimal asymptotic communication complexity for certain repeated multiparty group operations: The amount of information broadcast by the user in a single round is equal to the amount of information he inputs into the computation. We emphasise that our results are achieved in the random oracle model (though we discuss how they could be achieved in the common reference string model).

Our main technical contribution in this paper is a very interesting construction for pseudo-random function families with a useful algebraic structure, on which our constructions rely. While the specific construction in Sect. 4.1 has previously appeared in the literature [3] (independent our construction, which first appeared in [7]), the algebraic structure has not been noted before. These constructions also have interesting applications outside of the current problem domain.

The secondary contribution is the idea that our construction can be used for a certain multiparty computation, specifically elections.

This paper is structured as follows: Sect. 2 contains basic material on pseudo-random function families. Sect. 3 discusses requirements for single-round election schemes in general, and describes our proposed election scheme. Sect. 4 describes two pseudo-random function families with a useful algebraic structure, several useful tools for these function families, and some concrete examples. In Sect. 5 we show how to remove the trusted dealer used in Sect. 3. Finally, in Sect. 6 we make some concluding remarks.

1.1 Notation

For any distribution D and algorithm A, we denote by $A(D)$ the output distribution we get when we sample x from D and run A with input x.

For any set S, $x \xleftarrow{r} S$ denotes that x is sampled from the uniform distribution on S. Following the above notation, we denote by $A(S)$ the output distribution we get when we sample x from the uniform distribution on S and run A with input x.

2   Pseudo-random Function

Definition 1. *Let S_1 and S_2 be sets. A* pseudo-random function family (PRF) *F from S_1 to S_2 is a subset of $\mathrm{Map}(S_1, S_2)$ indexed by a key set K: $F = \{f_k : S_1 \to S_2 \mid k \in K\}$.*

A *l-distinguisher A* for F is an algorithm that is allowed to query a function chosen either uniformly at random from F or uniformly at random from $\mathrm{Map}(S_1, S_2)$ in at most l points of its choosing, and then output 0 or 1. The advantage of A in distinguishing functions in F from random functions is defined to be

$$\mathrm{Adv}_A = |\Pr[A^f = 1 \mid f \xleftarrow{r} F] - \Pr[A^f = 1 \mid f \xleftarrow{r} \mathrm{Map}(S_1, S_2)]|,$$

where A^f denotes that A is run with oracle access to the function f. We say that F is (t, ϵ, l)-*secure* if no l-distinguisher with advantage at least ϵ and run-time at most t exists.

A *weak l-distinguisher A* for F is an algorithm that is allowed to see a function chosen either uniformly at random from F or uniformly at random from $\mathrm{Map}(S_1, S_2)$ evaluated in l points chosen uniformly at random from S_1, and then output 0 or 1. The advantage of A in distinguishing functions in F from random functions is defined to be

$$\mathrm{Adv}_A = |\Pr[A((x_i, f(x_i))_{i=0}^{L-1}) = 1 \mid x_i \xleftarrow{r} S_1, f \xleftarrow{r} F]$$
$$- \Pr[A((x_i, f(x_i))_{i=0}^{L-1}) = 1 \mid x_i \xleftarrow{r} S_1, f \xleftarrow{r} \mathrm{Map}(S_1, S_2)]|.$$

We say that F is *weakly (t, ϵ, l)-secure* if no weak l-distinguisher with advantage at least ϵ and run-time at most t exists.

In the *random oracle model*, we can construct a secure PRF from S_0 to S_2 using a weakly secure PRF from S_1 to S_2. This result is well-known.

Theorem 1. *Let F be a pseudo-random function family from S_1 to S_2, let $h : S_0 \to S_1$ be a function chosen uniformly at random from $\mathrm{Map}(S_0, S_1)$, and let $F' = \{h \circ f_k \mid f_k \in F\}$, where $\circ$ denotes function composition. For any l-distinguisher A for F' in the random oracle model making at most L queries to the random oracle h, there exists a weak $L + l$-distinguisher for F with the same advantage.*

We also need a minor extension of this notion, where we only consider the indistinguishability of part of the function value. Consider a pseudo-random function family F from S_1 to G, where G is a group that has a subgroup J. Consider the set of functions $\tilde{F} = \{x \mapsto f_0(x)f_1(x) \mid f_0 \in F, f_1 \in \mathrm{Map}(S_1, J)\}$. We note that if $J = G$, then $\tilde{F} = \mathrm{Map}(S_1, G)$, and the following notions coincide with the above notions.

An *l-J-distinguisher* for F is an algorithm that is allowed to query a function chosen uniformly at random from either F or $\tilde{F}$ in at most l points of its choosing, and then output 0 or 1. The advantage of the distinguisher is defined as above.

A *weak l-J-distinguisher* for F is an algorithm that is allowed to see a function chosen either uniformly at random from F or uniformly at random from $\tilde{F}$

evaluated in l points chosen uniformly at random from S_1, and then output 0 or 1. The advantage of the distinguisher is defined as above, but can also be expressed as

$$\text{Adv}_A = |\Pr[A((x_i, f(x_i))_{i=0}^{l-1}) = 1 \mid x_i \xleftarrow{r} S_1, f \xleftarrow{r} F]$$
$$- \Pr[A((x_i, f(x_i)r_i)_{i=0}^{l-1}) = 1 \mid x_i \xleftarrow{r} S_1, r_i \xleftarrow{r} J, f \xleftarrow{r} F]|.$$

Again, we can construct a secure PRF from S_0 to G using a weakly secure PRF from S_1 to G, in the *random oracle model*. From F and $\tilde{F}$, we get $F' = \{h \circ f \mid f \in F\}$ and $\tilde{F}' = \{h \circ f \mid f \in \tilde{F}\}$, where $h \in \text{Map}(S_0, S_1)$.

Suppose that we have some group structure on the set F', written multiplicatively, such that for any $f_1, f_2 \in F'$ and any x, $(f_1 f_2)(x) = f_1(x) f_2(x)$. We now prove that in a certain situation, this property is of no help to an adversary trying to decide if a function comes from F'.

Theorem 2. *Let F' be a pseudo-random function family from S_0 to G with a group structure as described above. Let A be an adversary that gets $f \in F'$ as input and has oracle access to a pair of functions f_1 and f_2. The adversary may make at most l queries to its oracles, and must decide if f_1, f_2 are chosen uniformly at random from F', subject to $f_1 f_2 = f$, or uniformly at random from $\tilde{F}'$ subject to the condition that for any x, $f_1(x) f_2(x) = f(x)$. If A has advantage ϵ, then there exists an $2l$-J-distinguisher A' for F' with essentially the same run time as A and with advantage ϵ.*

Proof. The adversary A' has oracle access to a function f_1 that has either been sampled from F' or $\tilde{F}'$. A' samples f from the uniform distribution on F' and runs A with f as input. When A queries its oracle for f_1, A' queries its own oracle and returns the response. When A queries its oracle for $f_2(x)$, A' queries its own oracle for $f_1(x)$ and returns $f(x)/f_1(x)$. When A terminates with guess b, A' also terminates and outputs b as its own guess.

Since A' only runs A and does at most $2l$ queries to its oracles, as well as $2l$ group operations and function evaluations, A' has essentially the same run time as A.

If f_1 is sampled from F', then A' clearly simulates the oracles for f_1 and f_2 as if they were sampled from F'. Likewise, if f_1 is sampled from $\tilde{F}'$, A' will also simulate the oracles as if they were both sampled from $\tilde{F}'$, subject to $f_1 f_2 \in F'$, since f and f_1 are correctly distributed and f_2 satisfies the required condition. It is therefore clear that A' guesses correctly if and only if A guesses correctly, which shows that A' has the claimed advantage. □

3   The Election Scheme

We assume that there is a broadcast channel available. We want an election scheme that can be used for multiple sequential or parallel elections after some initial setup, and that satisfies the following functional requirements:

Constant-Round Setup. The number of rounds in the setup phase must be independent of the number of voters and the number of elections that are to be held.

Single Round per Election. Each of the multiple elections must require just one round. No voter must be required to decide on his vote before the start of the round, and after that round, every voter must be in possession of the result for that election round, unless some fault occurred.

Any election scheme must satisfy at least the following security requirements:

Privacy. Every vote must be as secret as possible in an election (e.g. if the result indicates that all votes were equal, no vote can possibly be private, regardless of the system).

Correctness. No voter should be able to submit incorrect votes.

Verifiability. Every voter should be able to verify that the tallying was performed correctly.

Usually, election schemes are also required to be *robust*, in the sense that a few voters cannot prevent the remaining voters from computing the result. Schemes that satisfy our functional requirements cannot be robust in this sense. If the scheme allows voters to compute the correct result when one or more votes are missing, any voter could first compute the correct result for all the votes, then pretend that vote i is missing and compute this result. That would reveal the ith vote, and privacy would be lost.

One might relax the functional requirements and say that the single round requirement should hold except in the presence of faults, when a fall-back election protocol should be used to compute the result. Unfortunately, this would allow an inside attacker that can read the ith vote, but prevent it from being broadcast to the other voters, to break the privacy of the ith voter. He could compute the complete result on his own, and the result without the ith vote together with the other voters.

It seems therefore that our functional election requirements forces us to accept schemes that are somewhat fragile in the presence of faults. Note that a denial of service attack is always possible, and against any election protocol, if the attacker controls the entire network.

3.1   The Scheme

We now describe our proposed election scheme, which takes the form of a *yes* or *no* election, encoded as 1 and 0, respectively. The t voters want to execute at most L elections, sequentially or partially in parallel. To focus on the interesting part, we assume that we have a trusted dealer available for now.

We assume that we have a pseudo-random function family F' from the set $\{0, 1, \ldots, L\}$ into the group G with a group structure on F' (written multiplicatively) such that for any $f_1, f_2 \in F'$ and any $0 \leq j < L$, $f_1(j)f_2(j) = (f_1 f_2)(j)$. We also need a one-out-of-two non-interactive zero-knowledge proof for F' (that is, a proof that a group element x equals either $f(i)$ or $f(i)z$ for some i and z) with certain added properties. We discuss constructions in Sect. 4.

Dealer. The dealer chooses an element g from G (either a generator for G or for a subgroup J), and for each user a function f_i uniformly at random from F'. He computes the function $f = f_1 f_2 \cdots f_t$. Then he sends f_i privately to the ith user, $i = 1, 2, \ldots, t$, and broadcasts $(g, f, y_1, y_2, \ldots, y_t)$ to every user, where $y_1, \ldots, y_t$ are commitments of some form to the private keys $f_1, \ldots, f_t$ that will allow other users to verify the non-interactive zero-knowledge proofs used later in the protocol. For the concrete constructions we use, $y_j = f_j(0)$.

Vote creation. In the jth election, voter i encrypts his vote $v_{i,j} \in \{0, 1\}$ as follows: First he computes $c_{i,j} = f_i(j)g^{v_{i,j}}$. He creates a one-out-of-two proof $p_{i,j}$ that proves that one of the values $c_{i,j}$ or $c_{i,j}/g$ is the correct value for $f_i(j)$. Then he broadcasts $(c_{i,j}, p_{i,j})$ to every user.

Tallying. The ith voter has the votes $\{c_{l,j}\}_l$ for the jth election, along the the proofs $\{p_{l,j}\}_l$. He verifies the one-out-of-two proofs (stopping if any proof fails), computes $r_j = (\prod_l c_{l,j})/f(j)$, and then computes the result v_j by computing the discrete logarithm of r_j to the base g.

Note that computing the vote count will always be feasible, since the number of votes is at most t.

Privacy. If at least two voters are honest, this scheme preserves the privacy of every honest vote, by Theorem 2. In order to get a proper reduction, we need the non-interactive zero knowledge proofs to be simulatable. The constructions in Sect. 4.3 are simulatable in the random oracle model.

Correctness. The non-interactive zero-knowledge proof ensures that every vote is correctly formed. Since every other action is performed either by the trusted dealer or the voter himself, this is sufficient to ensure correctness.

Verifiability. Again, since the votes are verified to be correct and every other action is performed either by the trusted dealer or the voter himself, every voter will know that the result is correct if the tallying procedure completes.

If we need to run something more complicated than a yes or no election, we could encode votes and use proofs as described in [5], although this would most likely require the Paillier-based [10] group structure, otherwise computing the discrete logarithm would be too expensive. For a multiparty computation, this amounts to computing integer sums modulo some (large) exponent.

4   The PRF Construction

We can construct a practical pseudo-random function family as follows: Let G be a cyclic group of order n (which may be prime or composite, known or unknown). Let $F = \mathrm{Hom}(G, G) = \{x \mapsto x^k \mid k \in \mathbb{Z}_n\} \subseteq \mathrm{Map}(G, G)$. The interesting thing about this is that even after we apply the construction in the previous section, the pseudo-random function family F' still has a group structure, namely that of $\mathrm{Hom}(G, G)$ induced by the group operation: $(x \mapsto x^k)(x \mapsto x^{k'}) = (x \mapsto x^{k+k'})$. As we shall see, this property is very useful.

When n is not prime, a random element of G may not be a generator. But if n has no small prime factors, the probability $\phi(n)/n$ that an element sampled uniformly at random from G is a generator is very close to 1.

If the group order n is unknown, but we know a reasonable bound on n, say $2^{N-1} < n < 2^N$ for some N, we can still efficiently sample 2^{-t}-close to uniformly from G if we have a generator g, simply by sampling k uniformly from $\{0, 1, \ldots, 2^{N+t} - 1\}$ and computing g^k. The cost of this is at most $2(N+t)$ group operations using a simple square-and-multiply algorithm, compared to $2N$ group operations if the order n is known.

Likewise, we can sample 2^{-t}-uniformly from $\mathrm{Hom}(G, G)$ by sampling uniformly from $\{0, 1, \ldots, 2^{n+t} - 1\}$. Evaluating such a function costs at most $2(N+t)$ group operations, compared to $2N$ group operations if the order n is known.

In the interest of simplicity, we shall in the following ignore the sampling error that comes from sampling not uniformly, but almost uniformly. The proofs of Theorems 3 and 4 are straight-forward and we skip them to save space.

4.1   Security Based on DDH

The Decision Diffie-Hellman problem for the group G is to distinguish tuples of the form (g, g^x, g^y, g^{xy}) from tuples of the form (g, g^x, g^y, g^{xy+z}), where g is a generator for the group G and x, y, z are chosen uniformly at random from $\mathbb{Z}_n$. The advantage of a DDH distinguisher A is defined to be

$$\mathrm{Adv}_A = |\Pr[A(g, g^x, g^y, g^{xy}) = 1 \mid x, y \xleftarrow{r} \mathbb{Z}_n]$$
$$- \Pr[A(g, g^x, g^y, g^{xy+z}) = 1 \mid x, y, z \xleftarrow{r} \mathbb{Z}_n]|$$

Any DDH adversary for G with advantage ϵ can trivially be turned into a weak 2-distinguisher for F with advantage ϵ.

Conversely, any weak distinguisher for F can be turned into a DDH distinguisher with essentially the same strength.

Theorem 3. *Let A be a weak L-distinguisher for F with advantage ϵ. Then there exists a DDH distinguisher A' for G with advantage $\epsilon - (1 - \phi(n)/n)$. The run time of A' is the run time of A plus $4L$ exponentiations and L multiplications in the group.*

If the group order of G is divisible by small primes, then the above theorem is no longer useful. However, in many cases a useful theorem can be recovered under reasonable assumptions, such as generators being indistinguishable from small powers of generators.

Prime Ordered Groups. The standard group structure for this construction is a group G of known prime order, say the group of rational points on an elliptic curve, or the multiplicative subgroup of a finite field. If the group itself is not of prime order, we can take G to be any prime-ordered subgroup such that the cofactor is relatively prime to the subgroup order.

4.2   Security Based on Subgroup Membership

If G has a proper, non-trivial subgroup H, the subgroup membership problem for G and H is to distinguish elements of H from elements of $G \setminus H$. The advantage of a distinguisher A for the subgroup membership problem is defined to be

$$\text{Adv}_A = |\Pr[A(H) = 1] - \Pr[A(G \setminus H)]|.$$

Now suppose G also has a proper, non-trivial subgroup J of order n' such that $J \cap H = \{1\}$ and $G = HJ$. We let $F = \text{Hom}(H, H)$, but consider F to be a pseudo-random function family from the subgroup H to G.

Any subgroup distinguisher with advantage ϵ can trivially be turned into a weak 1-distinguisher for F with advantage $\epsilon(n'-1)/n'$, since the output of F will always be in the subgroup H. Conversely, we can use a J-distinguisher for F to construct a distinguisher for the subgroup membership problem for G and H.

Theorem 4. *Suppose G has two disjoint subgroups H and J, such that $G = HJ$, and let A be a weak L-J-distinguisher for F with advantage ϵ. Then there exists a distinguisher for the subgroup membership problem for G and H with advantage at least $\epsilon/(2L) - (1 - \phi(n')/n')$. The run time of A' is the run time of A plus at most $L - 1$ samples from H, L samples from J, L exponentiations and multiplications in G.*

We remark that if we are willing to accept that both DDH and the Subgroup Membership problems are hard, we can get tighter bounds in the security proof for the family F. However, the bounds we have established are sufficient for our uses.

Paillier's Group. Another useful structure, especially for election schemes, is $\mathbb{Z}_{n^{s+1}}^*$ where n is a product of two prime numbers such that n is relatively prime to the order of $\mathbb{Z}_n^*$. As first described by Paillier [10] and elaborated on by Damgård and Jurik [2], $\mathbb{Z}_{n^{s+1}}^*$ contains a subgroup isomorphic to $\mathbb{Z}_n^*$ that is plausibly hard to distinguish (this would be our H), and a subgroup of order n^s where discrete logarithm computations are easy (this would be our J). This subgroup membership problem is know as the Decision Composite Residuosity problem.

Note that we have a nice map from $\mathbb{Z}_n^*$ into $\mathbb{Z}_{n^{s+1}}^*$ given by taking any representative r for the residue class x and taking it to the residue class y with representative r^{n^s}.

The most natural construction to apply in this situation is that of Theorem 4 (note that there are no small primes in the order of J). However, if we do not trust the hardness of the DCR problem, we could still use Theorem 3 and rely on Decision Diffie-Hellman, at a modest computational cost. In this case, it would be natural to require that n is a product of two safe primes and consider only the quadratic residues.

4.3 Useful Zero-Knowledge Proofs

In an election scheme, it is vital that every voter proves the correctness of his vote. To do this and still preserve zero latency, we must use non-interactive zero-knowledge proofs. Since we already employ the random oracle model, we can use standard honest-verifier zero-knowledge proofs since in the random oracle model, these can be converted to non-interactive zero-knowledge proofs.

The zero knowledge proofs in this section are all completely standard, and we skip the detailed protocol descriptions. For completeness, the protocols are included in Appendix A.

Correct Evaluation. The first proof is that a we have correctly evaluated $f \in F$, relative to a known function value, or alternatively, of equality of discrete logarithms. The prover P wants to prove that there exists a such that $h_0 = g_0^a$ and $h_1 = g_1^a$, for some g_0 and g_1. This amounts to showing that $(h_0, h_1) = (g_0, g_1)^a$ in the group $G \times G$, and we can do that by proving that we know a logarithm of (h_0, h_1) to the base (g_0, g_1) in the group $G \times G$.

One of Two is Correct. The second proof is that for a given $f \in F$, one out of two values correspond to the correct value of $f(x)$ for some x. Again, for our family F this corresponds to showing that one out of two pairs have the same discrete logarithm as a reference pair. We prove this by running the previous proof in parallel and tying the two runs together through the challenge. The prover fakes an accepting conversation for the incorrect value and then creates an accepting conversation for the correct.

This protocol can obviously be extended to one out of k by running k proofs in parallel, faking conversations for $k - 1$ of them and creating the correct proof for the final one.

5 Removing the Dealer

We would like to remove the trusted dealer from the scheme. The choice of the element g used to encode the votes as group elements is arbitrary. The pseudo-random function family hides any value in the subgroup equally well. The only requirement is that every user must be able to verify that g is really inside the proper subgroup. This is usually easy.

All that remains is for each player to choose f_i and to compute a joint representation for f without each player revealing their secret function, nor allowing any player to cheat. The solution depends on whether the group order is known or unknown.

5.1 Known Prime Group Order

When the group order n is known and prime, everything is simple and we do essentially a verifiable multiparty addition. At the start, every voter chooses their function f_i simply by choosing an exponent a_i uniformly at random from $\{0, 1, \ldots, n - 1\}$.

In the first round, every voter sends a share of a_i secretly to every other voter. He also commits to his choice by broadcasting $f_i(0)$ along with a non-interactive zero-knowledge proof of knowledge of possession of the key. (One possibility is essentially the proof given in Sect. 4.3 and Appendix A.1.)

Then every voter verifies every non-interactive zero knowledge proof, adds every secret key share he has received, and in the second round publishes the sum of all the shares. Finally, every voter adds together all the share sums, to get the number $a = \sum_i a_i$, and this number defines $f = \prod_i f_i$. Note that the correctness of this result can be verified by computing $f(0)$.

5.2   Modulo a Power of an RSA Modulus

Now we consider the case of $\mathbb{Z}^*_{n^{s+1}}$. Note that the modulus can be jointly generated using for example the protocol in [1], which supposedly can be made robust against cheating.

Let N be as in Sect 4.3, and set $Q = 2tN$. Every voter chooses his function f_i by choosing an exponent a_i uniformly at random from $\{0, 1, \ldots, N - 1\}$. Note that f_i is sampled almost uniformly at random. The idea is now to add the exponents together using multiparty addition modulo Q.

In the first round, every voter sends a share of a_i secretly to every other voter. He also commits to his choice by broadcasting $f_i(0)$ along with a non-interactive zero-knowledge proof of knowledge of possession of the key. (One possibility is essentially the proof given in Sect. 4.3 and Appendix A.1, except that since the known exponent a_i is very large, we must use even larger numbers in the proof to hide a_i.)

Then every voter verifies every non-interactive zero knowledge proof, adds every secret key share he has received, and in the second round publishes the sum of all the shares. Finally, every voter adds together all the share sums, to get the number $a = \sum_i a_i$ (which is the integer sum), and this number defines $f = \prod_i f_i$. Note that the correctness of this result can be verified by computing $f(0)$.

6   Concluding Remarks

We have described and motivated the problem of doing multiparty computations in a single communication round. We have also shown that this is possible, by giving an election scheme.

If we skip the vote verification parts of the election scheme, we are left with a general multiparty group operation protocol that is secure in the honest-but-curious model. With the construction from Sect. 4.1, this protocol achieves asymptotically optimal broadcast communication complexity: The user contributes one group element to the multiparty computation and broadcasts one group element per computation. It is impossible to achieve lower broadcast communication complexity.

Our election scheme is realised in the random oracle model. While this is a very good heuristic for security in the real world, many people would prefer schemes in some weaker cryptographic model. It is possible to realise our scheme in the common reference string model, where the common reference string replaces the random values derived from the random function in the construction of the PRF. Obviously, the size of the reference string will limit the number of possible rounds. The non-interactive zero-knowledge proofs would have to be replaced with proofs that work in the common reference string model (see for example [8], or [4] for a somewhat different model).

We have not yet considered the general problem of what kind of multiparty computations can at all be performed in a single broadcast round. This is currently an open problem.

Acknowledgements

Thanks to David Wagner and Ivan Damgård for useful discussions, and to the anonymous referees for helpful remarks.

References

1. Algesheimer, J., Camenisch, J., Shoup, V.: Efficient computation modulo a shared secret with application to the generation of shared safe-prime products. In: Yung, M. (ed.) CRYPTO 2002. LNCS, vol. 2442, pp. 417–432. Springer, Heidelberg (2002)
2. Damgård, I., Jurik, M.: A generalisation, a simplification and some applications of Paillier's probabilistic public-key system. In: Kim, K. (ed.) PKC 2001. LNCS, vol. 1992, pp. 119–136. Springer, Heidelberg (2001)
3. Damgård, I., Dupont, K., Pedersen, M.Ø.: Unclonable group identification. In: Vaudenay, S. (ed.) EUROCRYPT 2006. LNCS, vol. 4004, pp. 555–572. Springer, Heidelberg (2006)
4. Damgård, I., Fazio, N., Nicolosi, A.: Non-interactive zero-knowledge from homomorphic encryption. In: Halevi, S., Rabin, T. (eds.) TCC 2006. LNCS, vol. 3876, pp. 41–59. Springer, Heidelberg (2006)
5. Damgård, I., Groth, J., Salomonsen, G.: The theory and implementation of an electronic voting system. In: Gritzalis, D. (ed.) Secure Electronic Voting, Kluwer Academic Publishers, Dordrecht (2002)
6. Franklin, M., Yung, M.: Communication complexity of secure computation. In: Proceedings of the 24th ACM STOC (1992)
7. Gjøsteen, K.: Re: Conditional decryption. Posted to USENET (May 2005), http://sci.crypt, message-id d55ooi$922$1@orkan.itea.ntnu.no
8. Groth, J., Ostrovsky, R., Sahai, A.: Perfect non-interactive zero knowledge for NP. In: Vaudenay, S. (ed.) EUROCRYPT 2006. LNCS, vol. 4004, pp. 339–358. Springer, Heidelberg (2006)
9. Kiayias, A., Yung, M.: Self-tallying elections and perfect ballot secrecy. In: Naccache, D., Paillier, P. (eds.) PKC 2002. LNCS, vol. 2274, pp. 141–158. Springer, Heidelberg (2002)
10. Paillier, P.: Public-key cryptosystems based on composite degree residue classes. In: Stern, J. (ed.) EUROCRYPT 1999. LNCS, vol. 1592, pp. 223–238. Springer, Heidelberg (1999)

A   Zero Knowledge Protocols

We include detailed descriptions of the protocols referred to in Sect. 4.3. We note that proving completeness, honest verifier zero knowledge and soundness is straight-forward for the group structures discussed in Sect. 4 using standard proof techniques and assumptions.

We specify two parameters, t and N. The security parameter t determines how easy it is for a cheating prover to convince the verifier, it is chosen so that the probability 2^{-t} is sufficiently low. If the group order n is known, $N = n$. Otherwise, N is chosen so that the uniform distributions on $\{0, 1, \ldots, N-1\}$ and $\{ae, ae + 1, \ldots, N + ae - 1\}$ are statistically close for any $0 \le a < n$ and $0 \le e < 2^t$, say $N \approx 2^{2t}n$.

A.1   Correct Evaluation

The prover's private input is a, the public input is (g_0, g_1), (h_0, h_1) such that $(h_0, h_1) = (g_0, g_1)^a$.

1. The prover chooses x uniformly at random from $\{0, 1, \ldots, N-1\}$, computes $(z_0, z_1) = (g_0, g_1)^x$, and sends (z_0, z_1) to the verifier..
2. The verifier chooses e uniformly at random from $\{0, 1, \ldots, 2^t - 1\}$ and sends e to the prover.
3. The prover computes $y = x + ea$ and sends y to the verifier.

The verifier accepts if $(g_0, g_1)^y = (z_0, z_1)(h_0, h_1)^e$.

A.2   One of Two Is Correct

The prover's private input is a and b, the public input is (g_0, g_1), (h_{00}, h_{01}), (h_{10}, h_{11}) such that $(h_{b0}, h_{b1}) = (g_0, g_1)^a$.

1. The prover generates an accepting conversation $(z_{1-b,0}, z_{1-b,1}, s_{1-b}, y_{1-b})$ for $(h_{1-b,0}, h_{1-b,1})$ by choosing s_{1-b} uniformly at random from $\{0, 1, \ldots, 2^t - 1\}$, y_{1-b} uniformly at random from $\{0, 1, \ldots, N-1\}$ and computing $(z_{1-b,0}, z_{1-b,1}) = (g_0, g_1)^{y_{1-b}}(h_{1-b,0}, h_{1-b,1})^{-s_{1-b}}$.
   He then chooses x uniformly at random from $\{0, 1 \ldots, N-1\}$ and computes $(z_{b,0}, z_{b,1}) = (g_0, g_1)^x$.
   The prover then sends (z_{00}, z_{01}) and (z_{10}, z_{11}) to the verifier.
2. The verifier chooses e uniformly at random from $\{0, 1, \ldots, 2^t - 1\}$ and sends e to the prover.
3. The prover chooses s_b from $\{0, 1, \ldots, 2^t - 1\}$ such that $s_0 + s_1 \equiv e \bmod 2^t$ and computes $y_b = as_b + x$ and sends s_0, s_1, y_0, y_1 to the verifier.

The verifier accepts if $s_0 + s_1 \equiv e \pmod{2^t}$ and $(g_0, g_1)^{y_i} = (z_{i,0}, z_{i,1})(h_{i,0}, h_{i,1})^e$ for $i = 0, 1$.

Author Index

Printing: Mercedes-Druck, Berlin
Binding: Stein+Lehmann, Berlin

Lecture Notes in Computer Science

Sublibrary 4: Security and Cryptology

For information about Vols. 1– 3858
please contact your bookseller or Springer